GAME THEORY

GAME THEORY

Drew Fudenberg
Jean Tirole

The MIT Press
Cambridge, Massachusetts
London, England

Second Printing, 1992
© 1991 Massachusetts Institute of Technology

Set in Times Roman by Asco Trade Typesetting Ltd., Hong Kong.
Printed and bound in the United States of America.

Library of Congress Cataloging-in-Publication Data

Fudenberg, Drew.
  Game theory / Drew Fudenberg, Jean Tirole.
    p.   cm.
  Includes bibliographical references and index.
  ISBN 0-262-06141-4
  1. Game theory. 2. Economics, Mathematical. I. Tirole, Jean. II. Title.
HB144.F83   1991
658.4′0353—dc20   91-2301
                                                                             CIP

to our parents

Contents

Acknowledgments

Since this is a textbook, it is appropriate to begin by acknowledging our debts to our teachers. Eric Maskin introduced us to the modern literature on game theory in many long sessions in his office, at a time when that literature was not yet an integral part of the economics curriculum. We have been fortunate to work with Eric on several projects since then, and have continued to benefit from his insights.

David Kreps and Robert Wilson were very generous with their advice and encouragement early in our careers. Though they were not our instructors in a formal sense, they taught us quite a lot about how to use and interpret game-theoretic models. Drew Fudenberg has continued to learn from Kreps in the course of their ongoing collaborations.

Drew Fudenberg would also like to thank David Levine and Eddie Dekel-Tabak for years of helpful conversations. Jean Tirole would like to thank Roger Guesnerie and Jean-Jacques Laffont for many insights.

Many people sent us comments on earlier drafts of the present work. Ken Binmore, Larry Blume, and Bernard Caillaud made detailed comments on the whole manuscript. Several others sent detailed comments on specific chapters: Larry Ausubel and Ray Deneckere (chapter 10), Eddie Dekel-Tabak (chapters 1 and 14), Eric van Damme (chapters 1 and 11), Dov Samet (chapter 14), Lars Stole (chapters 7 and 11), and Jorgen Weibull (chapters 1 and 3).

We thank In-Koo Cho, Peter Cramton, Mathias Dewatripont, Robert Gibbons, Peter Hammer, Chris Harris, and Larry Samuelson for their comments as well.

We should also thank Lindsey Klecan, Fred Kofman, and Jim Ratliff for their superb research assistance; they read through the entire manuscript, pointed out mistakes, suggested alternative ways of introducing the material, and corrected typos. Glenn Ellison wrote the solutions to the exercises, and simultaneously rewrote the exercises so that the suggested answers were in fact correct. We were very fortunate to have all four work on the book. We are also grateful to our students at Berkeley and MIT; their questions helped show us how to present this material.

Several typists suffered through the many drafts. Special thanks here to Emily Gallagher, who did the biggest share and who succeeded in sorting out poor handwriting, inconsistent notation, and the renumbering of chapters, all with good cheer. Joel Gwynn drafted the figures. Terry Vaughn shepherded the book through The MIT Press in record time.

We acknowledge generous research grants from the National Science Foundation and the Guggenheim Foundation. Jean Tirole also enjoyed support from the Taussig Visiting Professorship at Harvard.

Finally, we thank our wives for their forbearance and support.

Introduction

Let us begin by describing a game-theoretic situation from the viewpoint of one of the participants. Consider the decision problem of a pie manufacturer—let's call it Piemax—that must choose between a high and a low price for today's batch of pies. In making this choice Piemax will want to think about what the prices of other pies and substitute products are likely to be. Piemax could simply optimize its pricing policy for some given exogenous beliefs about the competitors' prices, but it seems more satisfactory to try to predict these prices from some knowledge of the industry. In particular, Piemax knows that the other firms choose their own prices on the basis of their own predictions of the market environment, including Piemax's prices. The game-theoretic way for Piemax to use this knowledge is to build a model of the behavior of each individual competitor, and (perhaps) to look for behavior that forms an *equilibrium* of the model.

Leaving aside for now the questions of what we mean by an equilibrium and whether Piemax ought to believe that the market outcome will look like one, we are still left with the question of what kind of model Piemax should use. The simplest case is one where Piemax and all its competitors operate for only a single day, where all the firms know the demand for pies (and, more generally, desserts), and where each firm knows the production technologies of the others, as in the famous models of Antoine Augustin Cournot and Joseph Bertrand. This kind of situation can be studied with the tools of strategic-form games and Nash equilibrium that we will develop in part I.

If the industry will continue in operation for a longer period, then Piemax will want to consider some other objectives besides maximizing today's net profit. For example, low prices today may induce consumers to switch from a rival brand and thus may increase Piemax's market share in future periods. Or producing a large batch of pies may help the production staff gain experience and lower future costs. However, the future prices of Piemax's competitors may be influenced by the price Piemax sets today; one particular worry might be that a low price could trigger a price war. Chapter 3 develops the kind of extensive-form model used to capture such dynamic issues and introduces the solution concept of *subgame perfection*, which is commonly used in analyzing them. The other chapters of part II treat various classes of dynamic games in some detail.

Yet another complication arises if Piemax is uncertain of the cost functions or the long-run objectives of its competitors. Has Cupcake Corp. just made a breakthrough in large-batch production? Does Sweetstuff care more about market share than about current profits? And how much do these firms really know about Piemax? Part III shows how to analyze these situations of *incomplete information* in a static context.

Next, if the industry will continue in operation for several periods, Piemax ought to learn about Cupcake Corp. and Sweetstuff's private information from their current pricing behavior and use this information to improve its strategy in the future. Anticipating this, Cupcake Corp. and Sweetstuff may be reluctant to let their prices reveal information that enhances Piemax's competitive position. That is, they may try to manipulate Piemax's information. Part IV extends the analysis to games in which both dynamic issues and incomplete information are important.

We began this introduction with a story of oligopoly pricing because we expected it to be familiar to many of our readers. But game theory has a much broader scope. The theory of noncooperative games studies the behavior of agents in any situation where each agent's optimal choice may depend on his forecast of the choices of his opponents. Although the common usage of the word "games" refers to parlor games such as chess and poker, the Piemax example is far more typical of the kinds of games we will consider, in that the players' objectives are more complex than simply to beat the others: The firms have competing interests in market shares, but a common interest in high prices. The word "noncooperative" means that the players' choices are based only on their perceived self-interest, in contrast to the theory of cooperative games, which develops axioms meant in part to capture the idea of fairness. "Noncooperative" does not mean that the players do not get along, or that they always refuse to cooperate. As we explain in chapters 5 and 9, noncooperative players, motivated solely by self-interest, can exhibit "cooperative" behavior in some settings.

Although game theory has been applied to many fields, this book focuses on the kinds of game theory that have been most useful in the study of economic problems. (We have included some applications to political science as well.) The game-theoretic viewpoint is more useful in settings with a small number of players, for then each player's choice is more likely to matter to his opponents. For example, when the number of firms in a market is small, each firm's output is likely to have a large impact on the market price; thus, it is not reasonable for each firm to take the market price as given.

The first studies of games in the economics literature were the papers by Cournot (1838), Bertrand (1883), and Edgeworth (1925) on oligopoly pricing and production, but these were seen as special models that did little to change the way economists thought about most problems. The idea of a general theory of games was introduced by John von Neumann and Oskar Morgenstern in their famous 1944 book *Theory of Games and Economic Behavior*, which proposed that most economic questions should be analyzed as games. They introduced the ideas of the extensive-form and normal-form (or strategic-form) representations of a game, defined the minmax solution, and showed that this solution exists in all two-player zero-sum games. (In a zero-sum game, the interests of the players are directly opposed, with no common interests at all.)

Nash (1950) proposed what came to be known as "Nash equilibrium" as a way of extending game-theoretic analyses to non-zero-sum games. Nash equilibrium requires that each player's strategy be a payoff-maximizing response to the strategies that he forecasts that his opponents will use, and further that each player's forecast be correct. This is a natural generalization of the equilibria studied in specific models by Cournot and Bertrand, and it is the starting point for most economic analyses. Chapter 1 presents an introduction to Nash equilibrium and its properties. Chapter 2 defines an extension of Nash equilibrium called "correlated equilibrium," and also asks what predictions can be derived solely from the assumption that the players' rationality and the players' payoffs are "common knowledge." This question leads to the notions of iterated strict dominance and rationalizability.

In the models of Cournot and Bertrand, the strategies of the players are simply their choices of outputs or prices. One of the insights of von Neumann and Morgenstern was that the strategies of a game could also be more complex plans for contingent actions—for example, "I'll cut my price tomorrow if you cut yours today." Chapter 3 shows how to model games where the players use contingent plans of this kind.

Selten (1965) and Harsanyi (1967–68) introduced concepts that have been widely used in recent years. Selten argued that in games where the players choose contingent plans not all the Nash equilibria are equally reasonable, because some of them may rely on the ability of players to make "empty threats"—that is, contingent plans that would not in fact be optimal to carry out. (Suppose, for example, that Piemax used the contingent plan "If you do not let me have $\frac{3}{4}$ of the market today, I'll give away free pies for the next ten years.") Selten introduced his concept of *subgame perfection* to rule out the equilibria that rely on these kinds of threats. Chapter 3 defines and discusses this concept and the related issue of credible commitments. Chapters 4 and 5 analyze the subgame-perfect equilibria of several general classes of dynamic games. Chapter 4 is organized around three examples: the repeated prisoner's dilemma, the Rubinstein-Ståhl model of alternating-offer bargaining, and games of timing (including wars of attrition and preemption games) and their applications in the theory of industrial organization. Chapter 5 presents a systematic treatment of repeated games, beginning with the case of perfectly observed actions (as in the prisoner's dilemma) and proceeding to games in which the players' actions are imperfectly observed.

Harsanyi proposed a way to use standard game-theoretic techniques, which suppose that all players know the payoff functions of the others, to model situations of incomplete information where the players are unsure of one another's payoffs. His *Bayesian Nash equilibrium* is the cornerstone of many game-theoretic analyses. We introduce Harsanyi's ideas in chapter 6, and in chapter 7 we apply them to the problem of "mechanism design."

Applications include nonlinear price discrimination, optimal auctions, the revelation of preferences for public goods, and the inefficiency of bargaining when information is incomplete.

When a game has both incomplete information and dynamic aspects, the concept of Bayesian Nash equilibrium may seem too weak, for it allows empty threats of the same kind that Nash equilibrium allows in dynamic games of complete information. Chapter 8 introduces solution concepts that extend the idea of subgame perfection to incomplete-information games; in order of increasing restrictiveness these solution concepts are perfect Bayesian equilibrium, the sequential equilibrium of Kreps and Wilson (1982), and Selten's (1975) trembling-hand perfect equilibrium. We illustrate these concepts with applications to predation and to job-market signaling.

Chapter 9 uses these concepts to study the idea of "reputation effects," which is that players may be able to develop and maintain "reputations" for playing in certain ways. Chapter 10 discusses some papers that model bargaining between a buyer and a seller as a dynamic game of incomplete information; the bargaining is dynamic in that it can involve a sequence of offers and counteroffers, and the information is incomplete because neither player knows the value of an agreement to his opponent.

The last four chapters present topics that are of interest primarily to advanced students. Chapter 11 discusses several of the more restrictive equilibrium refinements that try to capture the idea of "forward induction," including "strategic stability," the "intuitive criterion," and "divinity." We apply these concepts to the signaling-game model introduced in chapter 8, and we discuss the sensitivity of the resulting conclusions to various changes that may and may not be viewed as "small." Chapter 12 presents three advanced topics having to do with the strategic form: generic properties, existence of equilibrium with a continuum of strategies, and supermodularity. Chapter 13 analyzes dynamic games of complete information using the concept of "Markov perfect equilibrium," which is more restrictive than subgame perfection and which requires that the players' current actions not depend on aspects of previous play which have no direct effect on current or future payoffs. Applications include strategic bequest games and games of resource extraction. Chapter 14 gives formal definitions of "common knowledge" and "almost common knowledge" and discusses how the equilibria of a game vary depending on the structure of common knowledge.

How to Use the Book

Although this book may be useful to established researchers who wish to learn more about game theory without taking a formal course and as a reference and guide to some portions of the literature, its primary role is to serve as a text for courses in game theory. We focus on presenting concepts and general results, with more "toy examples" than detailed applications. The applications we do develop are chosen to illustrate the power of the theory; we have not given a comprehensive survey of applications to any particular field. Most of our applications are drawn from the economics literature, and we expect that most of our readers will be economists; however, we have included a few examples from political science, and the book may be useful to political scientists as well.

The book is intended for use in a first course in game theory and in courses for more advanced students. No previous knowledge of game theory is assumed, and the key concepts of Nash equilibrium, subgame perfection, and incomplete information are developed relatively slowly. Most chapters progress from easier to more difficult material, to facilitate jumping from chapter to chapter. The level of mathematics is kept at that of Kreps 1990 and Varian 1984 except in the sections labeled "technical," and none of the technical material is needed to read the other sections.

A first course for advanced undergraduates or first-year graduate students could use most of the core chapters (1, 3, 6, and 8), with the technical sections omitted, and a few selected applications from the other chapters.

One pedagogical innovation in the book is that in chapter 3 we develop subgame perfection in the class of multi-stage games with observed actions without first developing the general extensive form. We do this in the belief that the extensive form involves more notation and more foundational issues (e.g., mixed versus behavior strategies) than are appropriate for the typical first-year course, in which class time is better spent on applications. Similarly, a first course might cover only perfect Bayesian equilibrium from chapter 8, leaving sequential equilibrium and trembling-hand perfection for a second course.

The median audience for this book is a class of first- or second-year graduate students who have already been informally exposed to the ideas of Nash equilibrium, subgame-perfect equilibrium, and incomplete information and who are now interested in a more formal treatment of these ideas and their implications. A one-semester course for these students could be built using all of chapters 1, 3, 6, and 8 and selections from the other chapters. (Sections 3.2 and 3.3, which concern perfection in multi-stage games, could be assigned as background reading but not discussed in class.) As a guide to the amount of material one might cover in a semester, the class here covers all of chapter 4, folk theorems and renegotiation from chapter 5, a little on reputation effects from chapter 9, bargaining from chapter 10, and a few lectures on the equilibrium refinements of chapter 11. Alternatively, the discussion of repeated games could be shortened to make time for Markov equilibrium (chapter 13). One could also incorporate a little on the formal treatment of "common knowledge" from chapter

14. The coverage of chapter 7, on mechanism design, might depend on what other courses the students have available. If another course is offered on contracts and mechanisms, chapter 7 can be skipped entirely. (Indeed, it might be a useful part of the other course.) If the students will not otherwise be exposed to optimal mechanisms, it may be desirable to work through the results on optimal auctions with risk-neutral buyers and on the necessity of disagreement in bargaining with incomplete information.

Some of the material would fit most naturally into an advanced-topics class for third-year students, either because of its difficulty or because it is of more specialized interest. Here we include chapter 12 (which presents more mathematically difficult results about strategic-form games), many of the variants of the repeated-games model covered in chapter 5, the identification of the "payoff-relevant state" in chapter 13, the optimal mechanisms covered in chapter 7, the refinements discussed in chapter 11, and chapter 14 on common knowledge. Of course, each instructor will have his or her own views on the relative importance of the various topics; we have tried to allow a great deal of flexibility as to which topics are covered.

We have used daggers to indicate the suitability of various sections to segments of the intended audience as follows:

†   advanced undergraduates and first-year graduate students
††   first- and second-year graduate students
†††   advanced students and researchers.

(In a few cases, certain subsections have been marked with more daggers than the section as a whole.) The difficulty of the material is closely related to the intended audience; however, not all of the "advanced" topics are difficult. A few of the sections are labeled "technical" to note the use of more powerful mathematical tools than are required in the rest of the book.

The exercises are indexed by degree of difficulty, from one asterisk to three. One-asterisk exercises should be appropriate for first-year graduate students; some of the three-asterisk exercises, as far as we know, have not yet been solved. A set of solutions to selected exercises, prepared by Glenn Ellison, is available to instructors from The MIT Press.

References

Bertrand, J. 1883. Théorie mathématique de la richesse sociale. *Journal des Savants* 499–508.

Cournot, A. 1838. *Recherches sur les Principes Mathematiques de la Theorie des Richesses.* English edition (ed. N. Bacon): *Researches into the Mathematical Principles of the Theory of Wealth* (Macmillan, 1897).

Edgeworth, F. 1897. La Teoria pura del monopolio. *Giornale degli Economisti* 13–31.

Harsanyi, J. 1967–68. Games with incomplete information played by Bayesian players. *Management Science* 14: 159–182, 320–334, 486–502.

Kreps, D. 1990. *A Course in Microeconomic Theory.* Princeton University Press.

Kreps, D., and R. Wilson. 1982. Sequential equilibrium. *Econometrica* 50: 863–894.

Nash, J. 1950. Equilibrium points in *N*-person games. *Proceedings of the National Academy of Sciences* 36: 48–49.

Selten, R. 1965. Spieltheoretische Behandlung eines Oligopolmodells mit Nachfrageträgheit. *Zeitschrift für die gesamte Staatswissenschaft* 12: 301–324.

Selten, R. 1975. Re-examination of the perfectness concept for equilibrium points in extensive games. *International Journal of Game Theory* 4: 25–55.

Varian, H. 1984. *Microeconomic Analysis*, second edition. Norton.

von Neumann, J., and O. Morgenstern. 1944. *Theory of Games and Economic Behavior*. Princeton University Press.

I STATIC GAMES OF COMPLETE INFORMATION

Games in Strategic Form and Nash Equilibrium

We begin with a simple, informal example of a game. Rousseau, in his *Discourse on the Origin and Basis of Equality among Men*, comments:

If a group of hunters set out to take a stag, they are fully aware that they would all have to remain faithfully at their posts in order to succeed; but if a hare happens to pass near one of them, there can be no doubt that he pursued it without qualm, and that once he had caught his prey, he cared very little whether or not he had made his companions miss theirs.[1]

To make this into a game, we need to fill in a few details. Suppose that there are only two hunters, and that they must decide simultaneously whether to hunt for stag or for hare. If both hunt for stag, they will catch one stag and share it equally. If both hunt for hare, they each will catch one hare. If one hunts for hare while the other tries to take a stag, the former will catch a hare and the latter will catch nothing. Each hunter prefers half a stag to one hare.

This is a simple example of a game. The hunters are the players. Each player has the choice between two strategies: hunt stag and hunt hare. The payoff to their choice is the prey. If, for instance, a stag is worth 4 "utils" and a hare is worth 1, then when both players hunt stag each has a payoff of 2 utils. A player who hunts hare has payoff 1, and a player who hunts stag by himself has payoff 0.

What prediction should one make about the outcome of Rousseau's game? Cooperation—both hunting stag—is an equilibrium, or more precisely a "Nash equilibrium," in that neither player has a unilateral incentive to change his strategy. Therefore, stag hunting seems like a possible outcome of the game. However, Rousseau (and later Waltz (1959)) also warns us that cooperation is by no means a foregone conclusion. If each player believes the other will hunt hare, each is better off hunting hare himself. Thus, the noncooperative outcome—both hunting hare—is also a Nash equilibrium, and without more information about the context of the game and the hunters' expectations it is difficult to know which outcome to predict.

This chapter will give precise definitions of a "game" and a "Nash equilibrium," among other concepts, and explore their properties. There are two nearly equivalent ways of describing games: the *strategic* (or *normal*) form and the *extensive* form.[2] Section 1.1 develops the idea of the strategic form and of dominated strategies. Section 1.2 defines the solution concept of Nash equilibrium, which is the starting point of most applications of game theory. Section 1.3 offers a first look at the question of when Nash equilibria exist; it is the one place in this chapter where powerful mathematics is used.

---

1. Quoted by Ordeshook (1986).
2. Historically, the term "normal form" has been standard, but many game theorists now prefer to use "strategic form," as this formulation treats the players' strategies as primitives of the model.

It may appear at first that the strategic form can model only those games in which the players act simultaneously and once and for all, but this is not the case. Chapter 3 develops the extensive-form description of a game, which explicitly models the timing of the players' decisions. We will then explain how the strategic form can be used to analyze extensive-form games.

1.1   Introduction to Games in Strategic Form and Iterated Strict Dominance[†]

1.1.1   Strategic-Form Games

A game in strategic (or normal) form has three elements: the set of players $i \in \mathscr{I}$, which we take to be the finite set $\{1, 2, \ldots, I\}$, the *pure-strategy space* S_i for each player i, and *payoff functions* u_i that give player i's von Neumann-Morgenstern utility $u_i(s)$ for each profile $s = (s_1, \ldots, s_I)$ of strategies. We will frequently refer to all players other than some given player i as "player i's opponents" and denote them by "$-i$." To avoid misunderstanding, let us emphasize that this terminology does not mean that the other players are trying to "beat" player i. Rather, each player's objective is to maximize his own payoff function, and this may involve "helping" or "hurting" the other players. For economists, the most familiar interpretations of strategies may be as choices of prices or output levels, which correspond to Bertrand and Cournot competition, respectively. For political scientists, strategies might be votes or choices of electoral platforms.

A two-player zero-sum game is a game such that $\sum_{i=1}^{2} u_i(s) = 0$ for all s. (The key feature of these games is that the sum of the utilities is a constant; setting the constant to equal 0 is a normalization.) In a two-player zero-sum game, whatever one player wins the other loses. This is the extreme case where the players are indeed pure "opponents" in the colloquial sense. Although such games are amenable to elegant analysis and have been widely studied in game theory, most games of interest in the social sciences are non-zero-sum.

It is helpful to think of players' strategies as corresponding to various "buttons" on a computer keyboard. The players are thought of as being in separate rooms, and being asked to choose a button without communicating with each other. Usually we also assume that all players know the structure of the strategic form, and know that their opponents know it, and know that their opponents know that they know, and so on *ad infinitum*. That is, the structure of the game is *common knowledge*, a concept examined more formally in chapter 14. This chapter uses common knowledge informally, to motivate the solution concept of Nash equilibrium and iterated strict dominance. As will be seen, common knowledge of payoffs on its own is in fact neither necessary nor sufficient to justify Nash equilibrium. In

	L	M	R
U	4,3	5,1	6,2
M	2,1	8,4	3,6
D	3,0	9,6	2,8

Figure 1.1

particular, for some justifications it suffices that the players simply know their *own* payoffs.

We focus our attention on finite games, that is, games where $S = \times_i S_i$ is finite; finiteness should be assumed wherever we do not explicitly note otherwise. Strategic forms for finite two-player games are often depicted as matrices, as in figure 1.1. In this matrix, players 1 and 2 have three pure strategies each: U, M, D (up, middle, and down) and L, M, R (left, middle, and right), respectively. The first entry in each box is player 1's payoff for the corresponding strategy profile; the second is player 2's.

A *mixed strategy* σ_i is a probability distribution over pure strategies. (We postpone the motivation for mixed strategies until later in this chapter.) Each player's randomization is statistically independent of those of his opponents, and the payoffs to a profile of mixed strategies are the expected values of the corresponding pure-strategy payoffs. (One reason we assume that the space of pure strategies is finite is to avoid measure-theoretic complications.) We will denote the space of player i's mixed strategies by Σ_i, where $\sigma_i(s_i)$ is the probability that σ_i assigns to s_i. The space of mixed-strategy profiles is denoted $\Sigma = \times_i \Sigma_i$, with element σ. The *support* of a mixed strategy σ_i is the set of pure strategies to which σ_i assigns positive probability. Player i's payoff to profile σ is

$$\sum_{s \in S} \left(\prod_{j=1}^{I} \sigma_j(s_j) \right) u_i(s),$$

which we denote $u_i(\sigma)$ in a slight abuse of notation. Note that player i's payoff to a mixed-strategy profile is a linear function of player i's mixing probability σ_i, a fact which has many important implications. Note also that player i's payoff is a polynomial function of the strategy profile, and so in particular is continuous. Last, note that the set of mixed strategies contains the pure strategies, as degenerate probability distributions are included. (We will speak of nondegenerate mixed strategies when we want to exclude pure strategies from consideration.)

For instance, in figure 1.1 a mixed strategy for player 1 is a vector $(\sigma_1(U), \sigma_1(M), \sigma_1(D))$ such that $\sigma_1(U)$, $\sigma_1(M)$, and $\sigma_1(D)$ are nonnegative and $\sigma_1(U) + \sigma_1(M) + \sigma_1(D) = 1$. The payoffs to profiles $\sigma_1 = (\frac{1}{3}, \frac{1}{3}, \frac{1}{3})$ and $\sigma_2 = (0, \frac{1}{2}, \frac{1}{2})$ are

$$u_1(\sigma_1, \sigma_2) = \tfrac{1}{3}(0 \cdot 4 + \tfrac{1}{2} \cdot 5 + \tfrac{1}{2} \cdot 6) + \tfrac{1}{3}(0 \cdot 2 + \tfrac{1}{2} \cdot 8 + \tfrac{1}{2} \cdot 3)$$
$$+ \tfrac{1}{3}(0 \cdot 3 + \tfrac{1}{2} \cdot 9 + \tfrac{1}{2} \cdot 2)$$
$$= \tfrac{11}{2}.$$

Similarly, $u_2(\sigma_1, \sigma_2) = \tfrac{27}{6}$.

1.1.2 Dominated Strategies

Is there an obvious prediction of how the game described in figure 1.1 should be played? Note that, no matter how player 1 plays, R gives player 2 a strictly higher payoff than M does. In formal language, strategy M is *strictly dominated*. Thus, a "rational" player 2 should not play M. Furthermore, if player 1 knows that player 2 will not play M, then U is a better choice than M or D. Finally, if player 2 knows that player 1 knows that player 2 will not play M, then player 2 knows that player 1 will play U, and so player 2 should play L.

The process of elimination described above is called *iterated dominance*, or, more precisely, *iterated strict dominance*.[3] In section 2.1 we give a formal definition of iterated strict dominance, as well as an application to an economic example. The reader may worry at this stage that the set of strategies that survive iterated strict dominance depends on the order in which strategies are eliminated, but this is not the case. (The key is that, if strategy s_i is strictly worse than strategy s_i' against all opponents' strategies in some set D, then strategy s_i is strictly worse than strategy s_i' against all opponents' strategies in any subset of D. Exercise 2.1 asks for a formal proof.)

Next, consider the game illustrated in figure 1.2. Here player 1's strategy M is not dominated by U, because M is better than U if player 2 moves R; and M is not dominated by D, because M is better than D when 2 moves L. However, if player 1 plays U with probability $\tfrac{1}{2}$ and D with probability $\tfrac{1}{2}$, he is guaranteed an expected payoff of $\tfrac{1}{2}$ regardless of how player 2 plays, which exceeds the payoff of 0 he receives from M. Hence, a pure strategy

	L	R
U	2,0	−1,0
M	0,0	0,0
D	−1,0	2,0

Figure 1.2

3. Iterated elimination of weakly dominated strategies has been studied by Luce and Raiffa (1957), Fahrquarson (1969), and Moulin (1979).

may be strictly dominated by a mixed strategy even if it is not strictly dominated by any pure strategy.

We will frequently wish to discuss varying the strategy of a single player i while holding the strategies of his opponents fixed. To do so, we let

$$s_{-i} \in S_{-i}$$

denote a strategy selection for all players but i, and write

$$(s_i', s_{-i})$$

for the profile

$$(s_1, \ldots, s_{i-1}, s_i', s_{i+1}, \ldots, s_I).$$

Similarly, for mixed strategies we let

$$(\sigma_i', \sigma_{-i}) = (\sigma_1, \ldots, \sigma_{i-1}, \sigma_i', \sigma_{i+1}, \ldots, \sigma_I).$$

Definition 1.1 Pure strategy s_i is *strictly dominated for player i* if there exists $\sigma_i' \in \Sigma_i$ such that

$$u_i(\sigma_i', s_{-i}) > u_i(s_i, s_{-i}) \text{ for all } s_{-i} \in S_{-i}. \tag{1.1}$$

The strategy s_i is *weakly dominated* if there exists a σ_i' such that inequality 1.1 holds with weak inequality, and the inequality is strict for at least one s_{-i}.

Note that, for a given s_i, strategy σ_i' satisfies inequality 1.1 for all pure strategies s_{-i} of the opponents if and only if it satisfies inequality 1.1 for all mixed strategies σ_{-i} as well, because player i's payoff when his opponents play mixed strategies is a convex combination of his payoffs when his opponents play pure strategies.

So far we have considered dominated pure strategies. It is easy to see that a mixed strategy that assigns positive probability to a dominated pure strategy is dominated. However, a mixed strategy may be strictly dominated even though it assigns positive probability only to pure strategies that are not even weakly dominated. Figure 1.3 gives an example. Playing U with probability $\frac{1}{2}$ and M with probability $\frac{1}{2}$ gives expected payoff

	L	R
U	1,3	−2,0
M	−2,0	1,3
D	0,1	0,1

Figure 1.3

	L	R
U	8,10	-100,9
D	7,6	6,5

Figure 1.4

$-\frac{1}{2}$ regardless of player 2's play and so is strictly dominated by playing D, even though neither U nor M is dominated.

When a game is solvable by iterated strict dominance in the sense that each player is left with a single strategy, as in figure 1.1, the unique strategy profile obtained is an obvious candidate for the prediction of how the game will be played. Although this candidate is often a good prediction, this need not be the case, especially when the payoffs can take on extreme values. When our students have been asked how they would play the game illustrated in figure 1.4, about half have chosen D even though iterated dominance yields (U, L) as the unique solution. The point is that although U is better than D when player 2 is certain not to use the dominated strategy R, D is better than U when there is a 1-percent chance that player 2 plays R. (The same casual empiricism shows that our students in fact do always play L.) If the loss to (U, R) is less extreme, say only − 1, then almost all players 1 choose U, as small fears about R matter less. This example illustrates the role of the assumptions that payoffs and the strategy spaces are common knowledge (as they were in this experiment) and that "rationality," in the sense of not playing a strictly dominated strategy, is common knowledge (as apparently was not the case in this experiment). The point is that the analysis of some games, such as the one illustrated in figure 1.4, is very sensitive to small uncertainties about the behavioral assumptions players make about each other. This kind of "robustness" test—testing how the theory's predictions change with small changes in the model—is an idea that will return in chapters 3, 8, and 11.

At this point we can illustrate a major difference between the analysis of games and the analysis of single-player decisions: In a decision, there is a single decision maker, whose only uncertainty is about the possible moves of "nature," and the decision maker is assumed to have fixed, exogenous beliefs about the probabilities of nature's moves. In a game, there are several decision makers, and the expectations players have about their opponents' play are not exogenous. One implication is that many familiar comparative-statics conclusions from decision theory do not extend once we take into account the way a change in the game may change the actions of *all* players.

Consider for example the game illustrated in figure 1.5. Here player 1's dominant strategy is U, and iterated strict dominance predicts that the

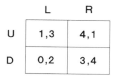

```
        L       R
    ┌───────┬───────┐
  U │  1,3  │  4,1  │
    ├───────┼───────┤
  D │  0,2  │  3,4  │
    └───────┴───────┘
```

Figure 1.5

```
        L       R
    ┌───────┬───────┐
  U │ -1,3  │  2,1  │
    ├───────┼───────┤
  D │  0,2  │  3,4  │
    └───────┴───────┘
```

Figure 1.6

solution is (U, L). Could it help player 1 to change the game and *reduce* his payoffs if U occurs by 2 utils, which would result in the game shown in figure 1.6? Decision theory teaches that such a change would not help, and indeed it would not *if we held player 2's action fixed at L.* Thus, player 1 would not benefit from this reduction in payoff if it were done without player 2's knowledge. However, if player 1 could arrange for this reduction to occur, and to become known to player 2 before player 2 chose his action, player 1 would indeed benefit, for then player 2 would realize that D is player 1's dominant choice, and player 2 would play R, giving player 1 a payoff of 3 instead of 1.

As we will see, similar observations apply to changes such as decreasing a player's choice set or reducing the quality of his information: Such changes cannot help a player in a fixed decision problem, but in a game they may have beneficial effects on the play of opponents. This is true both when one is making predictions using iterated dominance and when one is studying the equilibria of a game.

1.1.3 Applications of the Elimination of Dominated Strategies

In this subsection we present two classic games in which a *single* round of elimination of dominated strategies reduces the strategy set of each player to a single pure strategy. The first example uses the elimination of strictly dominated strategies, and the second uses the elimination of weakly dominated strategies.

Example 1.1: Prisoner's Dilemma
One round of the elimination of strictly dominated strategies gives a unique answer in the famous "prisoner's dilemma" game, depicted in figure 1.7. The story behind the game is that two people are arrested for a crime. The police lack sufficient evidence to convict either suspect and consequently

	C	D
C	1,1	− 1,2
D	2,− 1	0,0

Figure 1.7

need them to give testimony against each other. The police put each suspect in a different cell to prevent the two suspects from communicating with each other. The police tell each suspect that if he testifies against (doesn't cooperate with) the other, he will be released and will receive a reward for testifying, provided the other suspect does not testify against him. If neither suspect testifies, both will be released on account of insufficient evidence, and no rewards will be paid. If one testifies, the other will go to prison; if both testify, both will go to prison, but they will still collect rewards for testifying. In this game, both players simultaneously choose between two actions. If both players cooperate (C) (do not testify), they get 1 each. If they both play noncooperatively (D, for defect), they obtain 0. If one cooperates and the other does not, the latter is rewarded (gets 2) and the former is punished (gets − 1). Although cooperating would give each player a payoff of 1, self-interest leads to an inefficient outcome with payoffs 0. (To readers who feel this outcome is not reasonable, our response is that their intuition probably concerns a different game—perhaps one where players "feel guilty" if they defect, or where they fear that defecting will have bad consequences in the future. If the game is played repeatedly, other outcomes can be equilibria; this is discussed in chapters 4, 5, and 9.)

Many versions of the prisoner's dilemma have appeared in the social sciences. One example is moral hazard in teams. Suppose that there are two workers, $i = 1, 2$, and that each can "work" ($s_i = 1$) or "shirk" ($s_i = 0$). The total output of the team is $4(s_1 + s_2)$ and is shared equally between the two workers. Each worker incurs private cost 3 when working and 0 when shirking. With "work" identified with C and "shirk" with D, the payoff matrix for this moral-hazard-in-teams game is that of figure 1.7, and "work" is a strictly dominated strategy for each worker.

Exercise 1.7 gives another example where strict dominance leads to a unique solution: that of a mechanism for deciding how to pay for a public good.

Example 1.2: Second-Price Auction

A seller has one indivisible unit of an object for sale. There are I potential buyers, or bidders, with valuations $0 \leq v_1 \leq \cdots \leq v_I$ for the object, and these valuations are common knowledge. The bidders simultaneously submit bids $s_i \in [0, +\infty)$. The highest bidder wins the object and pays the second bid (i.e., if he wins ($s_i > \max_{j \neq i} s_j$), bidder i has utility $u_i =$

$v_i - \max_{j \neq i} s_j$), and the other bidders pay nothing (and therefore have utility 0). If several bidders bid the highest price, the good is allocated randomly among them. (The exact probability determining the allocation is irrelevant because the winner and the losers have the same surplus, i.e., 0.)

For each player i the strategy of bidding his valuation ($s_i = v_i$) weakly dominates all other strategies. Let $r_i \equiv \max_{j \neq i} s_j$. Suppose first that $s_i > v_i$. If $r_i \geq s_i$, bidder i obtains utility 0, which he would get by bidding v_i. If $r_i \leq v_i$, bidder i obtains utility $v_i - r_i$, which again is what he would get by bidding v_i. If $v_i < r_i < s_i$, then bidder i has utility $v_i - r_i < 0$; if he were to bid v_i, his utility would be 0. The reasoning is similar for $s_i < v_i$: When $r_i \leq s_i$ or $r_i \geq v_i$, the bidder's utility is unchanged when he bids v_i instead of s_i. However, if $s_i < r_i < v_i$, the bidder forgoes a positive utility by underbidding.

Thus, it is reasonable to predict that bidders bid their valuation in the second-price auction. Therefore, bidder I wins and has utility $v_I - v_{I-1}$. Note also that because bidding one's valuation is a dominant strategy, it does not matter whether the bidders have information about one another's valuations. Hence, if bidders know their own valuation but do not know the other bidders' valuations (see chapter 6), it is still a dominant strategy for each bidder to bid his valuation.

1.2   Nash Equilibrium[†]

Unfortunately, many if not most games of economic interest are not solvable by iterated strict dominance. In contrast, the concept of a Nash-equilibrium solution has the advantage of existing in a broad class of games.

1.2.1   Definition of Nash Equilibrium

A Nash equilibrium is a profile of strategies such that each player's strategy is an optimal response to the other players' strategies.

Definition 1.2   A mixed-strategy profile σ^* is a *Nash equilibrium* if, for all players i,

$$u_i(\sigma_i^*, \sigma_{-i}^*) \geq u_i(s_i, \sigma_{-i}^*) \text{ for all } s_i \in S_i. \tag{1.2}$$

A pure-strategy Nash equilibrium is a pure-strategy profile that satisfies the same conditions. Since expected utilities are "linear in the probabilities," if a player uses a nondegenerate mixed strategy in a Nash equilibrium (one that puts positive weight on more than one pure strategy) he must be indifferent between all pure strategies to which he assigns positive probability. (This linearity is why, in equation 1.2, it suffices to check that no player has a profitable pure-strategy deviation.)

A Nash equilibrium is *strict* (Harsanyi 1973b) if each player has a unique best response to his rivals' strategies. That is, s^* is a strict equilibrium if and only if it is a Nash equilibrium and, for all i and all $s_i \neq s_i^*$,

$$u_i(s_i^*, s_{-i}^*) > u_i(s_i, s_{-i}^*).$$

By definition, a strict equilibrium is necessarily a pure-strategy equilibrium. Strict equilibria remain strict when the payoff functions are slightly perturbed, as the strict inequalities remain satisfied.[4,5]

Strict equilibria may seem more compelling than equilibria where players are indifferent between their equilibrium strategy and a nonequilibrium response, as in the latter case we may wonder why players choose to conform to the equilibrium. Also, strict equilibria are robust to various small changes in the nature of the game, as is discussed in chapters 11 and 14. However, strict equilibria need not exist, as is shown by the "matching pennies" game of example 1.6 below: The unique equilibrium of that game is in (nondegenerate) mixed strategies, and no (nondegenerate) mixed-strategy equilibrium can be strict.[6] (Even pure-strategy equilibria need not be strict; an example is the profile (D, R) in figure 1.18 when $\lambda = 0$.)

To put the idea of Nash equilibrium in perspective, observe that it was implicit in two of the first games to have been studied, namely the Cournot (1838) and Bertrand (1883) models of oligopoly. In the Cournot model, firms simultaneously choose the quantities they will produce, which they then sell at the market-clearing price. (The model does not specify how this price is determined, but it is helpful to think of it being chosen by a Walrasian auctioneer so as to equate total output and demand.) In the Bertrand model, firms simultaneously choose prices and then must produce enough output to meet demand after the price choices become known. In each model, equilibrium is determined by the condition that all firms choose the action that is a best response to the anticipated play of their opponents. It is common practice to speak of the equilibria of these two models as "Cournot equilibrium" and "Bertrand equilibrium," respectively, but it is more helpful to think of them as the *Nash* equilibria of the two different games. We show below that the concepts of "Stackelberg equi-

4. Harsanyi called this "strong" equilibrium; we use the term "strict" to avoid confusion with "strong equilibrium" of Aumann 1959—see note 11.
5. An equilibrium is *quasi-strict* if each pure-strategy best response to one's rivals' strategies belongs to the support of the equilibrium strategy: $\{\sigma_i^*\}_{i \in \mathcal{I}}$ is a quasi-strict equilibrium if it is a Nash equilibrium and if, for all i and s_i,

$$u_i(s_i, \sigma_{-i}^*) = u_i(\sigma_i^*, \sigma_{-i}^*) \Rightarrow \sigma_i^*(s_i) > 0.$$

The equilibrium in matching pennies is quasi-strict, but some games have equilibria that are not quasi-strict. The game in figure 1.18b for $\lambda = 0$ has two Nash equilibria, (U, L) and (D, R). The equilibrium (U, L) is strict, but the equilibrium (D, R) is not even quasi-strict. Harsanyi (1973b) has shown that, for "almost all games," all equilibria are quasi-strict (that is, the set of all games that possess an equilibrium that is not quasi-strict is a closed set of measure 0 in the Euclidean space of strategic-form payoff vectors).
6. Remember that in a mixed-strategy equilibrium a player must receive the same expected payoff from every pure strategy he assigns positive probability.

librium" and "open-loop equilibrium" are also best thought of as shorthand ways of referring to the equilibria of different *games*.

Nash equilibria are "consistent" predictions of how the game will be played, in the sense that if all players predict that a particular Nash equilibrium will occur then no player has an incentive to play differently. Thus, a Nash equilibrium, and only a Nash equilibrium, can have the property that the players can predict it, predict that their opponents predict it, and so on. In contrast, a prediction that any fixed non-Nash profile will occur implies that at least one player will make a "mistake," either in his prediction of his opponents' play or (given that prediction) in his optimization of his payoff.

We do not maintain that such mistakes never occur. In fact, they may be likely in some special situations. But predicting them requires that the game theorist know more about the outcome of the game than the participants know. This is why most economic applications of game theory restrict attention to Nash equilibria.

The fact that Nash equilibria pass the test of being consistent predictions does not make them good predictions, and in situations it seems rash to think that a precise prediction is available. By "situations" we mean to draw attention to the fact that the likely outcome of a game depends on more information than is provided by the strategic form. For example, one would like to know how much experience the players have with games of this sort, whether they come from a common culture and thus might share certain expectations about how the game will be played, and so on.

When one round of elimination of strictly dominated strategies yields a unique strategy profile $s^* = (s_1^*, \ldots, s_I^*)$, this strategy profile is necessarily a Nash equilibrium (actually the unique Nash equilibrium). This is because any strategy $s_i \neq s_i^*$ is necessarily strictly dominated by s_i^*. In particular,

$$u_i(s_i, s_{-i}^*) < u_i(s_i^*, s_{-i}^*).$$

Thus, s^* is a pure-strategy Nash equilibrium (indeed a strict equilibrium). In particular, not cooperating is the unique Nash equilibrium in the prisoner's dilemma of example 1.1.[7]

We show in section 2.1 that the same property holds for iterated dominance. That is, if a single strategy profile survives iterated deletion of strictly dominated strategies, then it is the unique Nash equilibrium of the game.

Conversely, any Nash-equilibrium strategy profile must put weight only on strategies that are not strictly dominated (or, more generally, do not survive iterated deletion of strictly dominated strategies), because a player

7. The same reasoning shows that if there exists a single strategy profile surviving one round of deletion of weakly dominated strategies, this strategy profile is a Nash equilibrium. So, bidding one's valuation in the second-price auction (example 1.2) is a Nash equilibrium.

could increase his payoff by replacing a dominated strategy with one that dominates it. However, Nash equilibria may assign positive probability to weakly dominated strategies.

1.2.2 Examples of Pure-Strategy Equilibria

Example 1.3: Cournot Competition

We remind the reader of the Cournot model of a duopoly producing a homogeneous good. The strategies are quantities. Firm 1 and firm 2 simultaneously choose their respective output levels, q_i, from feasible sets $Q_i = [0, \infty)$, say. They sell their output at the market-clearing price $p(q)$, where $q = q_1 + q_2$. Firm i's cost of production is $c_i(q_i)$, and firm i's total profit is then

$$u_i(q_1, q_2) = q_i p(q) - c_i(q_i).$$

The feasible sets Q_i and the payoff functions u_i determine the strategic form of the game. The "Cournot reaction functions" $r_1 : Q_2 \to Q_1$ and $r_2 : Q_1 \to Q_2$ specify each firm's optimal output for each fixed output level of its opponent. If the u_i are differentiable and strictly concave, and the appropriate boundary conditions are satisfied,[8] we can solve for these reaction functions using the first-order conditions. For example, $r_2(\cdot)$ satisfies

$$p(q_1 + r_2(q_1)) + p'(q_1 + r_2(q_1))r_2(q_1) - c_2'(r_2(q_1)) = 0. \tag{1.3}$$

The intersections (if any exist) of the two reaction functions r_1 and r_2 are the Nash equilibria of the Cournot game: Neither firm can gain by a change in output, given the output level of its opponent.

For instance, for linear demand $(p(q) = \max(0, 1 - q))$ and symmetric, linear cost $(c_i(q_i) = cq_i$ where $0 \le c \le 1)$, firm 2's reaction function, given by equation 1.3, is (over the relevant range)

$$r_2(q_1) = (1 - q_1 - c)/2.$$

By symmetry, firm 1's reaction function is

$$r_1(q_2) = (1 - q_2 - c)/2.$$

The Nash equilibrium satisfies $q_2^* = r_2(q_1^*)$ and $q_1^* = r_1(q_2^*)$ or $q_1^* = q_2^* = (1 - c)/3$.

Example 1.4: Hotelling Competition

Consider Hotelling's (1929) model of differentiation on the line. A linear city of length 1 lies on the abscissa of a line, and consumers are uniformly

---

8. The "appropriate boundary conditions" refer to sufficient conditions for the optimal reaction of each firm to be in the interior of the feasible set Q_i. For example, if all positive outputs are feasible $(Q_i = [0, +\infty))$, it suffices that $p(q) - c_2'(0) > 0$ for all q (which, in general, implies that $c_2'(0) = 0$) for $r_2(q_1)$ to be strictly positive for all q_1, and $\lim_{q \to \infty} p(q) + p'(q)q - c_2'(q) < 0$ for $r_2(q_1)$ to be finite for all q_1.

distributed with density 1 along this interval. There are two stores (firms) located at the two extremes of the city, which sell the same physical product. Firm 1 is at $x = 0$, firm 2 at $x = 1$. The unit cost of each store is c. Consumers incur a transportation cost t per unit of distance. They have unit demands and buy one unit if and only if the minimum generalized price (price plus transportation cost) for the two stores does not exceed some large number $\bar{s}$. If prices are "not too high," the demand for firm 1 is equal to the number of consumers who find it cheaper to buy from firm 1. Letting p_i denote the price of firm i, the demand for firm 1 is given by

$$D_1(p_1, p_2) = x,$$

where

$$p_1 + tx = p_2 + t(1 - x)$$

or

$$D_1(p_1, p_2) = \frac{p_2 - p_1 + t}{2t}$$

and

$$D_2(p_1, p_2) = 1 - D_1(p_1, p_2).$$

Suppose that prices are chosen simultaneously. A Nash equilibrium is a profile (p_1^*, p_2^*) such that, for each player i,

$$p_i^* \in \arg\max_{p_i} \{(p_i - c)D_i(p_i, p_{-i}^*)\}.$$

For instance, firm 2's reaction curve, $r_2(p_1)$, is given (in the relevant range) by

$$D_2(p_1, r_2(p_1)) + [r_2(p_1) - c]\frac{\partial D_2}{\partial p_2}(p_1, r_2(p_1)) = 0.$$

In our example, the Nash equilibrium is given by $p_1^* = p_2^* = c + t$ (and the above analysis is valid as long as $c + 3t/2 \leq \bar{s}$).

Example 1.5: Majority Voting
There are three players, 1, 2, and 3, and three alternatives, A, B, and C. Players vote simultaneously for an alternative; abstaining is not allowed. Thus, the strategy spaces are $S_i = \{A, B, C\}$. The alternative with the most votes wins; if no alternative receives a majority, then alternative A is selected. The payoff functions are

$$u_1(A) = u_2(B) = u_3(C) = 2,$$

$$u_1(B) = u_2(C) = u_3(A) = 1,$$

and

$$u_1(C) = u_2(A) = u_3(B) = 0.$$

This game has three pure-strategy equilibrium *outcomes*: A, B, and C. There are more *equilibria* than this: If players 1 and 3 vote for outcome A, then player 2's vote does not change the outcome, and player 3 is indifferent about how he votes. Hence, the profiles (A, A, A) and (A, B, A) are both Nash equilibria whose outcome is A. (The profile (A, A, B) is not a Nash equilibrium, since if player 3 votes for B then player 2 would prefer to vote for B as well.)

1.2.3   Nonexistence of a Pure-Strategy Equilibrium

Not all games have pure-strategy Nash equilibria. Two examples of games whose only Nash equilibrium is in (nondegenerate) mixed strategies follow.

Example 1.6: Matching Pennies

A simple example of nonexistence is "matching pennies" (figure 1.8). Players 1 and 2 simultaneously announce heads (H) or tails (T). If the announcements match, then player 1 gains a util and player 2 loses a util. If the announcements differ, it is player 2 who wins the util and player 1 who loses. If the predicted outcome is that the announcements will match, then player 2 has an incentive to deviate, while player 1 would prefer to deviate from any prediction in which announcements do not match. The only "stable" situation is one in which each player randomizes between his two pure strategies, assigning equal probability to each. To see this, note that if player 2 randomizes $\frac{1}{2}$-$\frac{1}{2}$ between H and T, player 1's payoff is $\frac{1}{2} \cdot 1 + \frac{1}{2} \cdot (-1) = 0$ when playing H and $\frac{1}{2} \cdot (-1) + \frac{1}{2} \cdot 1 = 0$ when playing T. In this case player 1 is completely indifferent between his possible choices and is willing to randomize himself.

This raises the question of why a player should bother to play a mixed strategy when he knows that any of the pure strategies in its support would do equally well. In matching pennies, if player 1 knows that player 2 will randomize between H and T with equal probabilities, player 1 has expected value 0 from all possible choices. As far as his payoff goes, he could just as well play "heads" with certainty, but if this is anticipated by player 2 the equilibrium disintegrates. Subsection 1.2.5 mentions one defense of mixed strategies, which is that it represents a large population of players

	H	T
H	1,−1	−1,1
T	−1,1	1,−1

Figure 1.8

who use different pure strategies. If we insist that there is only one "player 1," though, this interpretation does not apply. Harsanyi (1973a) offered the alternative defense that the "mixing" should be interpreted as the result of small, unobservable variations in a player's payoffs. Thus, in our example, sometimes player 1 might prefer matching on T to matching on H, and conversely. Then, for each value of his payoff, player 1 would play a pure strategy. This "purification" of mixed-strategy equilibria is discussed in chapter 6.

Example 1.7: Inspection Game

A popular variant of the "matching pennies" game is the "inspection game," which has been applied to arms control, crime deterrence, and worker incentives. The simplest version of this game is depicted in figure 1.9. An agent (player 1) works for a principal (player 2). The agent can either shirk (S) or work (W). Working costs the agent g and produces output of value v for the principal. The principal can either inspect (I) or not inspect (NI). An inspection costs h to the principal but provides evidence of whether the worker shirks. The principal pays the agent a wage w unless he has evidence that the agent has shirked. (The principal is not allowed to condition the wage on the observed level of output.) If the agent is caught shirking, he gets 0 (because of limited liability). The two players choose their strategies simultaneously (in particular, the principal does not know whether the worker has chosen to shirk when he decides whether to inspect). To limit the number of cases to consider, assume that $g > h > 0$. To make things interesting we also assume that $w > g$ (otherwise working would be a weakly or strictly dominated strategy for the agent).

There is no pure-strategy equilibrium in the inspection game: If the principal does not inspect, the agent strictly prefers shirking, and therefore the principal is better off inspecting as $w > h$. On the other hand, if the principal inspects with probability 1 in equilibrium, the agent prefers working (as $w > g$), which implies that the principal is better off not inspecting. Thus, the principal must play a mixed strategy in equilibrium. Similarly, the agent must also randomize. Let x and y denote the probabilities that the agent shirks and the principal inspects, respectively. For the agent to be indifferent between shirking and working, it must be the case that the gain from shirking (g) equals the expected loss in income (yw). For the principal to be indifferent between inspecting and not inspecting, the

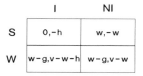

Figure 1.9

cost of inspection (h) must equal the expected wage savings (xw). Hence, $y = g/w$ and $x = h/w$ (both x and y belong to $(0, 1)$).[9]

1.2.4 Multiple Nash Equilibria, Focal Points, and Pareto Optimality

Many games have several Nash equilibria. When this is the case, the assumption that a Nash equilibrium is played relies on there being some mechanism or process that leads all the players to expect the same equilibrium.

One well-known example of a game with multiple equilibria is the "battle of the sexes," illustrated by figure 1.10a. The story that goes with the name "battle of the sexes" is that the two players wish to go to an event together, but disagree about whether to go to a football game or the ballet. Each player gets a utility of 2 if both go to his or her preferred event, a utility of 1 if both go to the other's preferred event, and 0 if the two are unable to agree and stay home or go out individually. Figure 1.10b displays a closely related game that goes by the names of "chicken" and "hawk-dove." (Chapter 4 discusses a related dynamic game that is also called "chicken.") One version of the story here is that the two players meet at a one-lane bridge and each must choose whether to cross or to wait for the other. If both play T (for "tough"), they crash in the middle of the bridge and get -1 each; if both play W (for "weak"), they wait and get 0; if one player chooses T and the other chooses W, then the tough player crosses first, receiving 2, and the weak one receives 1. In the bridge-crossing story, the term "chicken" is used in the colloquial sense of "coward." (Evolutionary biologists call this game "hawk-dove," because they interpret strategy T as "hawk-like" and strategy W as "dove-like.")

Though the different payoff matrices in figures 1.10a and 1.10b describe different sorts of situations, the two games are very similar. Each of them has three equilibria: two in pure strategies, with payoffs $(2, 1)$ and $(1, 2)$, and

9. Building on this result, one can compute the optimal contract, i.e., the w that maximizes the principal's expected payoff

$$v(1 - x) - w(1 - xy) - hy = v(1 - h/w) - w.$$

The optimal wage is thus $w = \sqrt{hv}$ (assuming $\sqrt{hv} > g$). Note that the principal would be better off if he could "commit" to an inspection level. To see this, consider the different game in which the principal plays first and chooses a probability y of inspection, and the agent, after observing y, chooses whether to shirk. For a given w ($>g$), the principal can choose $y = g/w + \varepsilon$, where ε is positive and arbitrarily small. The agent then works with probability 1, and the principal has (approximately) payoff

$$v - w - hg/w > v(1 - h/w) - w.$$

Technically, commitment eliminates the constraint $xw \geq h$, (i.e., that it is *ex post* worthwhile to inspect). (It is crucial that the principal is committed to inspecting with probability y. If the "toss of the coin" determining inspection is not public, the principal has an *ex post* incentive not to inspect, as he knows that the agent works.) This reasoning will become familiar in chapter 3. See chapters 5 and 10 for discussions of how repeated play might make the commitment credible whereas it would not be if the game was played only once.

	B	F
F	0,0	2,1
B	1,2	0,0

a

	T	W
T	-1,-1	2,1
W	1,2	0,0

b

Figure 1.10

one that is mixed. In the battle of the sexes, the mixed equilibrium is that player 1 plays F with probability $\frac{2}{3}$ (and B with probability $\frac{1}{3}$) and player 2 plays B with probability $\frac{2}{3}$ (and F with probability $\frac{1}{3}$). To obtain these probabilities, we solve out the conditions that the players be indifferent between their two pure strategies. So, if x and y denote the probabilities that player 1 plays F and player 2 plays B, respectively, player 1's indifference between F and B is equivalent to

$$0 \cdot y + 2 \cdot (1 - y) = 1 \cdot y + 0 \cdot (1 - y),$$

or

$$y = \tfrac{2}{3}.$$

Similarly, for player 2 to be indifferent between B and F it must be the case that

$$0 \cdot x + 2 \cdot (1 - x) = 1 \cdot x + 0 \cdot (1 - x),$$

or

$$x = \tfrac{2}{3}.$$

In the chicken game of figure 1.10b, the mixed-strategy equilibrium has players 1 and 2 play tough with probability $\frac{1}{2}$.

If the two players have not played the battle of the sexes before, it is hard to see just what the right prediction might be, because there is no obvious way for the players to coordinate their expectations. In this case we would not be surprised to see the outcome (B, F). (We would still be surprised if (B, F) turned out to be the "right" prediction, i.e., if it occurred almost every time.) However, Schelling's (1960) theory of "focal points" suggests that in some "real-life" situations players may be able to coordinate on a particular equilibrium by using information that is abstracted away by the strategic form. For example, the *names* of the strategies may have some commonly understood "focal" power. For example, suppose two players are asked to name an exact time, with the promise of a reward if their choices match. Here "12 noon" is focal; "1:43 P.M." is not. One reason that game theory abstracts away from such considerations is that the "focalness" of various strategies depends on the players' culture and past experiences. Thus, the focal point when choosing between "Left" and "Right" may vary across countries with the direction of flow of auto traffic.

Another example of multiple equilibria is the stag-hunt game we used to begin this chapter, where each player has to choose whether to hunt hare by himself or to join a group that hunts stag. Suppose now that there are I players, that choosing hare gives payoff 1 regardless of the other players' actions, and that choosing stag gives payoff 2 if all players choose stag and gives payoff 0 otherwise. This game has two pure-strategy equilibria: "all stag" and "all hare." Nevertheless, it is not clear which equilibrium should be expected. In particular, which equilibrium is more plausible may depend on the number of players. With only two players, stag is better than hare provided that the single opponent plays stag with probability $\frac{1}{2}$ or more, and given that "both stag" is efficient the opponent might be judged this likely to play stag. However, with nine players stag is optimal only if there is a probability of at least $\frac{1}{2}$ that all eight opponents play stag; if each opponent plays stag with probability p independent of the others, then this requires $p^8 \geq \frac{1}{2}$, or $p \gtrsim 0.93$. In the language of Harsanyi and Selten (1988), "all hare" *risk-dominates* "all stag."[10] (See Harsanyi and Selten 1988 for a formal definition. In a symmetric 2×2 game—that is, a symmetric two-player game with two strategies per player—if both players strictly prefer the same action when their prediction is that the opponent randomizes $\frac{1}{2}$-$\frac{1}{2}$, then the profile where both players play that action is the risk-dominant equilibrium.)

Although risk dominance then suggests that a Pareto-dominant equilibrium need not always be played, it is sometimes argued that players will

10. Very similar games have been discussed in the economics literature, where they are called "coordination failures." For example, Diamond (1982) considered a game where two players have to decide whether to produce one unit of a good that they cannot consume themselves in the hope of trading it for a good produced by the other player. Consumption yields 2 units of utility, and production costs 1 unit. Trade takes place only if both players have produced. Not producing yields 0; producing yields 1 if the opponent produces and -1 otherwise. This game is exactly "stag hunt" in the two-player case. With more players the two games can differ, as the payoff to producing might not equal 2 but might instead be

2(no. of opponents who produce)/(total no. of opponents) $- 1$,

assuming that a trader is matched randomly to another trader, who may or may not have produced. The literature on network externalities in adopting a new technology (e.g. Farrell and Saloner 1985) is a more recent study of coordination problems in economics. For example, all players gain if all switch to the new technology; but if less than half of the population is going to switch, each individual is better off staying with the old technology.

	L	R
U	9,9	0,8
D	8,0	7,7

Figure 1.11

in fact coordinate on the Pareto-dominant equilibrium (provided one exists) if they are able to talk to one another before the game is played. The intuition for this is that, even though the players cannot commit themselves to play the way they claim they will, the preplay communication lets the players reassure one another about the low risk of playing the strategy of the Pareto-dominant equilibrium. Although preplay communication may indeed make the Pareto-dominant equilibrium more likely in the stag-hunt game, it is not clear that it does so in general.

Consider the game illustrated in figure 1.11 (from Harsanyi and Selten 1988). This game has two pure-strategy equilibria ((U, L) with payoffs (9, 9) and (D, R) with payoffs (7, 7)) and a mixed equilibrium with even lower payoffs. Equilibrium (U, L) Pareto-dominates the others. Is this the most reasonable prediction of how the game will be played?

Suppose first that the players do not communicate before play. Then, while the Pareto efficiency of (U, L) may tend to make it a focal point, playing D is much safer for player 1, as it guarantees 7 regardless of how player 2 plays, and player 1 should play D if he assesses the probability of R to be greater than $\frac{1}{8}$ (so (D, R) is risk dominant). Moreover, player 1 knows that player 2 should play R if player 2 believes the probability of D is more than $\frac{1}{8}$. In this situation we are not certain what outcome to predict.

Does (U, L) become compelling if we suppose that the players are able to meet and communicate before they play? Aumann (1990) argues that the answer is no. Suppose that the players meet and assure each other that they plan to play (U, L). Should player 1 take player 2's assurances at face value? As Aumann observes, regardless of his own play, player 2 gains if player 1 plays U; thus, no matter how player 2 intends to play, he should tell player 1 that he intends to play L. Thus, it is not clear that the players should expect their assurances to be believed, which means that (D, R) might be the outcome after all. Thus, even with preplay communication, (U, L) does not seem like the necessary outcome, although it may seem more likely than when communication is not possible.

Another difficulty with the idea that the Pareto-dominant equilibrium is the natural prediction arises in games with more than two players. Consider the game illustrated in figure 1.12 (taken from Bernheim, Peleg, and Whinston 1987), where player 1 chooses rows, player 2 chooses columns, and

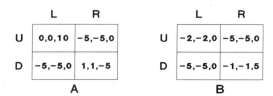

Figure 1.12

player 3 chooses matrices. (Harsanyi and Selten (1988) give a closely related example where player 3 moves before players 1 and 2.) This game has two pure-strategy Nash equilibria, (U, L, A) and (D, R, B), and an equilibrium in mixed strategies. Bernheim, Peleg, and Whinston do not consider mixed strategies, so we will temporarily restrict our attention to pure ones. The equilibrium (U, L, A) Pareto-dominates (D, R, B). Is (U, L, A) then the obvious focal point? Imagine that this was the expected solution, and hold player 3's choice fixed. This induces a two-player game between players 1 and 2. In this two-player game, (D, R) is the Pareto-dominant equilibrium! Thus, if players 1 and 2 expect that player 3 will play A, and if they can coordinate their play on their Pareto-preferred equilibrium in matrix A, they should do so, upsetting the "good" equilibrium (U, L, A).

In response to this example, Bernheim, Peleg, and Whinston propose the idea of a coalition-proof equilibrium, as a way of extending the idea of coordinating on the Pareto-dominant equilibrium to games with more than two players.[11]

To summarize our remarks on multiple equilibria: Although some games have focal points that are natural predictions, game theory lacks a general and convincing argument that a Nash outcome will occur.[12] However, equilibrium analysis has proved useful to economists, and we will focus attention on equilibrium in this book. (Chapter 2 discusses the "rationalizability" notion of Bernheim and Pearce, which investigates the predictions

11. The definition of a coalition-proof equilibrium proceeds by induction on coalition size. First one requires that no one-player coalition can deviate, i.e., that the given strategies are a Nash equilibrium. Then one requires that no two-player coalition can deviate, given that once such a deviation has "occurred" either of the deviating players (but none of the others) is free to deviate again. That is, the two-player deviations must be Nash equilibria of the two-player game induced by holding the strategies of the others fixed. And one proceeds in this way up to the coalition of all players. Clearly (U, L, A) in figure 1.12 is not coalition-proof; brief inspection shows that (D, R, B) is.

Coalition-proof equilibrium is a weakening of Aumann's (1959) "strong equilibrium," which requires that no subset of players, taking the actions of others as given, can jointly deviate in a way that increases the payoffs of all its members. Since this requirement applies to the grand coalition of all players, strong equilibria must be Pareto efficient, unlike coalition-proof equilibria. No strong equilibrium exists in the game of figure 1.12.

12. Aumann (1987) argues that the "Harsanyi doctrine," according to which all players' beliefs must be consistent with Bayesian updating from a common prior, implies that Bayesian rational players must predict a "correlated equilibrium" (a generalization of Nash equilibrium defined in section 2.2).

one can make without invoking equilibrium. As we will see, rationalizability is closely linked to the notion of iterated strict dominance.)

1.2.5   Nash Equilibrium as the Result of Learning or Evolution

To this point we have motivated the solution concepts of dominance,
iterated dominance, and Nash equilibrium by supposing that players make
their predictions of their opponents' play by introspection and deduction,
using their knowledge of the opponents' payoffs, the knowledge that the
opponents are rational, the knowledge that each player knows that the
others know these things, and so on through the infinite regress implied by
"common knowledge."

An alternative approach to introspection for explaining how players
predict the behavior of their opponents is to suppose that players extra-
polate from their past observations of play in "similar games," either with
their current opponents or with "similar"[13] ones. At the end of this subsection we will discuss how introspection and extrapolation differ in the nature
of their assumptions about the players' information about one another.

The idea of using learning-type adjustment processes to explain equilibrium goes back to Cournot, who proposed a process that might lead the
players to play the Cournot-Nash equilibrium outputs. In the Cournot
adjustment process, players take turns setting their outputs, and each
player's chosen output is a best response to the output his opponent chose
the period before. Thus, if player 1 moves first in period 0, and chooses
q_1^0, then player 2's output in period 1 is $q_2^1 = r_2(q_1^0)$, where r_2 is the Cournot reaction function defined in example 1.3. Continuing to iterate the
process,

$$q_1^2 = r_1(q_2^1) = r_1(r_2(q_1^0)),$$

and so on. This process may settle down to a steady state where the output
levels are constant, but it need not do so. *If* the process does converge to
(q_1^*, q_2^*), then $q_2^* = r_2(q_1^*)$ and $q_1^* = r_1(q_2^*)$, so the steady state is a Nash
equilibrium.

If the process converges to a particular steady state for all initial quantities sufficiently close to it, we say that the steady state is *asymptotically
stable*. As an example of an asymptotically stable equilibrium, consider the
Cournot game where $p(q) = 1 - q$, $c_i(q_i) = 0$, and the feasible sets are
$Q_i = [0, 1]$. The reaction curves for this game are $r_i(q_j) = (1 - q_j)/2$, and the
unique Nash equilibrium is at the intersection of the reaction curves, which
is the point $A = (\frac{1}{3}, \frac{1}{3})$. Figure 1.13 displays the path of the Cournot adjust-

---

13. Of course the distinction between introspection and extrapolation is not absolute. One
might suppose that introspection leads to the idea that extrapolation is likely to work, or
conversely that past experience has shown that introspection is likely to make the correct
prediction.

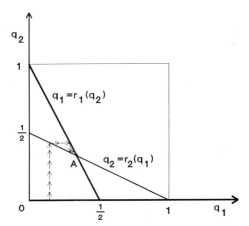

Figure 1.13

ment or *tâtonnement* process for the initial condition $q_1^0 = \frac{1}{6}$. The process converges to the Nash equilibrium from every starting point; that is, the Nash equilibrium is globally stable.

Now suppose that the cost and demand functions yield reaction curves as in figure 1.14 (we spare the reader the derivation of such reaction functions from a specification of cost and demand functions). The reaction functions in figure 1.14 intersect at three points, B, C, and D, all of which are Nash equilibria. Now, however, the intermediate Nash equilibrium, C, is not stable, as the adjustment process converges either to B or to D unless it starts at exactly C.

Comparing figures 1.13 and 1.14 may suggest that the question of asymptotic stability is related to the relative slopes of the reaction functions, and this is indeed the case. If the payoff functions are twice continuously differentiable, the slope of firm i's reaction function is

$$\frac{dr_i}{dq_j} = -\frac{\partial^2 u_i}{\partial q_i \partial q_j} \bigg/ \frac{\partial^2 u_i}{\partial q_i^2},$$

and a sufficient condition for an equilibrium to be asymptotically stable is that

$$\left| \frac{dr_1}{dq_2} \right| \left| \frac{dr_2}{dq_1} \right| < 1$$

or

$$\frac{\partial^2 u_1}{\partial q_1 \partial q_2} \frac{\partial^2 u_2}{\partial q_1 \partial q_2} < \frac{\partial^2 u_1}{\partial q_1^2} \frac{\partial^2 u_2}{\partial q_2^2}$$

in an open neighborhood of the Nash equilibrium.

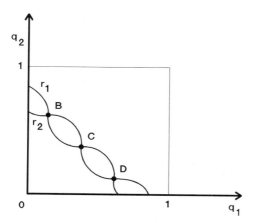

Figure 1.14

Technical aside   The condition for asymptotic stability when firms react *simultaneously*, instead of alternatively, to their opponent's most recent outputs is the same as the one just described. To see this, suppose that both players simultaneously adjust their quantities each period by choosing a best response to their opponent's output in the previous period. View this as a dynamic process

$$q^t = (q_1^t, q_2^t) = (r_1(q_2^{t-1}), r_2(q_1^{t-1})) \equiv f(q^{t-1}).$$

From the study of dynamical systems (Hirsch and Smale 1974), we know that a fixed point q^* of f is asymptotically stable in this process if all the eigenvalues of $\partial f(q^*)$ have real parts whose absolute value is less than 1. The condition on the slopes of the reaction functions is exactly sufficient to imply that this eigenvalue condition is satisfied. Classic references on the stability of the Cournot adjustment process include Fisher 1961, Hahn 1962, Seade 1980, and Dixit 1986; see Moulin 1986 for a discussion of more recent work and of subtleties that arise with more than two players.

One way to interpret Cournot's adjustment process with either alternating or simultaneous adjustment is that in each period the player making a move expects that his opponent's output in the future will be the same as it is now. Since output in fact changes every period, it may seem more plausible that players base their forecasts on the average value of their opponent's past play, which suggests the alternative dynamic process

$$q_i^t = r_i \left(\sum_{\tau=0}^{t-1} q_j^\tau / t \right).$$

This alternative has the added value of converging under a broader set

	L	M	R
U	0,0	4,5	5,4
M	5,4	0,0	4,5
D	4,5	5,4	0,0

Figure 1.15

of assumptions, which makes it more useful as a tool for computing equilibria.[14]

However, even when players do respond to the past averages of their opponents' play, the adjustment process need not converge, especially once we move away from games with one-dimensional strategy spaces and concave payoffs. The first example of cycles in this context is due to Shapley (1964), who considered the game illustrated here in figure 1.15.

Suppose first that, in each period, each player chooses a best response to the action his opponent played the period before. If play starts at the point (M, L), it will proceed to trace out the cycle (M, L), (M, R), (U, R), (U, M), (D, M), (D, L), (M, L). If instead players take turns reacting to one another's previous action, then once again play switches from one point to the next each period. If players respond to their opponents' average play, the play cycles increasingly (in fact, geometrically) slowly but never converges: Once (M, L) is played, (M, R) occurs for the next two periods, then player 1 switches to U; (U, R) occurs for the next four periods, then player 2 switches to M; after eight periods of (U, M), player 1 switches to D; and so on.

Thus, even assuming that behavior follows an adjustment process does not imply that play must converge to a Nash equilibrium. And the adjustment processes are not compelling as a description of players' behavior. One problem with all the processes we have discussed so far is that the players ignore the way that their current action will influence their opponent's action in the next period. That is, the adjustment process itself may not be an equilibrium of the "repeated game," where players know they face one another repeatedly.[15] It might seem natural that if the same two players face each other repeatedly they would come to recognize the dynamic effect of their choices. (Note that the effect is smaller if players react to past averages.)

14. For a detailed study of convergence when Cournot oligopolists respond to averages, see Thorlund-Petersen 1990.
15. If firms have perfect foresight, they choose their output taking into account its effect on their rival's future reaction. On this, see exercise 13.2. The Cournot tâtonnement process can be viewed as a special case of the perfect-foresight model where the firms have discount factor 0.

A related defense of Nash equilibrium supposes that there is a large group of players who are matched at random and asked to play a specific game. The players are not allowed to communicate or even to know who their opponents are. At each round, each player chooses a strategy, observes the strategy chosen by his opponent, and receives the corresponding payoff. If there are a great many players then a pair of players who are matched today are unlikely to meet again, and players have no reason to worry about how their current choice will affect the play of their future opponents. Thus, in each period the players should tend to play the strategy that maximizes that period's expected payoff. (We say "tend to play" to allow for the possibility that players may occasionally "experiment" with other choices.)

The next step is to specify how players adjust their expectations about their opponents' play in light of their experience. Many different specifications are possible, and, as with the Cournot process, the adjustment process need not converge to a stable distribution. However, if players observe their opponents' strategies at the end of each round, and players eventually receive a great many observations, then one natural specification is that each player's expectations about the play of his opponents converges to the probability distribution corresponding to the sample average of play he has observed in the past. In this case, *if* the system converges to a steady state, the steady state must be a Nash equilibrium.[16]

Caution   The assumption that players observe one another's strategies at the end of each round makes sense in games like the Cournot competition where strategies correspond to uncontingent choices of actions. In the general extensive-form games we introduce in chapter 3, strategies are contingent plans, and the observed outcome of play need not reveal the action a player would have used in a contingency that did not arise (Fudenberg and Kreps 1988).

The idea of a large population of players can also be used to provide an alternative interpretation of mixed strategies and mixed-strategy equilibria. Instead of supposing that individual players randomize among several strategies, a mixed strategy can be viewed as describing a situation in which different fractions of the population play different pure strategies. Once again a Nash equilibrium in mixed strategies requires that all pure strategies that receive positive probability are equally good responses, since if one pure strategy did better than the other we would expect more and more of the players to learn this and switch their play to the strategy with the higher payoff.

16. Recent papers on the explanation of Nash equilibrium as the result of learning include Gul 1989, Milgrom and Roberts 1989, and Nyarko 1989.

The large-population model of adjustment to Nash equilibrium has yet another application: It can be used to discuss the adjustment of population fractions by *evolution* as opposed to learning. In theoretical biology, Maynard Smith and Price (1973) pioneered the idea that animals are genetically programmed to play different pure strategies, and that the genes whose strategies are more successful will have higher reproductive fitness. Thus, the population fractions of strategies whose payoff against the current distribution of opponents' play is relatively high will tend to grow at a faster rate, and, any stable steady state must be a Nash equilibrium. (Non-Nash profiles can be unstable steady states, and not all Nash equilibria are locally stable.) It is interesting to note that there is an extensive literature applying game theory to questions of animal behavior and of the determination of the relative frequency of male and female offspring. (Maynard Smith 1982 is the classic reference.)

More recently, some economists and political scientists have argued that evolution can be taken as a metaphor for learning, and that evolutionary stability should be used more broadly in economics. Work in this area includes Axelrod's (1984) study of evolutionary stability in the repeated prisoner's dilemma game we discuss in chapter 4 and Sugden's (1986) study of how evolutionary stability can be used to ask which equilibria are more likely to become focal points in Schelling's sense.

To conclude this section we compare the informational assumptions of deductive and extrapolative explanations of Nash equilibrium and iterated strict dominance. The deductive justification of the iterated deletion of strictly dominated strategies requires that players are rational and know the payoff functions of all players, that they know their opponents are rational and know the payoff functions, that they know the opponents know, and so on for as many steps as it takes for the iterative process to terminate. In contrast, if players play one another repeatedly, then, even if players do not know their opponents' payoffs, they will eventually learn that the opponents do not play certain strategies, and the dynamics of the learning system will replicate the iterative deletion process. And for an extrapolative justification of Nash equilibrium, it suffices that players know their own payoffs, that play eventually converges to a steady state, and that if play does converge all players eventually learn their opponents' steady-state strategies. Players need not have *any* information about the payoff functions or information of their opponents.

Of course, the reduction in the informational requirements is made possible by the additional hypotheses of the learning story: Players must have enough experience to learn how their opponents play, and play must converge to a steady state. Moreover, we must suppose either that there is a large population of players who are randomly matched, or that, even though the same players meet one another repeatedly, they ignore any dynamic links between their play today and their opponents' play tomorrow.

1.3   Existence and Properties of Nash Equilibria (technical)[††]

We now tackle the question of the existence of Nash equilibria. Although some of the material in this section is technical, it is quite important for those who wish to read the formal game-theory literature. However, the section can be skipped in a first reading by those who are pressed for time and have little interest in technical detail.

1.3.1   Existence of a Mixed-Strategy Equilibrium

Theorem 1.1 (Nash 1950b)   Every finite strategic-form game has a mixed-strategy equilibrium.

Remark   Remember that a pure-strategy equilibrium is an equilibrium in degenerate mixed strategies. The theorem does not assert the existence of an equilibrium with nondegenerate mixing.

Proof   Since this is the archetypal existence proof in game theory, we will go through it in detail. The idea of the proof is to apply Kakutani's fixed-point theorem to the players' "reaction correspondences." Player i's *reaction correspondence*, r_i, maps each strategy profile σ to the set of mixed strategies that maximize player i's payoff when his opponents play σ_{-i}. (Although r_i depends only on σ_{-i} and not on σ_i, we write it as a function of the strategies of all players, because later we will look for a fixed point in the space Σ of strategy profiles.) This is the natural generalization of the Cournot reaction function we defined above. Define the correspondence $r: \Sigma \rightrightarrows \Sigma$ to be the Cartesian product of the r_i. A *fixed point* of r is a σ such that $\sigma \in r(\sigma)$, so that, for each player, $\sigma_i \in r_i(\sigma)$. Thus, a fixed point of r is a Nash equilibrium.

From Kakutani's theorem, the following are sufficient conditions for $r: \Sigma \rightrightarrows \Sigma$ to have a fixed point:

(1) Σ is a compact,[17] convex,[18] nonempty subset of a (finite-dimensional) Euclidean space.

(2) $r(\sigma)$ is nonempty for all σ.

(3) $r(\sigma)$ is convex for all σ.

---

17. A subset X of a Euclidean space is compact if any sequence in X has a subsequence that converges to a limit point in X. The definition of compactness for more general topological spaces uses the notion of "cover," which is a collection of open sets whose union includes the set X. X is compact if any cover has a finite subcover.
18. A set X in a linear vector space is convex if, for any x and x' belonging to X and any $\lambda \in [0, 1]$, $\lambda x + (1 - \lambda) x'$ belongs to X.

(4) $r(\cdot)$ has a closed graph: If $(\sigma^n, \hat{\sigma}^n) \to (\sigma, \hat{\sigma})$ with $\hat{\sigma}^n \in r(\sigma^n)$, then $\hat{\sigma} \in r(\sigma)$. (This property is also often referred to as *upper hemi-continuity*.[19])

Let us check that these conditions are satisfied.

Condition 1 is easy—each Σ_i is a simplex of dimension $(\# S_i - 1)$. Each player's payoff function is linear, and therefore continuous in his own mixed strategy, and since continuous functions on compact sets attain maxima, condition 2 is satisfied. If $r(\sigma)$ were not convex, there would be a $\sigma' \in r(\sigma)$, a $\sigma'' \in r(\sigma)$, and a $\lambda \in (0, 1)$ such that $\lambda\sigma' + (1 - \lambda)\sigma'' \notin r(\sigma)$. But for each player i,

$$u_i(\lambda\sigma'_i + (1 - \lambda)\sigma''_i, \sigma_{-i}) = \lambda u_i(\sigma'_i, \sigma_{-i}) + (1 - \lambda)u_i(\sigma''_i, \sigma_{-i}),$$

so that if both σ'_i and σ''_i are best responses to σ_{-i}, then so is their weighted average. This verifies condition 3.

Finally, assume that condition 4 is violated so there is a sequence $(\sigma^n, \hat{\sigma}^n) \to (\sigma, \hat{\sigma})$, $\hat{\sigma}^n \in r(\sigma^n)$, but $\hat{\sigma} \notin r(\sigma)$. Then $\hat{\sigma}_i \notin r_i(\sigma)$ for some player i. Thus, there is an $\varepsilon > 0$ and a σ'_i such that $u_i(\sigma'_i, \sigma_{-i}) > u_i(\hat{\sigma}_i, \sigma_{-i}) + 3\varepsilon$. Since u_i is continuous and $(\sigma^n, \hat{\sigma}^n) \to (\sigma, \hat{\sigma})$, for n sufficiently large we have

$$u_i(\sigma'_i, \sigma^n_{-i}) > u_i(\sigma'_i, \sigma_{-i}) - \varepsilon > u_i(\hat{\sigma}_i, \sigma_{-i}) + 2\varepsilon > u_i(\hat{\sigma}^n_i, \sigma^n_{-i}) + \varepsilon.$$

Thus, σ'_i does *strictly* better against σ^n_{-i} than $\hat{\sigma}^n_i$ does, which contradicts $\hat{\sigma}^n_i \in r_i(\sigma^n)$. This verifies condition 4. ∎

Once existence has been established, it is natural to consider the characterization of the equilibrium set. Ideally one would prefer there to be a unique equilibrium, but this is true only under very strong conditions. When several equilibria exist, one must see which, if any, seem to be reasonable predictions, but this requires examination of the entire Nash set. The reasonableness of one equilibrium may depend on whether there are others with competing claims. Unfortunately, in many interesting games the set of equilibria is difficult to characterize.

1.3.2   The Nash-Equilibrium Correspondence Has a Closed Graph

We now analyze how the set of Nash equilibria changes when the payoff functions change continuously with some parameters. The intuition for the results can be gleaned from the case of a single decision maker (see figure 1.16). Suppose that the decision maker gets payoff $1 + \lambda$ when playing L and $1 - \lambda$ when playing R. Let x denote the probability that the decision maker plays L, and consider the optimal x for each λ in $[-1, 1]$. This

---

19. The graph of a correspondence $f: X \rightrightarrows Y$ is the set of (x, y) such that $y \in f(x)$. Upper hemi-continuity requires that, for any x_0, and for any open set V that contains $f(x_0)$, there exists a neighborhood U of x_0 such that $f(x) \subseteq V$ if $x \in U$. In general this differs from the closed-graph notion, but the two concepts coincide if the range of f is compact and $f(x)$ is closed for each x—conditions which are generally satisfied when applying fixed-point theorems. See Green and Heller 1981.

Figure 1.16

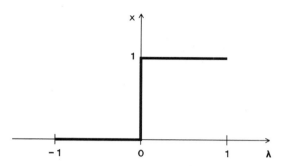

Figure 1.17

defines the Nash-equilibrium correspondence for this one-player game. In particular, for $\lambda = 0$, any $x \in [0, 1]$ is optimal. Figure 1.17, which exhibits the graph of the Nash correspondence (in bold), suggests its main properties. First, the correspondence has a closed graph (is upper hemicontinuous). For any sequence (λ^n, x^n) belonging to the graph of the correspondence and converging to some (λ, x), the limit (λ, x) belongs to the graph of correspondence.[20] Second, the correspondence may not be "lower hemi-continuous." That is, there may exist (λ, x) belonging to the graph of the correspondence and a sequence $\lambda^n \to \lambda$ such that there exists no x^n such that (λ^n, x^n) belongs to the graph of the correspondence and $x^n \to x$. Here, take $\lambda = 0$ and $x \in (0, 1)$. These two properties generalize to multi-player situations.[21]

One key step in the proof of existence of subsection 1.3.1 is verifying that when payoffs are continuous the reaction correspondences have closed graphs. The same argument applies to the set of Nash equilibria: Consider a family of strategic-form games with the same finite pure-strategy space S and payoffs $u_i(s, \lambda)$ that are continuous functions of λ. Let $G(\lambda)$ denote the game associated with λ and let $E(\cdot)$ be the Nash correspondence that associates with each λ the set of (mixed-strategy) Nash equilibria of $G(\lambda)$. Then, if the set of possible values Λ of λ is compact, the Nash correspondence has a closed graph and, in particular, $E(\lambda)$ is closed for each λ. The proof is as in the verification of condition (4) in the existence proof. Con-

20. This result is part of the "theorem of the maximum" (Berge 1963).
21. A correspondence $f: X \rightrightarrows Y$ is lower hemi-continuous if, for any $(x, y) \in X \times Y$ such that $y \in f(x)$, and any sequence $x^n \in X$ such that $x^n \to x$, there exists a sequence y^n in Y such that $y^n \to y$ and $y^n \in f(x^n)$ for each x^n.

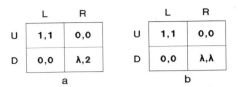

Figure 1.18

sider two sequences $\lambda^n \to \lambda$ and $\sigma^n \to \sigma$ such that $\sigma^n \in r(\sigma^n)$ and $\sigma \notin r(\sigma)$. That is, σ^n is a Nash equilibrium of $G(\lambda^n)$, but σ is not a Nash equilibrium of $G(\lambda)$. Then there is a player i and a $\hat{\sigma}_i$ that does strictly better than σ_i against σ_{-i}. Since payoffs are continuous in λ, for any λ^n near λ and any σ^n_{-i} near σ_{-i}, $\hat{\sigma}_i$ is a strictly better response to σ^n_{-i} than σ^n_i is—a contradiction.

It is important to note that this does not mean that the correspondence $E(\cdot)$ is continuous. Loosely speaking, a closed graph (plus compactness) implies that the set of equilibria cannot shrink in passing to the limit. If σ^n are Nash equilibria of $G(\lambda^n)$ and $\lambda^n \to \lambda$, then σ^n has a limit point $\sigma \in E(\lambda)$. However, $E(\lambda)$ can contain additional equilibria that are not limits of equilibria of "nearby" games. Thus, $E(\cdot)$ is not lower hemi-continuous, and hence is not continuous. We illustrate this with the two games in figure 1.18. In both of these games, (U, L) is the unique Nash equilibrium if $\lambda < 0$, while for $\lambda > 0$ there are three equilibria (U, L), (D, R), and an equilibrium in mixed strategies. While the equilibrium correspondence has a closed graph in both games, the two games have very different sets of equilibria at the point $\lambda = 0$.

First consider the game illustrated in figure 1.18a. For $\lambda > 0$, there are two pure-strategy equilibria and a unique equilibrium with nondegenerate mixing, as each player can be indifferent between his two choices only if the other player randomizes. If we let p denote the probability of U and q denote the probability of L, a simple computation shows that the unique mixed-strategy equilibrium is

$$(p, q) = \left(\frac{2}{3}, \frac{\lambda}{1 + \lambda} \right).$$

As required by a closed graph, the profiles $(p, q) = (1, 1), (0, 0)$, and $(\frac{2}{3}, 0)$ are all Nash equilibria at $\lambda = 0$. There are also additional equilibria for $\lambda = 0$ that are not limits of equilibria for any sequence $\lambda^n \to 0$, namely $(p, 0)$ for any $p \in [0, \frac{2}{3}]$. When $\lambda = 0$, player 1 is willing to randomize even if player 2 plays R with probability 1, and so long as the probability of U is not too large player 2 is still willing to play R. This illustrates how the equilibrium correspondence can fail to be lower hemi-continuous.

In the game of figure 1.18b, the equilibria for $\lambda > 0$ are $(1, 1), (0, 0)$, and $(\lambda/(1 + \lambda), \lambda/(1 + \lambda))$, whereas for $\lambda = 0$ there are only *two* equilibria:

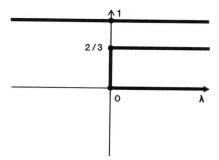

Figure 1.19

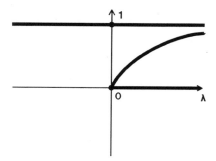

Figure 1.20

$(1,1)$ and $(0,0)$. (To see this, note that if p is greater than 0 then player 2 will set $q = 1$, and so p must equal 1, and $(1,1)$ is the only equilibrium with $q > 0$.)

At first sight a decrease in the number of equilibria might appear to violate the closed-graph property, but this is not the case: For λ positive but small, the mixed-strategy equilibrium $(\lambda/(1 + \lambda), \lambda/(1 + \lambda))$ is very close to the pure-strategy equilibrium $(0,0)$. Figures 1.19 and 1.20 display the equilibrium correspondences of these two games. More precisely, for each λ we display the set of p such that (p,q) is an equilibrium of $N(\lambda)$ for some q; this allows us to give a two-dimensional diagram.

Inspection of the diagrams reveals that each of these games has an odd number of Nash equilibria everywhere except $\lambda = 0$. Chapter 12 explains that this observation is generally true: If the strategy spaces are held fixed, there is an odd number of Nash equilibria for "almost all" payoff functions.

Finally, note that in figures 1.18a and 1.18b, although (D, R) is not a Nash equilibrium for $\lambda < 0$, it is an "ε-Nash equilibrium" in the sense of Radner (1980) if $\varepsilon \geq |\lambda|$: Each player's maximum gain to deviation is less than ε. More generally, an equilibrium of a given game will be an ε-Nash equilibrium for games "nearby"—a point developed and exploited by Fudenberg and Levine (1983, 1986), whose results are discussed in chapter 4.

1.3.3   Existence of Nash Equilibrium in Infinite Games with Continuous Payoffs

Economists often use models of games with an uncountable number of actions (as in the Cournot game of example 1.3 and the Hotelling game of example 1.4). Some might argue that prices or quantities are "really" infinitely divisible, while others might argue that "reality" is discrete and the continuum is a mathematical abstraction, but it is often easier to work with a continuum of actions rather than a large finite grid. Moreover, as Dasgupta and Maskin (1986) argue, when the continuum game does not have a Nash equilibrium, the equilibria corresponding to fine, discrete grids (whose existence was proved in subsection 1.3.1) could be very sensitive to exactly which finite grid is specified: If there were equilibria of the finite-grid version of the game that were fairly insensitive to the choice of the grid, one could take a sequence of finer and finer grids "converging" to the continuum, and the limit of a convergent subsequence of the discrete-action-space equilibria would be a continuum equilibrium under appropriate continuity assumptions. (To put it another way, one can pick equilibria of the discrete-grid version of the game that do not fluctuate with the grid if the continuum game has an equilibrium.)

Theorem 1.2 (Debreu 1952; Glicksberg 1952; Fan 1952)   Consider a strategic-form game whose strategy spaces S_i are nonempty compact convex subsets of an Euclidean space. If the payoff functions u_i are continuous in s and quasi-concave in s_i, there exists a pure-strategy Nash equilibrium.[22]

Proof   The proof is very similar to that of Nash's theorem: We verify that continuous payoffs imply nonempty, closed-graph reaction correspondences, and that quasi-concavity in players' own actions implies that the reaction correspondences are convex-valued.   ∎

Note that Nash's theorem is a special case of this theorem. The set of mixed strategies over a finite set of actions, being a simplex, is a compact, convex subset of an Euclidean space; the payoffs are polynomial, and therefore quasi-concave, in the player's own mixed strategy.

If the payoff functions are not continuous, the reaction correspondences can fail to have a closed graph and/or fail to be nonempty. The latter problem arises because discontinuous functions need not attain a maximum, as for example the function $f(x) = -|x|$, $x \neq 0$, $f(0) = -1$. To see how the reaction correspondence may fail to have a closed graph even when optimal reactions always exist, consider the following two-player game:

$$S_1 = S_2 = [0,1],$$

$$u_1(s_1,s_2) = -(s_1 - s_2)^2,$$

---

22. It is interesting to note that Debreu (1952) used a generalization of theorem 1.2 to prove that competitive equilibria exist when consumers have quasi-convex preferences.

$$u_2(s_1, s_2) = \begin{cases} -(s_1 - s_2 - \frac{1}{3})^2, & s_1 \geq \frac{1}{3} \\ -(s_1 - s_2 + \frac{1}{3})^2, & s_1 < \frac{1}{3}. \end{cases}$$

Here each player's payoff is strictly concave in his own strategy, and a best response exists (and is unique) for each strategy of the opponent. However, the game does not have a pure-strategy equilibrium: Player 1's reaction function is $r_1(s_2) = s_2$, while player 2's reaction function is $r_2(s_1) = s_1 - \frac{1}{3}$ for $s_1 \geq \frac{1}{3}, r_2(s_1) = s_1 + \frac{1}{3}$ for $s_1 < \frac{1}{3}$, and these reaction functions do not intersect.

Quasi-concavity is hard to satisfy in some contexts. For example, in the Cournot game the quasi-concavity of payoffs requires strong conditions on the second derivatives of the price and cost functions. Of course, Nash equilibria can exist even when the conditions of the existence theorems are not satisfied, as these conditions are sufficient but not necessary. However, in the Cournot case Roberts and Sonnenschein (1976) show that pure-strategy Cournot equilibria can fail to exist with "nice" preferences and technologies.

The absence of a pure-strategy equilibrium in some games should not be surprising, since pure-strategy equilibria need not exist in finite games, and these games can be approximated by games with real-valued action spaces but nonconcave payoffs. Figure 1.21 depicts the payoffs of player 1, who chooses an action s_1 in the interval $[\underline{s}_1, \bar{s}_1]$. Payoff function u_1 is continuous in s but not quasi-concave in s_1. This game is "almost" a game where player 1 has two actions, s_1' and s_1''. Suppose the same holds for player 2. Then the game is similar to a game with two actions per player, and we know (from "matching pennies," for instance) that such games may have no pure-strategy equilibrium.

When payoffs are continuous (but not necessarily quasi-concave), mixed strategies can be used to obtain convex-valued reactions, as in the following theorem.

Theorem 1.3 (Glicksberg 1952) Consider a strategic-form game whose strategy spaces S_i are nonempty compact subsets of a metric space. If the payoff functions u_i are continuous then there exists a Nash equilibrium in mixed strategies.

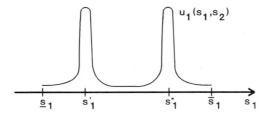

Figure 1.21

Here the mixed strategies are the (Borel) probability measures over the pure strategies, which we endow with the topology of weak convergence.[23] Once more, the proof applies a fixed-point theorem to the reaction correspondences. As we remarked above, the introduction of mixed strategies again makes the strategy spaces convex, the payoffs linear in own strategy and continuous in all strategies (when payoffs are continuous functions of the pure strategies, they are continuous in the mixed strategies as well[24]), and the reaction correspondences convex-valued. With infinitely many pure strategies, the space of mixed strategies is infinite-dimensional, so a more powerful fixed-point theorem than Kakutani's is required. Alternatively, one can approximate the strategy spaces by a sequence of finite grids. From Nash's theorem, each grid has a mixed-strategy equilibrium. One then argues that, since the space of probability measures is weakly compact, the sequence of these discrete equilibria has an accumulation point. Since the payoffs are continuous, it is easy to verify that the limit point is an equilibrium.

We have already seen that pure-strategy equilibria need not exist when payoffs are discontinuous. There are many examples to show that in this case mixed-strategy equilibria may fail to exist as well. (The oldest such example we know of is given in Sion and Wolfe 1957—see exercise 2.2 below.) Note: The Glicksberg theorem used above fails because when the pure-strategy payoffs are discontinuous the mixed-strategy payoffs are discontinuous too. Thus, as before, best responses may fail to exist for some of the opponents' strategies. Section 12.2 discusses the existence of mixed-strategy equilibria in discontinuous games and conditions that guarantee the existence of pure-strategy equilibria.

Exercises

Exercise 1.1*   This exercise asks you to work through the characterization of all the Nash equilibria of general two-player games in which each player has two actions (i.e., 2×2 matrix games). This process is time consuming but straightforward and is recommended to the student who is unfamiliar with the mechanics of determining Nash equilibria.

Let the game be as illustrated in figure 1.22.

The pure-strategy Nash equilibria are easily found by testing each cell of the matrix; e.g., (U, L) is a Nash equilibrium if and only if $a \geq e$ and $b \geq d$.

---

23. Fix a compact metric space A. A sequence of measures μ^n on A converges "weakly" to a limit μ if $\int f \, d\mu^n \to \int f \, d\mu$ for every real-valued continuous function f on A. The set of probability measures on A endowed with the topology of weak convergence is compact.
24. This is an immediate consequence of the definition of convergence we gave in note 23.

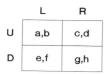

	L	R
U	a,b	c,d
D	e,f	g,h

Figure 1.22

	L	R
U	1,−1	3,0
D	4,2	0,−1

Figure 1.23

To determine the mixed-strategy equilibria requires more work. Let x be the probability player 1 plays U and let y be the probability player 2 plays L. We provide an outline, which the student should complete:

(i) Compute each player's reaction correspondence as a function of his opponent's randomizing probability.

(ii) For which parameters is player i indifferent between his two strategies regardless of the play of his opponent?

(iii) For which parameters does player i have a strictly dominant strategy?

(iv) Show that if neither player has a strictly dominant strategy, and the game has a unique equilibrium, the equilibrium must be in mixed strategies.

(v) Consider the particular example illustrated in figure 1.23.

(a) Derive the best-response correspondences graphically by plotting player i's payoff to his two pure strategies as a function of his opponent's mixed strategy.

(b) Plot the two reaction correspondences in the (x, y) space. What are the Nash equilibria?

Exercise 1.2* Find all the equilibria of the voting game of example 1.5.

Exercise 1.3 (Nash demand game)* Consider the problem of dividing a pie between two players. If we let x and y denote player 1's and player 2's payoffs, the vector (x, y) is feasible if and only if $x \geq x_0$, $y \geq y_0$, and $g(x, y) \leq 1$, where g is a differentiable function with $\partial g/\partial x > 0$ and $\partial g/\partial y > 0$ (for instance, $g(x, y) = x + y$). Assume that the feasible set is convex. The point (x_0, y_0) will be called the *status quo*. Nash (1950a) proposed axioms which implied that the "right" way to divide the pie is the allocation (x^*, y^*) that maximizes the product of the differences from the status quo $(x - x_0)(y - y_0)$ subject to the feasibility constraint $g(x, y) \leq 1$. In his 1953 paper, Nash looked for a game that would give this axiomatic bargaining solution as a Nash equilibrium.

(a) Suppose that both players simultaneously formulate demands x and y. If (x, y) is feasible, each player gets what he demanded. If (x, y) is infeasible, player 1 gets x_0 and player 2 gets y_0. Show that there exists a continuum of pure-strategy equilibria, and, more precisely, that any efficient division (x, y) (i.e., feasible and satisfying $g(x, y) = 1$) is a pure-strategy-equilibrium outcome.

(b)** Consider Binmore's (1981) version of the Nash "modified demand game." The feasible set is defined by $x \geq x_0$, $y \geq y_0$, and $g(x, y) \leq z$, where z has cumulative distribution F on $[z, \bar{z}]$ (suppose that $\forall z$, the feasible set is nonempty). The players do not know the realization of z before making demands. The allocation is made as previously, after the demands are made and z is realized. Derive the Nash-equilibrium conditions. Show that when F converges to a mass point at 1, any Nash equilibrium converges to the axiomatic bargaining solution.

Exercise 1.4 (Edgeworth duopoly)** There are two identical firms producing a homogeneous good whose demand curve is $q = 100 - p$. Firms simultaneously choose prices. Each firm has a capacity constraint of K. If the firms choose the same price they share the market equally. If the prices are unequal, $p_i < p_j$, the low-price firm, i, sells $\min(100 - p_i, K)$ and the high-price firm, j, sells $\min[\max(0, 100 - p_i - K), K]$. (There are many possible rationing rules, depending on the distribution of consumers' preferences and on how consumers are allocated to firms. If the aggregate demand represents a group of consumers each of whom buys one unit if the price p_i is less than his reservation price of r, and buys no units otherwise, and the consumer's reservation prices are uniformly distributed on $[0, 100]$, the above rationing rule says that the high-value consumers are allowed to purchase at price p_i before lower-value consumers are.) The cost of production is 10 per unit.

(a) Show that firm 1's payoff function is

$$u_1(p_1, p_2) = \begin{cases} (p_1 - 10)\min(100 - p_1, K), & p_1 < p_2 \\ (p_1 - 10)\min(50 - p_1/2, K), & p_1 = p_2 \\ (p_1 - 10)\min(100 - K - p_1, K), & p_1 > p_2, p_1 < 100 - K \\ 0, & \text{otherwise.} \end{cases}$$

(b) Suppose $30 < K < 45$. (Note that these inequalities are strict.) Show that this game does not have a pure-strategy Nash equilibrium by proving the following sequence of claims:

(i) If (p_1, p_2) is a pure-strategy Nash equilibrium, then $p_1 = p_2$. (Hint: If $p_1 \neq p_2$, then the higher-price firm has customers (Why?) and so the lower-price firm's capacity constraint is strictly binding. What happens if this firm charges a slightly higher price?)

(ii) If (p, p) is a pure-strategy Nash equilibrium, then $p > 10$.

(iii) If (p, p) is a pure-strategy Nash equilibrium, then p satisfies $p \leq 100 - 2K$.

(iv) If (p, p) is a pure-strategy Nash equilibrium, then $p = 100 - 2K$. (Hint: If $p < 100 - 2K$, is a deviation to a price between p and $100 - 2K$ profitable for either firm?)

(v) Since $K > 30$, there exists $\delta > 0$ such that a price of $100 - 2K + \delta$ earns a firm a higher profit than $100 - 2K$ when the other firm charges $100 - 2K$.

Note: The Edgeworth duopoly game does satisfy the assumptions of theorem 1.3 (restrict prices to the set $[0, 100]$) and so has a mixed-strategy equilibrium.

Exercise 1.5 (final-offer arbitration)*   Farber (1980) proposes the following model of final-offer arbitration. There are three players: a management $(i = 1)$, a union $(i = 2)$, and an arbitrator $(i = 3)$. The arbitrator must choose a settlement $t \in \mathbb{R}$ from the two offers, $s_1 \in \mathbb{R}$ and $s_2 \in \mathbb{R}$, made by the management and the union respectively. The arbitrator has exogenously given preferences $v_0 = -(t - s_0)^2$. That is, he would like to be as close to his "bliss point," s_0, as possible. The management and the union don't know the arbitrator's bliss point; they know only that it is drawn from the distribution P with continuous, positive density p on $[\underline{s}_0, \overline{s}_0]$. The management and the union choose their offers simultaneously. Their objective functions are $u_1 = -t$ and $u_2 = +t$, respectively.

Derive and interpret the first-order conditions for a Nash equilibrium. Show that the two offers are equally likely to be chosen by the arbitrator.

Exercise 1.6**   Show that the two-player game illustrated in figure 1.24 has a unique equilibrium. (Hint: Show that it has a unique pure-strategy equilibrium; then show that player 1, say, cannot put positive weight on both U and M; then show that player 1, say, cannot put positive weight on both U and D, but not on M, for instance.)

	L	M	R
U	1, -2	-2, 1	0, 0
M	-2, 1	1, -2	0, 0
D	0, 0	0, 0	1, 1

Figure 1.24

Exercise 1.7 (public good)*   Consider an economy with I consumers with "quasi-linear" utility functions,

$$u_i = V_i(x, \theta_i) + t_i,$$

where t_i is consumer i's income, x is a public decision (for instance, the quantity of a public good), $V_i(x, \theta_i)$ is consumer i's gross surplus for decision x, and θ_i is a utility parameter. The monetary cost of decision x is $C(x)$.
    The socially efficient decision is

$$x^*(\theta_1, \ldots, \theta_I) \in \arg\max_x \left\{ \sum_{i=1}^{I} V_i(x, \theta_i) - C(x) \right\}.$$

Assume (i) that the maximand in this program is strictly concave and (ii) that for all θ_{-i}, θ_i, and θ_i',

$$\theta_i' \neq \theta_i \Rightarrow x^*(\theta_{-i}, \theta_i') \neq x^*(\theta_{-i}, \theta_i).$$

Condition ii says that the optimal decision is responsive to the utility parameter of each consumer. (Condition i is satisfied if x belongs to $\mathbb{R}$, V_i is strictly concave in x, and C is strictly convex in x. Furthermore, if θ_i belongs to an interval of $\mathbb{R}$, V_i and C are twice differentiable, $\partial V_i / \partial x \partial \theta_i > 0$ or < 0, and x^* is an interior solution, then x^* is strictly increasing or strictly decreasing in θ_i, so that condition (ii) is satisfied as well.)
    Now consider the following "demand-revelation game": Consumers are asked to announce their utility parameters simultaneously. A pure strategy for consumer i is thus an announcement $\hat{\theta}_i$ of his parameter ($\hat{\theta}_i$ may differ from the true parameter θ_i). The realized decision is the optimal one for the announced parameters $x^*(\hat{\theta}_1, \ldots, \hat{\theta}_I)$, and consumer i receives a transfer from a "social planner" equal to

$$t_i(\hat{\theta}_1, \ldots, \hat{\theta}_I) = K_i + \sum_{j \neq i} V_j(x^*(\hat{\theta}_1, \ldots, \hat{\theta}_I), \hat{\theta}_j) - C(x^*(\hat{\theta}_1, \ldots, \hat{\theta}_I)),$$

when K_i is a constant.
    Show that telling the truth is dominant, in that any report $\hat{\theta}_i \neq \theta_i$ is strictly dominated by the truthful report $\hat{\theta}_i = \theta_i$.
    Because each player has a dominant strategy, it does not matter whether he knows the other players' utility parameters. Hence, even if the players do not know one another's payoffs (see chapter 6), it is still rational for them to tell the truth. This property of the dominant-strategy demand-revelation mechanism (called the *Groves mechanism*) makes it particularly interesting in a situation in which a consumer's utility parameter is known only to that consumer.

Exercise 1.8*   Consider the following model of bank runs, which is due to Diamond and Dybvig (1983). There are three periods ($t = 0, 1, 2$). There are many consumers—a continuum of them, for simplicity. All consumers are *ex ante* identical. At date 0, they deposit their entire wealth, $1, in a bank.

The bank invests in projects that yield $R each if the money is invested for two periods, where $R > 1$. However, if a project is interrupted after one period, it yields only $1 (it breaks even). Each consumer "dies" (or "needs money immediately") at the end of date 1 with probability x, and lives for two periods with probability $1 - x$. He learns which one obtains at the beginning of date 1. A consumer's utility is $u(c_1)$ if he dies in period 1 and $u(c_1 + c_2)$ if he dies in period 2, where $u' > 0$, $u'' < 0$, and c_1 and c_2 are the consumptions in periods 1 and 2.

An optimal insurance contract (c_1^*, c_2^*) maximizes a consumer's *ex ante* or expected utility. The consumer receives c_1^* if he dies at date 1, and otherwise consumes nothing at date 1 and receives c_2^* at date 2. The contract satisfies $xc_1^* + (1 - x)c_2^*/R = 1$ (the bank breaks even) and $u'(c_1^*) = R u'(c_2^*)$ (equality between the marginal rates of substitution). Note that $1 < c_1^* < c_2^*$. The issue is whether the bank can implement this optimal insurance scheme if it is unable to observe who needs money at the end of the first period. Suppose that the bank offers to pay $r_1 = c_1^*$ to consumers who want to withdraw their money in period 1. If $f \in [0, 1]$ is the fraction of consumers who withdraw at date 1, each withdrawing consumer gets r_1 if $fr_1 \leq 1$, and gets $1/f$ if $fr_1 > 1$. Similarly, consumers who do not withdraw at date 1 receive $\max\{0, R(1 - r_1 f)/(1 - f)\}$ in period 2.

(a) Show that it is a Nash equilibrium for each consumer to withdraw at date 1 if and only if he "dies" at that date.

(b) Show that another Nash equilibrium exhibits a bank run $(f = 1)$.

(c) Compare with the stag hunt.

Exercise 1.9* Suppose $p(q) = a - bq$ in the Cournot duopoly game of example 1.3.

(a) Check that the second-order and boundary conditions for equation (1.3) are satisfied. Compute the Nash equilibrium.

(b) Now suppose there are I identical firms, which all have cost function $c_i(q_i) = cq_i$. Compute the limit of the Nash equilibria as $I \to \infty$. Comment.

Exercise 1.10* Suppose there are I farmers, each of whom has the right to graze cows on the village common. The amount of milk a cow produces depends on the total number of cows, N, grazing on the green. The revenue produced by n_i cows is $n_i v(N)$ for $N < \bar{N}$, and $v(N) \equiv 0$ for $N \geq \bar{N}$, where $v(0) > 0$, $v' < 0$, and $v'' \leq 0$. Each cow costs c, and cows are perfectly divisible. Suppose $v(0) > c$. Farmers simultaneously decide how many cows to purchase; all purchased cows will graze on the common.

(a) Write this as a game in strategic form.

(b) Find the Nash equilibrium, and compare it against the social optimum.

(c) Discuss the relationship between this game and the Cournot oligopoly model.

(This exercise, constructed by R. Gibbons, is based on a discussion in Hume 1739.)

Exercise 1.11** We mentioned that theorem 1.3, which concerns the existence of a mixed-strategy Nash equilibrium when strategy spaces are nonempty, compact subsets of a metric space ($\mathbb{R}^n$, say) and when the payoff functions are continuous, can also be proved by taking a sequence of discrete approximations of the strategy spaces that "converge" to it. Go through the steps of the proof as carefully as you can.

Here is a sketch of the proof: Each discrete grid has a mixed-strategy equilibrium. By compactness, the sequence of discrete-grid equilibria has an accumulation point. Argue that this limit must be an equilibrium of the limit game with a continuum of actions. (This relies on the discrete grids becoming increasingly good approximations and the payoffs being continuous.)

Exercise 1.12* Consider a simultaneous-move auction in which two players simultaneously choose bids, which must be in nonnegative integer multiples of one cent. The higher bidder wins a dollar bill. If the bids are equal, neither player receives the dollar. Each player must pay his own bid, whether or not he wins the dollar. (The loser pays too.) Each player's utility is simply his net winnings; that is, the players are risk neutral. Construct a symmetric mixed-strategy equilibrium in which every bid less than 1.00 has a positive probability.

References

Aumann, R. 1959. Acceptable points in general cooperative *n*-person games. In *Contributions to the Theory of Games IV.* Princeton University Press.

Aumann, R. 1987. Correlated equilibrium as an extension of Bayesian rationality. *Econometrica* 55: 1–18.

Aumann, R. 1990. Communication need not lead to Nash equilibrium. Mimeo, Hebrew University of Jerusalem.

Axelrod, R. 1984. *The Evolution of Cooperation.* Basic Books.

Berge, C. 1963. *Topological Spaces.* Macmillan.

Bernheim, D. 1984. Rationalizable strategic behavior. *Econometrica* 52: 1007–1028.

Bernheim, D., D. Peleg, and M. Whinston. 1987. Coalition-proof Nash equilibria. I: Concepts. *Journal of Economic Theory.* 42: 1–12.

Bernheim, D., and M. Whinston. 1987. Coalition-proof Nash equilibria. II. Applications. *Journal of Economic Theory* 42: 13–22.

Bertrand, J. 1883. Théorie mathématique de la richesse sociale. *Journal des Savants* 499–508.

Binmore, K. 1981. Nash bargaining theory II. London School of Economics.

Cournot, A. 1838. *Recherches sur les Principes Mathematiques de la Theorie des Richesses.* English edition: *Researches into the Mathematical Principles of the Theory of Wealth,* ed. N. Bacon (Macmillan, 1897).

Dasgupta, P., and E. Maskin. 1986. The existence of equilibrium in discontinuous economic games. 1: Theory. *Review of Economic Studies* 53: 1–26.

Debreu, D. 1952. A social equilibrium existence theorem. *Proceedings of the National Academy of Sciences* 38: 886–893.

Diamond, D., and P. Dybvig. 1983. Bank runs, deposit insurance and liquidity. *Journal of Political Economy* 91: 401–419.

Diamond, P. 1982. Aggregate demand in search equilibrium. *Journal of Political Economy* 90: 881–894.

Dixit, A. 1986. Comparative statics for oligopoly. *International Economic Review* 27: 107–122.

Fahrquarson, R. 1969. *Theory of Voting*. Yale University Press.

Fan, K. 1952. Fixed point and minimax theorems in locally convex topological linear spaces. *Proceedings of the National Academy of Sciences* 38: 121–126.

Farber, H. 1980. An analysis of final-offer arbitration. *Journal of Conflict Resolution* 35: 683–705.

Farrell, J., and G. Saloner. 1985. Standardization, compatibility, and innovation. *Rand Journal of Economics* 16: 70–83.

Fisher, F. 1961. The stability of the Cournot oligopoly solution: The effects of speed of adjustment and increasing marginal costs. *Review of Economic Studies* 28: 125–135.

Fudenberg, D., and D. Kreps. 1988. A theory of learning, experimentation, and equilibrium in games. Mimeo, Stanford Graduate School of Business.

Fudenberg, D., and D. Levine. 1983. Subgame-perfect equilibria of finite and infinite horizon games. *Journal of Economic Theory* 31: 251–268.

Fudenberg, D., and D. Levine. 1986. Limit games and limit equilibria. *Journal of Economic Theory* 38: 261–279.

Glicksberg, I. L. 1952. A further generalization of the Kakutani fixed point theorem with application to Nash equilibrium points. *Proceedings of the National Academy of Sciences* 38: 170–174.

Green, J., and W. Heller. 1981. Mathematical analysis and convexity with applications to economics. In *Handbook of Mathematical Economics*, volume I, ed. K. Arrow and M. Intriligator. North-Holland.

Gul, F. 1989. Rational strategic behavior and the notion of equilibrium. Mimeo, Stanford Graduate School of Business.

Hahn, F. 1962. The stability of the Cournot oligopoly solution. *Review of Economic Studies* 29: 929–931.

Harsanyi, J. 1973a. Games with randomly disturbed payoffs: A new rationale for mixed strategy equilibrium points. *International Journal of Game Theory* 1: 1–23.

Harsanyi, J. 1973b. Oddness of the number of equilibrium points: A new proof. *International Journal of Game Theory* 2: 235–250.

Harsanyi, J., and R. Selten. 1988. *A General Theory of Equilibrium Selection in Games*. MIT Press.

Hirsch, M., and S. Smale. 1974. *Differential Equations, Dynamical Systems and Linear Algebra*. Academic Press.

Hotelling, H. 1929. Stability in competition. *Economic Journal* 39: 41–57.

Hume, D. 1739. *A Treatise on Human Nature* (Everyman edition: J. M. Dent, 1952).

Kuhn, H. 1953. Extensive games and the problem of information. *Annals of Mathematics Studies, No. 28*. Princeton University Press.

Luce, D., and H. Raiffa. 1957. *Games and Decisions*. Wiley.

Maynard Smith, J. 1982. *Evolution and the Theory of Games*. Cambridge University Press.

Maynard Smith, J., and G. R. Price. 1973. The logic of animal conflicts. *Nature* 246: 15–18.

Milgrom, P., and J. Roberts. 1989. Adaptive and sophisticated learning in repeated normal forms. Mimeo, Stanford University.

Moulin, H. 1979. Dominance solvable voting schemes. *Econometrica* 37: 1337–1353.

Moulin, H. 1984. Dominance solvability and Cournot stability. *Mathematical Social Sciences* 7: 83–102.

Moulin, H. 1986. *Game Theory for the Social Sciences*. New York University Press.

Nash, J. 1950a. The bargaining problem. *Econometrica* 18: 155–162.

Nash, J. 1950b. Equilibrium points in *n*-person games. *Proceedings of the National Academy of Sciences* 36: 48–49.

Nash, J. 1953. Two-person cooperative games. *Econometrica* 21: 128–140.

Novshek, W. 1985. On the existence of Cournot equilibrium. *Review of Economic Studies* 52: 85–98.

Nyarko, Y. 1989. Bayesian learning in games. Mimeo, New York University.

Ordeshook, P. 1986. *Game Theory and Political Theory: An Introduction*. Cambridge University Press.

Radner, R. 1980. Collusive behavior in non-cooperative epsilon equilibria of oligopolies with long but finite lives. *Journal of Economic Theory* 22: 136–154.

Roberts, J., and H. Sonnenschein. 1976. On the existence of Cournot equilibrium without concave profit functions. *Journal of Economic Theory* 13: 112–117.

Schelling, T. 1960. *The Strategy of Conflict*. Harvard University Press.

Seade, J. 1980. The stability of Cournot revisited. *Journal of Economic Theory* 23: 15–17.

Shapley, L. 1964. Some topics in two-person games. In *Contributions to the Theory of Games* (Princeton Annals of Mathematical Studies, no. 52).

Sion, M., and P. Wolfe. 1957. On a game without a value. In *Contributions to the Theory of Games*, Volume III (Princeton Annals of Mathematical Studies, no. 39).

Sugden, R. 1986. *The Economic of Rights, Cooperation, and Welfare*. Blackwell.

Thorlund-Petersèn, L. 1990. Iterative computation of Cournot equilibrium. *Games and Economic Behavior* 2: 61–95.

von Neumann, J. 1928. Zur Theorie der Gesellschaftsspiele. *Math. Annalen* 100: 295–320.

Waltz, K. 1959. *Man, the State, and War*. Columbia University Press.

2 Iterated Strict Dominance, Rationalizability, and Correlated Equilibrium

Most economic applications of game theory use the concept of Nash equilibrium or one of the more restrictive "equilibrium refinements" we introduce in later chapters. However, as we warned in chapter 1, in some situations the Nash concept seems too demanding. Thus, it is interesting to know what predictions one can make without assuming that a Nash equilibrium will occur. Section 2.1 presents the notions of iterated strict dominance and rationalizability, which derive predictions using only the assumptions that the structure of the game (i.e., the strategy spaces and the payoffs) and the rationality of the players are common knowledge. As we will see, these two notions are closely related, as rationalizability is essentially the contrapositive of iterated strict dominance.

Section 2.2 introduces the idea of a correlated equilibrium, which extends the Nash concept by supposing that players can build a "correlating device" that sends each of them a private signal before they choose their strategy.

2.1 Iterated Strict Dominance and Rationalizability[††]

We introduced iterated strict dominance informally at the beginning of chapter 1. We will now define it formally, derive some of its properties, and apply it to the Cournot model. We will then define rationalizability and relate the two concepts. As throughout, we restrict our attention to finite games except where we explicitly indicate otherwise.

2.1.1 Iterated Strict Dominance: Definition and Properties

Definition 2.1 The process of iterated deletion of strictly dominated strategies proceeds as follows: Set $S_i^0 \equiv S_i$ and $\Sigma_i^0 \equiv \Sigma_i$. Now define S_i^n recursively by

$$S_i^n = \{s_i \in S_i^{n-1} \mid \text{there is no } \sigma_i \in \Sigma_i^{n-1} \text{ such that}$$

$$u_i(\sigma_i, s_{-i}) > u_i(s_i, s_{-i}) \text{ for all } s_{-i} \in S_{-i}^{n-1}\}$$

and define

$$\Sigma_i^n = \{\sigma_i \in \Sigma_i \mid \sigma_i(s_i) > 0 \text{ only if } s_i \in S_i^n\}.$$

Set

$$S_i^\infty = \bigcap_{n=0}^\infty S_i^n.$$

S_i^∞ is the set of player i's pure strategies that survive iterated deletion of strictly dominated strategies. Set Σ_i^∞ to be all mixed strategies σ_i such that there is no σ_i' with $u_i(\sigma_i', s_{-i}) > u_i(\sigma_i, s_{-i})$ for all $s_{-i} \in S_{-i}^\infty$. This is the set of player i's mixed strategies that survive iterated strict dominance.

In words, S_i^n is the set of player i's strategies that are not strictly dominated when players $j \neq i$ are constrained to play strategies in S_j^{n-1} and Σ_i^n

is the set of mixed strategies over S_i^n. Note, however, that Σ_i^∞ may be smaller than the set of mixed strategies over S_i^∞. The reason for this, as was shown in figure 1.3, is that some mixed strategies with support S_i^∞ can be dominated. (In that example, $S_i^\infty = S_i$ for both players i because no pure strategy is eliminated in the first round of the process.)

Note that in a finite game the sequence of iterations defined above must cease to delete further strategies after a finite number of steps. The intersection S_i^∞ is simply the final set of surviving strategies. Note also that each step of the iteration requires one more level of the assumption "I know that you know... that I know the payoffs." For this reason, conclusions based on a large number of iterations tend to be less robust to small changes in the information players have about one another.

The reader may wonder whether the limit set $S^\infty = S_1^\infty \times \cdots \times S_I^\infty$ depends on the particular way that we have specified the process of deletion proceeds: We assumed that at each iteration all dominated strategies of each player are deleted simultaneously. Alternatively, we could have eliminated player 1's dominated strategies, than player 2's,..., then player I's, and started again with player 1,..., *ad infinitum*. Clearly there are many other iterative procedures that can be defined to eliminate strictly dominated strategies. Fortunately *all these procedures yield the same surviving strategies S^∞ and Σ^∞*, as is shown by exercise 2.1. (We will show in chapter 11 that this property does not hold for weakly dominated strategies; that is, which strategies survive in the limit may depend on the order of deletion.)

The reader may also wonder whether one could not delete all the dominated (pure and mixed) strategies at each round of the iterative process instead of first deleting only dominated pure strategies and then deleting mixed strategies at the end. The two ways to proceed actually yield the same sets Σ_i^∞. The reason is that a strategy is strictly dominated against all pure strategies of the opponents if and only if it is dominated against all of their mixed strategies, as we saw in subsection 1.1.2. Thus, whether a nondegenerate mixed strategy σ_i for player i is deleted at round n doesn't alter which strategies of player i's opponents are deleted at the next round. Thus, at each round, the sets of remaining *pure* strategies are the same under the two alternative definitions. Therefore, the undominated mixed strategies Σ_i^∞ are the same.

Definition 2.2 A game is solvable by iterated (strict) dominance if, for each player i, S_i^∞ is a singleton (i.e., a one-element set).

When the iterated deletion of strictly dominated strategies yields a unique strategy profile (as is the case in figure 1.1 or in the prisoner's dilemma of figure 1.7), this strategy profile is necessarily a Nash equilibrium (indeed, it is the unique Nash equilibrium). The proof goes as follows: Let $(s_1^*,\ldots,s_I^*)$ denote this strategy profile, and suppose that there exist i and $s_i \in S_i$ such that $u_i(s_i, s_{-i}^*) > u_i(s_i^*, s_{-i}^*)$. Then if one round of elimination of strictly dominated strategies has sufficed to yield this unique profile, s_i^* must dominate all other strategies in S_i, which is impossible as s_i is a better response to s_{-i}^* than s_i^*. More generally, suppose that in the iterated deletion s_i is strictly dominated at some round by s_i', which in turn is eliminated at a later round because it becomes strictly dominated by s_i'', ..., which is finally eliminated by s_i^*. Because s_{-i}^* belongs to the undominated strategies of player i's opponents at each round, by transitivity s_i^* must be a better response to s_{-i}^* than s_i—a contradiction. Conversely, it is easy to see that in any Nash equilibrium the players must play strategies that are not eliminated by iterated strict dominance.

It is also easy to see that if players repeatedly play the same game, and infer their opponents' behavior from past observations, eventually only strategies that survive iterated deletion of strictly dominated strategies will be played. First, because opponents won't play dominated strategies, players will learn that such strategies are not used. They will then use only strategies that are not strictly dominated, given that the dominated strategies of their opponents are not used. After more learning, this will be learned by the opponents, and so on.

2.1.2 An Application of Iterated Strict Dominance

Example 2.1: Iterated Deletion in the Cournot Model[1]

We now make stronger assumptions on the (infinite-action) Cournot model introduced in example 1.3: Suppose that u_i is strictly concave in $q_i (\partial^2 u_i/\partial q_i^2 < 0)$, that the cross-partial derivative is negative $(\partial^2 u_i/\partial q_i \partial q_j < 0$, which is the case if $p' < 0$ and $p'' \le 0$), and that the reaction curves r_1 and r_2 (which are continuous and downward-sloping from the previous two assumptions) intersect only once at a point N, at which r_1 is strictly steeper than r_2. This situation is depicted in figure 2.1. (Note that N is stable, in the terminology introduced in subsection 1.2.5.)

Let q_1^m and q_2^m denote the monopoly outputs: $q_1^m = r_1(0)$ and $q_2^m = r_2(0)$. The first round of deletion of strictly dominated strategies yields $S_i^1 = [0, q_i^m]$. The second round of deletion yields $S_i^2 = [r_i(q_j^m), q_i^m] \equiv [\underline{q}_i^2, q_i^m]$, as indicated in figure 2.1. Consider, for instance, firm 2. Knowing that firm 1 won't pick output greater than q_i^m, choosing output q_2 under $r_2(q_1^m) \equiv \underline{q}_2^2$ is strictly dominated by playing $\underline{q}_2^2$ by strict concavity of firm 2's payoff in its own output. And similarly for firm 1. The third round of deletion yields $S_i^3 = [\underline{q}_i^2, r_i(\underline{q}_j^2)] \equiv [\underline{q}_i^2, \overline{q}_i^3]$, and so on. More generally, iterated deletion yields a sequence of shrinking intervals around the outputs (q_1^*, q_2^*) corresponding to the intersection N of the reaction curves. For $n = 2k + 1$,

$$\underline{q}_i^{2k+1} = \underline{q}_i^{2k} \qquad \text{and} \qquad \overline{q}_i^{2k+1} = r_i(\underline{q}_j^{2k});$$

1. This example is inspired by Gabay and Moulin 1980. See also Moulin 1984.

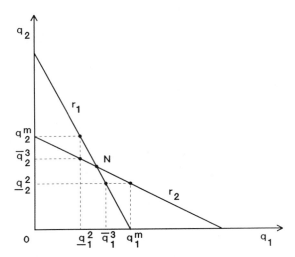

Figure 2.1

for $n = 2k$,

$$\underline{q}_i^{2k} = r_i(\overline{q}_j^{2k-1}) \qquad \text{and} \qquad \overline{q}_i^{2k} = \overline{q}_i^{2k-1}.$$

A difference between this process and the case of finite strategy spaces is that the process of deletion does not stop after a finite number of steps. Nevertheless, the process does converge, because the sequences $\underline{q}_i^n$ and $\overline{q}_i^n$ both converge to q_i^*, so that the process of iterated deletion of strictly dominated strategies yields N as the unique "reasonable" prediction. (Let $\underline{q}_i^\infty \equiv \lim \underline{q}_i^n \leq q_i^*$ and $\overline{q}_i^\infty \equiv \lim \overline{q}_i^n \geq q_i^*$. From the definition of $\underline{q}_i^n$ and $\overline{q}_i^n$ and by continuity of the reaction curves, one has $\overline{q}_i^\infty = r_i(\underline{q}_j^\infty)$ and $\underline{q}_j^\infty = r_j(\overline{q}_i^\infty)$. Hence, $\overline{q}_i^\infty = r_i(r_j(\overline{q}_i^\infty))$, which is possible only if $\overline{q}_i^\infty = q_i^*$; and similarly for $\underline{q}_i^\infty$.)

We conclude that this Cournot game is solvable by iterated strict dominance. This need not be the case for other specifications of the payoff functions; see exercise 2.4.

2.1.3 Rationalizability

The concept of rationalizability was introduced independently by Bernheim (1984) and Pearce (1984), and was used by Aumann (1987) and by Brandenberger and Dekel (1987) in their papers on the "Bayesian approach" to the choice of strategies.

Like iterated strict dominance, rationalizability derives restrictions on play from the assumptions that the payoffs and the "rationality" of the players are common knowledge. The starting point of iterated strict dominance is the observation that a rational player will never play a strictly dominated strategy. The starting point of rationalizability is the complementary question: What are *all* the strategies that a rational player could play? The answer is that a rational player will use only those strategies that are best responses to some beliefs he might have about the strategies of his opponents. Or, to use the contrapositive, a player cannot reasonably play a strategy that is not a best response to some beliefs about his opponents' strategies. Moreover, since the player knows his opponents' payoffs, and knows they are rational, he should not have arbitrary beliefs about their strategies. He should expect his opponents to use only strategies that are best responses to some beliefs that they might have. And these opponents' beliefs, in turn, should also not be arbitrary, which leads to an infinite regress. In the two-player case, the infinite regress has the form "I'm playing strategy σ_1 because I think player 2 is using σ_2, which is a reasonable belief because I would play it if I were player 2 and I thought player 1 was using σ_1', which is a reasonable thing for player 2 to expect because σ_1' is a best response to $\sigma_2', \ldots$."

Formally, rationalizability is defined by the following iterative process.

Definition 2.3 Set $\tilde{\Sigma}_i^0 \equiv \Sigma_i$, and for each i recursively define

$$\tilde{\Sigma}_i^n = \left\{ \sigma_i \in \tilde{\Sigma}_i^{n-1} \mid \exists \; \sigma_{-i} \in \underset{j \neq i}{\times} \text{ convex hull } (\tilde{\Sigma}_j^{n-1}) \text{ such that} \right.$$

$$\left. u_i(\sigma_i, \sigma_{-i}) \geq u_i(\sigma_i', \sigma_{-i}) \text{ for all } \sigma_i' \in \tilde{\Sigma}_i^{n-1} \right\}.$$

The *rationalizable strategies for player i* are $R_i = \bigcap_{n=0}^{\infty} \tilde{\Sigma}_i^n$.

In words, $\tilde{\Sigma}_{-i}^{n-1}$ are the strategies for player i's opponents that "survive" through round $(n-1)$, and $\tilde{\Sigma}_i^n$ is the set of i's surviving strategies that are best responses to some strategy in $\tilde{\Sigma}_{-i}^{n-1}$. The reason the convex hull operator appears in the definition is that player i might not be certain which of several strategies $\sigma_j \in \tilde{\Sigma}_j^{n-1}$ player j will use.[2] And it may be that, although both σ_j' and σ_j'' are in $\tilde{\Sigma}_j^{n-1}$, the mixture $(\frac{1}{2}\sigma_j', \frac{1}{2}\sigma_j'')$ is not. This is illustrated in figure 2.2. In the game of figure 2.2, player 2 has only two pure strategies: L and R. Then any pure strategy s_1 of player 1 is associated with two potential payoffs: $x \equiv u_1(s_1, L)$ and $y \equiv u_1(s_1, R)$. Figure 2.2a describes x and y for player 1's four pure strategies. Strategy A is a best response for player 1 to L and strategy B is a best response to R, but the mixed strategy $(\frac{1}{2}A, \frac{1}{2}B)$ is dominated by C and hence is not a best response to any strategy of player 2.

A strategy profile σ is *rationalizable* if σ_i is rationalizable for each player i. Note that every Nash equilibrium is rationalizable, since if σ^* is a Nash

---

2. The convex hull of a set X is the set of linear combinations of elements in this set:

$$\{x \mid \exists \, (y, z) \in X^2, \exists \, \lambda \in [0, 1] \text{ s.t. } x = \lambda y + (1 - \lambda)z\}.$$

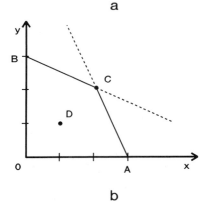

s_1	x	y
A	3	0
B	0	3
C	2	2
D	1	1

a

b

Figure 2.2

equilibrium then $\sigma_i^* \in \tilde{\Sigma}_i^n$ for each n. Thus, the set of rationalizable strategies is nonempty.

Theorem 2.1 (Bernheim 1984; Pearce 1984)   The set of rationalizable strategies is nonempty and contains at least one pure strategy for each player. Further, each $\sigma_i \in R_i$ is (in Σ_i) a best response to an element of

$$\times_{j \neq i} \text{ convex hull } (R_j).$$

Sketch of Proof   The proof shows inductively that the $\tilde{\Sigma}_i^n$ in the definition of rationalizability are closed, nonempty, and nested and that they contain a pure strategy. Their infinite intersection is thus nonempty and contains a pure strategy. The existence of an element of $\times_{j \neq i}$ convex hull (R_j) to which $\sigma_i \in R_i$ is a best response is obtained by induction on n.   ∎

2.1.4   Rationalizability and Iterated Strict Dominance (technical)

The condition of not being a best response, which is used in defining rationalizability, looks very close to that of being strictly dominated. In fact these two conditions are equivalent in two-player games.

It is clear that, with any number of players, a strictly dominated strategy is never a best response: If σ_i' strictly dominates σ_i relative to Σ_{-i}, then σ_i' is a strictly better response than σ_i to every σ_{-i} in Σ_{-i}. Thus, *in general games, the set of rationalizable strategies is contained in the set that survives*

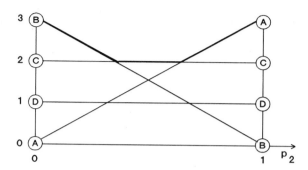

Figure 2.3

iterated strict dominance. The converse in two-player games is a consequence of the separating hyperplane theorem.

To gain some intuition, consider the game in figure 2.2. Figure 2.2b plots player 1's payoff possibilities (x, y) corresponding to different strategies of player 1. A strategy σ_1 is strictly dominated if there is another strategy that gives player 1 a strictly higher payoff no matter how player 2 plays—i.e., if σ_1 yields payoff possibility (x, y) and there is a strategy that yields (x', y'), with $x' > x$ and $y' > y$. It is clear that the pure strategy D is dominated by C and that the only undominated strategies correspond to payoffs on the line segments $\overline{AC}$ and $\overline{BC}$. (Note that mixtures of A and B are dominated, even though neither A nor B is dominated as a pure strategy.)

It is also easy to see from the diagram which strategies are not best responses to any strategy of player 2: A strategy $(p_2, 1 - p_2)$ for player 2 corresponds to weights on L and R, and thus to a family of lines which we can interpret as player 1's "indifference curves." Player 1's best responses to $(p_2, 1 - p_2)$ are those strategies yielding the maximal payoffs with these weights, which are exactly the points of the efficient set $\overline{AC} \cup \overline{BC}$ with which the indifference curves are "tangent" (subtangent) (see figure 2.3). Any strategy of player 1 that is a best response to some strategy of player 2 thus corresponds to a point on the efficient frontier (the bold line in the figure), so any strategy that is not a best response must lie in the interior and hence be dominated. This is the intuition for the following theorem.

Theorem 2.2 (Pearce 1984) Rationalizability and iterated strict dominance coincide in two-player games.

Proof Let S^n denote the set of pure strategies remaining after n rounds of the deletion of strictly dominated strategies, let Σ^n be the corresponding mixed strategies, and let $\tilde{\Sigma}^n$ be the set of mixed strategies that survive n rounds of the iteration in the definition of rationalizability. Clearly the set Σ^0 of mixed strategies corresponding to S^0 equals $\tilde{\Sigma}^0$. Assume that $\Sigma^n = \tilde{\Sigma}^n$. For any finite set A, let $\Delta(A)$ denote the space of probability distributions over A. Any s_i in S_i^{n+1} is undominated in $\Delta(S_i^{n+1})$ given that σ_j belongs to

Σ_j^n; otherwise it would have been eliminated. Now consider the vectors

$$\vec{u}_i(\sigma_i) = \{u_i(\sigma_i, s_j)\}_{s_j \in S_j^n}$$

for each $\sigma_i \in \Sigma_i^n$. The set of such vectors is convex, and, by the definition of iterated dominance, S_i^{n+1} contains exactly the s_i such that $\vec{u}_i(s_i)$ is undominated in this set. Fix $\tilde{s}_i$ in S_i^{n+1}. By the separating hyperplane theorem, there exists

$$\sigma_j = \{\sigma_j(s_j)\}_{s_j \in S_j^n}$$

such that, for all $\sigma_i \in \Sigma_i^n$,

$$\sigma_j \cdot (\vec{u}_i(\tilde{s}_i) - \vec{u}_i(\sigma_i)) \geq 0$$

(where a dot denotes the inner product), or

$$u_i(\tilde{s}_i, \sigma_j) \geq u_i(\sigma_i, \sigma_j) \; \forall \; \sigma_i \in \Sigma_i^n = \tilde{\Sigma}_i^n.$$

This means that $\tilde{s}_i$ is a best response in $\tilde{\Sigma}_i^n$ to a strategy σ_j in convex hull ($\tilde{\Sigma}_j^n$). Thus, $\tilde{s}_i \in \tilde{\Sigma}_i^{n+1}$, and we conclude that $\Sigma^{n+1} = \tilde{\Sigma}^{n+1}$. ∎

Remark Pearce gives a different proof based on the existence of the minmax value in finite two-player zero-sum games.[3] The minmax theorem, in turn, is usually proved with the separating hyperplane theorem.

The equivalence between being strictly dominated and not being a best response breaks down in games with three or more players (see exercise 2.7). The point is that, since mixed strategies assume independent mixing, the set of mixed strategies is not convex. In figure 2.2, the problem becomes that the mixed strategies no longer correspond to the set of all tangents to the efficient surface, so a strategy might be on the efficient surface without being a best response to a mixed strategy. However, allowing for correlation in the definition of rationalizability restores equivalence: A strategy is strictly dominated if and only if it is not a best response to a correlated mixed strategy of the opponents. (A correlated mixed strategy for player i's opponents is a general probability distribution on S_{-i}, i.e., an element of $\Delta(S_{-i})$, while a mixed-strategy profile for player i's opponents is an element of $\times_{j \neq i} \Delta(S_j)$.) This gives rise to the notion of *correlated rationalizability*, which is equivalent to iterated strict dominance.

To see this, modify the proof above, replacing the subscript j with the subscript $-i$. The separating hyperplane theorem shows that if $\tilde{s}_i \in S_i^{n+1}$,

---

3. A two-person, zero-sum game with strategy spaces S_1 and S_2 has a (minmax) value if

$$\sup_{s_1 \in S_1} \inf_{s_2 \in S_2} u_1(s_1, s_2) = \inf_{s_2 \in S_2} \sup_{s_1 \in S_1} u_1(s_1, s_2).$$

If a game has a value u_1^* and if there exists (s_1^*, s_2^*) such that $u_1(s_1^*, s_2^*) = u_1^*$, then (s_1^*, s_2^*) is called a *saddle point*. Von Neumann (1928) and Fan (1952, 1953) have given sufficient conditions for the existence of a saddle point.

there is a vector

$$\sigma_{-i} = \{\sigma(s_1, \ldots, s_{i-1}, s_{i+1}, \ldots, s_I)\}_{s_{-i} \in \tilde{S}^n_{-i}}$$

such that $u_i(\tilde{s}_i, \sigma_{-i}) \geq u_i(\sigma_i, \sigma_{-i})$ for all $\sigma_i \in \tilde{\Sigma}^n_i$. However, σ_{-i} is an arbitrary probability distribution over S^n_{-i}, and in general it cannot be interpreted as a mixed strategy, as it may involve player i's rivals' correlating their randomizations.

2.1.5 Discussion

Rationalizability, by design, makes very weak predictions; it does not distinguish between any outcomes that cannot be excluded on the basis of common knowledge of rationality. For example, in the battle of the sexes (figure 1.10a), rationalizability allows the prediction that the players are certain to end up at (F, B), where both get 0. (F, B) is not a Nash equilibrium; in fact, both players can gain by deviating. We can see how it might nevertheless occur: Player 1 plays F, expecting player 2 to play F, and 2 plays B expecting 1 to play B. Thus, we might be unwilling to say (F, B) *wouldn't* happen, especially if these players haven't played each other before. In some special cases, such as if we know that player 2's past play with other opponents has led him to expect (B, B) while player 1's has led him to expect (F, F), (F, B) might even be the most likely outcome. However, such situations seem rare; most often we might hesitate to predict that (F, B) has high probability. Rabin (1989) formalizes this idea by asking how likely each player can consider a given outcome. If player 1 is choosing a best response to his subjective beliefs $\hat{\sigma}_2$ about player 2's strategy, then for any value of $\hat{\sigma}_2$ player 1 must assign (F, B) a probability no greater than $\frac{1}{3}$: If he assigns a probability greater than $\frac{1}{3}$ to player 2's playing B, player 1 will play B. Similarly, player 2 cannot assign (F, B) a probability greater than $\frac{2}{3}$. Thus, Rabin argues that we ought to be hesitant to assign (F, B) a probability greater than the maximum of the two probabilities (that is, $\frac{2}{3}$).

2.2 Correlated Equilibrium[††]

The concept of Nash equilibrium is intended to be a minimal necessary condition for "reasonable" predictions in situations where the players must choose their strategies independently. Now consider players who may engage in preplay discussion, but then go off to isolated rooms to choose their strategies. In some situations, both players might gain if they could build a "signaling device" that sent signals to the separate rooms. Aumann's (1974) notion of a correlated equilibrium captures what could be achieved with any such signals. (See Myerson 1986 for a fuller introduction to this concept, and for a discussion of its relationship to the theory of mechanism design.)

	L	R
U	5,1	0,0
D	4,4	1,5

Figure 2.4

To motivate this concept, consider Aumann's example, presented in figure 2.4. This game has three equilibria: (U, L),(D, R), and a mixed-strategy equilibrium in which each player puts equal weight on each of his pure strategies and that gives each player 2.5. If they can jointly observe a "coin flip" (or sunspots, or any other publicly observable random variable) before play, they can achieve payoffs (3, 3) by a joint randomization between the two pure-strategy equilibria. (For example, flip a fair coin, and use the strategies "player 1 plays U if heads and D if tails; player 2 plays L if heads and R if tails"). More generally, by using a publicly observable random variable, the players can obtain any payoff vector in the convex hull of the set of Nash-equilibrium payoffs. Conversely, the players cannot obtain any payoff vector outside the convex hull of Nash payoffs by using publicly observable random variables.

However, the players can do even better (still without binding contracts) if they can build a device that sends *different but correlated signals* to each of them. This device will have three equally likely states: A, B, and C. Suppose that if A occurs player 1 is perfectly informed, but if the state is B or C player 1 does not know which of the two prevails. Player 2, conversely, is perfectly informed if the state is C, but he cannot distinguish between A and B. In this transformed game, the following is a Nash equilibrium: Player 1 plays U when told A, and D when told (B, C); player 2 plays R when told C, and L when told (A, B). Let's check that player 1 does not want to deviate. When he observes A, he knows that player 2 observes (A, B), and thus that player 2 will play L; in this case U is player 1's best response. If player 1 observes (B, C), then conditional on his information he expects player 2 to play L and R with equal probability. In this case player 1 will average 2.5 from either of his choices, so he is willing to choose D. So player 1 is choosing a best response; the same is easily seen to be true for player 2. Thus, we have constructed an equilibrium in which the players' choices are correlated: The outcomes (U, L),(D, L), and (D, R) are chosen with probability $\frac{1}{3}$ each, and the "bad" outcome (U, R) never occurs. In this new equilibrium the expected payoffs are $3\frac{1}{3}$ each, which is outside the convex hull of the equilibrium payoffs of the original game without the signaling device. (Note that adding the signaling device does not remove the "old" equilibria: Since the signals do not influence payoffs, if player 1 ignores his signal, player 2 may as well ignore hers.)

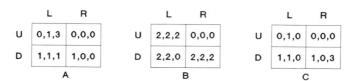

Figure 2.5

The next example of a correlated equilibrium illustrates the familiar game-theoretic point that a player may *gain* from limiting his own information *if the opponents know he has done so*, because this may induce the opponents to play in a desirable fashion.

In the game illustrated in figure 2.5, player 1 chooses rows, player 2 chooses columns, and player 3 chooses matrices. In this game the unique Nash equilibrium is (D, L, A), with payoffs (1, 1, 1).

Now imagine that the players build a correlating device with two equally likely outcomes, H ("heads") and T ("tails"), and that they arrange for the outcome to be perfectly revealed to players 1 and 2, while player 3 receives no information at all. In this game, a Nash equilibrium is for player 1 to play U if H and D if T, player 2 to play L if H and R if T, and player 3 to play B. Player 3 now faces a distribution of $\frac{1}{2}$(U, L) and $\frac{1}{2}$(D, R), which makes B a best response. Note the importance of players 1 and 2 knowing that player 3 does not know whether heads or tails prevailed when choosing the matrix. If the random variable were publicly observable and players 1 and 2 played the above strategies, then player 3 would choose matrix A if H and matrix C if T, and thus players 1 and 2 would deviate as well. As we observed, the equilibrium would then give player 3 a payoff of 1.

With these examples as an introduction, we turn to a formal definition of correlated equilibrium. There are two equivalent ways to formulate the definition.

The first definition explicitly defines strategies for the "expanded game" with a correlating device and then applies the definition of Nash equilibrium to the expanded game. Formally, we identify a correlating device with a triple $(\Omega, \{H_i\}, p)$. Here Ω is a (finite) state space corresponding to the outcomes of the device (e.g., H or T in our discussion of figure 2.5), and p is a probability measure on the state space Ω.

Player i's information about which $\omega \in \Omega$ occurred is represented by the *information partition* H_i; if the true state is ω, player i is told that the state lies in $h_i(\omega)$. In our discussion of figure 2.4, player 1's information partition is ((A), (B, C)) and player 2's partition is ((A, B), (C)). In the discussion of figure 2.5, players 1 and 2 have the partition ((H), (T)); player 3's partition is the one-element set (H, T).

More generally, a *partition* of a finite set Ω is a collection of disjoint subsets of Ω whose union is Ω. An *information partition* H_i assigns an $h_i(\omega)$

to each ω in such a way that $\omega \in h_i(\omega)$ for all ω. The set $h_i(\omega)$ consists of those states that player i regards as possible when the truth is ω; the requirement that $\omega \in h_i(\omega)$ means that player i is never "wrong" in the weak sense that he never regards the true state as impossible. However, player i may be poorly informed. If his partition is the one-element set $h_i(\omega) = \Omega$ for all ω, he has no information at all beyond his prior. (This is called the "trivial partition.")

For all h_i with positive prior probability, player i's posterior beliefs about Ω are given by Bayes' law: $p(\omega|h_i) = p(\omega)/p(h_i)$ for ω in h_i, and $p(\omega|h_i) = 0$ for ω not in h_i.

Given a correlating device $(\Omega, \{H_i\}, p)$, the next step is to define strategies for the expanded game where players can condition their play on the signal the correlating device sends them. A pure strategy for the expanded game can be viewed as a function ∂_i that maps elements h_i of H_i—the possible signals that player i receives—to pure strategies $s_i \in S_i$ of the game without the correlating device. Note that if $\omega' \in h_i(\omega)$, then necessarily ∂_i prescribes the same actions in states ω and ω'. Instead of defining strategies in this way as maps from information sets to elements of S_i, it will be more convenient for our analysis to use an equivalent formulation: We will define pure strategies ∂_i as maps from Ω to S_i with the additional property that $\partial_i(\omega) = \partial_i(\omega')$ if $\omega' \in h_i(\omega)$. The formal term for this is that the strategies are *adapted* to the information structure. (Mixed strategies can be defined in the obvious way, but they will be irrelevant if we take the state space Ω to be sufficiently large. For example, instead of player 1 playing $(\frac{1}{2}U, \frac{1}{2}D)$ when given signal h_i, we could construct an expanded state space $\hat{\Omega}$ where each $\omega \in h_i$ is replaced by two equally likely states, ω' and ω'', and player 1 is told both "h_i" and whether the state is of the single-prime or the double-prime kind. Then player i can use the pure strategy "play U if told h_i and single-prime, play D if told h_i and double-prime." This will be equivalent to the original mixed strategy.)

Definition 2.4A   A *correlated equilibrium* ∂ relative to information structure $(\Omega, \{H_i\}, p)$ is a Nash equilibrium in strategies that are adapted to this information structure. That is, $(\partial_1, \ldots, \partial_I)$ is a correlated equilibrium if, for every i and every adapted strategy $\tilde{\partial}_i$,

$$\sum_{\omega \in \Omega} p(\omega)u_i(\partial_i(\omega), \partial_{-i}(\omega)) \geq \sum_{\omega \in \Omega} p(\omega)u_i(\tilde{\partial}_i(\omega), \partial_{-i}(\omega)). \tag{2.1}$$

This definition, where the distribution p over Ω is the same for all players, is sometimes called an "objective correlated equilibrium" to distinguish it from "subjective correlated equilibria" where players may disagree on prior beliefs and each player i is allowed to have different beliefs p_i. We say more about subjective correlated equilibrium in section 2.3.

Definition 2.4A, which requires that ∂_i maximize player i's "ex ante" payoff—her expected payoff before knowing which h_i contains the true state—implies that a_i maximizes player i's payoff *conditional on h_i* for each h_i that player i assigns positive prior probability (this conditional payoff is often called an "interim" payoff). That is, (2.1) is equivalent to the condition that, for all players i, information sets h_i with $p(h_i) > 0$, and all s_i,

$$\sum_{\{\omega | h_i(\omega)=h_i\}} p(\omega | h_i)u_i(a_i(\omega), a_{-i}(\omega)) \geq \sum_{\{\omega | h_i(\omega)=h_i\}} p(\omega | h_i)u_i(s_i, a_{-i}(\omega)). \quad (2.2)$$

When all players have the same prior, any h_i with $p(h_i) = 0$ is irrelevant, and all states $\omega \in h_i$ can be omitted from the specification of Ω. New issues arise when the priors are different, as we will see when we discuss Brandenburger and Dekel 1987.

An awkward feature of this definition is that it depends on the particular information structure specified, yet there are an infinite number of possible state spaces Ω and many information structures possible for each. Fortunately there is a more concise way to define correlated equilibrium. This alternative definition is based on the realization that any joint distribution over actions that forms a correlated equilibrium for some correlating device can be attained as an equilibrium with the "universal device" whose signals to each player constitute a recommendation of how that player should play. In the example of figure 2.4, player 1 would be told "play D" instead of "the state is (B, C)," and player 1 would be willing to follow this recommendation so long as, when he is told to play D, the conditional probability of player 2 being instructed to play R is $\frac{1}{2}$. (Those familiar with the literature on mechanism design will recognize this observation as a version of the "revelation principle"; see chapter 7.)

Definition 2.4B A *correlated equilibrium* is any probability distribution $p(\cdot)$ over the pure strategies $S_1 \times \cdots \times S_I$ such that, for every player i and every function $d_i(\cdot)$ that maps S_i to S_i,

$$\sum_{s \in S} p(s)u_i(s_i, s_{-i}) \geq \sum_{s \in S} p(s)u_i(d_i(s_i), s_{-i}).$$

Just as with definition 2.4A, there is an equivalent version of the definition stated in terms of maximization conditional on each recommendation: $p(\cdot)$ is a correlated equilibrium if, for every player i and every s_i with $p(s_i) > 0$,

$$\sum_{s_{-i} \in S_{-i}} p(s_{-i} | s_i)u_i(s_i, s_{-i}) \geq \sum_{s_{-i} \in S_{-i}} p(s_{-i} | s_i)u_i(s_i', s_{-i}) \quad \forall s_i' \in S_i.$$

That is, player i should not be able to gain by disobeying the recommendation to play s_i if every other player obeys his recommendation.

Let us explain why the two definitions of correlated equilibrium are equivalent. Clearly an equilibrium in the sense of definition 2.4B is an equilibrium according to definition 2.4A—just take $\Omega = S$, and $h_i(s) = \{s' | s_i' = s_i\}$.

Conversely, if σ is an equilibrium relative to some $(\Omega, \{H_i\}, \tilde{p})$ as in definition 2.4A, set $p(s)$ to be the sum of $\tilde{p}(\omega)$ over all $\omega \in \Omega$ such that $\sigma_i(\omega) = s_i$ for all players i. Let us check that no player i can gain by disobeying any recommendation $s_i \in S_i$. (The only reason this isn't completely obvious is that there may have been several information sets h_i where player i played s_i, in which case his information has been reduced to s_i alone.) Set

$$J_i(s_i) = \{\omega \mid \sigma_i(\omega) = s_i\},$$

so that $\tilde{p}(J_i(s_i)) = p(s_i)$ is the probability that player i is told to play s_i. If we view each pure-strategy profile $\sigma_{-i}(\omega)$ as a degenerate mixed strategy that places probability 1 on $s_{-i} = \sigma_{-i}(\omega)$, then the probability distribution on opponents' strategies that player i believes he faces, conditional on being told to play s_i, is

$$\sum_{\omega \in J_i(s_i)} \frac{\tilde{p}(\omega)\sigma_{-i}(\omega)}{\tilde{p}(J_i(s_i))},$$

which is a convex combination of the distributions conditional on each h_i such that $\sigma_i(h_i) = s_i$. Since player i could not gain by deviating from σ_i at any such h_i, he cannot gain by deviating when this finer information structure is replaced by the one that simply tells him his recommended strategy.

A pure-strategy Nash equilibrium is a correlated equilibrium in which the distribution $p(\cdot)$ is degenerate. Mixed-strategy Nash equilibria are also correlated equilibria: Just take $p(\cdot)$ to be the joint distribution implied by the equilibrium strategies, so that the recommendations made to each player convey no information about the play of his opponents.

Inspection of the definition shows that the set of correlated equilibria is convex, so the set of correlated equilibria is at least as large as the convex hull of the Nash equilibria. This convexification could be attained by using only public correlating devices. But, as we have seen, nonpublic (imperfect) correlation can lead to equilibria outside the convex hull of the Nash set.

Since Nash equilibria exist in finite games, correlated equilibria do too. Actually, the existence of correlated equilibria would seem to be a simpler problem than the existence of Nash equilibria, because the set of correlated equilibria is defined by a system of linear inequalities and is therefore convex; indeed, Hart and Schmeidler (1989) have provided an existence proof that uses only linear methods (as opposed to fixed-point theorems). One might also like to know when the set of correlated equilibria differs "greatly" from the convex hull of the Nash equilibria, but this question has not yet been answered.

One may take the view that the correlation in correlated equilibria should be thought of as the result of the players receiving "endogenous" correlated signals, so that the notion of correlated equilibrium is particularly appropriate in situations with preplay communication, for then the players might be able to design and implement a procedure for obtaining correlated, private signals.[4] When players do not meet and design particular correlated devices, it is plausible that they may still observe exogenous random signals (i.e., "sunspots" or "moonspots") on which they can condition their play. If the signals are publicly observed they can only serve to convexify the set of Nash equilibrium payoffs. But if the signals are observed privately and yet are correlated, they also allow imperfectly correlated equilibria, which may have payoffs outside the convex hull of Nash equilibria, such as $(3\frac{1}{3}, 3\frac{1}{3})$ in figure 2.4. (Aumann (1987) argues that Bayesian rationality, broadly construed, implies that play *must* correspond to a correlated equilibrium, though not necessarily to a Nash equilibrium.)

2.3 Rationalizability and Subjective Correlated Equilibria[†††]

In matching pennies (figure 1.10a), rationalizability allows player 1 to be sure he will outguess player 2, and player 2 to be sure he'll outguess player 1; the players' strategic beliefs need not be consistent. It is interesting to note that this kind of inconsistency in beliefs can be modeled as a kind of correlated equilibrium with inconsistent beliefs. We mentioned the possibility of inconsistent beliefs when we defined subjective correlated equilibrium, which generalizes objective correlated equilibrium by allowing each player i to have different beliefs $p_i(\cdot)$ over the joint recommendation $s \in S$. That notion is weaker than rationalizability, as is shown by figure 2.6 (which is drawn from Brandenburger and Dekel 1987). One subjective correlated equilibrium for this game has player 1's beliefs assign probability 1 to (U, L) and player 2's beliefs assign probability $\frac{1}{2}$ each to (U, L) and (D, L). Given his beliefs, player 2 is correct to play L. However, that

	L	R
U	2,0	1,1
D	1,1	0,0

Figure 2.6

4. Barany (1988) shows that if there are at least four players $(I \geq 4)$, any *correlated* equilibrium of a strategic-form game coincides with a *Nash* equilibrium of an extended game in which the players engage in costless conversations (cheap talk) before they play the strategic-form game in question. If there are only two players, then the set of Nash equilibria with cheap talk coincides with the subset of correlated equilibria induced by perfectly correlated signals (i.e., publicly observed randomizing devices.)

strategy is deleted by iterated dominance, and so we see that subjective correlated equilibrium is less restrictive than rationalizability.

The point is that subjective correlated equilibrium allows each player's beliefs about his opponents to be completely arbitrary, and thus cannot capture the restrictions implied by common knowledge of the payoffs. Brandenburger and Dekel introduce the idea of an *a posteriori* equilibrium, which does capture these restrictions.

Although this equilibrium concept, like correlated equilibrium, can be defined either with reference to explicit correlating devices or in a "direct version," it is somewhat simpler here to make the correlating device explicit.

Given state space Ω, partition H_i, and priors $p_i(\cdot)$, we now require, for each ω (even those with $p_i(\omega) = 0$),[5] that player i have well-defined conditional beliefs $p_i(\omega' | h_i(\omega))$, satisfying $p_i(h_i(\omega) | h_i(\omega)) = 1$.

Definition 2.5 The adapted strategies $(\sigma_1, \ldots, \sigma_I)$ are an *a posteriori* equilibrium if, for all $\omega \in \Omega$, all players i, and all s_i,

$$\sum_{\omega' \in h_i(\omega)} p_i(\omega' | h_i(\omega)) u_i(\sigma_i(\omega), \sigma_{-i}(\omega'))$$

$$\geq \sum_{\omega' \in h_i(\omega)} p_i(\omega' | h_i(\omega)) u_i(s_i, \sigma_{-i}(\omega')).$$

Thus, player i's strategy is required to be optimal for all ω, even those to which he assigns prior probability 0.

Brandenburger and Dekel show that the set of correlated rationalizable payoffs is precisely the set of interim payoffs to *a posteriori* equilibria; that is, they are the payoffs player i can expect to receive conditional on a particular $\omega \in \Omega$.

Exercises

Exercise 2.1**
(a) Consider an alternative definition of iterated strict dominance that proceeds as in section 2.1 except that, at each state n, only the strictly dominated pure strategies of players $I(n) \subseteq I$ are deleted. Suppose that, for each player i, there exists an infinite number of steps n such that $i \in I(n)$. If the game is finite, show that the resulting limit set is S^∞ (as given in definition 2.1), so that there is no loss of generality in taking $I(n) = I$ for all n. Hint: The intuition is that if a strategy s_i is strictly dominated at step n but is not eliminated because $i \notin I(n)$, then it will be eliminated at the next step $n' > n$ such that $i \in I(n')$, as (i) the set of strategies s_{-i} remaining at step n' is no larger than the set of strategies s_{-i} remaining at step n and (ii) if

---

5. Note that we do not require priors to be absolutely continuous with respect to each other—that is, they may disagree on which ω's have positive probability.

strategy s_i is strictly dominated relative to a set Σ'_{-i} of opponents' mixed strategies it is strictly dominated relative to any subset $\Sigma_{-i} \subseteq \Sigma'_{-i}$. Show by induction on n that any strategy that is deleted at stage n under the maximal-deletion process $I(k) = I$ for all k is deleted in a finite number of steps (no fewer than n) when deletion is not required to be maximal.

(b) Verify that, in a finite game, the two definitions of iterated deletion of dominated mixed strategies given in section 2.1 are equivalent.

Exercise 2.2* Prove that if a game is solvable by iterated strict dominance, it has a unique Nash equilibrium.

Exercise 2.3** Consider an arbitrary two-player game with action spaces $A_1 = A_2 = [0, 1]$ and payoff functions that are twice continuously differentiable and concave in own action. Say that the game is *locally solvable by iterated strict dominance at* a^* if there is a rectangle N containing a^* such that when players are restricted to choosing actions in N, the successive elimination of strictly dominated strategies yields the unique point a^*. Relate the conditions for local solvability by iterated strict dominance of the simultaneous-move process to those for local stability of the alternating-move Cournot adjustment process. (The answer is in Gabay and Moulin 1980.)

Exercise 2.4* Show that in the Cournot game with three Nash equilibria with the reaction curves depicted in figure 1.14, the strategies that survive iterated deletion of strictly dominated strategies are the outputs that belong to the interval whose boundaries are the projections of B and D.

Exercise 2.5** A competitive economy may be described as a game with a continuum of players. Concepts such as iterated dominance, rationalizability, and Nash equilibrium can, with minor adjustments, be applied to such situations. Consider the following "wheat market": There is a continuum of farmers indexed by a parameter i distributed with a density $f(i)$ on $[\underline{i}, \bar{i}]$, where $\underline{i} > 0$. They must choose the size of their crop $q(i)$ before the market for wheat opens. The cost function of farmer i is $C(q, i) = q^2/2i$. The farmer's utility function is thus $u_i = p\,q(i) - q(i)^2/2i$, where p is the price of wheat. Let $O(p)$ denote the aggregate supply function when farmers perfectly predict p:

$$O(p) = \left(\int_{\underline{i}}^{\bar{i}} i f(i)\, di \right) p \equiv kp.$$

The demand curve is $D(p) = a - bp$ for $0 \leq p \leq a/b$ and 0 otherwise. The timing is such that the farmers simultaneously choose the size of their crop, then the price clears the market:

$$\int_{\underline{i}}^{\bar{i}} q(i) f(i)\, di = D(p).$$

A perfect foresight (or Nash equilibrium) is a price p^* such that $O(p^*) = D(p^*)$ (more correctly, it is a strategy profile $q^*(\cdot)$ such that $q^*(i) = ip^*$.) Note that $p^* = a/(b + k)$.

Apply iterated strict dominance in this game among farmers. Show that if $b > k$, the game is solvable by iterated strict dominance, which yields the perfect-foresight equilibrium. When $b \leq k$, determine the interval of prices that correspond to outputs that survive iterated strict dominance. Draw the link with the stability of the "cobweb" tatônnement in which the market is repeated over time, and farmers have point price expectations equal to the last period's price. (This exercise is drawn from Guesnerie 1989, which also addresses production and demand uncertainty, price floors and ceilings, and sequential timing of crop planting.)

Exercise 2.6** Consider the two-player game in figure 2.7. This is matching pennies with an outside option α for player 1. Suppose that $\alpha \in (0, 1)$.

(a) Show that the set of mixed strategies for player 1 surviving iterated deletion of strictly dominated strategies consists of two "edges of the strategy simplex": the set of mixed strategies with support (H, α) and the set of mixed strategies with support (T, α).

(b) Show directly (that is, without applying theorem 2.2) that the set of rationalizable strategies for player 1 is also composed of these two edges. (Hint: Use a diagram similar to figure 2.3.)

	H	T
H	1,−1	−1,1
T	−1,1	1,−1
α	α,0	α,0

Figure 2.7

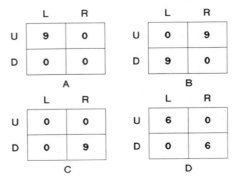

Figure 2.8

Exercise 2.7** Consider the game in figure 2.8, where player 1 chooses rows, player 2 chooses columns, and player 3 chooses matrices. Player 3's payoffs are given in the figure. Show that action D is not a best response to any mixed strategies of players 1 and 2, but that D is *not* dominated. Comment.

Exercise 2.8*** Find *all* the correlated equilibria of the games illustrated in figures 2.4 and 2.5.

References

Aumann, R. 1974. Subjectivity and correlation in randomized strategies. *Journal of Mathematical Economics* 1: 67–96.

Aumann, R. 1987. Correlated equilibrium as an extension of Bayesian rationality. *Econometrica* 55: 1–18.

Barany, I. 1988. Fair distribution protocols, or how the players replace fortune. Mimeo, University College, London.

Bernheim, D. 1984. Rationalizable strategic behavior. *Econometrica* 52: 1007–1028.

Brandenburger, A., and E. Dekel. 1987. Rationalizability and correlated equilibria. *Econometrica* 55: 1391–1402.

Fan, K. 1952. Fixed point and minimax theorems in locally convex topological linear spaces. *Proceedings of the National Academy of Sciences* 38: 121–126.

Fan, K. 1953. Minimax theorems. *Proceedings of the National Academy of Sciences* 39: 42–47.

Gabay, D., and H. Moulin. 1980. On the uniqueness of Nash equilibrium in noncooperative games. In *Applied Stochastic Control in Econometric and Management Science*, ed. Bensoussan, Kleindorfer, and Tapien. North-Holland.

Guesnerie, R. 1989. An exploration of the eductive justifications of the rational expectations hypothesis. Mimeo, EHESS, Paris.

Hart, S., and D. Schmeidler, 1989. Existence of correlated equilibria. *Mathematics of Operations Research* 14: 18–25.

Moulin, H. 1984. Dominance solvability and Cournot stability. *Mathematical Social Sciences* 7: 83–102.

Myerson, R. 1985. Bayesian equilibrium and incentive compatibility: An introduction, in *Social Goals and Social Organization: Essays in Honor of Elizha Pazner*, ed. L. Hurwicz, D. Schmeidler, and H. Sonnenschein. Cambridge University Press.

Pearce, D. 1984. Rationalizable strategic behavior and the problem of perfection. *Econometrica* 52: 1029–1050.

Rabin, M. 1989. Predictions and solution concepts in noncooperative games. Ph.D. dissertation, Department of Economics, MIT.

von Neumann, J. 1928. Zur Theorie der Gesellschaftsspiele. *Math. Annalen* 100: 295–320.

II DYNAMIC GAMES OF COMPLETE INFORMATION

Extensive-Form Games

3.1 Introduction[†]

In the examples we examined in part I, such as the stag hunt, the prisoner's dilemma, and the battle of the sexes, the players choose their actions simultaneously. Much of the recent interest in the economic applications of game theory has been in situations with an important dynamic structure, such as entry and entry deterrence in industrial organization and the "time-consistency" problem in macroeconomics. Game theorists use the concept of a *game in extensive form* to model such dynamic situations. The extensive form makes explicit the order in which players move, and what each player knows when making each of his decisions. In this setting, strategies correspond to contingent plans instead of uncontingent actions. As we will see, the extensive form can be viewed as a multi-player generalization of a decision tree. Not surprisingly, many results and intuitions from decision theory have game-theoretic analogs. We will also see how to build up the strategic-form representation of a game from its extensive form. Thus, we will be able to apply the concepts and results of part I to dynamic games.

As a simple example of an extensive-form game, consider the idea of a "Stackelberg equilibrium" in a duopoly. As in the Cournot model, the actions of the firms are choices of output levels, q_1 for player 1 and q_2 for player 2. The difference is that we now suppose that player 1, the "Stackelberg leader," chooses her output level q_1 first, and that player 2 observes q_1 before choosing his own output level. To make things concrete, we suppose that production is costless, and that demand is linear, with $p(q) = 12 - q$, so that player i's payoff is $u_i(q_1, q_2) = [12 - (q_1 + q_2)]q_i$. How should we extend the idea of Nash equilibrium to this setting? And how should we expect the players to play?

Since player 2 observes player 1's choice of output q_1 before choosing q_2, in principle player 2 could condition his choice of q_2 on the observed level of q_1. And since player 1 moves first, she cannot condition her output on player 2's. Thus, it is natural that player 2's *strategies* in this game should be maps of the form $s_2: Q_1 \rightarrow Q_2$ (where Q_1 is the space of feasible q_1's and Q_2 is the space of feasible q_2's), while player 1's strategies are simply choices of q_1. Given a (pure) strategy profile of this form, the outcome is the output vector $(q_1, s_2(q_1))$, with payoffs $u_i(q_1, s_2(q_1))$.

Now that we have identified strategy spaces and the payoff functions, we can define a Nash equilibrium of this game in the obvious way: as a strategy profile such that neither player can gain by switching to a different strategy. Let's consider two particular Nash equilibria of this game.

The first equilibrium gives rise to the Stackelberg output levels normally associated with this game. In this equilibrium, player 2's strategy s_2 is to choose, for each q_1, the level of q_2 that solves $\max_{q_2'} u_2(q_1, q_2')$, so that s_2 is identically equal to the Cournot reaction function r_2 defined in chapter 1. With the payoffs we have specified, $r_2(q_1) = 6 - q_1/2$.

Nash equilibrium requires that player 1's strategy maximize her payoff given that $s_2 = r_2$, so that player 1's output level q_1^* is the solution to $\max_{q_1} u_1(q_1, r_2(q_1))$, which with the payoffs we specified gives $q_1^* = 6$.

The output levels $(q_1^*, r_2(q_1^*))$ (here equal to $(6, 3)$) are called the *Stackelberg outcome* of the game; this is the outcome economics students are taught to expect. In the usual case, r_2 is a decreasing function, and so player 1 can decrease player 2's output by increasing her own. As a result, player 1's Stackelberg output level and payoff are typically higher than in the Cournot equilibrium where both players move simultaneously, and player 2's output and payoff are typically lower. (In our case the unique Cournot equilibrium is $q_1^C = q_2^C = 4$, with payoffs of 16 each; in the Stackelberg equilibrium the leader's payoff is 18 and the follower's is 9.)

Though the Stackelberg outcome may seem the natural prediction in this game, there are many other Nash equilibria. One of them is the profile "$q_1 = q_1^C; s_2(q_1) = q_2^C$ for all q_1." These strategies really are a Nash equilibrium: Given that player 2's output will be q_2^C independent of q_1, player 1's problem is to maximize $u_1(q_1, q_2^C)$, and by definition this maximization is solved by the Cournot output q_1^C. And given that $q_1 = q_1^C$, player 2's payoff will be $u_2(q_1^C, s_2(q_1^C))$, which is maximized by *any* strategy s_2 such that $s_2(q_1^C) = q_2^C$, including the constant strategy $s_2(\cdot) \equiv q_2^C$. Note, though, that this strategy is *not* a best response to other output levels that player 1 might have chosen but did not; i.e., q_2^C is not in general a best response to q_1 for $q_1 \neq q_1^C$.

So we have identified two Nash equilibria for the game where player 1 chooses her output first: one equilibrium with the "Stackelberg outputs" and one where the output levels are the same as if the players moved simultaneously. Why is the first equilibrium more reasonable, and what is wrong with the second one? Most game theorists would answer that the second equilibrium is "not credible," as it relies on an "empty threat" by player 2 to hold his output at q_2^C regardless of player 1's choice. This threat is empty because if player 1 were to present player 2 with the *fait accompli* of choosing the Stackelberg output q_1^*, player 2 would do better to choose a different level of q_2—in particular, $q_2 = r_2(q_1^*)$. Thus, if player 1 knows player 2's payoffs, the argument goes, she should not believe that player 2 would play q_2^C no matter what player 1's output. Rather, player 1 should predict that player 2 will play an optimal response to whatever q_1 player 1 actually chooses, so that player 1 should predict that whatever level of q_1 she chooses, player 2 will choose the optimal response $r_2(q_1)$. This argument picks out the "Stackelberg equilibrium" as the unique credible outcome. A more formal way of putting this is that the Stackelberg equilibrium is consistent with *backward induction*, so called because the idea is to start by solving for the optimal choice of the last mover for each possible situation he might face, and then work backward to compute the optimal choice for the player before. The ideas of credibility and backward induction are clearly present in the textbook analysis of the Stackelberg game; they were informally applied by Schelling (1960) to the analysis of commitment in a number of settings. Selten (1965) formalized the intuition with his concept of a *subgame-perfect equilibrium*, which extends the idea of backward induction to extensive games where players move simultaneously in several periods, so the backward-induction algorithm is not applicable because there are several "last movers" and each of them must know the moves of the others to compute his own optimal choice.

This chapter will develop the formalism for modeling extensive games and develop the solution concepts of backward induction and subgame perfection. Although the extensive form is a fundamental concept in game theory, its definition may be a bit detailed for readers who are more interested in applications of games than in mastering the general theory. With such readers in mind, section 3.2 presents a first look at dynamic games by treating a class of games with a particularly simple structure: the class of "multi-stage games with observed actions." These games have "stages" such that (1) in each stage every player knows all the actions taken by any player, including "Nature," at any previous stage, and (2) players move "simultaneously" within each stage.

Though very special, this class of games includes the Stackelberg example we have just discussed, as well as many other examples from the economics literature. We use multi-stage games to illustrate the idea that strategies can be contingent plans, and to give a first definition of subgame perfection. As an illustration of the concepts, subsection 3.2.3 discusses how to model the idea of commitment, and addresses the particular example called the "time-consistency problem" in macroeconomics. Readers who lack the time or interest for the general extensive-game model are advised to skip from the end of section 3.2 to section 3.6, which gives a few cautions about the potential drawbacks of the ideas of backward induction and subgame perfection.

Section 3.3 introduces the concepts involved in defining an extensive form. Section 3.4 discusses strategies in the extensive form, called "behavior strategies," and shows how to relate them to the strategic-form strategies discussed in chapters 1 and 2. Section 3.5 gives the general definition of subgame perfection. We postpone discussion of more powerful equilibrium refinements to chapters 8 and 11 in order to first study several interesting classes of games which can be fruitfully analyzed with the tools we develop in this chapter.

Readers who already have some informal understanding of dynamic games and subgame perfection probably already know the material of section 3.2, and are invited to skip directly to section 3.3. (Teaching note: When planning to cover all of this chapter, it is probably not worth taking the time to teach section 3.2 in class; you may or may not want to ask the students to read it on their own.)

3.2 Commitment and Perfection in Multi-Stage Games with Observed Actions[†]

3.2.1 What Is a Multi-Stage Game?

Our first step is to give a more precise definition of a "multi-stage game with observed actions." Recall that we said that this meant that (1) all players knew the actions chosen at all previous stages $0, 1, 2, \ldots, k-1$ when choosing their actions at stage k, and that (2) all players move "simultaneously" in each stage k. (We adopt the convention that the first stage is "stage 0" in order to simplify the notation concerning discounting when stages are interpreted as periods.) Players move simultaneously in stage k if each player chooses his or her action at stage k without knowing the stage-k action of any other player. Common usage to the contrary, "simultaneous moves" does not exclude games where players move in alternation, as we allow for the possibility that some of the players have the one-element choice set "do nothing." For example, the Stackelberg game has two stages: In the first stage, the leader chooses an output level (and the follower "does nothing"). In the second stage, the follower knows the leader's output and chooses an output level of his own (and the leader "does nothing"). Cournot and Bertrand games are one-stage games: All players choose their actions at once and the game ends. Dixit's (1979) model of entry and entry deterrence (based on work by Spence (1977)) is a more complex example: In the first stage of this game, an incumbent invests in capacity; in the second stage, an entrant observes the capacity choice and decides whether to enter. If there is no entry, the incumbent chooses output as a monopolist in the third stage; if entry occurs, the two firms choose output simultaneously as in Cournot competition.

Often it is natural to identify the "stages" of the game with time periods, but this is not always the case. A counterexample is the Rubinstein-Ståhl model of bargaining (discussed in chapter 4), where each "time period" has two stages. In the first stage of each period, one player proposes an agreement; in the second stage, the other player either accepts or rejects the proposal. The distinction is that time periods refer to some physical measure of the passing of time, such as the accumulation of delay costs in the bargaining model, whereas the stages need not have a direct temporal interpretation.

In the first stage of a multi-stage game (stage 0), all players $i \in \mathscr{I}$ simultaneously choose actions from choice sets $A_i(h^0)$. (Remember that some of the choice sets may be the singleton "do nothing." We let $h^0 = \varnothing$ be the "history" at the start of play.) At the end of each stage, all players observe

the stage's action profile. Let $a^0 \equiv (a_1^0, \ldots, a_I^0)$ be the stage-0 action profile. At the beginning of stage 1, players know history h^1, which can be identified with a^0 given that h^0 is trivial. In general, the actions player i has available in stage 1 may depend on what has happened previously, so we let $A_i(h^1)$ denote the possible second-stage actions when the history is h^1. Continuing iteratively, we define h^{k+1}, the history at the end of stage k, to be the sequence of actions in the previous periods,

$$h^{k+1} = (a^0, a^1, \ldots, a^k),$$

and we let $A_i(h^{k+1})$ denote player i's feasible actions in stage $k + 1$ when the history is h^{k+1}. We let $K + 1$ denote the total number of stages in the game, with the understanding that in some applications $K = +\infty$, corresponding to an infinite number of stages; in this case the "outcome" when the game is played will be an infinite history, h^∞. Since each h^{K+1} by definition describes an entire sequence of actions from the beginning of the game on, the set H^{K+1} of all "terminal histories" is the same as the set of possible outcomes when the game is played.

In this setting, a *pure strategy for player i* is simply a contingent plan of how to play in each stage k for possible history h^k. (We will postpone discussion of mixed strategies until section 3.3, as they will not be used in the examples we discuss here.) If we let H^k denote the set of all stage-k histories, and let

$$A_i(H^k) = \bigcup_{h^k \in H^k} A_i(h^k),$$

a pure strategy for player i is a sequence of maps $\{s_i^k\}_{k=0}^K$, where each s_i^k maps H^k to the set of player i's feasible actions $A_i(H^k)$ (i.e., satisfies $s_i^k(h^k) \in A_i(h^k)$ for all h^k). It should be clear how to find the sequence of actions generated by a profile of such strategies: The stage-0 actions are $a^0 = s^0(h^0)$, the stage-1 actions are $a^1 = s^1(a^0)$, the stage-2 actions are $a^2 = s^2(a^0, a^1)$, and so on. This is called the *path* of the strategy profile. Since the terminal histories represent an entire sequence of play, we can represent each player i's payoff as a function $u_i: H^{K+1} \to \mathbb{R}$. In most applications the payoff functions are additively separable over stages (i.e., each player's overall payoff is some weighted average of single-stage payoffs $g_i(a^k)$, $k = 0, \ldots, K$), but this restriction is not necessary.

Since we can assign an outcome in H^{K+1} to each strategy profile, and a payoff vector to each outcome, we can now compute the payoff to any strategy profile; in an abuse of notation, we will represent the payoff vector to profile s as $u(s)$. A (pure-strategy) *Nash equilibrium* in this context is simply a strategy profile s such that no player i can do better with a different strategy, which is the familiar condition that $u_i(s_i, s_{-i}) \geq u_i(s_i', s_{-i})$ for all s_i'.

The Cournot and Bertrand "equilibria" discussed in chapter 1 are trivial examples of Nash equilibria of multi-stage (actually one-stage) games. We saw two other examples of Nash equilibria when we discussed the Stackelberg game at the beginning of this chapter. We also saw that some of these Nash equilibria may rely on "empty threats" of suboptimal play at histories that are not expected to occur—that is, at histories off the path of the equilibrium.

3.2.2 Backward Induction and Subgame Perfection

In the Stackelberg game, it was easy to see how player 2 "ought" to play, because once q_1 was fixed player 2 faced a simple decision problem. This allowed us to solve for player 2's optimal second-stage choice for each q_1 and then work backward to find the optimal choice for player 1. This algorithm can be extended to other games where only one player moves at each stage. We say that a multi-stage game has *perfect information* if, for every stage k and history h^k, exactly one player has a nontrivial choice set—a choice set with more than one element—and all the others have the one-element choice set "do nothing." A simple example of such a game has player 1 moving in stages 0, 2, 4, etc. and player 2 moving in stages 1, 3, 5, and so on. More generally, some players could move several times in a row, and which player gets to move in stage k could depend on the previous history. The key thing is that only one player moves at each stage k. Since we have assumed that each player knows the past choices of all rivals, this implies that the single player on move at k is "perfectly informed" of all aspects of the game except those which will occur in the future.

Backward induction can be applied to any finite game of perfect information, where *finite* means that the number of stages is finite and the number of feasible actions at any stage is finite, too.[1] The algorithm begins by determining the optimal choices in the final stage K for each history h^K—that is, the action for the player on move, given history h^K, that maximizes that player's payoff conditional on h^K being reached. (There may be more than one maximizing choice; in this case backward induction allows the player to choose any of the maximizers.) Then we work back to stage $K-1$, and determine the optimal action for the player on move there, given that the player on move at stage K with history h^K will play the action we determined previously. The algorithm proceeds to "roll back," just as in solving decision problems, until the initial stage is reached. At this point we have constructed a strategy profile, and it is easy to verify that this profile is a Nash equilibrium. Moreover, it has the nice property that each player's actions are optimal at every possible history.

The argument for the backward-induction solution in the two-stage Stackelberg game—that player 1 should be able to forecast player 2's second-stage play—strikes us as quite compelling. In a three-stage game,

---

1. Section 4.6 extends backward induction to infinite games of perfect information, where there is no last period from which to work backward.

the argument is a bit more complex: The player on move at stage 0 must forecast that the player on move at stage 1 will correctly forecast the play of the player on move at stage 2, which clearly is a more demanding hypothesis. And the arguments for backward induction in longer games require correspondingly more involved hypotheses. For this reason, backward-induction arguments may not be compelling in "long" games. For the moment, though, we will pass over the arguments against backward induction; section 3.6 discusses its limitations in more detail.

As defined above, backward induction applies only to games of perfect information. It can be extended to a slightly larger class of games. For instance, in a multi-stage game, if all players have a dominant strategy in the last stage, given the history of the game (or, more generally, if the last stage is solvable by iterated strict dominance), one can replace the last-stage strategies by the dominant strategies, then consider the penultimate stage and apply the same reasoning, and so on. However, this doesn't define backward induction for games that cannot be solved by this backward-induction version of dominance solvability. Yet one would think that the backward-induction idea of predicting what the players are likely to choose in the future ought to carry over to more general games. Suppose that a firm—call it firm 1—has to decide whether or not to invest in a new cost-reducing technology. Its choice will be observed by its only competitor, firm 2. Once the choice is made and observed, the two firms will choose output levels simultaneously, as in Cournot competition. This is a two-stage game, but not one of perfect information. How should firm 1 forecast the second-period output choice of its opponent? In the spirit of equilibrium analysis, a natural conjecture is that the second-period output choices will be those of a Cournot equilibrium for the prevailing cost structure of the industry. That is, each history h^1 generates a simultaneous-move game between the two firms, and firm 1 forecasts that play in this game will correspond to an equilibrium for the payoffs prevailing under h^1. This is exactly the idea of Selten's (1965) *subgame-perfect equilibrium.*

Defining subgame perfection requires a few preliminary steps. First, since all players know the history h^k of moves before stage k, we can view the game from stage k on with history h^k as a game in its own right, which we will denote $G(h^k)$. To define the payoff functions in this game, note that if the actions in stages k through K are a^k though a^K, the final history will be $h^{K+1} = (h^k, a^k, a^{k+1}, \ldots, a^K)$, and so the payoffs will be $u_i(h^{K+1})$. Strategies in $G(h^k)$ are defined in the obvious way: as maps from histories to actions, where the only histories we need consider are those consistent with h^k. So now we can speak of the Nash equilibria of $G(h^k)$.

Next, any strategy profile s of the whole game induces a strategy profile $s|h^k$ on any $G(h^k)$ in the obvious way: For each player i, $s_i|h^k$ is simply the restriction of s_i to the histories consistent with h^k.

Definition 3.1 A strategy profile s of a multi-stage game with observed actions is a *subgame-perfect equilibrium* if, for every h^k, the restriction $s|h^k$ to $G(h^k)$ is a Nash equilibrium of $G(h^k)$.

This definition reduces to backward induction in finite games of perfect information, for the only Nash equilibrium in game $G(h^K)$ at the final stage is for the player on move to choose (one of) his preferred action(s) as in backward induction, the only Nash-equilibrium choice in the next-to-last stage given Nash play at the last stage is as in backward induction, and so on.

Example 3.1

To illustrate the ideas of this section, consider the following model of strategic investment in a duopoly: Firm 1 and firm 2 currently both have a constant average cost of 2 per unit. Firm 1 can install a new technology with an average cost of 0 per unit; installing the technology costs f. Firm 2 will observe whether or not firm 1 invests in the new technology. Once firm 1's investment decision is observed, the two firms will simultaneously choose output levels q_1 and q_2 as in Cournot competition. Thus, this is a two-stage game.

To define the payoffs, we suppose that the demand is $p(q) = 14 - q$ and that each firm's goal is to maximize its net revenue minus costs. Firm 1's payoff is then $[12 - (q_1 + q_2)]q_1$ if it does not invest, and $[14 - (q_1 + q_2)]q_1 - f$ if it does; firm 2's payoff is $[12 - (q_1 + q_2)]q_2$.

To find the subgame-perfect equilibria, we work backward. If firm 1 does not invest, both firms have unit cost 2, and hence their reaction functions are $r_i(q_j) = 6 - q_j/2$. These reaction functions intersect at the point $(4, 4)$, with payoffs of 16 each. If firm 1 does invest, its reaction becomes $\tilde{r}_1(q_2) = 7 - q_2/2$, the second-stage equilibrium is $(\frac{16}{3}, \frac{10}{3})$, and firm 1's total payoff is $256/9 - f$. Thus, firm 1 should make the investment if $256/9 - f > 16$, or $f < 112/9$.

Note that making the investment increases firm 1's second-stage profit in two ways. First, firm 1's profit is higher at any fixed pair of outputs, because its cost of production has gone down. Second, firm 1 gains because firm 2's second-stage output is decreased. The reason firm 2's output is lower is because by lowering its cost firm 1 altered its own second-period incentives, and in particular made itself "more aggressive" in the sense that $\tilde{r}_1(q_2) > r_1(q_2)$ for all q_2. We say more about this kind of "self-commitment" in the next subsection. Note that firm 2's output would not decrease if it continued to believe that firm 1's cost equaled 2.

3.2.3 The Value of Commitment and "Time Consistency"

One of the recurring themes in the analysis of dynamic games has been that in many situations players can benefit from the opportunity to make a binding commitment to play in a certain way. In a one-player game—i.e., a decision problem—such commitments cannot be of value, as any payoff that a player could attain while playing according to the commitment could be attained by playing in exactly the same way without being committed to do so. With more than one player, though, commitments can be of value, since by committing himself to a given sequence of actions a player may be able to alter the play of his opponents. This "paradoxical" value of commitment is closely related to our observation in chapter 1 that a player can gain by reducing his action set or decreasing his payoff to some outcomes, provided that his opponents are aware of the change. Indeed, some forms of commitment can be represented in exactly this way.

The way to model the possibility of commitments (and related moves like "promises") is to explicitly include them as actions the players can take. (Schelling (1960) was an early proponent of this view.) We have already seen one example of the value of commitment in our study of the Stackelberg game, which describes a situation where one firm (the "leader") can commit itself to an output level that the follower is forced to take as given when making its own output decision. Under the typical assumption that each firm's optimal reaction $r_i(q_j)$ is a decreasing function of its opponent's output, the Stackelberg leader's payoff is higher than in the "Cournot equilibrium" outcome where the two firms choose their output levels simultaneously.

In the Stackelberg example, commitment is achieved simply by moving earlier than the opponent. Although this corresponds to a different extensive form than the simultaneous moves of Cournot competition, the set of "physical actions" is in some sense the same. The search for a way to commit oneself can also lead to the use of actions that would not otherwise have been considered. Classic examples include a general burning his bridges behind him as a commitment not to retreat and Odysseus having himself lashed to the mast and ordering his sailors to plug their ears with wax as a commitment not to go to the Sirens' island. (Note that the natural way to model the Odysseus story is with two "players," corresponding to Odysseus before and Odysseus after he is exposed to the Sirens.) Both of these cases correspond to a "total commitment": Once the bridge is burned, or Odysseus is lashed to the mast and the sailors' ears are filled with wax, the cost of turning back or escaping from the mast is taken to be infinite. One can also consider partial commitments, which increase the cost of, e.g., turning back without making it infinite.

As a final example of the value of commitment, we consider what is known as the "time-consistency problem" in macroeconomics. This problem was first noted by Kydland and Prescott (1977); our discussion draws on the survey by Mankiw (1988). Suppose that the government sets the inflation rate π, and has preferences over inflation and output y represented by $u_g(\pi, y) = y - \pi^2$, so that it is prepared to tolerate inflation if doing so increases the output level. The working of the macroeconomy is such that only unexpected inflation changes output:

$$y = y^* + (\pi - \hat{\pi}),\tag{3.1}$$

where y^* is the "natural level" of output and $\hat{\pi}$ is the expected inflation.[2]

Regardless of the timing of moves, the agents' expectations of inflation are correct in any pure-strategy equilibrium, and so output is at its natural level. (In a mixed-strategy equilibrium the expectations need only be correct on average.) The variable of interest is thus the level of inflation. Suppose first that the government can commit itself to an inflation rate, i.e., the government moves first and chooses a level of π that is observed by the agents. Then output will equal y^* regardless of the chosen level of π, so the government should choose $\pi = 0$.

As Kydland and Prescott point out, this solution to the commitment game is not "time consistent," meaning that if the agents mistakenly believe that π is set equal to 0 when in fact the government is free to choose any level of π it wishes, then the government would prefer to choose a different level of π. That is, the commitment solution is not an equilibrium of the game without commitment.

If the government cannot commit itself, it will choose the level of inflation that equates the marginal benefit from increased output to the marginal cost of increased inflation. The government's utility function is such that this tradeoff is independent of the level of output or the level of expected inflation, and the government will choose $\pi = \frac{1}{2}$. Since output is the same in the two cases, the government does strictly worse without commitment. In the context of monetary policy, the "commitment path" can be interpreted as a "money growth rule," and noncommitment corresponds to a "discretionary policy"; hence the conclusion that "rules can be better than discretion."[3]

As a gloss on the time-consistency problem, let us consider the analogous questions in relation to Stackelberg and Cournot equilibria. If we think of the government and the agents as both choosing output levels, the commitment solution corresponds to the Stackelberg outcome (q_1^*, q_2^*). This outcome is not an equilibrium of the game where the government cannot commit itself, because in general q_1^* is not a best response to q_2^* when q_2^*

2. Equation 3.1 is a reduced form that incorporates the way that the agents' expectations influence their production decisions and in turn influence output. Since the actions of the agents have been suppressed, the model does not directly correspond to an extensive-form game, but the same intuitions apply. Here is an artificial extensive-form game with the same qualitative properties: The government chooses the money supply m, and a single agent chooses a nominal price p. Aggregate demand is $y = \max(0, m - p)$, and the agent is constrained to supply all demanders. The agent's utility is $p - p^2/2m$, and the government's utility is $y - (m - 1)^2$. This does not quite give equation 3.1, but the resulting model has very similar properties.

3. In the extensive-game model where the agent chooses prices (see note 2), the agent chooses $p = m$ and the commitment solution is to set $m = 1$. Without commitment this is not an equilibrium, since for fixed p the government could gain by choosing a larger value of m.

is held fixed. The no-commitment solution $\pi = \frac{1}{2}$ derived above corresponds to a situation of simultaneous moves—that is, to the Cournot outcome.

Whether and when a commitment to a monetary rule is credible have been important topics of theoretical and applied research in macroeconomics. This research has started from the observation that decisions about the money supply are not made once and for all, but rather are made repeatedly. Chapter 5, on repeated games, and chapter 9, on reputation effects, discuss game-theoretic analyses of the question of when repeated play makes commitments credible.

Finally, note that a player does not always do better when he moves first (and his choice of action is observed) than when players move simultaneously: In "matching pennies" (example 1.6) each player's equilibrium payoff is 0, whereas if one player moves first his equilibrium payoff is -1.

3.3 The Extensive Form[††]

This section gives a formal development of the idea of an extensive-form game. The extensive form is a fundamental concept in game theory and one to which we will refer frequently, particularly in chapters 8 and 11, but the details of the definitions are not essential for much of the material in the rest of the book. Thus, readers who are primarily interested in applications of the theory should not be discouraged if they do not master all the fine points of the extensive-form methodology. Instead of dwelling on this section, they should proceed along, remembering to review this material before beginning section 8.3.

3.3.1 Definition

The extensive form of a game contains the following information:

(1) the set of players
(2) the order of moves—i.e., who moves when
(3) the players' payoffs as a function of the moves that were made
(4) what the players' choices are when they move
(5) what each player knows when he makes his choices
(6) the probability distributions over any exogenous events.

The set of players is denoted by $i \in \mathscr{I}$; the probability distributions over exogenous events (point 6) are represented as moves by "Nature," which is denoted by N. The order of play (point 2) is represented by a *game tree*, T, such as the one shown in figure 3.1.[4] A tree is a finite collection of ordered

---

4. Our development of the extensive form follows that of Kreps and Wilson 1982 with a simplification suggested by Jim Ratliff. Their assumptions (and ours) are equivalent to those of Kuhn 1953.

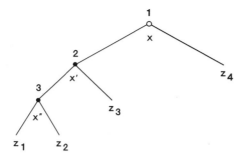

Figure 3.1

nodes $x \in X$ endowed with a precedence relation denoted by $\succ$; $x \succ x'$ means "x is before x'." We assume that the precedence relation is transitive (if x is before x' and x' is before x'', then x is before x'') and asymmetric (if x is before x', then x' is not before x). These assumptions imply that the precedence relation is a *partial order*. (It is not a complete order, because two nodes may not be comparable: In figure 3.1, z_3 is not before x'', and x'' is not before z_3.) We include a single initial node $\circ \in X$ that is before all other nodes in X; this node will correspond to a move by nature if any. Figure 3.1 describes a situation where "nature's move" is trivial, as nature simply gives the move to player 1. As in this figure, we will suppress nature's move whenever it is trivial, and begin the tree with the first "real" choice. The initial node will be depicted with $\circ$ to distinguish it from the others. In figure 3.1, the precedence order is from the top of the diagram down. Given the assumptions we will impose, the precedence ordering will be clear in most diagrams; when the intended precedence is not clear we will use arrows ($\rightarrow$) to connect a node to its immediate successors.

The assumption that precedence is a partial order rules out cycles of the kind shown in figure 3.2a: If $x \succ x' \succ x'' \succ x$, then by transitivity $x'' \succ x'$. Since we already have $x' \succ x''$, this would violate the asymmetry condition. However, the partial ordering does not rule out the situation shown in figure 3.2b, where both x and x' are immediate predecessors of node x''.

We wish to rule out the situation in figure 3.2b, because each node of the tree is meant to be a complete description of all events that preceded it, and not just of the "physical situation" at a given point in time. For example, in figure 3.2c, a firm in each of two markets, A and B, might have entered A and then B (node x and then x'') or B and then A (node x' and then x''), but we want our formalism to distinguish between these two sequences of events instead of describing them by a single node x''. (Of course, we are free to specify that both sequences lead to the same payoff for the firm.) In order to ensure that there is only one path through the tree to a given node, so that each node is a complete description of the path

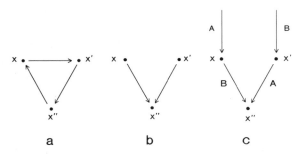

Figure 3.2

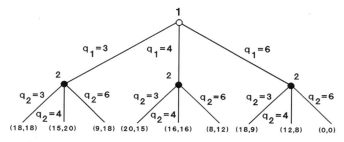

Figure 3.3

preceding it, we require that each node x (except the initial node o) have exactly one immediate predecessor—that is, one node $x' \succ x$ such that $x'' \succ x$ and $x'' \neq x'$ implies $x'' \succ x'$. Thus, if x and x' are both predecessors of x'', then either x is before x' or x' is before x. (This makes the pair $(X, \succ)$ an *arborescence*.)

The nodes that are not predecessors of any other node are called "terminal nodes" and denoted by $z \in Z$. Because each z completely determines a path through the tree, we can assign payoffs to sequences of moves using functions $u_i: Z \to \mathbb{R}$, with $u_i(z)$ being player i's payoff if terminal node z is reached. In drawing extensive forms, the payoff vectors (point 3 in the list above) are displayed next to the corresponding terminal nodes, as in figures 3.3 and 3.4. To complete the specification of point 2 (who moves when), we introduce a map $i: X/Z \to \mathscr{I}$ with the interpretation that player $i(x)$ moves at node x. Next we must describe what player $i(x)$'s choices are, which was point 4 of our list. To do so, we introduce a finite set A of actions and a function ℓ that labels each noninitial node x with the last action taken to reach it. We require that ℓ be one-to-one on the set of immediate successors of each node x, so that different successors correspond to different actions, and let $A(x)$ denote the set of feasible actions at x. (Thus $A(x)$ is the range of ℓ on the set of immediate successors of x.)

Point 5, the information players have when choosing their actions, is the most subtle of the six points. This information is represented using *information sets* $h \in H$, which *partition* the nodes of the tree—that is, every node is in exactly one information set.[5] The interpretation of the information set $h(x)$ containing node x is that the player who is choosing an action at x is uncertain if he is at x or at some other $x' \in h(x)$. We require that if $x' \in h(x)$ the same player move at x and x'. Without this requirement, the players might disagree about who was supposed to move. Also, we require that if $x' \in h(x)$ then $A(x') = A(x)$, so the player on move has the same set of choices at each node of this information set. (Otherwise he might "play" an infeasible action.) Thus, we can let $A(h)$ denote the action set at information set h.

A special case of interest is that of *games of perfect information*, in which all the information sets are singletons. In a game of perfect information, players move one at a time, and each player knows all previous moves when making his decision. The Stackelberg game we discussed at the start of this chapter is a game of perfect information. Figure 3.3 displays a tree for this

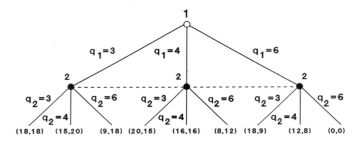

a

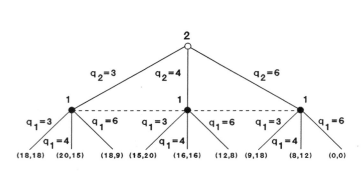

b

Figure 3.4

5. Note that we use the same notation, h, for information sets and for histories in multi-stage games. This should not cause too much confusion, especially as information sets can be viewed as a generalization of the idea of a history.

game on the assumption that each player has only three possible output levels: 3, 4, and 6. The vectors at the end of each branch of the tree are the payoffs of players 1 and 2, respectively.

Figure 3.4a displays an extensive form for the Cournot game, where players 1 and 2 choose their output levels simultaneously. Here player 2 does not know player 1's output level when choosing his own output. We model this by placing the nodes corresponding to player 1's three possible actions in the same information set for player 2. This is indicated in the figure by the broken line connecting the three nodes. (Some authors use "loops" around the nodes instead.) Note well the way simultaneous moves are represented: As in figure 3.3, player 1's decision comes "before" player 2's in terms of the precedence ordering of the tree; the difference is in player 2's information set. As this shows, the precedence ordering in the tree need not correspond to calendar time. To emphasize this point, consider the extensive form in figure 3.4b, which begins with a move by player 2. Figures 3.4a and 3.4b describe exactly the same strategic situation: Each player chooses his action not knowing the choice of his opponent. However, the situation represented in figure 3.3, where player 2 observed player 1's move before choosing his own, can only be described by an extensive form in which player 1 moves first.

Almost all games in the economics literature are games of *perfect recall*: No player ever forgets any information he once knew, and all players know the actions they have chosen previously. To impose this formally, we first require that if x and x' are in the same information set then neither is a predecessor of the other. This is not enough to ensure that a player never forgets, as figure 3.5 shows. To rule out this situation, we require that if $x'' \in h(x')$, if x is a predecessor of x', and if the same player i moves at x and at x' (and thus at x''), then there is a node $\hat{x}$ (possibly x itself) that is in the same information set as x, that $\hat{x}$ is a predecessor of x'', and that the action taken at x along the path to x' is the same as the action taken at $\hat{x}$ along the path to x''. Intuitively, the nodes x' and x'' are distinguished by information the player doesn't have, so he can't have had it when he was at information set $h(x)$; x' and x'' must be consistent with the same action at $h(x)$, since the player remembers his action there.

When a game involves moves by Nature, the exogenous probabilities are displayed in *brackets*, as in the two-player extensive form of figure 3.6. In

Figure 3.5

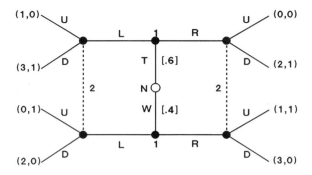

Figure 3.6

figure 3.6, Nature moves first and chooses a "type" or "private information" for player 1. With probability 0.6 player 1 learns that his type is "tough" (T), and with probability 0.4 he learns that his type is "weak" (W). Player 1 then plays left (L) or right (R). Player 2 observes player 1's action but not his type, and chooses between up (U) and down (D). Note that we have allowed both players' payoffs to depend on the choice by Nature even though this choice is initially observed only by player 1. (Player 2 will be able to infer Nature's move from his payoffs.) Figure 3.6 is an example of a "signaling game," as player 1's action may reveal information about his type to player 2. Signaling games, the simplest games of incomplete information, will be studied in detail in chapters 8 and 11.

3.3.2 Multi-Stage Games with Observed Actions

Many of the applications of game theory to economics, political science, and biology have used the special class of extensive forms that we discussed in section 3.2: the class of "*multi-stage games with observed actions.*"[6] These games have "stages" such that (1) in each stage k every player knows all the actions, including those by Nature, that were taken at any previous stage; (2) each player moves at most once within a given stage; and (3) no information set contained in stage k provides any knowledge of play in that stage. (Exercise 3.4 asks you to give a formal definition of these conditions in terms of a game tree and information sets.)

In a multi-stage game, all past actions are common knowledge at the beginning of stage k, so there is a well-defined "history" h^k at the start of each stage k. Here a pure strategy for player i is a function s_i that specifies an action $a_i \in A_i(h^k)$ for each k and each history h^k; mixed strategies specify probability mixtures over the actions in each stage.

Caution Although the idea of a multi-stage game seems natural and intrinsic, it suffers from the following drawback: There may be two exten-

6. Such games are also often called "games of almost-perfect information."

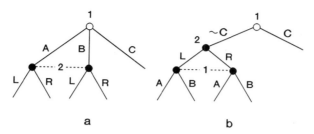

Figure 3.7

sive forms that seem to represent the same real game, with one of them a multi-stage game and the other one not. Consider for example figure 3.7. The extensive form on the left is not a multi-stage game: Player 2's information set is not a singleton, and so it must belong to the first stage and not to a second one. However, player 2 does have some information about player 1's first move (if player 2's information set is reached, then player 1 did not play C), so player 2's information set cannot belong to the first stage either. However, the extensive form on the right is a two-stage game, and the two extensive forms seem to depict the same situation: When player 2 moves, he knows that player 1 is choosing A or B but not C; player 1 chooses A or B without knowing player 2's choice of L or R. The question as to which extensive forms are "equivalent" is still a topic of research—see Elmes and Reny 1988. We will have more to say about this topic when we discuss recent work on equilibrium refinements in chapter 11.

Before proceeding to the next section, we should point out that in applications the extensive form is usually described without using the apparatus of the formal definition, and game trees are virtually never drawn except for very simple "toy" examples. The test of a good informal description is whether it provides enough information to construct the associated extensive form; if the extensive form is not clear, the model has not been well specified.

3.4 Strategies and Equilibria in Extensive-Form Games[††]

3.4.1 Behavior Strategies

This section defines strategies and equilibria in extensive-form games and relates them to strategies and equilibria of the strategic-form model. Let H_i be the set of player i's information sets, and let $A_i \equiv \bigcup_{h_i \in H_i} A(h_i)$ be the set of all actions for player i. A *pure strategy* for player i is a map $s_i: H_i \to A_i$, with $s_i(h_i) \in A(h_i)$ for all $h_i \in H_i$. Player i's space of pure strategies, S_i, is simply the space of all such s_i. Since each pure strategy is a map from information sets to actions, we can write S_i as the Cartesian product of the action spaces at each h_i:

$$S_i = \underset{h_i \in H_i}{\times} A(h_i).$$

In the Stackelberg example of figure 3.3, player 1 has a single information set and three actions, so that he has three pure strategies. Player 2 has three information sets, corresponding to the three possible choices of player 1, and player 2 has three possible actions at each information set, so player 2 has 27 pure strategies in all. More generally, the number of player i's pure strategies, $\#S_i$, equals

$$\prod_{h_i \in H_i} \#(A(h_i)).$$

Given a pure strategy for each player i and the probability distribution over Nature's moves, we can compute a probability distribution over outcomes and thus assign expected payoffs $u_i(s)$ to each strategy profile s. The information sets that are reached with positive probability under profile s are called the *path* of s.

Now that we have defined the payoffs to each pure strategy, we can proceed to define a pure-strategy Nash equilibrium for an extensive-form game as a strategy profile s^* such that each player i's strategy s_i^* maximizes his expected payoff given the strategies s_{-i}^* of his opponents. Note that since the definition of Nash equilibrium holds the strategies of player i's opponents fixed in testing whether player i wishes to deviate, it is as if the players choose their strategies simultaneously. This does *not* mean that in Nash equilibrium players necessarily choose their *actions* simultaneously. For example, if player 2's fixed strategy in the Stackelberg game of figure 3.3 is the Cournot reaction function $\hat{s}_2 = (4,4,3)$, then when player 1 treats player 2's strategy as fixed he does not presume that player 2's action is unaffected by his own, but rather that player 2 will respond to player 1's action in the way specified by $\hat{s}_2$.

To fill in the details missing from our discussion of the Stackelberg game in the introduction: The "Stackelberg equilibrium" of this game is the outcome $q_1 = 6$, $q_2 = 3$. This outcome corresponds to the Nash-equilibrium strategy profile $s_1 = 6$, $s_2 = \hat{s}_2$. The Cournot outcome is (4,4); this is the outcome of the Nash equilibrium $s_1 = 4$, $s_2 = (4,4,4)$.

The next order of business is to define mixed strategies and mixed-strategy equilibria for extensive-form games. Such strategies are called *behavior strategies* to distinguish them from the strategic-form mixed strategies we introduced in chapter 1. Let $\Delta(A(h_i))$ be the probability distributions on $A(h_i)$. A *behavior strategy for player i*, denoted b_i, is an element of the Cartesian product $\times_{h_i \in H_i} \Delta(A(h_i))$. That is, a behavior strategy specifies a probability distribution over actions at each h_i, and the probability distributions at different information sets are independent. (Note that a pure strategy is a special kind of behavior strategy in which the distribution at each information set is degenerate.) A profile $b = (b_1, \ldots, b_I)$ of behavior strategies generates a probability distribution over outcomes in the obvious way, and hence gives rise to an expected payoff for each player. A *Nash equilibrium in behavior strategies* is a profile such that no player can increase his expected payoff by using a different behavior strategy.

3.4.2 The Strategic-Form Representation of Extensive-Form Games

Our next step is to relate extensive-form games and equilibria to the strategic-form model. To define a strategic form from an extensive form, we simply let the pure strategies $s \in S$ and the payoffs $u_i(s)$ be exactly those we defined in the extensive form. A different way of saying this is that the same pure strategies can be interpreted as either extensive-form or strategic-form objects. With the extensive-form interpretation, player i "waits" until h_i is reached before deciding how to play there; with the strategic-form inter-pretation, he makes a complete contingent plan in advance.

Figure 3.8 illustrates this passage from the extensive form to the strategic form in a simple example. We order player 2's information sets from left to right, so that, for example, the strategy $s_2 = (L, R)$ means that he plays L after U and R after D.

As another example, consider the Stackelberg game illustrated in figure 3.3. We will again order player 2's information sets from left to right, so that player 2's strategy $\hat{s}_2 = (4,4,3)$ means that he plays 4 in response to $q_1 = 3$, plays 4 in response to 4, and plays 3 in response to 6. (This strategy happens to be player 2's Cournot reaction function.) Since player 2 has three information sets and three possible actions at each of these sets, he has 27 pure strategies. We trust that the reader will forgive our not display-ing the strategic form in a matrix diagram!

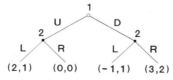

a. Extensive Form

	(L,L)	(L,R)	(R,L)	(R,R)
U	2,1	2,1	0,0	0,0
D	−1,1	3,2	−1,1	3,2

b. Strategic Form

Figure 3.8

There can be several extensive forms with the same strategic form, as the example of simultaneous moves shows: Figures 3.4a and 3.4b both correspond to the same strategic form for the Cournot game.

At this point we should note that the strategy space as we have defined it may be unnecessarily large, as it may contain pairs of strategies that are "equivalent" in the sense of having the same consequences regardless of how the opponents play.

Definition 3.2 Two pure strategies s_i and s_i' are *equivalent* if they lead to the same probability distribution over outcomes for all pure strategies of the opponents.

Consider the example in figure 3.9. Here player 1 has four pure strategies: $(a, c), (a, d), (b, c)$, and (b, d). However, if player 1 plays b, his second information set is never reached, and the strategies (b, c) and (b, d) are equivalent.

Definition 3.3 The *reduced strategic form* (or reduced normal form) of an extensive-form game is obtained by identifying equivalent pure strategies (i.e., eliminating all but one member of each equivalence class).

Once we have derived the strategic form from the extensive form, we can (as in chapter 1) define mixed strategies to be probability distributions over pure strategies in the reduced strategic form. Although the extensive form and the strategic form have exactly the same pure strategies, the sets of mixed and behavior strategies are different. With behavior strategies, player i performs a different randomization at each information set. Luce and Raiffa (1957) use the following analogy to explain the relationship between mixed and behavior strategies: A pure strategy is a book of instructions, where each page tells how to play at a particular information set. The strategy space S_i is like a library of these books, and a mixed strategy is a probability measure over books—i.e., a random way of making a selection from the library. A given behavior strategy, in contrast, is a single book, but it prescribes a random choice of action on each page.

The reader should suspect that these two kinds of strategies are closely related. Indeed, they are equivalent in games of perfect recall, as was proved

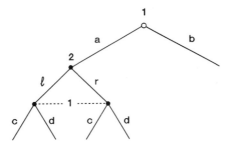

Figure 3.9

by Kuhn (1953). (Here we use "equivalence" as in our earlier definition: Two strategies are equivalent if they give rise to the same distributions over outcomes for all strategies of the opponents.)

3.4.3 The Equivalence between Mixed and Behavior Strategies in Games of Perfect Recall

The equivalence between mixed and behavior strategies under perfect recall is worth explaining in some detail, as it also helps to clarify the workings of the extensive-form model. Any mixed strategy σ_i of the strategic form (not of the reduced strategic form) generates a unique behavior strategy b_i as follows: Let $R_i(h_i)$ be the set of player i's pure strategies that do not preclude h_i, so that for all $s_i \in R_i(h_i)$ there is a profile s_{-i} for player i's opponents that reaches h_i. If σ_i assigns positive probability to some s_i in $R_i(h_i)$, define the probability that b_i assigns to $a_i \in A(h_i)$ as

$$b_i(a_i|h_i) = \sum_{\{s_i \in R_i(h_i) \text{ and } s_i(h_i)=a_i\}} \sigma_i(s_i) \Bigg/ \sum_{\{s_i \in R_i(h_i)\}} \sigma_i(s_i).$$

If σ_i assigns probability 0 to all $s_i \in R_i(h_i)$, then set

$$b_i(a_i|h_i) = \sum_{\{s_i(h_i)=a_i\}} \sigma_i(s_i).^{[7]}$$

In either case, the $b_i(\cdot|\cdot)$ are nonnegative, and

$$\sum_{a_i \in A(h_i)} b_i(a_i|h_i) = 1,$$

because each s_i specifies an action for player i at h_i.

Note that in the notation $b_i(a_i|h_i)$, the variable h_i is redundant, as $a_i \in A(h_i)$, but the conditioning helps emphasize that a_i is an action that is feasible at information set h_i.

It is useful to work through some examples to illustrate the construction of behavior strategies from mixed strategies. In figure 3.10, a single player (player 1) moves twice. Consider the mixed strategy $\sigma_1 = (\frac{1}{2}(L, \ell), \frac{1}{2}(R, \imath))$.

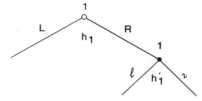

Figure 3.10

7. Since h_i cannot be reached under σ_i, the behavior strategies at h_i are arbitrary in the same sense that Bayes' rule does not determine posterior probabilities after probability-0 events. Our formula is one of many possible specifications.

The strategy plays $\imath$ with probability 1 at information set h_1', as only $(R, \imath) \in R_1(h_1')$.

Figure 3.11 gives another example. Player 2's strategy σ_2 assigns probability $\frac{1}{2}$ each to $s_2 = (L, L', R'')$ and $\tilde{s}_2 = (R, R', L'')$. The equivalent behavior strategy is

$$b_2(L \mid h_2) = b_2(R \mid h_2) = \tfrac{1}{2};$$

$$b_2(L' \mid h_2') = 0 \text{ and } b_2(R' \mid h_2') = 1,$$

and

$$b_2(L'' \mid h_2'') = b_2(R'' \mid h_2'') = \tfrac{1}{2}.$$

Many different mixed strategies can generate the same behavior strategy. This can be seen from figure 3.12, where player 2 has four pure strategies: $s_2 = (A, C)$, $s_2' = (A, D)$, $s_2'' = (B, C)$, and $s_2''' = (B, D)$.

Now consider two mixed strategies: $\sigma_2 = (\frac{1}{4}, \frac{1}{4}, \frac{1}{4}, \frac{1}{4})$, which assigns probability $\frac{1}{4}$ to each pure strategy, and $\hat{\sigma}_2 = (\frac{1}{2}, 0, 0, \frac{1}{2})$, which assigns probability $\frac{1}{2}$ to s_2 and $\frac{1}{2}$ to s_2'''. Both of these mixed strategies generate the behavior strategy b_2, where $b_2(A \mid h) = b_2(B \mid h) = \frac{1}{2}$ and $b_2(C \mid h') = b_2(D \mid h') = \frac{1}{2}$.

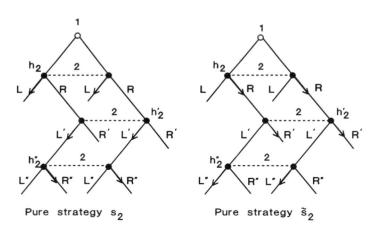

Pure strategy s_2 Pure strategy $\tilde{s}_2$

Figure 3.11

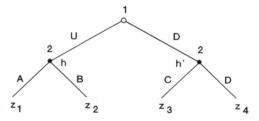

Figure 3.12

Moreover, for any strategy σ_1 of player 1, σ_2, $\hat{\sigma}_2$, and b_2 all lead to the same probability distribution over terminal nodes; for example, the probability of reaching node z_1 equals the probability of player 1's playing U times $b_2(A|h)$.

The relationship between mixed and behavior strategies is different in the game illustrated in figure 3.13, which is not a game of perfect recall. (Exercise 3.2 asks you to verify this using the formal definition.) Here, player 1 has four strategies in the strategic form:

$$s_1 = (A, C),\ s_1' = (A, D),\ s_1'' = (B, C),\ s_1''' = (B, D).$$

Now consider the mixed strategy $\sigma_1 = (\frac{1}{2}, 0, 0, \frac{1}{2})$. As in the last example, this generates the behavior strategy $b_1 = \{(\frac{1}{2}, \frac{1}{2}), (\frac{1}{2}, \frac{1}{2})\}$, which says that player 1 mixes $\frac{1}{2}$-$\frac{1}{2}$ at each information set. But b_1 is *not* equivalent to the σ_1 that generated it. Consider the strategy $s_2 = L$ for player 2. Then (σ_1, L) generates a $\frac{1}{2}$ probability of the terminal node corresponding to (A, L, C), and a $\frac{1}{2}$ probability of (B, L, D). However, since behavior strategies describe independent randomizations at each information set, (b_1, L) assigns probability $\frac{1}{4}$ to each of the four paths (A, L, C), (A, L, D), (B, L, C), and (B, L, D). Since both A vs. B and C vs. D are choices made by player 1, the strategic-form strategy σ_1 can have the property that both A and B have positive probability but C is played wherever A is. Put differently, the strategic-form strategies, where player 1 makes all his decisions at once, allow the decisions at different information sets to be *correlated*. Behavior strategies can't produce this correlation in the example, because when it comes time to choose between C and D player 1 has forgotten whether he chose A or B. This forgetfulness means that there is not perfect recall in this game. If we change the extensive form so that there is perfect recall (by partitioning player 1's second information set into two, corresponding to his choice of A or B), it is easy to see that every mixed strategy is indeed equivalent to the behavior strategy it generates.

Theorem 3.1 (Kuhn 1953)   In a game of perfect recall, mixed and behavior strategies are equivalent. (More precisely: Every mixed strategy is equiv-

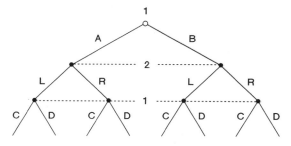

Figure 3.13

alent to the unique behavior strategy it generates, and each behavior strategy is equivalent to every mixed strategy that generates it.)

We will restrict our attention to games of perfect recall throughout this book, and will use the terms "mixed strategy" and "Nash equilibrium" to refer to the mixed and behavior formulations interchangeably. This leads us to the following important *notational convention*: In the rest of part II and in most of part IV (except in sections 8.3 and 8.4), we will be studying behavioral strategies. Thus, when we speak of a mixed strategy of an extensive form, we will mean a behavior strategy unless we state otherwise. Although the distinction between the mixed strategy σ_i and the behavior strategy b_i was necessary to establish their equivalence, we will follow standard usage by denoting both objects by σ_i (thus, the notation b_i is not used in the rest of the book). In a multi-stage game with observed actions, we will let $\sigma_i(a_i^k | h^k)$ denote player i's probability of playing action $a_i^k \in A_i(h^k)$ given the history of play h^k at stage k. In general extensive forms (with perfect recall), we let $\sigma_i(a_i | h_i)$ denote player i's probability of playing action a_i at information set h_i.

3.4.4   Iterated Strict Dominance and Nash Equilibrium

If the extensive form is finite, so is the corresponding strategic form, and the Nash existence theorem yields the existence of a mixed-strategy equilibrium. The notion of iterated strict dominance extends to extensive-form games as well; however, as we mentioned above, this concept turns out to have little force in most extensive forms. The point is that a player cannot strictly prefer one action over another at an information set that is not reached given his opponents' play.

Consider figure 3.14. Here, player 2's strategy R is not strictly dominated, as it is as good as L when player 1 plays U. Moreover, this fact is not "pathological." It obtains for all strategic forms whose payoffs are derived from an extensive form with the tree on the left-hand side of the figure. That is, for any assignment of payoffs to the terminal nodes of the tree, the payoffs to (U, L) and (U, R) must be the same, as both strategy profiles lead to the same terminal node. This shows that the set of strategic-form payoffs of a fixed game tree is of lower dimension than the set of all payoffs of the corresponding strategic form, so theorems based on generic strategic-form payoffs (see chapter 12) do not apply. In particular, there can be an even number of Nash equilibria for an open set of extensive-form payoffs. The game illustrated in figure 3.14 has two Nash equilibria, (U, R) and (D, L), and this number is not changed if the extensive-form payoffs are slightly perturbed. The one case where the odd-number theorem of chapter 12 applies is to a simultaneous-move game such as that of figure 3.4; in such a game, each terminal node corresponds to a unique strategy profile. Put

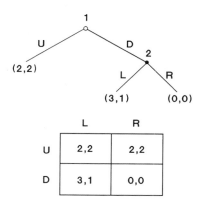

Figure 3.14

differently: In simultaneous-move games, every strategy profile reaches every information set, and so no player's strategy can involve a choice that is not implemented given his opponents' play.

Recall that a game of perfect information has all its information sets as singletons, as in the games illustrated in figures 3.3 and 3.14.

Theorem 3.2 (Zermelo 1913; Kuhn 1953)   A finite game of perfect information has a pure-strategy Nash equilibrium.

The proof of this theorem constructs the equilibrium strategies using "Zermelo's algorithm," which is a many-player generalization of backward induction in dynamic programming. Since the game is finite, it has a set of penultimate nodes—i.e., nodes whose immediate successors are terminal nodes. Specify that the player who can move at each such node chooses whichever strategy leads to the successive terminal node with the highest payoff for him. (In case of a tie, make an arbitrary selection.) Now specify that each player at nodes whose immediate successors are penultimate nodes chooses the action that maximizes her payoff over the feasible successors, given that players at the penultimate nodes play as we have just specified. We can now roll back through the tree, specifying actions at each node. When we are done, we will have specified a strategy for each player, and it is easy to check that these strategies form a Nash equilibrium. (In fact, the strategies satisfy the more restrictive concept of subgame perfection, which we will introduce in the next section.)

Zermelo's algorithm is not well defined if the hypotheses of the theorem are weakened. First consider infinite games. An infinite game necessarily has either a single node with an infinite number of successors (as do games with a continuum of actions) or a path consisting of an infinite number of nodes (as do multi-stage games with an infinite number of stages). In the first case, an optimal choice need not exist without further restrictions on the payoff functions[8]; in the second, there need not be a penultimate node on a given path from which to work backward. Finally, consider a game of imperfect information in which some of the information sets are not singletons, as in figure 3.4a. Here there is no way to define an optimal choice for player 2 at his information set without first specifying player 2's belief about the previous choice of player 1; the algorithm fails because it presumes that such an optimal choice exists at every information set given a specification of play at its successors.

We will have much more to say about this issue when we treat equilibrium refinements in detail. We conclude this section with one caveat about the assertion that the Nash equilibrium is a minimal requirement for a "reasonable" point prediction: Although the Nash concept can be applied to any game, the assumption that each player correctly forecasts his opponents' strategy may be less plausible when the strategies correspond to choices of contingent plans than when the strategies are simply choices of actions. The issue here is that when some information sets may not be reached in the equilibrium, Nash equilibrium requires that players correctly forecast their opponents' play at information sets that have 0 probability according to the equilibrium strategies. This may not be a problem if the forecasts are derived from introspection, but if the forecasts are derived from observations of previous play it is less obvious why forecasts should be correct at the information sets that are not reached. This point is examined in detail in Fudenberg and Kreps 1988 and in Fudenberg and Levine 1990.

3.5 Backward Induction and Subgame Perfection[††]

As we have seen, the strategic form can be used to represent arbitrarily complex extensive-form games, with the strategies of the strategic form being complete contingent plans of action in the extensive form. Thus, the concept of Nash equilibrium can be applied to all games, not only to games where players choose their actions simultaneously. However, many game theorists doubt that Nash equilibrium is the right solution concept for

---

8. The existence of an optimal choice from a compact set of actions requires that payoffs be upper semi-continuous in the choice made. (A real-valued function $f(x)$ is upper semi-continuous if $x^n \to x$ implies $\lim_{n \to \infty} f(x^n) \leq f(x)$.)

Assuming that payoffs u_i are continuous in s does not guarantee that an optimal action exists at each node. Although the last mover's payoff is continuous and therefore an optimum exists if his action set is compact, the last mover's optimal action need not be a continuous function of the action chosen by the previous player. In this case, when we replace that last mover by an arbitrary specification of an optimal action on each path, the next-to-last mover's derived payoff function need not be upper semi-continuous, even though that player's payoff is a continuous function of the actions chosen at each node. Thus, the simple backward-induction algorithm defined above cannot be applied. However, subgame-perfect equilibria do exist in infinite-action games of perfect information, as shown by Harris (1985) and by Hellwig and Leininger (1986).

general games. In this section we will present a first look at "equilibrium refinements," which are designed to separate the "reasonable" Nash equilibria from the "unreasonable" ones. In particular, we will discuss the ideas of backward induction and "subgame perfection." Chapters 4, 5 and 13 apply these ideas to some classes of games of interest to economists.

Selten (1965) was the first to argue that in general extensive games some of the Nash equilibria are "more reasonable" than others. He began with the example illustrated here in figure 3.14. This is a finite game of perfect information, and the backward-induction solution (that is, the one obtained using Kuhn's algorithm) is that player 2 should play L if his information set is reached, and so player 1 should play D. Inspection of the strategic form corresponding to this game shows that there is another Nash equilibrium, where player 1 plays U and player 2 plays R. The profile (U, R) *is* a Nash equilibrium because, given that player 1 plays U, player 2's information set is not reached, and player 2 loses nothing by playing R. But Selten argued, and we agree, that this equilibrium is suspect. After all, if player 2's information set is reached, then, as long as player 2 is convinced that his payoffs are as specified in the figure, player 2 should play L. And if we were player 2, this is how we would play. Moreover, if we were player 1, we would expect player 2 to play L, and so we would play D.

In the now-familiar language, the equilibrium (U, R) is not "credible," because it relies on an "empty threat" by player 2 to play R. The threat is "empty" because player 2 would never wish to carry it out.

The idea that backward induction gives the right answer in simple games like that of figure 3.14 was implicit in the economics literature before Selten's paper. In particular, it is embodied in the idea of Stackelberg equilibrium: The requirement that player 2's strategy be the Cournot reaction function is exactly the idea of backward induction, and all other Nash equilibria of the game are inconsistent with backward induction. So we see that the expression "Stackelberg equilibrium" does not simply refer to the extensive form of the Stackelberg game, but instead is shorthand for "the backward-induction solution to the sequential quantity-choice game." Just as with "Cournot equilibrium," this shorthand terminology can be convenient when no confusion can arise. However, our experience suggests that the terminology can indeed lead to confusion, so we advise the student to use the more precise language instead.

Consider the game illustrated in figure 3.15. Here neither of player 2's choices is dominated at his last information set, and so backward induction does not apply. However, given that one accepts the logic of backward induction, the following argument seems compelling as well: "The game beginning at player 1's second information set is a zero-sum simultaneous-move game ('matching pennies') whose unique Nash equilibrium has expected payoffs (0, 0). Player 2 should choose R only if he expects that there is

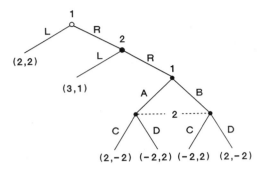

Figure 3.15

probability $\frac{3}{4}$ or better that he will outguess player 1 in the simultaneous-move subgame and end up with $+2$ instead of -2. Since player 2 assumes that player 1 is as rational as he is, it would be very rash of player 2 to expect to get the better of player 1, especially to such an extent. Thus, player 2 should go L, and so player 1 should go R." This is the logic of subgame perfection: Replace any "proper subgame" of the tree with one of its Nash-equilibrium payoffs, and perform backward induction on the reduced tree. (If the subgame has multiple Nash equilibria, this requires that all players agree on which of them would occur; we will come back to this point in subsection 3.6.1.) Once the subgame starting at player 1's second information set is replaced by its Nash-equilibrium outcome, the games described in figures 3.14 and 3.15 coincide.

To define subgame perfection formally we must first define the idea of a proper subgame. Informally, a proper subgame is a portion of a game that can be analyzed as a game in its own right, like the simultaneous-move game embedded in figure 3.15. The formal definition is not much more complicated:

Definition 3.4 A *proper subgame* G of an extensive-form game T consists of a *single* node and all its successors in T, with the property that if $x' \in G$ and $x'' \in h(x')$ then $x'' \in G$. The information sets and payoffs of the subgame are inherited from the original game. That is, x' and x'' are in the same information set in the subgame if and only if they are in the same information set in the original game, and the payoff function on the subgame is just the restriction of the original payoff function to the terminal nodes of the subgame.

Here the word "proper" here does not mean strict inclusion, as it does in the term "proper subset." Any game is always a proper subgame of itself. Proper subgames are particularly easy to identify in the class of deterministic multi-stage games with observed actions. In these games, all

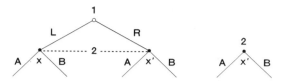

Figure 3.16

Figure 3.17

previous actions are known to all players at the start of each stage, so each stage begins a new proper subgame. (Checking this is part of exercise 3.4.)

The requirements that all the successors of x be in the subgame and that the subgame not "chop up" any information set ensure that the subgame corresponds to a situation that could arise in the original game. In figure 3.16, the game on the right isn't a subgame of the game on the left, because on the right player 2 knows that player 1 didn't play L, which he did not know in the original game.

Together, the requirements that the subgame begin with a single node x and that the subgame respect information sets imply that in the original game x must be a singleton information set, i.e., $h(x) = \{x\}$. This ensures that the payoffs in the subgame, conditional on the subgame being reached, are well defined. In figure 3.17, the "game" on the right has the problem that player 2's optimal choice depends on the relative probabilities of nodes x and x', but the specification of the game does not provide these probabilities. In other words, the diagram on the right cannot be analyzed as an independent game; it makes sense only as a component of the game on the left, which is needed to provide the missing probabilities.

Since payoffs conditional on reaching a proper subgame are well defined, we can test whether strategies yield a Nash equilibrium when restricted to the subgame in the obvious way. That is, if σ_i is a behavior strategy for player i in the original game, and $\hat{H}_i$ is the collection of player i's information sets in the proper subgame, then the restriction of σ_i to the subgame is the map $\hat{\sigma}_i$ such that $\hat{\sigma}_i(\cdot | h_i) = \sigma_i(\cdot | h_i)$ for every $h_i \in \hat{H}_i$.

We have now developed the machinery needed to define subgame perfection.

Definition 3.5 A behavior-strategy profile σ of an extensive-form game is a *subgame-perfect equilibrium* if the restriction of σ to G is a Nash equilibrium of G for every proper subgame G.

Because any game is a proper subgame of itself, a subgame-perfect equilibrium profile is necessarily a Nash equilibrium. If the only proper subgame is the whole game, the sets of Nash and subgame-perfect equilibria coincide. If there are other proper subgames, some Nash equilibria may fail to be subgame perfect.

It is easy to see that subgame perfection coincides with backward induction in finite games of perfect information. Consider the penultimate nodes of the tree, where the last choices are made. Each of these nodes begins a trivial one-player proper subgame, and Nash equilibrium in these subgames requires that the player now make a choice that maximizes his payoff; thus, any subgame-perfect equilibrium must coincide with a backward-induction solution at every penultimate node, and we can continue up the tree by induction. But subgame perfection is more general than backward induction; for example, it gives the suggested answer in the game of figure 3.15.

We remarked above that in multi-stage games with observed actions every stage begins a new proper subgame. Thus, in these games, subgame perfection is simply the requirement that the restrictions of the strategy profile yield a Nash equilibrium from the start of each stage k for each history h^k. If the game has a fixed finite number of stages ($K + 1$), then we can characterize the subgame-perfect equilibria using backward induction: The strategies in the last stage must be a Nash equilibrium of the corresponding one-shot simultaneous-move game, and for each history h^K we replace the last stage by one of its Nash-equilibrium payoffs. For each such assignment of Nash equilibria to the last stage, we then consider the set of Nash equilibria beginning from each stage h^{K-1}. (With the last stage replaced by a payoff vector, the game from h^{K-1} on is a one-shot simultaneous-move game.) The characterization proceeds to "roll back the tree" in the manner of the Kuhn-Zermelo algorithm. Note that even if two different stage-K histories lead to the "same game" in the last stage (that is, if there is a way of identifying strategies in the two games that preserves payoffs), the two histories still correspond to different subgames, and subgame perfection allows us to specify a different Nash equilibrium for each history. This has important consequences, as we will see in section 4.3 and in chapter 5.

3.6 Critiques of Backward Induction and Subgame Perfection[††]

This section discusses some of the limitations of the arguments for backwards induction and subgame perfection as necessary conditions for reasonable play. Although these concepts seem compelling in simple two-stage games of perfect information, such as the Stackelberg game we discussed at the start of the chapter, things are more complicated if there are many players or if each player moves several times; in these games, equilibrium refinements are less compelling.

3.6.1 Critiques of Backward Induction

Consider the I-player game illustrated in figure 3.18, where each player $i < I$ can either end the game by playing "D" or play "A" and give the move to player $i + 1$. (To readers who skipped sections 3.3–3.5: Figure 3.18 depicts a "game tree." Though you have not seen a formal definition of such trees, we trust that the particular trees we use in this subsection will be clear.) If player i plays D, each player gets $1/i$; if all players play A, each gets 2.

Since only one player moves at a time, this is a game of perfect information, and we can apply the backward-induction algorithm, which predicts that all players should play A. If I is small, this seems like a reasonable prediction. If I is very large, then, as player 1, we ourselves would play D and not A on the basis of a "robustness" argument similar to the one that suggested the inefficient equilibrium in the stag-hunt game of subsection 1.2.4.

First, the payoff 2 requires that all $I - 1$ other players play A. If the probability that a given player plays A is $p < 1$, independent of the others, the probability that all $I - 1$ other players play A is p^{I-1}, which can be quite small even if p is very large. Second, we would worry that player 2 might have these same concerns; that is, player 2 might play D to safeguard against either "mistakes" by future players or the possibility that player 3 might intentionally play D.

A related observation is that longer chains of backward induction presume longer chains of the hypothesis that "player 1 knows that player 2 knows that player 3 knows... the payoffs." If $I = 2$ in figure 3.18, backward induction supposes that player 1 knows player 2's payoff, or at least that player 1 is fairly sure that player 2's optimal choice is A. If $I = 3$, not only must players 1 and 2 know player 3's payoff, in addition, player 1 must know that player 2 knows player 3's payoff, so that player 1 can forecast player 2's forecast of player 3's play. If player 1 thinks that player 2 will forecast player 3's play incorrectly, then player 1 may choose to play D. Traditionally, equilibrium analysis is motivated by the assumption that

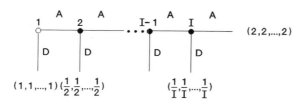

Figure 3.18

payoffs are "common knowledge," so that arbitrarily long chains of "i knows that j knows that k knows" are valid, but conclusions that require very long chains of this form are less compelling than conclusions that require less of the power of the common-knowledge assumption. (In part this is because longer chains of backward induction are more sensitive to small changes in the information structure of the game, as we will see in chapter 9.)

The example in figure 3.18 is most troubling if I is very large. A second complication with backward induction arises whenever the same player can move several times in succession. Consider the game illustrated in figure 3.19. Here the backward-induction solution is that at every information set the player who has the move plays D. Is this solution compelling? Imagine that it is, that you are player 2, and that, contrary to expectation, player 1 plays A_1 at his first move. How should you play? Backward induction says to play D_2 because player 1 will choose D_3 if given a chance, but backward induction also says that player 1 should have played D_1. In this game, unlike the simple examples we started with, player 2's best action if player 1 deviates from the predicted play A_1 depends on how player 2 expects player 1 to play in the future: If player 2 thinks there is at least a 25 percent chance that player 1 will play A_3, then player 2 should play A_2. How should player 2 form these beliefs, and what beliefs are reasonable? In particular, how should player 2 predict how player 1 will play if, contrary to backward induction, player 1 decides to play A_1? In some contexts, playing A_2 may seem like a good gamble.

Most analyses of dynamic games in the economics literature continue to use backward induction and its refinements without reservations, but recently the skeptics have become more numerous. The game depicted in figure 3.19 is based on an example provided by Rosenthal (1981), who was one of the first to question the logic of backward induction. Basu (1988, 1990), Bonanno (1988), Binmore (1987, 1988), and Reny (1986) have argued that reasonable theories of play should not try to rule out any behavior once an event to which the theory assigns probability 0 has occurred, because the theory provides no way for players to form their predictions conditional on these events. Chapter 11 discusses the work of Fudenberg, Kreps, and Levine (1988), who propose that players interpret unexpected deviations as being due to the payoffs' differing from those that were

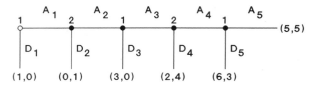

Figure 3.19

originally thought to be most likely. Since any observation of play can be explained by some specification of the opponents' payoffs, this approach sidesteps the difficulty of forming beliefs conditional on probability-0 events, and it recasts the question of how to predict play after a "deviation" as a question of which alternative payoffs are most likely given the observed play. Fudenberg and Kreps (1988) extend this to a methodological principle: They argue that any theory of play should be "complete" in the sense of assigning positive probability to any possible sequence of play, so that, using the theory, the players' conditional forecasts of subsequent play are always well defined.

Payoff uncertainty is not the only way to obtain a complete theory. A second family of complete theories is obtained by interpreting any extensive-form game as implicitly including the fact that players sometimes make small "mistakes" or "trembles" in the sense of Selten 1975. If, as Selten assumes, the probabilities of "trembling" at different information sets are independent, then no matter how often past play has failed to conform to the predictions of backward induction, a player is justified in continuing to use backward induction to predict play in the current subgame. Thus, interpreting "trembles" as deviations is a way to defend backward induction. The relevant question is how likely players view this "trembles" explanation of deviations as opposed to others. In figure 3.19, if player 2 observes A_1, should she (or will she) interpret this as a "tremble," or as a signal that player 1 is likely to play A_3?

3.6.2 Critiques of Subgame Perfection

Since subgame perfection is an extension of backward induction, it is vulnerable to the critiques just discussed. Moreover, subgame perfection requires that players all agree on the play in a subgame even if that play cannot be predicted from backward-induction arguments. This point is emphasized by Rabin (1988), who proposes alternative, weaker equilibrium refinements that allow players to disagree about which Nash equilibrium will occur in a subgame off the equilibrium path.

To see the difference this makes, consider the following three-player game. In the first stage, player 1 can either play L, ending the game with payoffs $(6, 0, 6)$, or play R, which gives the move to player 2. Player 2 can then either play R, ending the game with payoffs $(8, 6, 8)$, or play L, in which case players 1 and 3 (but not player 2) play a simultaneous-move "coordination game" in which they each choose F or G. If their choices differ, they each receive 7 and player 2 gets 10; if the choices match, all three players receive 0. This game is depicted in figure 3.20.

The coordination game between players 1 and 3 at the third stage has three Nash equilibria: two in pure strategies with payoffs $(7, 10, 7)$ and a mixed-strategy equilibrium with payoffs $(3\frac{1}{2}, 5, 3\frac{1}{2})$. If we specify an equilibrium in which players 1 and 3 successfully coordinate, then player 2 plays

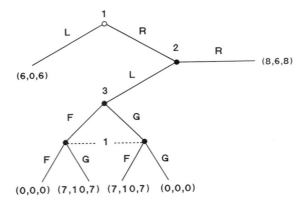

Figure 3.20

L and so player 1 plays R, expecting a payoff of 7. If we specify the inefficient mixed equilibrium in the third stage, then player 2 will play R and again player 1 plays R, this time expecting a payoff of 8. Thus, in all subgame-perfect equilibria of this game, player 1 plays R.

As Rabin argues, it may nevertheless be reasonable for player 1 to play L. He would do so if he saw no way to coordinate in the third stage, and hence expected a payoff of $3\frac{1}{2}$ conditional on that stage being reached, but feared that player 2 would believe that play in the third stage would result in coordination on an efficient equilibrium.

The point is that subgame perfection supposes not only that the players expect Nash equilibria in all subgames but also that all players expect the *same* equilibria. Whether this is plausible depends on the reason one thinks an equilibrium might arise in the first place.

Exercises

Exercise 3.1* Players 1 and 2 must decide whether or not to carry an umbrella when leaving home. They know that there is a 50-50 chance of rain. Each player's payoff is -5 if he doesn't carry an umbrella and it rains, -2 if he carries an umbrella and it rains, -1 if he carries an umbrella and it is sunny, and 1 if he doesn't carry an umbrella and it is sunny. Player 1 learns the weather before leaving home; player 2 does not, but he can observe player 1's action before choosing his own. Give the extensive and strategic forms of the game. Is it dominance solvable?

Exercise 3.2.* Verify that the game in figure 3.13 does not meet the formal definition of a game of perfect recall.

Exercise 3.3* Player 1, the "government," wishes to influence the choice of player 2. Player 2 chooses an action $a_2 \in A_2 = \{0, 1\}$ and receives a transfer $t \in T = \{0, 1\}$ from the government, which observes a_2. Player 2's objective is to maximize the expected value of his transfer, minus the cost of his action, which is 0 for $a_2 = 0$ and $\frac{1}{2}$ for $a_2 = 1$. Player 1's objective is to minimize the sum $2(a_2 - 1)^2 + t$. Before player 2 chooses his action, the government can announce a transfer rule $t(a_2)$.

(a) Draw the extensive form for the case where the government's announcement is not binding and has no effect on payoffs.

(b) Draw the extensive form for the case where the government is constrained to implement the transfer rule it announced.

(c) Give the strategic forms for both games.

(d) Characterize the subgame-perfect equilibria of the two games.

Exercise 3.4** Define a deterministic multi-stage game with observed actions using conditions on the information sets of an extensive form. Show that in these games the start of each stage begins a proper subgame.

Exercise 3.5** Show that subgame-perfect equilibria exist in finite multi-stage games.

Exercise 3.6* There are two players, a seller and a buyer, and two dates. At date 1, the seller chooses his investment level $I \geq 0$ at cost I. At date 2, the seller may sell one unit of a good and the seller has cost $c(I)$ of supplying it, where $c'(0) = -\infty$, $c' < 0$, $c'' > 0$, and $c(0)$ is less than the buyer's valuation. There is no discounting, so the socially optimal level of investment, I^*, is given by $1 + c'(I^*) = 0$.

(a) Suppose that at date 2 the buyer observes the investment I and makes a take-it-or-leave-it offer to the seller. What is this offer? What is the perfect equilibrium of the game?

(b) Can you think of a contractual way of avoiding the inefficient outcome of (i)? (Assume that contracts cannot be written on the level of I.)

Exercise 3.7* Consider a voting game in which three players, 1, 2, and 3, are deciding among three alternatives, A, B, and C. Alternative B is the "status quo" and alternatives A and C are "challengers." At the first stage, players choose which of the two challengers should be considered by casting votes for either A or C, with the majority choice being the winner and abstentions not allowed. At the second stage, players vote between the status quo B and whichever alternative was victorious in the first round, with majority rule again determining the winner. Players vote simultaneously in each round. The players care only about the alternative that is finally selected, and are indifferent as to the sequence of votes that leads to a given selection. The payoff functions are $u_1(A) = 2$, $u_1(B) = 0$, $u_1(C) = 1$; $u_2(A) = 1$, $u_2(B) = 2$, $u_2(C) = 0$; $u_3(C) = 2$, $u_3(B) = 1$, $u_3(A) = 0$.

(a) What would happen if at each stage the players voted for the alternative they would most prefer as the final outcome?

(b) Find the subgame-perfect equilibrium outcome that satisfies the additional condition that no strategy can be eliminated by iterated weak dominance. Indicate what happens if dominated strategies are allowed.

(c) Discuss whether different "agendas" for arriving at a final decision by voting between two alternatives at a time would lead to a different equilibrium outcome.

(This exercise is based on Eckel and Holt 1989, in which the play of this game in experiments is reported.)

Exercise 3.8* Subsection 3.2.3 discussed a player's "strategic incentive" to alter his first-period actions in order to change his own second-period incentives and thus alter the second-period equilibrium. A player may also have a strategic incentive to alter the second-period incentives of others. One application of this idea is the literature on strategic trade policies (e.g. Brander and Spencer 1985; Eaton and Grossman 1986—see Helpman and Krugman 1989, chapters 5 and 6, for a clear review of the arguments). Consider two countries, A and B, and a single good which is consumed only in country B. The inverse demand function is $p = P(Q)$, where Q is the total output produced by firms in countries A and B. Let c denote the constant marginal cost of production and Q_m the monopoly output (Q_m maximizes $Q(P(Q) - c)$).

(a) Suppose that country B does not produce the good. The I (≥ 1) firms in country A are Cournot competitors. Find conditions under which an optimal policy for the government of country A is to levy a unit export *tax* equal to $-P'(Q_m)(I - 1)Q_m/I$. Give an externality interpretation.

(b) Suppose now that there are two producers, one in each country. The game has two periods. In period 1, the government of country A chooses an export tax or subsidy (per unit of exports); in period 2, the two firms, which have observed the government's choice, simultaneously choose quantities. Suppose that the Cournot reaction curves are downward sloping and intersect only once, at a point at which country A's firm's reaction curve is steeper than country B's firm's reaction curve in the (q_A, q_B) space. (The objective of country A's government is to maximize the sum of its own receipts and the profit of its firm.) Show that an export *subsidy* is optimal.

(c) What would happen in question (b) if there were more than one firm in country A? If the strategic variables of period 2 gave rise to upward-sloping reaction curves? Caution: The answer to the latter depends on a "stability condition" of the kind discussed in subsection 1.2.5.

Exercise 3.9** Consider the three-player extensive-form game depicted in figure 3.21.

(a) Show that (A, A) is not the outcome of a Nash equilibrium.

(b) Consider the nonequilibrium situation where player 1 expects player 3 to play R, player 2 expects player 3 to play L, and consequently players

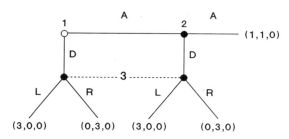

Figure 3.21

1 and 2 both play A. When might this be a fixed point of a learning process like those discussed in chapter 1? When might learning be expected to lead players 1 and 2 to have the same beliefs about player 3's action, as required for Nash equilibrium? (Give an informal answer.) For more on this question see Fudenberg and Kreps 1988 and Fudenberg and Levine 1990.

Exercise 3.10***   In the class of zero-sum games, the sets of outcomes of Nash and subgame-perfect equilibria are the same. That is, for every outcome (probability distribution over terminal nodes) of a Nash-equilibrium strategy profile, there is a perfect equilibrium profile with the same outcome. This result has limited interest, because most games in the social sciences are not zero-sum; however, its proof, which we give in the context of a multi-stage game with observed actions, is a nice way to get acquainted with the logic of perfect equilibrium. Consider a two-person game and let $u_1(\sigma_1, \sigma_2)$ denote player 1's expected payoff (by definition of a zero-sum game, $u_2 = -u_1$). Let $u_1(\sigma_1, \sigma_2 \mid h^t)$ denote player 1's expected payoff conditional on history h^t having been reached at date t (for simplicity, we identify "stages" with "dates"). Last, let $\sigma_i / \hat{\sigma}_i^{h^t}$ denote player i's strategy σ_i, except that if h^t is reached at date t, player i adopts strategy $\hat{\sigma}_i^{h^t}$ in the subgame associated with history h^t (henceforth called "the subgame").

(a) Let (σ_1, σ_2) be a Nash equilibrium. If (σ_1, σ_2) is not perfect, there is a date t, a history h^t, and a player (say player 1) such that this player does not maximize his payoff conditional on history h^t being reached. (Of course, this history h^t must have probability 0 of being reached according to strategies (σ_1, σ_2); otherwise player 1 will not be maximizing his unconditional payoff $u_1(\sigma_1, \sigma_2)$ given σ_2.)

Let $\hat{\sigma}_1^{h^t}$ denote the strategy that maximizes $u_1(\sigma_1 / \hat{\sigma}_1^{h^t}, \sigma_2 \mid h^t)$. Last, let $(\sigma_1^{*h^t}, \sigma_2^{*h^t})$ denote a Nash equilibrium of the subgame. Show that for *any* $\tilde{\sigma}_1$

$$u_1(\tilde{\sigma}_1 / \hat{\sigma}_1^{h^t}, \sigma_2 \mid h^t) \geq u_1(\tilde{\sigma}_1 / \sigma_1^{*h^t}, \sigma_2 / \sigma_2^{*h^t} \mid h^t).$$

(Hint: Use the facts that $\sigma_2^{*h^t}$ is a best response to $\sigma_1^{*h^t}$ in subgame h^t, that the game is a zero-sum game, and that $\hat{\sigma}_1^{h^t}$ is an optimal response to σ_2 in the subgame.)

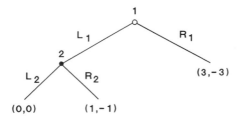

Figure 3.22

(b) Show that the strategy profile $(\sigma_1/\sigma_1^{*h^t}, \sigma_2/\sigma_2^{*h^t})$ is also a Nash equilibrium. (Hint: Use the fact that subgame h^t is not reached under (σ_1, σ_2) and the definition of Nash behavior in the subgame.)

(c) Conclude that the Nash-equilibrium outcome (the probability distribution on terminal nodes generated by σ_1 and σ_2) is also a perfect-equilibrium outcome.

Note that although *outcomes* coincide, the Nash-equilibrium *strategies* need not be perfect-equilibrium strategies—as is demonstrated in figure 3.22, where (R_1, R_2) is a Nash, but not a perfect, equilibrium.

Exercise 3.11* Consider the agenda-setter model of Romer and Rosenthal (1978) (see also Shepsle 1981). The object of the game is to make a one-dimensional decision. There are two players. The "agenda-setter" (player 1, who may stand for a committee in a closed-rule voting system) offers a point $s_1 \in \mathbb{R}$. The "voter" (player 2, who may stand for the median voter in the legislature) can then accept s_1 or refuse it; in the latter case, the decision is the *status quo* or reversion point s_0. Thus, $s_2 \in \{s_0, s_1\}$. The adopted policy is thus s_2. The voter has quadratic preferences $-(s_2 - \hat{s}_2)^2$, where $\hat{s}_2$ is his bliss point.

(a) Suppose that the agenda setter's objective is s_2 (she prefers higher policy levels). Show that, in perfect equilibrium, the setter offers $s_1 = s_0$ if $s_0 \geq \hat{s}_2$ and $s_1 = 2\hat{s}_2 - s_0$ if $s_0 < \hat{s}_2$.

(b) Suppose that the agenda setter's objective function is quadratic as well: $-(s_2 - \hat{s}_1)^2$. Fixing $\hat{s}_1$ and $\hat{s}_2$ ($\hat{s}_1 \gtrsim \hat{s}_2$), depict how the perfect-equilibrium policy varies with the reversion s_0.

Exercise 3.12** Consider the twice-repeated version of the agenda-setter model developed in the previous exercise. The new status quo in period 2 is whatever policy (agenda setter's proposal or initial status quo) was adopted in period 1. Suppose that the objective function of the agenda setter is the sum of the two periods' policies, and that the voter's preferences are $-(s_2^1 - 4)^2 - (s_2^2 - 12)^2$ (that is, his bliss point is 4 for the first-period policy and 12 for the second-period one). The initial status quo is 2.

(a) Suppose first that the voter is myopic (acts as if his discount factor were 0 instead of 1), but that the agenda setter is not. Show that the agenda setter offers 6 in period 1, and that the payoffs are 24 for the agenda setter and 40 for the voter. Assume in this exercise that the voter chooses the higher acceptable policy when indifferent. If you are courageous, show that this policy is uniquely optimal when her discount factor is slightly less than 1 instead of 1.

(b) Suppose now that both players are rational. Show that the agenda setter's utility is higher and the voter's utility is lower than in question (a). What point does this comparison illustrate? (See Ingberman 1985 and Rosenthal 1990.)

References

Basu, K. 1988. Strategic irrationality in extensive games. *Mathematical Social Sciences* 15: 247–260.

Basu, K. 1990. On the non-existence of a rationality definition for extensive games. *International Journal of Game Theory*.

Binmore, K. 1987. Modeling rational players: Part I. *Economics and Philosophy* 3: 179–214.

Binmore, K. 1988. Modeling rational players: Part II. *Economics and Philosophy* 4: 9–55.

Bonanno, G. 1988. The logic of rational play in extensive games of perfect information. Mimeo, University of California, Davis.

Brander, J., and B. Spencer. 1985. Export subsidies and market share rivalry. *Journal of International Economics* 18: 83–100.

Dixit, A. 1979. A model of duopoly suggesting a theory of entry barriers. *Bell Journal of Economics* 10: 20–32.

Eaton, J., and G. Grossman. 1986. Optimal trade and industrial policy under oligopoly. *Quarterly Journal of Economics* 101: 383–406.

Eckel, C., and C. Holt. 1989. Strategic voting in agenda-controlled committee experiments. *American Economic Review* 79: 763–773.

Elmes, S., and P. Reny. 1988. The equivalence of games with perfect recall. Mimeo.

Fudenberg, D., and D. Kreps. 1988. A theory of learning, experimentation, and equilibrium in games. Mimeo, MIT.

Fudenberg, D., D. Kreps, and D. Levine. 1988. On the robustness of equilibrium refinements. *Journal of Economic Theory* 44: 354–380.

Fudenberg, D., and D. Levine. 1990. Steady-state learning and self-confirming equilibrium. Mimeo.

Harris, C. 1985. Existence and characterization of perfect equilibrium in games of perfect information. *Econometrica* 53: 613–627.

Hellwig, M., and W. Leininger. 1986. On the existence of subgame-perfect equilibrium in infinite-action games of perfect information. Mimeo, Universität Bonn.

Helpman, E., and P. Krugman. 1989. *Trade Policy and Market Structure*. MIT Press.

Ingberman, D. 1985. Running against the status-quo. *Public Choice* 146: 14–44.

Kreps, D., and R. Wilson. 1982. Sequential equilibria. *Econometrica* 50: 863–894.

Kuhn, H. 1953. Extensive games and the problem of information. *Annals of Mathematics Studies*, no. 28. Princeton University Press.

Kydland, F., and E. Prescott. 1977. Rules rather than discretion: The inconsistency of optimal plans. *Journal of Political Economy* 85: 473–491.

Luce, R., and H. Raiffa. 1957. *Games and Decisions*. Wiley.

Mankiw, G. 1988. Recent developments in macroeconomics: A very quick refresher course. *Journal of Money, Credit, and Banking* 20: 436–459.

Rabin, M. 1988. Consistency and robustness criteria for game theory. Mimeo, MIT.

Reny, P. 1986. Rationality, common knowledge, and the theory of games. Ph.D. Dissertation, Princeton University.

Romer, T., and H. Rosenthal. 1978. Political resource allocation, controlled agendas, and the status-quo. *Public Choice* 33: 27–44.

Rosenthal, H. 1990. The setter model. In *Readings in the Spatial Theory of Elections*, ed. Enelow and Hinich. Cambridge University Press.

Rosenthal, R. 1981. Games of perfect information, predatory pricing and the chain-store paradox. *Journal of Economic Theory* 25: 92–100.

Schelling, T. 1960. *The Strategy of Conflict*. Harvard University Press.

Selten, R. 1965. Spieltheoretische Behandlung eines Oligopolmodells mit Nachfrageträgheit. *Zeitschrift für die gesamte Staatswissenschaft* 12: 301–324.

Selten, R. 1975. Re-examination of the perfectness concept for equilibrium points in extensive games. *International Journal of Game Theory* 4: 25–55.

Shepsle, K. 1981. Structure-induced equilibrium and legislative choice. *Public Choice* 37: 503–520.

Spence, A. M. 1977. Entry, capacity, investment and oligopolistic pricing. *Bell Journal of Economics* 8: 534–544.

Zermelo, E. 1913. Über eine Anwendung der Mengenlehre auf der Theorie des Schachspiels. In Proceedings of the Fifth International Congress on Mathematics.

4.1   Introduction[†]

In chapter 3 we introduced a class of extensive-form games that we called "multi-stage games with observed actions," where the players move simultaneously within each stage and know the actions that were chosen in all past stages. Although these games are very special, they have been used in many applications in economics, political science, and biology. The repeated games we study in chapter 5 belong to this class, as do the games of resource extraction, preemptive investment, and strategic bequests discussed in chapter 13. This chapter develops a basic fact about dynamic optimization and presents a few interesting examples of multi-stage games. The chapter concludes with discussions of what is meant by "open-loop" and "closed-loop" equilibria, of the notion of iterated conditional dominance, and of the relationship between equilibria of finite-horizon and infinite-horizon games.

Recall that in a multi-stage game with observed actions the history h^t at the beginning of stage t is simply the sequence of actions $(a^0, a^1, \ldots, a^{t-1})$ chosen in previous periods, and that a pure strategy s_i for player i is a sequence of maps s_i^t from histories h^t to actions a_i^t in the feasible sets $A_i(h^t)$. Player i's payoff u_i is a function of the terminal history h^{T+1}, i.e., of the entire sequence of actions from the initial stage 0 through the terminal stage T, where T is sometimes taken to be infinite. In some of the examples of this section, payoffs take the special form of the discounted sum $\sum_{t=0}^{T} \delta_i^t g_i^t(a^t)$ of per-period payoffs $g_i^t(a^t)$.

Section 4.3 presents a first look at the subclass of repeated games, where the payoffs are given by averages as above and where the sets of feasible actions at each stage and the per-period payoffs are independent of previous play and time, so that the "physical environment" of the game is memoryless. Nevertheless, the fact that the game is repeated means that the players can condition their current play on the past play of their opponents, and indeed there can be equilibria in strategies of this kind. Section 4.3 considers only a few examples of repeated games, and does not try to characterize all the equilibria of the examples it examines; chapter 5 gives a more thorough treatment.

In this chapter we consider mostly games with an infinite horizon as opposed to a horizon that is long but finite. Games with a long but finite horizon represent a situation where the horizon is long but well foreseen; infinite-horizon games describe a situation where players are fairly uncertain as to which period will be the last. This latter assumption seems to be a better model of many situations with a large number of stages; we will say more about this point when discussing some of the examples.

When the horizon is infinite, the set of subgame-perfect equilibria cannot be determined by backward induction from the terminal date, as it can be in the finitely repeated prisoner's dilemma and in any finite game of perfect information. As we will see, however, subgame perfection does lead to very strong predictions in some infinite-horizon games with a great many Nash equilibria, such as the bargaining model of Rubinstein (1982) and Ståhl (1972). A key feature of this model and of some of the others we will discuss is that, although the horizon is not *a priori* bounded, there are some actions, such as accepting an offer or exiting from a market, that effectively "end the game." These games have been applied to the study of exit from a declining industry, noncooperative bargaining, and the introduction of new technology, among other topics. Section 4.4 discusses the Rubinstein-Ståhl alternating-offer bargaining game, where there are many ways the game can end, corresponding to the various possible agreements the players can reach. Section 4.5 discusses the class of simple timing games, where the players' only decision is when to stop the game and not the "way" to stop it. We have not tried to give a thorough survey of the many applications of games with absorbing states; our purpose is to introduce some of the flavor of the ideas involved.

Section 4.6 introduces the concept of "iterated conditional dominance," which extends the concept of backward induction to games with a potentially infinite number of stages. As we will see, the unique subgame-perfect equilibria of several of the examples we discuss in this chapter can be understood as the consequence of there being a single strategy profile that survives the weaker condition of iterated conditional dominance. Section 4.7 discusses the relationship between open-loop equilibria and closed-loop equilibria, which are the equilibria of two different information structures for the same "physical game." Section 4.8 discusses the relationship between the equilibria of finite- and infinite-horizon versions of the "same game."

The last two sections are more technical than the rest of the chapter, and could be skipped in a first course. Sections 4.3–4.6 are meant to be examples of the uses of the theory we developed in chapter 3. Most courses would want to cover at least one of these sections, but it is unnecessary to do all of them. Section 4.2, though, is used in chapters 5 and 13; it develops a fact that is very useful in determining whether a strategy profile is subgame perfect.

4.2   The Principle of Optimality and Subgame Perfection[†]

To verify that a strategy profile of a multi-stage game with observed actions is subgame perfect, it suffices to check whether there are any histories h^t where some player i can gain by deviating from the actions prescribed by s_i at h^t and conforming to s_i thereafter. Since this "one-stage-deviation principle" is essentially the principle of optimality of dynamic programming, which is based on backward induction, it helps illustrate how subgame perfection extends the idea of backward induction. We split the observation into two parts, corresponding to finite- and infinite-horizon games; some readers may prefer to read the first proof and take the second one on faith, although both are quite simple. For notational simplicity, we state the principle for pure strategies; the mixed-strategy counterpart is straightforward.

Theorem 4.1 (one-stage-deviation principle for finite-horizon games)   In a finite multi-stage game with observed actions, strategy profile s is subgame perfect if and only if it satisfies the one-stage-deviation condition that no player i can gain by deviating from s in a single stage and conforming to s thereafter. More precisely, profile s is subgame perfect if and only if there is no player i and no strategy $\hat{s}_i$ that agrees with s_i except at a single t and h^t, and such that $\hat{s}_i$ is a better response to s_{-i} than s_i conditional on history h^t being reached.[1]

Proof The necessity of the one-stage-deviation condition ("only if") follows from the definition of subgame perfection. (Note that the one-stage-deviation condition is not necessary for *Nash* equilibrium, as a Nash-equilibrium profile may prescribe suboptimal responses at histories that do not occur when the profile is played.) To see that the one-stage-deviation condition is sufficient, suppose to the contrary that profile s satisfies the condition but is not subgame perfect. Then there is a stage t and a history h^t such that some player i has a strategy $\hat{s}_i$ that is a better response to s_{-i} than s_i is in the subgame starting at h^t. Let $\hat{t}$ be the largest t' such that, for some $h^{t'}$, $\hat{s}_i(h^{t'}) \neq s_i(h^{t'})$. The one-stage-deviation condition implies $\hat{t} > t$, and since the game is finite, $\hat{t}$ is finite as well. Now consider an alternative strategy $\tilde{s}_i$ that agrees with $\hat{s}_i$ at all $t < \hat{t}$ and follows s_i from stage $\hat{t}$ on. Since $\hat{s}_i$ agrees with s_i from $\hat{t} + 1$ on, the one-stage-deviation condition implies that $\tilde{s}_i$ is as good a response as $\hat{s}_i$ in every subgame starting at $\hat{t}$, so $\tilde{s}_i$ is as good a response as $\hat{s}_i$ in the subgame starting at t with history h^t. If $\hat{t} = t + 1$, then $\tilde{s}_i = s_i$, which contradicts the hypothesis that $\hat{s}_i$ improves on s_i. If $\hat{t} > t + 1$, we construct a strategy that agrees with $\hat{s}_i$ until $\hat{t} - 2$, and argue that it is as good a response as $\hat{s}_i$, and so on: The alleged sequence of improving deviations unravels from its endpoint.   ■

What if the horizon is infinite? The proof above leaves open the possibility that player i could gain by some infinite sequence of deviations, even though he cannot gain by a single deviation in any subgame. Just as in dynamic programming, this possibility can be excluded if the payoff functions take the form of a discounted sum of per-period payoffs. More generally, the key condition is that the payoffs be "continuous at infinity." To make this precise, let h denote an infinite-horizon history, i.e., an

---

1. Even more precisely, there cannot be a history h^t such that the restriction of $\hat{s}_i$ to the subgame $G(h^t)$ is a better response than the restriction of s_i is.

outcome of the infinite-horizon game. For a fixed infinite-horizon history h, let h^t denote *the restriction of h* to the first t periods.

Definition 4.1   A game is *continuous at infinity* if for each player i the utility function u_i satisfies

$$\sup_{h, \tilde{h} \text{ s.t. } h^t = \tilde{h}^t} |u_i(h) - u_i(\tilde{h})| \to 0 \text{ as } t \to \infty.$$

This condition says that events in the distant future are relatively unimportant. It will be satisfied if the overall payoffs are a discounted sum of per-period payoffs $g_i^t(a^t)$ and the per-period payoffs are uniformly bounded, i.e, there is a B such that

$$\max_{t, a^t} |g_i^t(a^t)| < B.$$

Theorem 4.2 (one-stage deviation principle for infinite-horizon games)   In an infinite-horizon multi-stage game with observed actions that is continuous at infinity, profile s is subgame perfect if and only if there is no player i and strategy $\hat{s}_i$ that agrees with s_i except at a single t and h^t, and such that $\hat{s}_i$ is a better response to s_{-i} than s_i conditional on history h^t being reached.

Proof   The proof of the last theorem establishes necessity, and also shows that if s satisfies the one-stage-deviation condition then it cannot be improved by any finite sequence of deviations in any subgame. Suppose to the contrary that s were not subgame perfect. Then there would be a stage t and a history h^t where some player i could improve on his utility by using a different strategy $\hat{s}_i$ in the subgame starting at h^t. Let the amount of this improvement be $\varepsilon > 0$. Continuity at infinity implies that there is a t' such that the strategy s_i' that agrees with $\hat{s}_i$ at all stages before t' and agrees with s_i at all stages from t' on must improve on s_i by at least $\varepsilon/2$ in the subgame starting at h^t. But this contradicts the fact that no finite sequence of deviations can make any improvement at all.   ∎

This theorem and its proof are essentially the principle of optimality for discounted dynamic programming.

4.3   A First Look at Repeated Games[†]

4.3.1   The Repeated Prisoner's Dilemma

This section discusses the way in which repeated play introduces new equilibria by allowing players to condition their actions on the way their opponents played in previous periods. We begin with what is probably the best-known example of a repeated game: the celebrated "prisoner's dilemma," whose static version we discussed in chapter 1. Suppose that the per-period payoffs depend only on current actions $(g_i(a^t))$ and are as shown

	Cooperate	Defect
Cooperate	1,1	−1,2
Defect	2,−1	0,0

Figure 4.1

in figure 4.1, and suppose that the players discount future payoffs with a common discount factor δ. We will wish to consider how the equilibrium payoffs vary with the horizon T. To make the payoffs for different horizons comparable, we normalize to express them all in the units used for the per-period payoffs, so that the utility of a sequence $\{a^0, \dots, a^T\}$ is

$$\frac{1-\delta}{1-\delta^{T+1}} \sum_{t=0}^{T} \delta^t g_i(a^t).$$

This is called the "average discounted payoff." Since the normalization is simply a rescaling, the normalized and present-value formulations represent the same preferences. The normalized versions make it easier to see what happens as the discount factor and the time horizon vary, by measuring all payoffs in terms of per-period averages. For example, the present value of a flow of 1 per period from date 0 to date T is $(1 - \delta^{T+1})/(1 - \delta)$; the average discounted value of this flow is simply 1.

We begin with the case in which the game is played only once. Then cooperating is strongly dominated, and the unique equilibrium is for both players to defect. If the game is repeated a finite number of times, subgame perfection requires both players to defect in the last period, and backward induction implies that the unique subgame-perfect equilibrium is for both players to defect in every period.[2]

If the game is played infinitely often, then "both defect every period" remains a subgame-perfect equilibrium. Moreover, it is the only equilibrium with the property that the play at each stage does not vary with the actions played at previous stages. However, if the horizon is infinite and $\delta > \frac{1}{2}$, then the following strategy profile is a subgame-perfect equilibrium as well: "Cooperate in the first period and continue to cooperate so long as no player has ever defected. If any player has ever defected, then defect for the rest of the game." With these strategies, there are two classes of subgames: class A, in which no player has defected, and class B, in which defect i on has occurred. If a player conforms to the strategies in every subgame in class A, his average discounted payoff is 1; if he deviates at time t and conforms to the (class B) strategies thereafter, his (normalized) payoff

---

2. This conclusion can be strengthened to hold for Nash equilibria as well. See section 5.2.

is

$$(1 - \delta)(1 + \delta + \cdots + \delta^{t-1} + 2\delta^t + 0 + \cdots) = 1 - \delta^t(2\delta - 1),$$

which is less than 1 as $\delta > \frac{1}{2}$. For any h^t in class B, the payoff from conforming to the strategies from period t on is 0; deviating once and then conforming gives -1 at period t and 0 in the future. Thus, in every subgame, no player can gain by deviating a single time from the specified strategy and then conforming, and so from the one-stage-deviation principle these strategies form a subgame-perfect equilibrium.

Depending on the size of the discount factor, there can be many other perfect equilibria. The next chapter presents the "folk theorem": Any feasible payoffs above the minmax levels (defined in chapter 5; in this example the minmax levels are 0) can be supported for a discount factor close enough to 1.[3] Thus, repeated play with patient players not only makes "cooperation"—meaning efficient payoffs—possible, it also leads to a large set of other equilibrium outcomes. Several methods have been proposed to reduce this multiplicity of equilibria; however, none of them has yet been widely accepted, and the problem remains a topic of research. We discuss one of the methods—"renegotiation-proofness"—in chapter 5.

Besides emphasizing the way that repeated play expands the set of equilibrium outcomes, the repeated prisoner's dilemma shows that the sets of equilibria of finite-horizon and infinite-horizon versions of the "same game" can be quite different, and in particular that new equilibria can arise when the horizon is allowed to be infinite. We return to this point at the end of this chapter.

4.3.2   A Finitely Repeated Game with Several Static Equilibria

The finitely repeated prisoner's dilemma has the same set of equilibria as the static version, but this is not always the case. Consider the multi-stage game corresponding to two repetitions of the stage game in figure 4.2. In the first stage of this game, players 1 and 2 simultaneously choose among

	L	M	R
U	0,0	3,4	6,0
M	4,3	0,0	0,0
D	0,6	0,0	5,5

Figure 4.2

3. The reason this holds only for large discount factors is that for small discount factors the short-term gain from deviating (for instance, deviating from cooperation in the prisoner's dilemma) necessarily exceeds any long-term losses that this behavior might create. See chapter 5.

U, M, D and L, M, R, respectively. At the end of the first stage the players observe the actions that were chosen, and in the second stage the players play the stage game again. As above, suppose that each player's payoff function in the multi-stage game is the discounted average of his or her payoffs in the two periods.

If this game is played once, there are three equilibria: (M, L), (U, M), and a mixed equilibrium ((3/7 U, 4/7 M), (3/7 L, 4/7 M)), with payoffs (4,3), (3,4), and (12/7, 12/7) respectively; the efficient payoff (5,5) is not attainable by an equilibrium. However, in the two-stage game, the following strategy profile is a subgame-perfect equilibrium if $\delta > 7/9$: "Play (D, R) in the first stage. If the first-stage outcome is (D, R), then play (M, L) in the second stage; if the first-stage outcome is not (D, R), then play ((3/7 U, 4/7 M), (3/7 L, 4/7 M)) in the second stage."

By construction, these strategies specify a Nash equilibrium in the second stage. Deviating in the first stage increases the current payoff by 1, and lowers the continuation payoffs for players 1 and 2 respectively from 4 or 3 to 12/7. Thus, player 1 will not deviate if $1 > (4 - 12/7)\delta$ or $\delta < 7/16$, and player 2 will not deviate if $1 > (3 - 12/7)\delta$ or $\delta < 7/9$.

4.4 The Rubinstein-Ståhl Bargaining Model[†]

In the model of Rubinstein 1982, two players must agree on how to share a pie of size 1. In periods 0, 2, 4, etc., player 1 proposes a sharing rule $(x, 1 - x)$ that player 2 can accept or reject. If player 2 accepts any offer, the game ends. If player 2 rejects player 1's offer in period $2k$, then in period $2k + 1$ player 2 can propose a sharing rule $(x, 1 - x)$ that player 1 can accept or reject. If player 1 accepts one of player 2's offers, the game ends; if he rejects, then he can make an offer in the subsequent period, and so on. This is an infinite-horizon game of perfect information. Note that the "stages" in our definition of a multi-stage game are not the same as "periods"—period 1 has two stages, corresponding to player 1's offer and player 2's acceptance or refusal.

We will specify that the payoffs if $(x, 1 - x)$ is accepted at date t are $(\delta_1^t x, \delta_2^t (1 - x))$, where x is players 1's share of the pie, and δ_1 and δ_2 are the two players' discount factors. (Rubinstein considered a somewhat larger class of preferences that allowed for a fixed per-period cost of bargaining in addition to the delay cost represented by the discount factors, and also allowed for utility functions that are not linear in the player's share of the pie.)

4.4.1 A Subgame-Perfect Equilibrium

Note that there are a great many *Nash* equilibria in this game. In particular, the strategy profile "player 1 always demands $x = 1$, and refuses all smaller shares; player 2 always offers $x = 1$ and accepts any offer" is a Nash equilibrium. However, this profile is not subgame perfect: If player 2 rejects player 1's first offer, and offers player 1 a share $x > \delta_1$, then player 1 should accept, because the best possible outcome if he rejects is to receive the entire pie tomorrow, which is worth only δ_1.

Here is a subgame-perfect equilibrium of this model: "Player i always demands a share $(1 - \delta_j)/(1 - \delta_i\delta_j)$ when it is his turn to make an offer. He accepts any share equal to or greater than $\delta_i(1 - \delta_j)/(1 - \delta_i\delta_j)$ and refuses any smaller share." Note that player i's demand of

$$\frac{1 - \delta_j}{1 - \delta_i\delta_j} = 1 - \frac{\delta_j(1 - \delta_i)}{1 - \delta_i\delta_j}$$

is the highest share for player i that is accepted by player j. Player i cannot gain by making a lower offer, for it too will be accepted. Making a higher (and rejected) offer and waiting to accept player j's offer next period hurts player i, as

$$\delta_i\left(1 - \frac{1 - \delta_i}{1 - \delta_i\delta_j}\right) = \delta_i^2 \frac{1 - \delta_j}{1 - \delta_i\delta_j} < \frac{1 - \delta_j}{1 - \delta_i\delta_j}.$$

Similarly, it is optimal for player i to accept any offer of at least $\delta_i(1 - \delta_j)/(1 - \delta_i\delta_j)$ and to reject lower shares, since if he rejects he receives the share $(1 - \delta_j)/(1 - \delta_i\delta_j)$ next period.

Rubinstein's paper extends the work of Ståhl (1972), who considered a finite-horizon version of the same game. With a finite horizon, the game is easily solved by backward induction: The unique subgame-perfect equilibrium in the last period is for the player who makes the offer (let's assume it is player 1) to demand the whole pie, and for his opponent to accept this demand. In the period before, the last offerer (player 1) will refuse all offers that give him less than δ_1, for he can ensure $\delta_1 \cdot 1$ by refusing. And so on.

The finite-horizon model has two potential drawbacks relative to the infinite-horizon model. First, the solution depends on the length of the game, and on which player gets to make the last offer; however, this dependence becomes small as the number of periods grows to infinity, as is shown in exercise 4.5. Second, and more important, the assumption of a last period means that if the last offer is rejected the players are not allowed to continue to try to reach an agreement. In situations in which there is no outside opportunity and no per-period cost of bargaining, it is natural to assume that the players keep on bargaining as long as they haven't reached an agreement. Thus the only way to dismiss the suspicion that it matters whether one prohibits further bargaining after the exogenous finite horizon is to prove uniqueness in the infinite-horizon game.

4.4.2 Uniqueness of the Infinite-Horizon Equilibrium

Let us now demonstrate that the infinite-horizon bargaining game has a unique equilibrium. The following proof, by Shaked and Sutton (1984), uses the stationarity of the game to obtain an upper bound and a lower bound on each player's equilibrium payoff and then shows that the upper and lower bounds are equal. Section 4.6 gives an alternative proof of uniqueness that, although slightly longer, clarifies the uniqueness result through a generalization of the concept of iterative strict dominance.

To exploit the stationarity of the game, we define *the continuation payoffs of a strategy profile in a subgame starting at t* to be the utility in time-*t* units of the outcome induced by that profile. For example, the continuation payoff of player 1 at period 2 of a profile that leads to player 1's getting the whole pie at date 3 is δ_1, whereas this outcome has utility δ_1^3 in time-0 units.

Now we define $\underline{v}_1$ and $\bar{v}_1$ to be player 1's lowest and highest continuation payoffs of player 1 in any perfect equilibrium of any subgame that begins with player 1 making an offer. (More formally, $\underline{v}_1$ is the infimum or greatest lower bound of these payoffs, and $\bar{v}_1$ is the supremum.) Similarly, let $\underline{w}_1$ and $\bar{w}_1$ be player 1's lowest and highest perfect-equilibrium continuation payoffs in subgames that begin with an offer by player 2. Also, let $\underline{v}_2$ and $\bar{v}_2$ be player 2's lowest and highest perfect-equilibrium continuation payoffs in subgames beginning with an offer by player 2, and let $\underline{w}_2$ and $\bar{w}_2$ be player 2's lowest and highest perfect-equilibrium continuation payoffs in subgames beginning with an offer by player 1.

When player 1 makes an offer, player 2 will accept any x such that player 2's share (of $1 - x$) exceeds $\delta_2 \bar{v}_2$, since player 2 cannot expect more than $\bar{v}_2$ in the continuation game following his refusal. Hence, $\underline{v}_1 \geq 1 - \delta_2 \bar{v}_2$. By the symmetric argument, player 1 accepts all shares above $\delta_1 \bar{v}_1$, and $\underline{v}_2 \geq 1 - \delta_1 \bar{v}_1$.

Since player 2 will never offer player 1 a share greater than $\delta_1 \bar{v}_1$, player 1's continuation payoff when player 2 makes an offer, $\bar{w}_1$, is at most $\delta_1 \bar{v}_1$.

Since player 2 can obtain at least $\underline{v}_2$ in the continuation game by rejecting player 1's offer, player 2 will reject any x such that $1 - x \leq \delta_2 \underline{v}_2$. Therefore, player 1's highest equilibrium payoff when making an offer, $\bar{v}_1$, satisfies

$$\bar{v}_1 \leq \max(1 - \delta_2 \underline{v}_2, \delta_1 \bar{w}_1) = \max(1 - \delta_2 \underline{v}_2, \delta_1^2 \bar{v}_1).$$

Next, we claim that

$$\max(1 - \delta_2 \underline{v}_2, \delta_1^2 \bar{v}_1) = 1 - \delta_2 \underline{v}_2:$$

If not, then we would have $\bar{v}_1 \leq \delta_1^2 \bar{v}_1$, implying $\bar{v}_1 \leq 0$, but then $1 - \delta_2 \underline{v}_2 > \delta_1^2 \bar{v}_1$, as neither δ_2 nor $\underline{v}_2$ can exceed 1. Thus, $\bar{v}_1 \leq 1 - \delta_2 \underline{v}_2$. By symmetry, $\bar{v}_2 \leq 1 - \delta_1 \underline{v}_1$. Combining these inequalities, we have

$$\underline{v}_1 \geq 1 - \delta_2 \bar{v}_2 \geq 1 - \delta_2(1 - \delta_1 \underline{v}_1),$$

or

$$\underline{v}_1 \geq \frac{1 - \delta_2}{1 - \delta_1 \delta_2},$$

and

$$\bar{v}_1 \leq 1 - \delta_2(1 - \delta_1 \bar{v}_1),$$

or

$$\bar{v}_1 \leq \frac{1 - \delta_2}{1 - \delta_1 \delta_2};$$

because $\underline{v}_1 \leq \bar{v}_1$, this implies $\underline{v}_1 = \bar{v}_1$. Similarly,

$$\underline{v}_2 = \bar{v}_2 = \frac{1 - \delta_1}{1 - \delta_1 \delta_2},$$

$$\underline{w}_1 = \bar{w}_1 = \frac{\delta_1(1 - \delta_2)}{1 - \delta_1 \delta_2},$$

and

$$\underline{w}_2 = \bar{w}_2 = \frac{\delta_2(1 - \delta_1)}{1 - \delta_1 \delta_2}.$$

This shows that the perfect-equilibrium continuation payoffs are unique. To see that there is a unique perfect-equilibrium strategy profile, consider a subgame that begins with an offer by player 1. The argument above shows that player 1 must offer exactly $x = \underline{v}_1$. Although player 2 is indifferent between accepting and rejecting this offer, perfect equilibrium requires that he accept with probability 1: If player 2's strategy is to accept all $x < \underline{v}_1$ with probability 1, but to accept $\underline{v}_1$ with probability less than 1, then no best response for player 1 exists. Hence, this randomization by player 2 is inconsistent with equilibrium. A similar argument applies in subgames that begin with an offer by player 2.

4.4.3 Comparative Statics

Note that as $\delta_1 \to 1$ for fixed δ_2, $v_1 \to 1$ and player 1 gets the whole pie, whereas player 2 gets the whole pie if $\delta_2 \to 1$ for fixed δ_1. Player 1 also gets the whole pie if $\delta_2 = 0$, since a myopic player 2 will accept any positive amount today rather than wait one period. Note also that even if $\delta_1 = 0$ player 2 does not get the whole pie if $\delta_2 < 1$: Due to his first-mover advantage, even a myopic player 1 receives a positive share. The first-mover advantage also explains why player 1 does better than player 2 even if the discount factors are equal: If $\delta_1 = \delta_2 = \delta$, then

$$v_1 = \frac{1}{1+\delta} > \tfrac{1}{2}.$$

As one would expect, this first-mover advantage disappears if we take the time periods to be arbitrarily short. To see this, let Δ denote the length of the time period, and set $\delta_1 = \exp(-r_1\Delta)$ and $\delta_2 = \exp(-r_2\Delta)$. Then, for Δ close to 0, δ_i is approximately $1 - r_i\Delta$, and v_1 converges to $r_2/(r_1 + r_2)$, so the relative patience of the players determines their shares. In particular, if $r_1 = r_2$ the players have equal shares in the limit. (See Binmore 1981 for a very thorough discussion of the Rubinstein-Ståhl model.)

4.5 Simple Timing Games[††]

4.5.1 Definition of Simple Timing Games

In a simple timing game, each player's only choice is when to choose the action "stop," and once a player stops he has no effect on future play. That is, if player i has not stopped at any $\tau < t$, his action set at t is

$A_i(t) = \{\text{stop, don't stop}\};$

if player i has stopped at some $\tau < t$, then $A_i(t)$ is the null action "don't move." Few situations can be exactly described in this way, because players typically have a wider range of choices. (For example, firms typically do not simply choose a time to enter a market; they also decide on the scale of entry, the quality level, etc.) But economists often abstract away from such details to study the timing question in isolation.

We will consider only two-player timing games, and restrict our attention to the subgame-perfect equilibria. Once one player has stopped, the remaining player faces a maximization problem that is easily solved. Thus, when considering subgame-perfect equilibria, we can first "fold back" subgames where one player has stopped and then proceed to subgames where neither player has yet stopped.[4] This allows us to express both players' payoffs as functions of the time

$\hat{t} = \min\{t \mid a_i^t = \text{stop for at least one } i\}$

at which the first player stops (with the strategies we will consider, this minimum is well defined); if no player ever stops, we set $\hat{t} = \infty$. We describe these payoffs using the functions L_i, F_i, and B_i: If only player i stops at $\hat{t}$, then player i is the "leader"; he receives $L_i(\hat{t})$, and his opponent receives "follower" payoff $F_j(\hat{t})$. If both players stop simultaneously at $\hat{t}$, the payoffs are $B_1(\hat{t})$ and $B_2(\hat{t})$. We will assume that

---

4. Although the one-player maximization problem will typically have a unique solution, this need not be the case. If there are multiple solutions, then one must consider the implications of each of them.

$$\lim_{\hat{t} \to +\infty} L_i(\hat{t}) = \lim_{\hat{t} \to +\infty} F_i(\hat{t}) = \lim_{\hat{t} \to +\infty} B_i(\hat{t}),$$

which will be the case if payoffs are discounted.

The last step in describing these games is to specify the strategy spaces. We begin with the technically simpler case where time is discrete, as it has been in our development so far. Since the feasible actions at each date until some player stops are simply {stop, don't stop}, and since once a player stops the game effectively ends (remember that we have folded back any subsequent play), the history at date t is simply the fact that the game is still going on then. Thus, pure strategies s_i are simply maps from the set of dates t to {stop, don't stop}, behavior strategies b_i specify a conditional probability $b_i(t)$ of stopping at t if no player has stopped before, and mixed strategies σ_i are probability distributions over the pure strategies s_i.

For some games, the set of equilibria is easier to compute in a model where time is continuous. The pure strategies, as in discrete time, are simply maps from times t to {stop, don't stop}, but two complications arise in dealing with mixed strategies. First, the formal notion of behavior strategies becomes problematic when players have a continuum of information sets. (This was first noted by Aumann (1964).) We will sidestep the question of behavior strategies by working only with the mixed (i.e., strategic-form) strategies. The second problem is that, as we will see, the set of continuous-time mixed strategies as they are usually defined is too small to ensure that the continuous-time model will capture the limits of discrete-time equilibria with short time intervals, although it does capture the short-time-interval limits of some classes of games.

Putting these problems aside, we introduce the space of continuous-time mixed strategies that we will use in most of this section. Given that pure strategies are stopping times, it is natural to identify mixed strategies as cumulative distribution functions G_i on $[0, \infty)$. In other words, $G_i(t)$ is the probability that player i stops at or before time t. (To be cumulative distribution functions, the G_i must take values in the interval $[0, 1]$ and be nondecreasing and right-continuous.[5]) The functions G_i need not be continuous; let

$$\alpha_i(t) = G_i(t) - \lim_{\tau \uparrow t} G_i(\tau)$$

be the size of the jump at t. When $\alpha_i(t)$ is nonzero, player i stops with probability $\alpha_i(t)$ at exactly time t; this is called an "atom" of the probability distribution. Where G_i is differentiable, its derivative dG_i is the probability density function; the probability that player i stops between times t and

---

5. The function $G(\cdot)$ is right-continuous at t if

$$\lim_{\tau \downarrow t} G(\tau) = G(t).$$

It is right-continuous if it is right-continuous at each t.

$t + \varepsilon$ is approximately $\varepsilon dG_i(t)$. Player i's payoff function is then

$$u_i(G_1, G_2) = \int_0^\infty [L_i(s)(1 - G_j(s)) \, dG_i(s) + F_i(s)(1 - G_i(s)) \, dG_j(s)]$$

$$+ \sum_s \alpha_i(s)\alpha_j(s)B_i(s).$$

That is, there is probability $dG_i(s)$ that player i stops at date s. If player j hasn't stopped yet, which has probability $1 - G_j(s)$, player i is the leader and obtains $L_i(s)$. And similarly for the other terms.

We now develop two familiar games of timing: the war of attrition and the preemption game.[6]

4.5.2 The War of Attrition

A classic example of a timing game is the war of attrition, first analyzed by Maynard Smith (1974).[7]

Stationary War of Attrition
In the *discrete-time* version of the stationary war of attrition, two animals are fighting for a prize whose current value at any time $t = 0, 1, \ldots$ is $v > 1$; fighting costs 1 unit per period. If one animal stops fighting in period t, his opponent wins the prize without incurring a fighting cost that period, and the choice of the second stopping time is irrelevant. If we introduce a per-period discount factor δ, the (symmetric) payoff functions are

$$L(\hat{t}) = -(1 + \delta + \cdots + \delta^{i-1}) = -\frac{1 - \delta^i}{1 - \delta}$$

and

$$F(\hat{t}) = -(1 + \delta + \cdots + \delta^{i-1}) + \delta^i v = L(\hat{t}) + \delta^i v.$$

If both animals stop simultaneously, we specify that neither wins the prize, so that

$$B_1(\hat{t}) = B_2(\hat{t}) = L(\hat{t}).$$

(Exercise 4.1 asks you to check that other specifications with $B(\hat{t}) < F(\hat{t})$ yield similar conclusions when the time periods are sufficiently short.) Figure 4.3 depicts $L(\cdot)$ and $F(\cdot)$ for the continuous-time version of this game.

This stationary game has several Nash equilibria. Here is one: Player 1's strategy is "never stop" and player 2's is "always stop." There is a unique

6. See Katz and Shapiro 1986 for hybrids of these two games.
7. Another name for the war of attrition is "chicken." The classic game of chicken is played in automobiles. In one version the cars head toward a cliff side by side, and the first driver to stop loses; in the other version the cars head toward each other, and the first driver to swerve out of the way to avoid the collision loses. We do not recommend experimental studies of either version of this game.

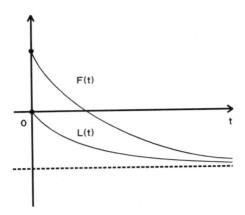

Figure 4.3

symmetric equilibrium, which is stationary and involves mixed strategies: For any p, let "always p" be the behavior strategy "if the other player has not stopped before t, then stop at t with probability p." The equivalent strategic-form mixed strategy assigns probability $(1-p)^t p$ to the pure strategy "stop at t if the other player hasn't stopped before then." For the stationary symmetric profile (always p, always p) to be an equilibrium, it is necessary that, for any t, the payoff to stopping at t conditional on the opponent's not having stopped previously, which is $L(t)$, is equal to

$$p[F(t)] + (1-p)[L(t+1)],$$

the payoff to staying in until $t+1$ and dropping out then unless the opponent drops out today. (If the opponent does drop out before t, then the strategies "stop at t" and "stop at $t+1$" yield the same payoff.) Equating these terms gives $p^* = 1/(1+v)$, which ranges from 1 to 0 as v ranges from 0 to infinity. Another way to arrive at this conclusion is to note that by staying in for one more period, a player gains v with probability p and loses the fighting cost 1 during that period with probability $1-p$. For him to be indifferent between staying in for one more period and stopping now, it must be the case that $pv = 1 - p$, which yields the above expression for p^*.

Thus, "always p^*" is the only candidate for a stationary symmetric equilibrium. To check that it is indeed an equilibrium, note that if player 1 plays "always p^*" the payoff to each possible stopping time t for player 2 is 0.

At this stage, the reader may wonder whether the Nash equilibria are subgame perfect. If the players are free to quit when they want and are not committed to abide by their date-0 choice of stopping time, do they want to deviate? The answer is No: All stationary Nash equilibria (i.e. equilibria with strategies that are independent of calendar time) are subgame perfect.

(To see this, note that the stationarity of the payoffs implies that all subgames where both players are still active are isomorphic.)

The *continuous-time* formulation is very convenient for wars of attrition. Consider the continuous-time version of the example considered above, where the terms δ^t are replaced by $\exp(-rt)$ and r is the rate of interest. Let $G_i(t)$ denote the probability that player i stops at or before t (that is, $G_i(\cdot)$ is a cumulative distribution function). As in the discrete-time version of the game, there is a stationary symmetric equilibrium G with the property that at each date the players are indifferent between stopping at time t and waiting a bit longer, until $t + \varepsilon$, to see if the opponent stops first. Conditional on not stopping before t, to the first order in ε,[8] the marginal cost of waiting ε longer is ε and the expected reward from doing so is $v\,dG/(1 - G)$. Equating these terms yields $dG/(1 - G) = 1/v$, so that G is the exponential distribution $G^*(t) = 1 - \exp(-t/v)$. (As in the discrete-time case, to verify that these are equilibrium strategies, note that if palyer 1 uses G^* then player 2's expected payoff to any strategy is 0.)

Thus, the war of attrition does have a symmetric equilibrium in the kind of continuous-time strategies we introduced above. Moreover, this equilibrium is the limit of the symmetric equilibria of the discrete-time game as the interval Δ between periods goes to 0, as we will now show. To make the discrete- and continuous-time formulations comparable, we assume that fighting costs 1 per unit of *real* time. Hence, if in discrete time the real length of each period is Δ (so that there are $1/\Delta$ periods per unit of time), the fighting cost is Δ per period. The value of the prize v does not need to be adjusted when the period length changes, as v was taken to be a stock rather than a flow in both formulations. The discrete-time equilibrium strategy is now given by $p^*v = (1 - p^*)\Delta$ or $p^* = \Delta/(\Delta + v)$. Fix a real time t, and let $n = t/\Delta$ be the number of (discrete-time) periods between 0 and t. The probability that a player does not stop before t in the discrete-time formulation is

$$1 - G(t) = (1 - p^*)^n = \left(\frac{v}{v + \Delta}\right)^{t/\Delta} = \exp\left[-\frac{t}{\Delta}\ln\left(1 + \frac{\Delta}{v}\right)\right]$$

$$\simeq \exp\left(-\frac{t}{\Delta}\frac{\Delta}{v}\right) = \exp\left(-\frac{t}{v}\right) \text{ for } \Delta \text{ small.}$$

Thus, the symmetric discrete-time equilibrium converges to the symmetric continuous-time equilibrium when Δ tends to 0.

Nonstationary Wars of Attrition

More generally, games satisfying the following (discrete- or continuous-time) conditions can be viewed as wars of attrition: For all players i and

---

8. We will say that a term $f(\varepsilon)$ is not of order ε if $\lim_{\varepsilon \to 0} f(\varepsilon)/\varepsilon = 0$.

all dates t,

    (i) $F_i(t) \geq F_i(\tau)$ for $\tau > t$,

    (ii) $F_i(t) \geq L_i(\tau)$ for $\tau > t$,

    (iii) $L_i(t) = B_i(t)$,

    (iv) $L_i(0) > L_i(+\infty)$,

    (v) $L_i(+\infty) = F_i(+\infty)$.

Condition i states that if player i's opponent is going to stop first in the subgame starting at t, then player i prefers that his opponent stop immediately at t. Condition ii says that each player i prefers his opponent stopping first at any time t to any outcome where player i stops first at some $\tau > t$. The motivation for this condition is that if player j stops at t, player i can always stay in until $\tau > t$ and quit at τ and obtain $L_i(\tau)$ plus the economized fighting costs between t and τ.

Condition iii asserts that when a player stops, it does not matter if the opponent stops or stays; this assumption simplifies the study of the discrete-time formulation and is irrelevant under continous time. Condition iv states that fighting forever is costly—each player would rather quit immediately than fight forever. Condition v is automatically satisfied if players discount their payoffs and the payoffs are bounded.

Two nonstationary variants satisfying these conditions and stronger assumptions have appeared frequently in the literature: "eventual continuation" and "eventual stopping." (We state these further assumptions for the discrete-time framework in order not to discuss continuous-time strategies.) In either case, subgame perfection uniquely pins down equilibrium behavior.

Eventual Continuation   Make assumptions i–v and the following additional assumptions:

    (ii') $F_i(t + 1) > L_i(t)$ for all i and t.

    (vi) For all i, there exists $T_i > 0$ such that $L_i(t) > L_i(+\infty)$ for $t < T_i$ and $L_i(t) < L_i(+\infty)$ for $t > T_i$.

    (vii) For all i, there exists $\tilde{T}_i$ such that $L_i(\cdot)$ is strictly decreasing before $\tilde{T}_i$ and increasing after $\tilde{T}_i$.

Condition ii', which states that fighting for one period is worthwhile if successful, is a strengthening of condition ii. (To see that conditions i and ii' imply condition ii, note that for $\tau > t$, $F_i(t) \geq F_i(\tau + 1) > L_i(\tau)$.) Condition vi states that even though at the start of the game it is better to quit than to fight forever, things get better later on so that, ignoring past sunk costs, the player would rather continue fighting than quit. Condition vii states that $L_i(\cdot)$ has a single peak. Note that, necessarily, $\tilde{T}_i \geq T_i$. In an industrial-organization context, conditions vi and vii correspond to the market growing or the technology improving over time (for exogenous reasons or because of learning by doing).

Example of eventual continuation Fudenberg et al. (1983) study an example of eventual continuation: Two firms are engaged in a patent race, and "stopping" means abandoning the race. The expected productivity of research is initially low, so that if both firms do R&D until one of them makes a discovery then both firms have a negative expected value. However, the productivity of R&D increases over time, so that there are times T_1 and T_2 such that, if both firms are still active at T_i, then it is a dominant strategy for firm i to never stop.

For simplicity, we give the continuous-time version of the patent-race game. Suppose that the patent has value v. If firm i has not quit before date t, it pays $c_i \, dt$ and makes a discovery with probability $x_i(t) \, dt$ between t and $t + dt$. The instantaneous flow profit is thus $[x_i(t)v - c_i] \, dt$. (The probability that firm j discovers between t and $t + dt$ is infinitesimal.) Suppose that $dx_i/dt > 0$ (due to learning).

In this game,

$$L_i(t) = \int_0^t [x_i(\tau)v - c_i] \exp\left(-\int_0^\tau [x_1(s) + x_2(s)] \, ds\right) \exp(-r\tau) \, d\tau,$$

where r is the rate of interest. The probability that no one has discovered at date τ conditional on both players having stayed in the race is

$$\exp\left(-\int_0^\tau [x_1(s) + x_2(s)] \, ds\right).$$

We assume that an R&D monopoly is viable:

$$0 < \int_0^\infty [x_i(\tau)v - c_i] \exp\left(-\int_0^\tau x_i(s) \, ds\right) \exp(-r\tau) \, d\tau = F_i(0),$$

and that a duopoly is not viable at date 0: $L_i(\infty) < 0$. (Recall that $L_i(\infty)$ is the date-0 payoff if neither firm ever stops.) Because $x_i(\cdot)$ is increasing, if a monopoly is viable at date 0 then it is viable from any date $t > 0$ on.[9] Therefore, it is optimal for each player to stay in until discovery once his opponent has quit. The follower's payoff is thus

$$F_i(t) = \int_0^t [x_i(\tau)v - c_i] \exp\left(-\int_0^\tau [x_1(s) + x_2(s)] \, ds\right) \exp(-r\tau) \, d\tau$$

$$+ \int_t^\infty [x_i(\tau)v - c_i] \exp\left[-\left(\int_0^\tau x_i(s) \, ds + \int_0^t x_j(s) \, ds\right)\right]$$

$$\times \exp(-r\tau) \, d\tau.$$

9. Note also that $F_i(0) > 0$ and $x_i(\cdot)$ increasing imply that there exists a time such that $x_i(t)v > c_i$ after that time. Therefore L_i is first decreasing and then increasing. The time $\widetilde{T}_i$ defined in condition vii is given by the equation $x_i(\widetilde{T}_i)v = c_i$.

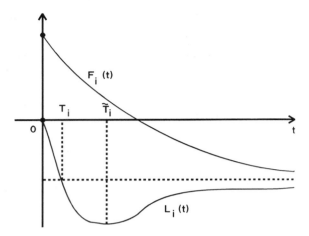

Figure 4.4

The leader's and the follower's payoffs in the continuous-time patent race are depicted in figure 4.4. It is clear that assumptions i–v, ii', vi, and vii are satisfied in the discrete-time version of the game as long as the interval between the periods is small (i.e., when discrete time is close to continuous time).

Uniqueness under eventual continuation Condition vi guarantees that player i never quits after T_i: By quitting at date $t > T_i$ he obtains

$$L_i(t) < L_i(+\infty) = F_i(+\infty) \le F_i(\tau)$$

for all τ. Thus, he always gets more by never quitting. Thus, quitting is a (conditionally) strictly dominated strategy at date $t > T_i$. Let us now assume that the times (T_1, T_2) defined in condition vi satisfy $T_1 + 1 < T_2$. If the time interval between the periods is short and the players are not quite the same (for instance, if $c_1 \ne c_2$ or $x_1(\cdot) \ne x_2(\cdot)$ in the patent race), this condition (or the symmetric condition) is likely to be satisfied. We claim that the war of attrition has a unique equilibrium, and that in this equilibrium player 2 quits at date 0 (there is no war).

Uniqueness is proved by backward induction. At date $T_1 + 1$, if both players are still fighting, player 2 knows that player 1 will never quit. Because $T_1 + 1 < T_2$,

$$L_2(T_1 + 1) > L_2(+\infty).$$

Furthermore, because $L_2(\cdot)$ has a single peak (condition vii),

$$L_2(T_1 + 1) > L_2(t)$$

for all $t > T_1 + 1$. Hence, it is optimal for player 2 to quit at $T_1 + 1$. Consider now date T_1. Because $F_1(T_1 + 1) > L_1(T_1)$ (condition ii'), player

1 does not quit. And by the same reasoning as before, $L_2(T_1) > L_2(t)$ for all $t > T_1$. Hence, player 2 quits at T_1 if both players are still around at that date. The same reasoning shows that player 2 quits and player 1 stays in at any date $t < T_1$. There exists a unique subgame-perfect equilibrium.

One reason why $T_1 + 1$ might be less than T_2 in the patent-race example is that firm 1 entered the patent race $k \geq 2$ periods before firm 2. Then, if the two firms have the same technology ($x_2(t) = x_1(t - k)$ and $c_1 = c_2$),

$$T_1 = T_2 - k.$$

If periods are short, then the case $T_1 = T_2 - 2$ seems like a small advantage for player 1, yet it is sufficient to make firm 1 the "winner" without a fight. In the terminology of Dasgupta and Stiglitz (1980), this game exhibits "ε-preemption," as an ε advantage proves decisive. Harris and Vickers (1985) develop related ε-preemption arguments. Hendricks and Wilson (1989) characterize the equilibria of a large family of discrete-time wars of attrition; Hendricks et al. (1988) do the same for the continuous-time version.

Eventual Stopping   Make assumptions i–v and the following additional assumptions:

(viii) There exists $T_2 > 0$ such that $\forall t < T_2, L_2(t) < F_2(t), \forall t > T_2, F_2(t) \leq L_2(t)$, and $\forall t \leq T_2, F_1(t + 1) > L_1(t)$.

(ix) For all i, $L_i(\cdot)$ has a single peak. It increases strictly before some time $\tilde{T}_i$ and decreases strictly after time $\tilde{T}_i$. Furthermore, $\tilde{T}_2 \leq \tilde{T}_1$.

Assumption viii states that after some date T_2, player 2 is better off exiting than staying even if the other player has quit. The following example illustrates these conditions. Note that necessarily $\tilde{T}_2 < T_2$.

Example of eventual stopping   As with eventual continuation, we give the example and draw the payoff functions in continuous time; the proof of uniqueness is performed in the discrete-time framework. Two firms wage duopoly competition in a market. If one quits, the other becomes a monopoly. Suppose that the firms differ only in their flow fixed cost, $f_1 < f_2$. The gross flow profits are $\Pi^m(t)$ for a monopolist and $\Pi^d(t)$ for a duopolist, where $\Pi^m(t) > \Pi^d(t)$ for all t. Demand is declining, so $\Pi^m(\cdot)$ and $\Pi^d(\cdot)$ are strictly decreasing. Suppose that there exist $\tilde{T}_2$ and T_2 such that $0 < \tilde{T}_2 < T_2 < +\infty$, $\Pi^d(\tilde{T}_2) = f_2$ (firm 2 stops making profit as a duopolist at date $\tilde{T}_2$), and $\Pi^m(T_2) = f_2$ (firm 2 is no longer viable as a monopolist after date T_2). The payoffs $F_2(\cdot)$ and $L_2(\cdot)$ are represented in figure 4.5 (firm 1's payoffs have similar shapes). After T_2 it is optimal for firm 2 to quit immediately, so in continuous time $F_2(t) = L_2(t)$ for $t > T_2$. (In discrete time and with short time periods, $F_2(t)$ is slightly lower than $L_2(t)$, since firm 2, as a follower, stays one period longer than the leader and loses money during that period.) In discrete time, this example satisfies assump-

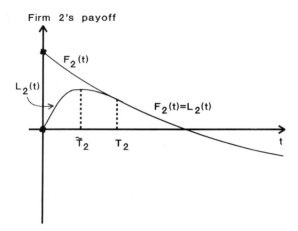

Figure 4.5

tions i–v, viii, and ix for sufficiently short time intervals between the periods.

Uniqueness under Eventual Stopping First we claim that player 2 quits at any date $t > T_2$. By stopping it gets $L_2(t)$. Because $L_2(t) \geq F_2(t)$ from condition viii, and because $L_2(\cdot)$ is strictly decreasing from condition ix, $L_2(t) > L_2(\tau) \geq F_2(\tau)$ for $\tau > t$. Hence, it is a strictly dominant strategy for player 2 to quit at any $t > T_2$. Therefore, $F_1(T_2 + 1) > L_1(T_2)$ implies that player 1 stays in at T_2. Because $L_2(T_2) > L_2(T_2 + 1)$, player 2 quits at T_2. The same holds by induction for any t greater than $\tilde{T}_2$. Before $\tilde{T}_2$, neither player quits since $L_i(\cdot)$ is increasing. Hence, the unique equilibrium of the game has player 2 quit first at $\tilde{T}_2$, and the two players' payoffs are $F_1(\tilde{T}_2)$ and $L_2(\tilde{T}_2)$.

Ghemawat and Nalebuff (1985) and Fudenberg and Tirole (1986) give examples of declining industries that fit the eventual-stopping example. Ghemawat and Nalebuff argue that if exit is an all-or-nothing choice, as in the simple model we have been considering, then a big firm will become unprofitable before a smaller one will, and so the big firm will be forced to exit first. Moreover, foreseeing this eventual exit, the small firm will stay in, and backward induction implies that the big firm exits once the market shrinks enough that staying in earns negative flow profits. (Whinston (1986) shows that this conclusion need not obtain if the big firm is allowed to shed capacity in small units.)

4.5.3 Preemption Games

Preemption games are a rough opposite to the war of attrition, with $L(\hat{t}) > F(\hat{t})$ for some range of times $\hat{t}$. Here the specification of the payoff to simultaneous stopping, $B(\cdot)$, is more important than in the war of attrition, as if L exceeds F we might expect both players to stop simultaneously. One example of a preemption game is the decision of whether and when to build a new plant or adopt a new innovation when the market is big enough to support only one such addition (Reinganum 1981a,b; Fudenberg and Tirole 1985). In this case $B(t)$ is often less than $F(t)$, as it can be better to let an opponent have a monopoly than to incur duopoly losses.

One very stylized preemption game is "grab the dollar." In this stationary game, time is discrete ($t = 0, 1, \ldots$) and there is a dollar on the table, which either or both of the players can try to grab. If only one player grabs, that player receives 1 and the other 0; if both try to grab at once, the dollar is destroyed and both pay a fine of 1; if neither player grabs, the dollar remains on the table. The players use the common discount factor δ, so that $L(t) = \delta^t$, $F(t) = 0$, and $B(t) = -\delta^t$ for all t. Like the war of attrition, this game has asymmetric equilibria, where one player "wins" with probability 1, and also a symmetric mixed-strategy equilibrium, where each player grabs the dollar with probability $p^* = \frac{1}{2}$ in each period. (It is easy to check that this yields a symmetric equilibrium; to see that it is the only one, note that each player must be indifferent between stopping—i.e., grabbing—at date t, which yields payoff $\delta^t((1 - p^*(t)) - p^*(t))$ if the other has not stopped before date t and 0 otherwise, and never stopping, which yields payoff 0, so that $p^*(t)$ must equal $\frac{1}{2}$ for all t.) The payoffs in the symmetric equilibrium are $(0, 0)$, and the distribution over outcomes is that the probability that player 1 alone stops first at t, the probability that player 2 alone stops first at t, and the probability that both players stop simultaneously at t are all equal to $(\frac{1}{4})^{t+1}$. Note that these probabilities are independent of the per-period discount factor, δ, and thus of the period length, Δ, in contrast to the war of attrition, where the probabilities were proportional to the period length. This makes finding a continuous-time representation of this game more difficult.

To understand the difficulties, let t denote a fixed "real time" after the initial time 0, define the number of periods between time 0 and time t when the real time length of the period is Δ as $n(t, \Delta) = t/\Delta$, and consider what happens as $\Delta \to 0$. The probability that at least one player has stopped by t is $1 - (\frac{1}{4})^{n(t, \Delta)}$, which converges to 1 as $\Delta \to 0$. The limit of the equilibrium distribution over outcomes is probability $\frac{1}{3}$ that player 1 wins the dollar at time 0, probability $\frac{1}{3}$ that player 2 wins at time 0, probability $\frac{1}{3}$ that both grab simultaneously at time 0, and probability 0 that the game continues beyond time 0. Fudenberg and Tirole (1985) observed that this limiting distribution cannot be expressed as an equilibrium in continuous-time strategies of the kind we have considered so far: If the game ends with probability 1 at time 0, then, for at least one player i, $G_i(0)$ must equal 1; but then there would be probability 0 that player i's opponent wins the dollar. The problem is that different sequences of discrete-time strategy profiles converge to a limit in which the game ends with probability 1 at the start, including "stop with probability 1 at time 0" and "stop with conditional probability $p > 0$ at each period." The usual continuous-time strategies implicitly associate an atom of size 1 in continuous time with an atom of that size in discrete time, and this cannot represent the limit of the discrete-time equilibria, where the atom at time 0 corresponds not to probability 1 of stopping at exactly time 0 but rather to an "interval of atoms" at all times just after 0. We proposed an extended version of the continuous-time strategies and payoff functions to capture the limit of discrete-time equilibria in the particular preemption game we were analyzing. Simon (1988) has generalized our approach to a much broader class of games.

As another example of a preemption game, suppose that $L(t) = 14 - (t - 7)^2$, $F(t) = 0$, and $B(t) < 0$. These payoffs are meant to describe a situation where either of two firms can introduce a new product. The product will have no effect on their existing business, which is why $F(t) = 0$, and the combination of fixed costs and aggressive duopoly pricing implies that both firms will lose money if both develop the product. Thus, once one firm introduces the product the other firm never will, and the only case in which both firms would introduce is if they did so simultaneously.

To avoid the need to consider the possibility of such mistakes and the associated mixed strategies, let us follow Gilbert and Harris (1984) and Harris and Vickers (1985) and make the simplifying assumption that player 1 can stop only in even-numbered periods ($t = 0, 2, \ldots$) and player 2 can stop only in odd-numbered periods ($t = 1, 3, \ldots$), so that the game is one of perfect information. There are three Pareto-efficient outcomes for the players—either player 1 stops at $t = 6$ or $t = 8$ or player 2 stops at $t = 7$—with two Pareto-efficient payoffs: $(13, 0)$ and $(0, 14)$. In the unique subgame-perfect equilibrium, firm 1 stops at $t = 4$, which is the first t where $L(t) > F(t)$. This is an example of "rent dissipation": Although there are possible rents to be made from introducing the product later, in equilibrium the race to be first forces the introduction of the product at the first time when rents are nonnegative.

4.6 Iterated Conditional Dominance and the Rubinstein Bargaining Game[†††]

The last two sections presented several examples of infinite-horizon games with unique equilibria. The uniqueness arguments there can be strengthened, in that these games have a unique profile that satisfies a weaker concept than subgame perfection.

Definition 4.2 In a multi-stage game with observed actions, action a_i^t is *conditionally dominated at stage t given history h^t* if, in the subgame beginning at h^t, every strategy for player i that assigns positive probability to a_i^t is strictly dominated. Iterated conditional dominance is the process that, at each round, deletes every conditionally dominated action in every subgame, given the opponents' strategies that have survived the previous rounds.

It is easy to check that iterated conditional dominance coincides with subgame perfection in finite games of perfect information. In these games it also coincides with Pearce's (1984) extensive-form rationalizability. In general multi-stage games, any action ruled out by iterated conditional dominance is also ruled out by extensive-form rationalizability, but the exact relationship between the two concepts has not been determined.

In a game of imperfect information, iterated conditional dominance can be weaker than subgame perfection, as it does not assume that players forecast that an equilibrium will occur in every subgame. To illustrate this point, consider a one-stage, simultaneous-move game. Then iterated conditional dominance coincides with iterated strict dominance, subgame perfection coincides with Nash equilibrium, and iterated strict dominance is in general weaker than Nash equilibrium.

Theorem 4.3   In a finite- or infinite-horizon game of perfect information, no subgame-perfect strategy profile is removed by iterated conditional dominance.

Proof   Proving this theorem is exercise 4.7.

Iterated conditional dominance is quite weak in some games. For example, in a repeated game with discount factor near 1, no action is conditionally dominated. (It is always worthwhile to play any fixed action today if playing the action induces the opponents to play cooperatively in every future period and if not playing the action causes the opponents to play something you do not like.) However, in games that "end" when certain actions are played, iterated conditional dominance has more bite, since if a player's current action ends the game then his opponents will not be able to "punish" him in the future. One such example is the infinite-horizon version of the bargaining model in section 4.4.

Let us see how iterated conditional dominance gives a unique solution. Note first that a player never accepts an offer that gives him a negative share (i.e., such that he loses money): Accepting such an offer is strictly dominated by the strategy "reject any offer, including this one, and make only offers that, if accepted, give a positive share." Next, in any subgame where player 2 has just made an offer, it is conditionally dominated for player 1 to refuse if the offer has player 1's share of the pie, x, exceeding δ_1; similarly, player 2 must accept any $x < 1 - \delta_2$. These are all the actions that are removed at the first round of iteration. At the second round, we conclude that

(i) player 2 will never offer player 1 more than δ_1,

(ii) player 2 will reject any $x > 1 - \delta_2(1 - \delta_1)$, because he can get $\delta_2(1 - \delta_1)$ by waiting one period,

(iii) player 1 never offers $x < 1 - \delta_2$, and

(iv) player 1 rejects all $x < \delta_1(1 - \delta_2)$.

To continue on, imagine that, after k rounds of elimination of conditionally dominated strategies, we have concluded that player 1 accepts all $x > x^k$, and player 2 accepts all $x < y^k$, with $x^k > y^k$. Then, after one more round, we conclude that

(i) player 2 never offers $x > x^k$,

(ii) player 2 rejects all $x > 1 - \delta_2(1 - x^k)$,

(iii) player 1 never offers $x < y^k$, and

(iv) player 1 rejects all $x < \delta_1 y^k$.

At the next round of iteration, we claim that player 1 must accept all $x > x^{k+1} \equiv \delta_1(1 - \delta_2) + \delta_1\delta_2 x^k$, and that player 2 accepts all $x < y^{k+1} = 1 - \delta_2 + \delta_1\delta_2 y^k$, where $x^{k+1} > y^{k+1}$. We will check this claim for player 1: If player 1 refuses an offer of player 2 in some subgame, one of three things can happen. Either (a) no agreement is ever reached, which has payoff 0, or (b) player 2 accepts one of player 1's offers, which has a current value of at most $\delta_1(1 - \delta_2(1 - x^k))$ (since the soonest this can happen is next period, and player 2 refuses x above $(1 - \delta_2(1 - x^k))$), or (c) player 1 accepts one of player 2's offers, which has a current value of at most $\delta_1^2 x^k$. Simple algebra shows that, for all discount factors δ_1, δ_2, the payoff in case b is largest, so player 1 accepts all $x > \delta_1(1 - \delta_2) + \delta_1\delta_2 x^k$.

The x^k and y^k are monotonic sequences, with limits $x^\infty = \delta_1(1 - \delta_2)/(1 - \delta_1\delta_2)$ and $y^\infty = (1 - \delta_2)/(1 - \delta_1\delta_2)$. Iterated conditional dominance shows that player 2 rejects all $x > 1 - \delta_2(1 - x^\infty) = y^\infty$ and accepts all $x < y^\infty$, so the unique equilibrium outcome is for player 1 to offer exactly y^∞ and for player 2 to accept. (There is no equilibrium profile in which player 2 rejects y^∞ with positive probability, for then player 1 would want to offer "just below" y^∞, so that no best response for player 1 exists.)

4.7   Open-Loop and Closed-Loop Equilibria[††]

4.7.1   Definitions

The terms *closed-loop* and *open-loop* are used to distinguish between two different information structures in multi-stage games. Our definition of a multi-stage game with observed actions corresponds to the closed-loop information structure, where players can condition their play at time t on the history of play until that date. In the terminology of the literature on optimal control, the corresponding strategies are called *closed-loop strate-*

gies or feedback strategies, while *open-loop* strategies are functions of calendar time alone.

Determining which are the appropriate strategies to consider is the same as determining the information structure of the game. Suppose first that the players never observe any history other than their own moves and time, or that at the beginning of the game they must choose time paths of actions that depend only on calendar time. (These two situations are equivalent from the extensive-form viewpoint, as the role of information sets is to describe what information players can use in choosing their actions.) In this case all strategies are open-loop, and all Nash equilibria (which in this case coincide with perfect equilibria) are in open-loop strategies. An equilibrium in open-loop strategies is called an *open-loop equilibrium*. (As with "Cournot" and "Stackelberg" equilibria, this is not really a new equilibrium concept but rather a way of describing the equilibria of a particular class of games.)

If the players can condition their strategies on other variables in addition to calendar time, they may prefer not to use open-loop strategies in order to react to exogenous moves by nature, to the realizations of mixed strategies by their rivals, and to possible deviations by their rivals from the equilibrium strategies. That is, they may prefer to use closed-loop strategies. When closed-loop strategies are feasible, subgame-perfect equilibria will typically not be in open-loop strategies, as subgame perfection requires players to respond optimally to the realizations of random variables as well as to unexpected deviations; in particular, for open-loop strategies to meet this condition requires that it be optimal to play the same actions whether or not an opponent has deviated in the past. The term *closed-loop equilibrium* usually means a subgame-perfect equilibrium of the game where players can observe and respond to their opponents' actions at the end of each period. Of course, games with this information structure can have Nash equilibria that are not perfect. In particular, a pure-strategy open-loop equilibrium is a Nash equilibrium in the closed-loop information structure if the game is deterministic (admits no moves by nature) and the players' action spaces depend only on time (exercise 4.10); yet it will typically not be perfect.

It is typically much easier to characterize the open-loop equilibria of a given situation than the closed-loop ones, in part because the closed-loop strategy space is so much larger. This tractability is one explanation for the use of open-loop equilibria in economic analyses. A second reason for interest in open-loop equilibria, discussed in the next subsection, is that they serve as a useful benchmark for discussing the effects of strategic incentives in the closed-loop information structure, i.e., the incentives to change current play so as to influence the future play of opponents. A third reason, discussed in subsection 4.7.3, is that the open-loop equilibria may be a good approximation to the closed-loop ones if there are very many "small" players. Intuitively, if players are small, an unexpected deviation by an opponent might have little influence on a player's optimal play.

4.7.2 A Two-Period Example

The use of open-loop equilibria as benchmarks for measuring strategic effects can be illustrated most easily in a game with continuous action spaces. Consider a two-player two-stage game where in the first stage players $i = 1, 2$ simultaneously choose actions $a_i \in A_i$, and in the second stage they simultaneously choose actions $b_i \in B_i$, where each of these action sets is an interval of real numbers. Suppose that the payoff functions u_i are differentiable and that each player's payoff is concave in his own actions.

An open-loop equilibrium is a time path (a^*, b^*) satisfying, for $i = 1, 2$,

a_i^* maximizes $u_i((a_i, a_{-i}^*), b^*)$

and

b_i^* maximizes $u_i(a^*, (b_i, b_{-i}^*))$.

Since payoffs are concave, an interior solution must satisfy the first-order conditions

$$\frac{\partial u_i}{\partial a_i} = 0 = \frac{\partial u_i}{\partial b_i}. \tag{4.1}$$

In a closed-loop equilibrium (supposing that one exists), the second-stage actions $b^*(a)$ after any first-period actions a are required to be a Nash equilibrium of the second-stage game. That is, for each $a = (a_1, a_2)$, $b_i^*(a)$ maximizes $u_i(a, b_i, b_{-i}^*(a))$. Moreover, the players recognize that the second-period actions will depend on the first-period ones according to the function b^* when choosing the first-period actions. Thus, the first-order condition for an optimal (interior) choice of a_i (assuming that $b^*(\cdot)$ is differentiable) is now

$$\frac{\partial u_i}{\partial a_i} + \frac{\partial u_i}{\partial b_{-i}} \frac{\partial b_{-i}^*}{\partial a_i} = 0. \tag{4.2}$$

Compared with the corresponding open-loop equation (4.1), there is now an extra term corresponding to player i's "strategic incentive" to alter a_i to influence b_{-i}. (The change in player i's utility due to the induced change in his own second-stage action is 0 by the "envelope theorem.") For example, if player i prefers decreases in b_{-i}, and $\partial b_{-i}^*/\partial a_i$ is negative at the open-loop equilibrium a^*, then player i's "strategic incentive," at least locally, is to increase a_i beyond a_i^*.

To make these observations more concrete, suppose that the actions are choices of outputs, as in Cournot competition, and there is "learning by doing" so that a firm's second-stage marginal cost is decreasing in its first-stage output. Then the second-period equilibrium, $b^*(a)$, is simply the Cournot equilibrium given the first-period costs. Since a firm's Cournot-equilibrium output level is increasing in its opponent's marginal cost (at least if the stability condition discussed in chapter 1 is satisfied), and since increasing a_i lowers firm i's second-stage costs, $\partial b^*_{-i}/\partial a_i$ is negative. Finally, in Cournot competition each firm prefers its opponent's output to be low. Thus the strategic incentives in this example favor additional investment in learning beyond what a firm would choose in an open-loop equilibrium.

As a final gloss on this point, note that if the second-period equilibrium actions are increasing in the first-period actions, and firms prefer their opponents' second-period actions to be low, then strategic incentives tend to reduce first-period actions from the open-loop levels. And note that by changing the sign of $\partial u_i/\partial b_{-i}$ we obtain two analogous cases.[10]

4.7.3 Open-Loop and Closed-Loop Equilibria in Games with Many Players[†††]

We remarked above that one defense of open-loop equilibria is that they may approximate the closed-loop ones if there are many small players. We now examine this intuition in a bit more detail. First consider the limit case where players are infinitesimal. That is, suppose that the game has a continuum of nonatomic individuals of each player type—a continuum of player 1s, a continuum of player 2s, and so on. (For concreteness, let the set of individuals be copies of the unit interval, endowed with Lebesgue measure.) Suppose further that each player i's payoff is independent of the actions of any subset of opponents with measure 0. Then if one individual player j deviates, and all players $k \neq i,j$ ignore j's deviation, it is clearly optimal for player i to ignore the deviation as well. Thus, the outcome of an open-loop equilibrium is subgame perfect.[11]

However, even in this nonatomic model, there can be equilibria in which all players do respond to a deviation by a single player, because the deviation shifts play from one continuation equilibrium to another one. This is easiest to see in a continuum-of-individuals version of the two-stage game in figure 4.2. Suppose that each player 1's payoff to strategy s_1 is the average of his payoffs against the distribution of strategies played by the continuum of player 2's payoff:

$$u_1 = \sum_{s_2} p(s_2)u_1(s_1, s_2),$$

10. Bulow, Geanakoplos, and Klemperer (1985) and Fudenberg and Tirole (1984, 1986) develop taxonomies along these lines and apply them to various problems in industrial organization.
11. The open-loop *strategies* are not perfect, as they ignore deviations by subsets of positive measure.

where $p(s_2)$ is the proportion of player 2's using strategy s_2; define player 2's payoff analogously as the average against the population distribution of s_1's. If no player can observe the action of any individual opponent, then each player's second-stage payoff is independent of how he plays in the first stage, and the efficient first-stage payoff $(5, 5)$ cannot occur in equilibrium. However, if players do observe the play of each individual opponent, then the first-stage payoff $(5, 5)$ can be enforced with the same strategies as in the two-player version of the game.

The key to enforcing $(5, 5)$ in this example is for all players to respond to the deviation of a single opponent, even though this deviation does not directly influence their payoffs. Economists fairly often rule out such "atomic" or "irregular" equilibria by requiring that players cannot observe the actions of measure-zero subsets.[12] However, these atomic equilibria are not pathologies of the continuum-of-players model. Fudenberg and Levine (1988) give an example of a sequence of two-period finite-player games, each of which has one open-loop equilibrium and two closed-loop ones, and such that every second-period subgame has a unique equilibrium for every finite number of players. As the number of players grows to infinity, one of the closed-loop equilibria has the same limiting path as the open-loop equilibrium, while the other closed-loop equilibrium converges to an atomic equilibrium of the limit game. To obtain the intuitive conclusion that open-loop and closed-loop equilibria are close together in T-period games with a large finite number of players requires (strong) conditions on the first $T + 1$ partial derivatives of the payoff functions.

It may be that the intuition that the actions of small players should be ignored corresponds to a continuum-of-players model that is the limit of finite-players models with a noise term that is large enough to mask the actions of any individual player, yet vanishes as the number of players grows to infinity so that the limit game is deterministic. We have not seen formal results along this line.

4.8   Finite-Horizon and Infinite-Horizon Equilibria (technical)[††]

Since continuity at infinity (see section 4.2) implies that events after t (for t large) have little effect, one would expect that under this condition the sets of equilibria of finite-horizon and infinite-horizon versions of the "same game" would be closely related. This is indeed the case, but it is not true that all infinite-horizon equilibria are limits of equilibria of the corresponding finite-horizon game. (That is, there is typically a failure of lower hemi-continuity in passing to the infinite-horizon limit.) We have already seen an example of this in the repeated prisoner's dilemma of figure 4.1.

12. For a recent example see Gul, Sonnenschein, and Wilson 1986.

Radner (1980) observed that cooperation can be restored as an equilibrium of the finite-horizon game (with time averaging, i.e., $\delta = 1$) if one relaxes the assumption that players exactly maximize their payoffs.

Definition 4.3 Profile σ^* is an ε-*Nash equilibrium* if, for all players i and strategies σ_i,

$$u_i(\sigma_i^*, \sigma_{-i}^*) \geq u_i(\sigma_i, \sigma_{-i}^*) - \varepsilon.$$

The profile is an ε-perfect equilibrium if no player can gain more than ε by deviating in any subgame.

In our opinion, the concept of ε-equilibrium is best viewed as a useful device to relate the structures of large finite and infinite horizons. Though it is sometimes proposed as a description of boundedly rational behavior, its rationality requirements are very close to those of Nash equilibrium. For instance, the players must have the correct beliefs about their opponents' strategies and must correctly compute the expected payoff to each action; for some unspecified reason, they may voluntarily sacrifice ε utils.[13] (However, ε-optimality might be a *necessary* condition for certain boundedly rational policies to "survive.")

Radner's use of ε-equilibria smoothes the finite-to-infinite-horizon limit when utilities are continuous at infinity. Fudenberg and Levine (1983) showed this, beginning with an infinite-horizon game G^∞, which is then "approximated" by a sequence of T-period "truncated" games G^T that are created by choosing an arbitrary strategy $\tilde{s}$ in G^∞, and specifying that play follows $\tilde{s}$ after T. (In a repeated game, the most natural truncated games to consider are those in which $\tilde{s}$ specifies the same strategy in every period, independent of the history; these truncations correspond to the finitely repeated version of the game. In more general multi-stage games, such constant strategies may not be feasible.) A strategy s^T for the game G^T specifies play in all periods up to and including T; play follows strategy $\tilde{s}$ in the subsequent periods. In an abuse of notation, we use the same notation s^T to denote both the strategy of a truncated game G^T and the corresponding strategy in G^∞; when we speak of truncated-game strategies converging we will view them as elements of G^∞. Given this embedding, the payoff functions of G^∞ induce payoff functions in G^T in the obvious way.

To characterize equilibria of G^∞ in terms of limits of strategy profiles in G^T, one must specify a topology on the space of strategies of G^∞. Recall that a behavior strategy σ_i for player i in G^∞ is a sequence $\sigma_i(\cdot \mid h^0)$, $\sigma_i(\cdot \mid h^1)$, etc. Fudenberg and Levine use a complicated metric to topologize these

---

13. Another difficulty with the concept as a descriptive model is that, although ε is small relative to total utility, it may be large relative to a given period's utility. Thus, for instance, cooperating in the last period of a finitely repeated prisoner's dilemma may entail a substantial cost in that period even though the cost is negligible overall.

sequences.[14] In games with a finite number of possible actions per period ("finite-action games"), their topology reduces to the product topology (also known as the topology of pointwise convergence), which is much easier to work with.[15]

Theorem 4.4 (Fudenberg and Levine 1983)   Consider an infinite-horizon finite-action game G^∞ whose payoffs are continuous at infinity. Then

(i) σ^* is a subgame-perfect equilibrium of G^∞ if and only if it is the limit (in the product topology) of a sequence σ^T of ε_T-perfect equilibria of a sequence of truncated games G^T with $\varepsilon_T \to 0$. Moreover,

(ii) the set of subgame-perfect equilibria of G^∞ is nonempty and compact.

Remark   To gain some intuition for the theorem, consider approximating the finite-horizon prisoner's dilemma of figure 4.1, with $\delta > \frac{1}{2}$, by truncated games in which players are required to defect in all periods after T. Although "always defect" is the only subgame-perfect equilibrium of these finite games, cooperation can occur in an ε-perfect equilibrium: If the opponent's strategy is to cooperate until defecting occurs and to defect thereafter, the best response is to cooperate until the last period T, and then to defect at T, which yields average utility

$$1 + \frac{\delta^T(1-\delta)}{1-\delta^{T+1}}.$$

Cooperating in every period yields utility 1, and the difference between this strategy and the optimum goes to 0 as $T \to \infty$. More generally, continuity at infinity implies that players lose very little (in *ex ante* payoff) by not optimizing at a far-distant horizon.

Proof   First note that if $\sigma^n \to \sigma$ in the product topology, then the continuation payoffs $u(\sigma^n|h^t)$ under σ^n in the subgame starting with h^t converge to the payoffs $u(\sigma|h^t)$ under σ. To see this, recall that $\sigma^n \to \sigma$ implies that

$$\sigma^n(a_i^t|h^t) \to \sigma(a_i^t|h^t) \text{ for all } t, h^t, a_i^t.$$

Thus, conditional on h^t, the probability that a^t is played in period t and

---

14. Harris (1985) shows that the complicated metric used by Fudenberg and Levine can be replaced by a simple one that sets the distance between two strategy profiles equal to $1/k$, where k is the largest number such that the two profiles prescribe exactly the same actions in every period $t \le k$ for every history h^t. This topology allows Harris to dispense with an additional continuity requirement that Fudenberg and Levine required for these games. (They required that payoffs as a function of the infinite history h^∞ be continuous in the product topology, which implies continuity in each period's realized action.) Borges (1989) shows that the *outcomes* of infinite-horizon pure-strategy equilibria coincide with the limits (in the product topology) of the outcomes of finite-horizon pure-strategy ε-equilibria.
15. A sequence σ^n converges to σ in the product topology (or topology of pointwise convergence) if and only if, for all i, all h^t, and all $a_i^t \in A_i(h^t)$,

$$\sigma_i^n(a_i^t|h^t) \to \sigma_i(a_i^t|h^t),$$

which implies that $\sigma^n(a|h^t) \to \sigma(a|h^t)$.

a^{t+1} is played in period $(t + 1)$ is

$$\sigma^n(a^t|h^t)\sigma^n(a^{t+1}|h^t, a^t) \to \sigma(a^t|h^t)\sigma(a^{t+1}|h^t, a^t),$$

and the distribution over outcomes at each date $\tau > t$ converges pointwise to that generated by σ. Thus, for any $\varepsilon > 0$ and T there is an N such that for all $n > N$ the distribution of actions from period t through period $(t + T)$ generated by σ^n, conditional on h^t, is within ε of the distribution generated by σ. Since outcomes after $(t + T)$ are unimportant for large T, continuation payoffs under σ^n converge to those under σ.[16]

To prove (i), note that continuity at infinity says that there is a sequence $\eta_T \to 0$ such that events after period T matter no more than η_T for each player. If σ is a subgame-perfect equilibrium of G^∞, the projection σ^T of σ onto G^T must be a $2\eta_T$-perfect equilibrium of G^T: For any i, h^t, and $\tilde{\sigma}_i$,

$$u_i(\sigma_i, \sigma_{-i}|h^t) \geq u_i(\tilde{\sigma}_i, \sigma_{-i}|h^t),$$

so from continuity at infinity

$$u_i(\sigma_i^T, \sigma_{-i}^T|h^t) + \eta_T \geq u_i(\tilde{\sigma}_i^T, \sigma_{-i}^T|h^t) - \eta_T.$$

Hence, σ is the limit of $2\eta_T$-equilibria of G^T.

Conversely, suppose $\sigma^T \to \sigma$ is a sequence of ε_T-perfect equilibria of G^T with $\varepsilon_T \to 0$. Continuity at infinity implies that each σ^T is an $(\varepsilon_T + \eta_T)$-perfect equilibrium of G^∞. If σ is not subgame perfect, there must be a time t and a history h^t where some player i can gain at least $2\varepsilon > 0$ by playing some $\hat{\sigma}_i$ instead of σ_i against σ_{-i}. But since $\sigma^T \to \sigma$ and payoffs are continuous, for T sufficiently large player i could gain at least ε by playing $\hat{\sigma}_i$ instead of σ_i^T against σ_{-i}^T, which contradicts $\varepsilon_T \to 0$.

To prove (ii), note first that for fixed $\tilde{s}$ each G^T is a finite multi-stage game and hence has a subgame-perfect equilibrium σ^T. (See exercise 3.5 and chapter 8.) Because the space of infinite-horizon strategies is compact (this is Tychonov's theorem[17]), the sequence σ^T has an accumulation point; this accumulation point is a perfect equilibrium of G^∞ from (i). Because payoffs are continuous, a standard argument shows that the set of subgame-perfect equilibria is closed, and closed subsets of compact sets are compact. ∎

Radner considered the repeated prisoner's dilemma with time averaging, and observed that "both cooperate" *is* an ε-perfect equilibrium outcome of the finitely repeated game, with the required $\varepsilon \to 0$ as the number of periods tends to infinity. That Radner obtained this result with time

---

16. With time averaging, payoffs are not continuous in the product topology because they are not continuous at infinity. Consider the one-player game where the feasible actions each period are 0 and 1 and the player's stage-game payoff equals his action. Then the sequence of strategies σ^n given by $\sigma^n(0|h^t) = 1$ for $t < n$ and $\sigma^n(1|h^t) = 1$ for $t \geq n$ converges to $\sigma(0|h^t) = 1$ for all t and h^t in the product topology, and the discounted payoffs converge to 0 as well, but under time averaging the payoff of each σ^n is 1.
17. See, for example, Munkres 1975.

averaging is somewhat misleading, as, in general, games with time averaging are not well behaved: There can be exact equilibrium payoffs of a finitely repeated stochastic game that are not even ε-equilibrium payoffs of the infinitely repeated version.[18]

Exercises

Exercise 4.1

(a)** Consider the following modification of the stationary symmetric war of attrition developed in subsection 4.5.2: $L(\hat{t}) = -(1 - \delta^t)/(1 - \delta)$, $F(\hat{t}) = L(\hat{t}) + \delta^t v$, and $B(\hat{t}) = L(\hat{t}) + \delta^t qv$, with $q \leq \frac{1}{2}$, which corresponds to the assumption that if both animals stop fighting simultaneously then each has probability q of winning the prize. Characterize the symmetric stationary equilibrium. Compute the limit as the time period shrinks and show it is independent of q.

(b)*** Characterize the entire set of perfect-equilibrium outcomes.

Exercise 4.2**
Consider the following modification of the perfect-information preemption game developed in the text. Players now choose two times: a time s to do a feasibility study and a time t to build a plant. In order to build a plant, the player must have done a feasibility study in some $s < t$, with s and t required to be odd for player 2 and even for player 1. Doing a feasibility study costs ε in present value, where ε is small,

---

18. Here is a one-player game where the (unique) infinite-horizon equilibrium payoff is less than the limit of the payoffs of finite-horizon equilibria: The player must decide when to chop down a tree, which grows by one unit per period from an initial size of 0 at date 0. If the tree is chopped down at the beginning of period t, the player receives a flow payoff of 0 each period before t, a flow of 1 in periods t through $2t - 1$, and 0 thereafter. If the player discounts flows at rate δ, the player's strategy in the infinite-horizon game is to chop down the tree at the time t^* that maximizes

$$\delta^t(1 + \delta + \cdots + \delta^{t-1}) = \frac{\delta^t(1 - \delta^t)}{1 - \delta};$$

so t^* is such that δ^{t^*} is as close as possible to $\frac{1}{2}$. Note that t^* is independent of the (finite or infinite) horizon as long as the horizon is large enough.

However, if the player's utility is the average payoff, the optimal strategy with finite horizon T is to cut at the first time t where $t \geq T/2$, yielding an average payoff that converges to $\frac{1}{2}$, yet *no* policy in the infinite-horizon problem yields a strictly positive payoff. Here the problem is that the sequence of finite-horizon strategies "cut at $T/2$" converges (in the product topology) to the limit strategy "never cut," but the limit of the corresponding payoffs is not equal to the payoff of the limit strategy, so that the payoff is not a continuous function of the strategy. Sorin (1986) identifies a different way the finite-to-infinite-horizon limit can be badly behaved. In his example, the finite-horizon equilibria involve one player using a behavior strategy that assigns probability roughly t/T to an action that sends the game to an absorbing state, so that the state is reached with probability close to 1 with a long horizon T. These strategies once again have a limit that assigns probability 0 to stopping (reaching the absorbing state) in each period, and indeed the equilibrium payoffs of the infinite-horizon game do not include the finite-horizon limits. Lehrer and Sorin (1989) provide conditions for the finite-to-infinite-horizon limit to be well behaved in one-player games with time averaging.

but this cost is recouped except for lost interest payments if only that player builds a plant. Thus, the payoff is $-\varepsilon$ if a player does a feasibility study and never builds, $-1-\varepsilon$ if a player does a feasibility study and both build, and $14-(t-7)^2-\varepsilon(1-\delta^{t-s})$ if a player does a study, builds, and his opponent does not build. Show that the equilibrium outcome that survives iterated conditional dominance is for player 1 to pay for a study in period 2 and wait until period 6 to build. Explain why player 1 is now able to postpone building, and why player 2 cannot preempt by doing a study at $s=1$. Relate this to "ε-preemption." (This exercise was provided by R. Wilson.)

Exercise 4.3* Consider a variant of Rubinstein's infinite-horizon bargaining game where partitions are restricted to be integer multiples of 0.01, that is, x can be 0, 0.01, 0.02, ..., 0.99, or 1. Characterize the set of subgame-perfect equilibria for $\delta=\frac{1}{2}$ and for δ very close to 1.

Exercise 4.4* Consider the I-player version of Rubinstein's game, which Moulin (1986) attributes to Dutta and Gevers. At dates $1, I+1, 2I+1, \ldots$, player 1 offers a division $(x_1, \ldots, x_I)$ of the pie with $x_i \ge 0$ for all i, and $\sum_{i=1}^{I} x_i \le 1$. At dates $2, I+2, 2I+2, \ldots$, player 2 offers a division, and so on. When player i offers a division, the other players simultaneously accept or veto the division. If all accept, the pie is divided; if at least one vetoes, player $i+1$ (player 1 if $i=I$) offers a division in the following period. Assuming that the players have common discount factor δ, show that, for all i, player i offering division

$$\left(\frac{1}{1+\cdots+\delta^{I-1}}, \frac{\delta}{1+\cdots+\delta^{I-1}}, \ldots, \frac{\delta^{I-1}}{1+\cdots\delta^{I-1}}\right)$$

for players $i, i+1, \ldots, i-1$ at each date $(kI+i)$ and the other players' accepting is a subgame-perfect equilibrium outcome.

Exercise 4.5* Solve Ståhl's finite-horizon bargaining problem for T even and then for T odd, and show that the outcomes of the two cases converge to a common limit as $T \to \infty$.

Exercise 4.6** Admati and Perry (1988) consider the following model of infinite-horizon, perfect-information joint investment in a public good: Players $i=1,2$ take turns making investments $x_i(t)$ in the project, which is "finished" at the first date T at which

$$\sum_{i=1,2} \sum_{t=0}^{T} x_i(t) \ge K.$$

Players receive no benefits from the project until it is completed; if it is completed at date T, player i receives benefit $\delta_i^T V$. Players have a convex cost of investment $c_i(x_i)$, with $c_i(0)=0$; thus, player i's total payoff is

$$\delta_i^T V - \sum_{t=0}^{T} \delta_i^t c_i(x_i(t)).$$

Use iterated conditional dominance to show that the game has a unique subgame-perfect equilibrium. Hint: First show that there is an $\bar{x}_1$ such that, if the investment to date exceeds $K - \bar{x}_1$, it is conditionally dominant for the player on move to finish the project. Then argue that the second round of conditional dominance implies that there is an $\bar{x}_2$ such that, if investment to date $K(t)$ exceeds $K - \bar{x}_1 - \bar{x}_2$ but does not exceed $K - \bar{x}_1$, the player on move should not invest less than $K - K(t) - \bar{x}_1$.

Exercise 4.7** Prove that, in a game of perfect information, no subgame-perfect strategy profile is removed by iterated conditional dominance.

Exercise 4.8**

(a) Consider the two-person Rubinstein-Ståhl model of section 4.4. The two players bargain to divide a pie of size 1 and take turns making offers. The discount factor is δ. Introduce "outside options" in the following way: At each period, the player whose turn it is to make the offer makes the offer; the other player then has the choice among (1) accepting the offer, (2) exercising his outside option instead, and (3) continuing bargaining (making an offer the next period). Let x_0 denote the value of the outside option. Show that, if $x_0 \leq \delta/(1 + \delta)$, the outside option has no effect on the equilibrium path. Comment. What happens if $x_0 > \delta/(1 + \delta)$?

(b) Consider an alternative way of formalizing outside options in bargaining. Suppose that there is an "exogenous risk of breakdown" of renegotiation (Binmore et al. 1986). At each period t, assuming that bargaining has gone on up to date t, there is probability $(1 - x)$ that bargaining breaks down at the end of period t if the period-t offer is turned down. The players then get x_0 each. Show that the "outside opportunity" x_0 matters even if it is small, and compute the subgame-perfect equilibrium.

(c) In their study of supply assurance, Bolton and Whinston (1989) consider a situation in which the outside option is endogenous. Suppose that there are three players: two buyers ($i = 1, 2$) and a seller ($i = 3$). The seller has one indivisible unit of a good for sale. Each buyer has a unit demand. The seller's cost of departing from the unit is 0 (the unit is already produced). The buyers have valuations v_1 and v_2, respectively. Without loss of generality, assume that $v_1 \geq v_2$. Bolton and Whinston consider a generalization of the Rubinstein-Ståhl process. At dates $0, 2, \ldots, 2k, \ldots$, the seller makes offers; at dates $1, 3, \ldots, 2k + 1, \ldots$, the buyers make offers. Buyers' offers are prices at which they are willing to buy and among which the seller may choose. The seller can make an offer to only a single buyer, as she has only one unit for sale (alternatively one could consider a situation in which the seller organizes an auction in each even period). Consider a stationary equilibrium and show that, if parties have the same discount factor and as the time between offers tends to 0, the parties' perfect-equilibrium payoffs converge to $v_1/2$ for both the seller and buyer 1 and to 0 for buyer 2 if $v_1/2 > v_2$, and to v_2 for the seller, $v_1 - v_2$ for buyer 1, and 0 for buyer 2 if $v_1/2 < v_2$. (For a uniqueness result see Bolton and Whinston 1989.)

Exercise 4.9** As shown by Rubinstein (see section 4.4 above), the alternating-move bargaining process between two players has a unique equilibrium. Shaked has pointed out that with $I \geq 3$ players there are many (subgame-) perfect equilibria (see Herrero 1985 for more details). Prove that with $I = 3$ players, and for discount factor $\delta > \frac{1}{2}$, any partition of the pie is the outcome of a perfect equilibrium.

The game is as follows: Three players bargain over the division of a pie of size 1. A division is a triple (x_1, x_2, x_3) of shares for each player, where $x_i \geq 0$, $\sum_{i=1}^{3} x_i = 1$. At dates $3k + 1, k = 0, 1, \ldots$, player 1 offers a division; if players 2 and 3 both accept, the game is over. If one or both of them veto, bargaining goes on. Similarly, at dates $3k + 2$ (respectively, $3k$), player 2 (respectively, player 3) makes the offer. The game stops once an offer by one player has been accepted by the other two players.

Show that, if $\delta > \frac{1}{2}$, any partition can be supported as a (subgame-) perfect equilibrium.

Exercise 4.10*

(a) Show that a pure-strategy open-loop Nash equilibrium of a deterministic game in which action spaces depend only on time is a closed-loop Nash equilibrium. Hint: Use the analogy with a control (single decision maker) problem.

(b) Does this result hold when the open-loop Nash equilibrium is in mixed strategies? When the players learn stochastic moves by nature?

References

Admati, A., and M. Perry. 1988. Joint projects without commitment. Mimeo.

Aumann, R. 1964. Mixed vs. behavior strategies in infinite extensive games. *Annals of Mathematics Studies* 52: 627–630.

Binmore, K. 1981. Nash bargaining theory 1–3. London School of Economics Discussion Paper.

Binmore, K., A. Rubinstein, and A. Wolinsky. 1986. The Nash bargaining solution in economic modeling. *Rand Journal of Economics* 17: 176–188.

Bolton, P., and M. Whinston. 1989. Incomplete contracts, vertical integration, and supply assurance. Mimeo, Harvard University.

Borges, T. 1989. Perfect equilibrium histories of finite and infinite horizon games. *Journal of Economic Theory* 47: 218–227.

Bulow, J., J. Geanakoplos, and P. Klemperer. 1985. Multimarket oligopoly: Strategic substitutes and complements. *Journal of Political Economy* 93: 488–511.

Dasgupta, P., and J. Stiglitz. 1980. Uncertainty, industrial structure, and the speed of R&D. *Bell Journal of Economics* 11: 1–28.

Fudenberg, D., R. Gilbert, J. Stiglitz, and J. Tirole. 1983. Preemption, leapfrogging, and competition in patent races. *European Economic Review* 22: 3–31.

Fudenberg, D., and D. Levine. 1983. Subgame-perfect equilibria of finite and infinite horizon games. *Journal of Economic Theory* 31: 227–256.

Fudenberg, D., and D. Levine. 1988. Open-loop and closed-loop equilibria in dynamic games with many players. *Journal of Economic Theory* 44: 1–18.

Fudenberg, D., and J. Tirole. 1984. The fat cat effect, the puppy dog ploy and the lean and hungry look. *American Economic Review, Papers and Proceedings* 74: 361–368.

Fudenberg, D., and J. Tirole. 1985. Preemption and rent equalization in the adoption of new technology. *Review of Economic Studies* 52: 383–402.

Fudenberg, D., and J. Tirole. 1986. *Dynamic Models of Oligopoly*. Harwood.

Ghemawat, P., and B. Nalebuff. 1985. Exit. *Rand Journal of Economics* 16: 184–194.

Gilbert, R., and R. Harris. 1984. Competition with lumpy investment. *Rand Journal of Economics* 15: 197–212.

Gul, F., H. Sonnenschein, and R. Wilson. 1986. Foundations of dynamic monopoly and the Coase conjecture. *Journal of Economic Theory* 39: 155–190.

Harris, C. 1985. A characterization of the perfect equilibria of infinite horizon games. *Journal of Economic Theory* 37: 99–127.

Harris, C., and C. Vickers. 1985. Perfect equilibrium in a model of a race. *Review of Economic Studies* 52: 193–209.

Hendricks, K., A. Weiss, and C. Wilson. 1988. The war of attrition in continuous time with complete information. *International Economic Review* 29: 663–680.

Hendricks, K., and C. Wilson. 1989. The war of attrition in discrete time. Mimeo, State University of New York, Stony Brook.

Herrero, M. 1985. A strategic bargaining approach to market institutions. Ph.D. thesis, London School of Economics.

Katz, M., and C. Shapiro. 1986. How to license intangible property. *Quarterly Journal of Economics* 101: 567–589.

Kuhn, H. 1953. Extensive games and the problem of information. *Annals of Mathematics Studies* 28. Princeton University Press.

Lehrer, E., and S. Sorin. 1989. A uniform Tauberian theorem in dynamic programming. Mimeo.

Maynard Smith, J. 1974. The theory of games and evolution in animal conflicts. *Journal of Theoretical Biology* 47: 209–221.

Moulin, H. 1986. *Game Theory for the Social Sciences*. New York University Press.

Munkres, I. 1975. *Topology: A First Course*. Prentice-Hall.

Pearce, D. 1984. Rationalizable strategic behavior and the problem of perfection. *Econometrica* 52: 1029–1050.

Radner, R. 1980. Collusive behavior in non-cooperative epsilon equilibria of oligopolies with long but finite lives. *Journal of Economic Theory* 22: 121–157.

Reinganum, J. 1981a. On the diffusion of a new technology: A game-theoretic approach. *Review of Economic Studies* 48: 395–405.

Reinganum, J. 1981b. Market structure and the diffusion of new technology. *Bell Journal of Economics* 12: 618–624.

Rubinstein, A. 1982. Perfect equilibrium in a bargaining model. *Econometrica* 50: 97–110.

Shaked, A., and J. Sutton. 1984. Involuntary unemployment as a perfect equilibrium in a bargaining game. *Econometrica* 52: 1351–1364.

Simon, L. 1988. Simple timing games. Mimeo, University of California, Berkeley.

Sorin, S. 1986. Asymptotic properties of a non zero-sum stochastic game. Mimeo, Université de Strasbourg.

Ståhl, I. 1972. *Bargaining Theory*. Stockholm School of Economics.

Whinston, M. 1986. Exit with multiplant firms. HIER DP 1299, Harvard University.

Repeated Games

The best-understood class of dynamic games is that of repeated games, in which players face the same "stage game" or "constituent game" in every period, and the player's overall payoff is a weighted average of the payoffs in each stage. If the players' actions are observed at the end of each period, it becomes possible for players to condition their play on the past play of their opponents, which can lead to equilibrium outcomes that do not arise when the game is played only once. One example of this in the repeated prisoner's dilemma of section 4.3 is the "unrelenting" strategy "cooperate until the opponent defects; if ever the opponent defects, then defect in every subsequent period." The profile where both players use this unrelenting strategy is a subgame-perfect equilibrium of the infinitely repeated game if the discount factor is sufficiently close to 1: even though each player could do better in the short run by defecting instead of cooperating, for a patient player this short-run gain is outweighed by the prospect of unrelenting future "punishment." Section 4.3 considers this equilibrium as well as the one where players defect each period; there are other equilibria as well. Our goal in this chapter is to present a more systematic treatment of general repeated games. (Surveys of the literature on repeated games have been published by Aumann (1986, 1989), Mertens (1987), and Sorin (1988). Mertens, Sorin, and Zamir (1990) give a detailed exposition of repeated games, with emphasis on "large" action spaces, and discuss the related topic of stochastic games.)

Because repeated games do not allow for past play to influence the feasible actions or payoff functions in the current period, they cannot be used to model such important phenomena as investment in productive machinery and learning about the physical environment. Nevertheless, repeated games may be a good approximation of some long-term relationships in economics and political science—particularly those where "trust" and "social pressure" are important, such as when informal agreements are used to enforce mutually beneficial trades without legally enforced contracts. There are many variations on this theme, including Chamberlin's (1929) informal argument that oligopolists may use repeated play to implicitly collude on higher prices[1] and Macaulay's (1963) observation that relations between a firm and its suppliers are often based on "reputation" and the threat of the loss of future business.[2] Chapter 9 discusses an alternative way of modeling long-run relationships, where past actions serve to signal a player's future intentions by providing information about his payoffs.

---

1. Fisher (1898) gave an earlier critique of the static Cournot model that can be interpreted as favoring a repeated-game model. He asserted that, contrary to the Cournot assumption that outputs are chosen once and for all, "no business man assumes ... that his opponents' output or price will remain constant" (quoted in Scherer 1980).
2. Recent economic applications of repeated games to explain trust and cooperation include Greif 1989, Milgrom, North, and Weingast 1989, Porter 1983a, and Rotemberg and Saloner 1986. For some recent applications to political science, see the essays in Oye 1986.

The reason repeated play introduces new equilibrium outcomes is that players can condition their play on the information they have received in previous stages. Thus, one would expect that a key issue in analyzing repeated games is just what form this information takes. In the prisoner's-dilemma example in chapter 4, the players perfectly observed the actions that had been played. Sections 5.1–5.4 discuss general repeated games with this information structure, which we call *repeated games with observed actions*. (Note that this is a special case of the multi-stage games with observed actions introduced in chapter 3.)

Section 5.1 analyzes the equilibria of infinite-horizon games, focusing on the "folk theorems," which describe the equilibria when players are either completely patient or almost so. Section 5.2 presents the parallel results for finitely repeated games, and section 5.3 discusses various extensions to models where not all the players play the game every period. Examples include a long-run firm that faces a different short-run consumer each period; in that case the firm must decide whether to produce high-quality or low-quality goods (Dybvig and Spatt 1980; Shapiro 1982), and an organization composed of overlapping generations of workers must decide whether to exert effort on joint production (Crémer 1986).

Section 5.4 discusses the ideas of Pareto perfection and renegotiation-proofness, which have been proposed as a way to restrict the large set of repeated game equilibria when players are patient.

Sections 5.5–5.7 consider repeated games in which the players observe imperfect signals of their opponents' play. One example of this sort of game is the oligopoly model of Green and Porter (1984), wherein firms choose quantities each period and observe the realized market price but not the outputs of their opponents. Since the market price is stochastic, a low price could be due either to unexpectedly low demand or to some rival's having produced an unexpectedly high output. A second example is the repeated partnership in which each player observes the realized level of production but not the effort level of his partner (Radner 1986; Radner, Myerson, and Maskin 1986).

5.1 Repeated Games with Observable Actions[††]

5.1.1 The Model

The building block of a repeated game, the game which is repeated, is called the *stage game*. Assume that the stage game is a finite I-player simultaneous-move game with finite action spaces A_i and stage-game payoff functions $g_i: A \to \mathbb{R}$, where $A = \times_{i \in \mathscr{I}} A_i$. Let $\mathscr{A}_i$ be the space of probability distributions over A_i.

To define the repeated game, we must specify the players' strategy spaces and payoff functions. This section considers games in which the players observe the realized actions at the end of each period. Thus, let $a^t \equiv (a_1^t, \ldots, a_I^t)$ be the actions that are played in period t. Suppose that the game begins in period 0, with the null history h^0. For $t \geq 1$, let $h^t = (a^0, a^1, \ldots, a^{t-1})$ be the realized choices of actions at all periods before t, and let $H^t = (A)^t$ be the space of all possible period-t histories.

Since all players observe h^t, a pure strategy s_i for player i in the repeated game is a sequence of maps s_i^t—one for each period t—that map possible period-t histories $h^t \in H^t$ to actions $a_i \in A_i$. (Remember that a strategy must specify play in all contingencies, even those that are not expected to occur.) A mixed (behavior) strategy σ_i in the repeated game is a sequence of maps σ_i^t from H^t to mixed actions $\alpha_i \in \mathscr{A}_i$. Note that a player's strategy cannot depend on the past values of his opponents' randomizing probabilities α_{-i}; it can depend only on the past values of a_{-i}. Note also that each period of play begins a proper subgame. Moreover, since moves are simultaneous in the stage game, these are the only proper subgames, a fact that we will use in testing for subgame perfection.[3]

This section considers infinitely repeated games; section 5.2 considers games with a fixed finite horizon. With a finite horizon, the set of subgame-perfect equilibria is determined by backward-induction arguments that do not apply to the infinite-horizon model. The infinite-horizon case is a better description of situations where the players always think the game extends one more period with high probability; the finite-horizon model describes a situation where the terminal date is well and commonly foreseen.[4]

There are several alternative specifications of payoff functions for the infinitely repeated game. We will focus on the case where players discount future utilities using discount factor $\delta < 1$. In this game, denoted $G(\delta)$, player i's objective function is to maximize the normalized sum

$$u_i = \mathrm{E}_\sigma (1 - \delta) \sum_{t=0}^{\infty} \delta^t g_i(\sigma^t(h^t)),$$

where the operator E_σ denotes the expectation with respect to the distribution over infinite histories that is generated by strategy profile σ. The normalization factor $(1 - \delta)$ serves to measure the stage-game and repeated-game payoffs in the same units: The normalized value of 1 util per period is 1.

3. Although it seems that many of the results in this chapter should extend to stage games in which the moves are not simultaneous, as far as we know no one has yet checked the details.
4. The importance of a common forecast of the terminal date is shown by Neyman (1989), who considers a repeated prisoner's dilemma where both players know the horizon is finite, and where both players know the true length of the game to within ± 1 period, but the length of the game is not common knowledge between them (it is not even "almost common knowledge," as defined in chapter 14 below). He shows that this game has "cooperative" equilibria of the sort that arise in the infinite-horizon model but that are ruled out by backward induction with a known finite horizon.

To recapitulate the notation: As in the rest of the book, u_i, s_i, and σ_i denote the payoffs and the pure and mixed strategies of the overall game. The payoffs and strategies of the stage game are denoted g_i, a_i, and α_i.

As in the games of chapter 4, the discount factor δ can be thought of as representing pure time preference: This interpretation corresponds to $\delta = e^{-r\Delta}$, where r is the rate of time preference and Δ is the length of the period. The discount factor can also represent the possibility that the game may terminate at the end of each period: Suppose that the rate of time preference is r, the period length is Δ, and there is probability μ of continuing from one period to the next. Then 1 util tomorrow, to be collected only if the game lasts that long, is worth nothing with probability $1 - \mu$ and worth $\delta = e^{-r\Delta}$ utils with probability μ, for an expected discounted value of $\delta' = \mu\delta$. Thus, the situation is the same as if $\mu' = 1$ and $r' = r - \ln(\mu)/\Delta$. This shows that infinitely repeated games can represent games that terminate in finite time with probability 1. The key is that the conditional probability of continuing one more period should be bounded away from 0.[5]

Since each period begins a proper subgame, for any strategy profile σ and history h^t we can compute the players' expected payoffs from period t on. We will call these the "continuation payoffs," and renormalize so that the continuation payoffs from time t are measured in time-t units. Thus, the continuation payoff from time t on is

$$(1 - \delta) \sum_{\tau=t}^{\infty} \delta^{\tau-t} g_i(\sigma^\tau(h^\tau)).$$

With this renormalization, the continuation payoff of a player who will receive 1 util per period from period t on is 1 unit for any period t. This renormalization will be convenient, as it exploits the stationary structure of the game.

Although we will focus on the case where players discount future payoffs, we will also discuss the case where players are "completely patient," corresponding to the limit model $\delta = 1$. Several different specifications of the payoffs have been proposed to model complete patience. The simplest is the time-average criterion, where each player i's objective is to maximize

$$\lim_{T\to\infty} \inf \, \mathrm{E}\,(1/T) \sum_{t=0}^{T} g_i(\sigma^t(h^t)).$$

The lim inf in this expression is in response to the fact that some infinite sequences of utilities do not have well-defined average values.[6] (See Lehrer

5. In unpublished notes, B. D. Bernheim has shown that if the stage game has a continuum of actions, cooperative equilibria can arise even if the continuation probability does converge to 0 over time, provided it does so sufficiently slowly.

1988 for a discussion of the difference between this notion of a time average and the analogous one using the lim sup.)

Any form of time-average criterion implies that players are unconcerned not only about the timing of payoffs but also about their payoff in any finite number of periods, so that, for example, the sequences $(1, 0, 0, \ldots)$ and $(0, 0, \ldots)$, which both have average 0, are equally attractive. The overtaking criterion is an alternative specification of "patience" where improvement in a single period matters. This criterion, which is not representable by a utility functional, says that the sequence $g = (g^0, g^1, \ldots)$ is preferred to $\tilde{g} = (\tilde{g}^0, \tilde{g}^1, \ldots)$ if and only if there exists a time T' such that for all $T > T'$ the partial sum $\sum_{t=0}^{T} g^t$ strictly exceeds the partial sum $\sum_{t=0}^{T} \tilde{g}^t$. If g is not preferred to $\tilde{g}$, and $\tilde{g}$ is not preferred to g, then the two sequences are judged to be equally attractive. Note that if g has a higher time average than $\tilde{g}$, then g is necessarily preferred to $\tilde{g}$ under the overtaking criterion.[7]

Now that we have specified strategy spaces and payoff functions for the repeated game, our description of the model is complete. We conclude this subsection with a simple but useful observation.

Observation If α^* is a Nash equilibrium of the stage game (that is, a "static equilibrium"), then the strategies "each player i plays α_i^* from now on" are a subgame-perfect equilibrium. Moreover, if the game has m static equilibria $\{\alpha^j\}_{j=1}^{m}$, then for any map $j(t)$ from time periods to indices the strategies "play $\alpha^{j(t)}$ in period t" are a subgame-perfect equilibrium as well.

To see that this observation is correct, note that with these strategies the future play of player i's opponents is independent of how he plays today, so his optimal response is to play to maximize his current period's payoff, i.e., to play a static best response to $\alpha_{-i}^{j(t)}$. Note also that these are "open-loop" strategies of the type discussed in section 4.7.

The observation shows that repeated play of a game does not decrease the set of equilibrium payoffs. Further, since the only reason not to play a static best response is concern about the future, if the discount factor is small enough, then the only Nash equilibria of the repeated game are strategies that specify a static equilibrium at every history to which the equilibrium gives positive probability. (Proving this is exercise 5.2. Note that the same static equilibrium need not occur in every period. In games with infinite strategy spaces the conclusion must be modified slightly, since

---

6. Recall that $\lim_{t \to \infty} \inf x^t = \sup_T \inf_{t > T} x^t$ is the greatest lower bound on the sequence's accumulation points. Thus, if $\liminf_{t \to \infty} x^t = \underline{x}$, then for all $x > \underline{x}$ and all T there is a $\tau > T$ with $x^\tau < x$.

7. It is not obvious how to extend the overtaking criterion to probability distributions over sequences. One formulation requires that with probability 1 the realized sequence of utilities under one distribution be preferred to that under the other, but with this formulation the overtaking criterion is no longer a refinement of time averaging.

even a small future punishment can induce players to forgo a sufficiently small current gain.)

Another important fact about repeated games with observed actions is that the set of Nash-equilibrium continuation payoff vectors is the same in every subgame. Proving this is exercise 5.3.

5.1.2   The Folk Theorem for Infinitely Repeated Games

The "folk theorems" for repeated games assert that if the players are sufficiently patient then any feasible, individually rational payoffs can be enforced by an equilibrium. Thus, in the limit of extreme patience, repeated play allows virtually any payoff to be an equilibrium outcome.

To make this assertion precise, we must define "feasible" and "individually rational." Define player i's *reservation utility* or *minmax value* to be

$$\underline{v}_i = \min_{\alpha_{-i}} \left[\max_{\alpha_i} g_i(\alpha_i, \alpha_{-i}) \right]. \tag{5.1}$$

This is the lowest payoff player i's opponents can hold him to by any choice of α_{-i}, provided that player i correctly foresees α_{-i} and plays a best response to it. Let m^i_{-i} be a strategy for player i's opponents that attains the minimum in equation 5.1. We call m^i_{-i} the *minmax profile* against player i. Let m^i_i be a strategy for player i such that $g_i(m^i_i, m^i_{-i}) = \underline{v}_i$.

To illustrate this definition, we compute the minmax values for the game in figure 5.1. To compute player 1's minmax value, we first compute his payoffs to U, M, and D as a function of the probability q that player 2 assigns to L; in the obvious notation, these payoffs are $v_U(q) = -3q + 1$, $v_M(q) = 3q - 2$, and $v_D(q) = 0$. Since player 1 can always attain a payoff of 0 by playing D, his minmax payoff is at least this large; the question is whether player 2 can hold player 1's maximized payoff to 0 by some choice of q. Since q does not enter into v_D, we can pick q to minimize the maximum of v_U and v_M, which occurs at the point where the two expressions are equal, i.e., $q = \frac{1}{2}$. Since $v_U(\frac{1}{2}) = v_M(\frac{1}{2}) = -\frac{1}{2}$, player 1's minmax value is the zero payoff he can achieve by playing D. (Note that $\max(v_U(q), v_M(q)) \le 0$ for any $q \in [\frac{1}{3}, \frac{2}{3}]$, so we can take player 2's minmax strategy against player 1, m^1_2, to be any q in this range.)

Similarly, to find player 2's minmax value, we first express player 2's payoff to L and R as a function of the probabilities p_U and p_M that player 1 assigns to U and M:

	L	R
U	-2,2	1,-2
M	1,-2	-2,2
D	0,1	0,1

Figure 5.1

$$v_L = 2(p_U - p_M) + (1 - p_U - p_M), \tag{5.2}$$

$$v_R = -2(p_U - p_M) + (1 - p_U - p_M). \tag{5.3}$$

Player 2's minmax payoff is then determined by

$$\min_{p_U, p_M} \max [2(p_U - p_M) + (1 - p_U - p_M),$$

$$-2(p_U - p_M) + (1 - p_U - p_M)].$$

By inspection (or plotting equations 5.2 and 5.3) we see that player 2's minmax payoff is 0, which is attained by the profile $(\frac{1}{2}, \frac{1}{2}, 0)$. Here, unlike the minmax against player 1, the minmax profile is uniquely determined: If $p_U > p_M$, the payoff to L is positive, if $p_M > p_U$ the payoff to R is positive, and if $p_U = p_M < \frac{1}{2}$, then both L and R have positive payoffs.

Note that if we restricted attention to pure strategies in equation 5.1, player 1's and player 2's minmax values will both be 1. Clearly, minimizing over a smaller set in equation 5.1 cannot give a lower value; the figure shows that the restriction can give values that are strictly higher.

At this point, the reader might question our identifying the minmax payoffs as the reservation utilities. This terminology is justified by the following observation.

Observation Player i's payoff is at least $\underline{v}_i$ in any static equilibrium and in any Nash equilibrium of the repeated game, regardless of the level of the discount factor.

Proof In a static equilibrium $\hat{\alpha}$, $\hat{\alpha}_i$ is a best response to $\hat{\alpha}_{-i}$, and so $g_i(\hat{\alpha}_i, \hat{\alpha}_{-i})$ is no less than the minimum defined in equation 5.1. Now consider a Nash equilibrium $\hat{\sigma}$ of the repeated game. One feasible, though not necessarily optimal, strategy for player i is the myopic one that chooses each period's action $a_i(h^t)$ to maximize the expected value of $g_i(a_i, \hat{\sigma}_{-i}(h^t))$. (This may not be optimal, because it ignores the possibility that the future play of i's opponents may depend on how he plays today.) The key is that, because all players have the same information at the start of each period t, the probability distribution over the opponents' period-t actions given player i's information corresponds to *independent* randomizations by player i's opponents. (This is not necessarily true when actions are imperfectly observed, as we discuss in section 5.5.) Thus, the myopic strategy for player i yields at least $\underline{v}_i$ in each period, and so $\underline{v}_i$ is a lower bound on player i's equilibrium payoff in the repeated game. ∎

Thus, we know on *a priori* grounds that no equilibrium of the repeated game can give any player a payoff lower than his minmax value.

Next we introduce a definition of the feasible payoffs. Here we encounter the following subtlety: The sets of feasible payoffs in the stage game, and thus in the repeated game for small discount factors, need not be convex.

The problem is that "many" convex combinations of pure-strategy payoffs correspond to correlated strategies, and cannot be obtained by independent randomizations. For example, in the "battle of the sexes" game (figure 1.10a), the payoffs $(\frac{3}{2}, \frac{3}{2})$ cannot be obtained by independent mixing.

As Sorin (1986) has shown, this nonconvexity does not occur when the discount factor is near enough to 1, as any convex combination of pure-strategy payoffs can be obtained by a time-varying deterministic path. This is easiest to see in the time-averaging limit: The payoffs $(\frac{3}{2}, \frac{3}{2})$ in the battle of the sexes can be obtained by playing (B, B) in even-numbered periods and (F, F) in odd-numbered ones.

To avoid the need to use such time-varying paths, Fudenberg and Maskin (1986a) convexify the feasible payoffs of the stage game by assuming that all players observe the outcome of a public randomizing device at the start of each period. Sorin's result on its own suggests, but does not imply, that these public randomizations are innocuous when the discount factor is near enough to 1; Fudenberg and Maskin (1990a) subsequently proved a stronger version of Sorin's result and used it to extend their proof of the perfect folk theorem to games without public randomizations.[8] To avoid the complications this involves, we will use the assumption of public randomizations in our proofs. Formally, let $\{\omega^0, \ldots, \omega^t \ldots\}$ be a sequence of independent draws from a uniform distribution on $[0, 1]$, and assume that the players observe ω^t at the beginning of period t. The history is now

$$h^t \equiv (a^0, \ldots, a^{t-1}, \omega^0, \ldots, \omega^t).$$

A pure strategy s_i for player i is then a sequence of maps s_i^t from histories h^t into A_i.

In this case, the set of feasible payoffs for any discount factor is

$$V = \text{convex hull}\{v \mid \exists a \in A \text{ with } g(a) = v\}.$$

This set is illustrated in figure 5.2. The shaded region in the figure is the set of all feasible payoffs that Pareto dominate the minmax payoffs, which are 0 for both players. The set of feasible, strictly individually rational payoffs is the set $\{v \in V \mid v_i > \underline{v}_i \ \forall i\}$. Figure 5.2 depicts these sets for the game of figure 5.1, in which the minmax payoffs are $(0, 0)$.

Theorem 5.1 (folk theorem)[9] For every feasible payoff vector v with $v_i > \underline{v}_i$ for all players i, there exists a $\underline{\delta} < 1$ such that for all $\delta \in (\underline{\delta}, 1)$ there is a Nash equilibrium of $G(\delta)$ with payoffs v.

---

8. For small discount factors public randomizations can allow equilibrium payoffs that are not in the convex hull of the set of equilibrium payoffs without public randomization. See Forges 1986, Myerson 1986, and our exercise 5.5.
9. This is called the "folk theorem" because it was part of game theory's oral tradition or "folk wisdom" long before it was recorded in print.

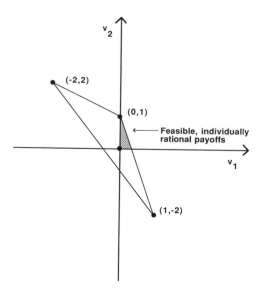

Figure 5.2

Remark The intuition for this theorem is simply that when the players are patient, any finite one-period gain from deviation is outweighed by even a small loss in utility in every future period. The strategies we construct in the proof are "unrelenting": A player who deviates will be minmaxed in every subsequent period.

Proof Assume first that there is a pure action profile a such that $g(a) = v$, and consider the following strategies for each player i: "Play a_i in period 0, and continue to play a_i so long as either (i) the realized action in the previous period was a or (ii) the realized action in the previous period differed from a in two or more components. If in some previous period player i was the only one not to follow profile a, then each player j plays m_j^i for the rest of the game."

Can player i gain by deviating from this strategy profile? In the period in which he deviates he receives at most $\max_a g_i(a)$, and since his opponents will minmax him forever afterward he can obtain at most $\underline{v}_i$ in periods after his first deviation. Thus, if player i's first deviation is in period t, he obtains at most

$$(1 - \delta^t)v_i + \delta^t(1 - \delta) \max_a g_i(a) + \delta^{t+1}\underline{v}_i, \tag{5.4}$$

which is less than v_i as long as δ exceeds the critical level $\underline{\delta}_i$ defined by

$$(1 - \underline{\delta}_i) \max_a g_i(a) + \underline{\delta}_i \underline{v}_i = v_i. \tag{5.5}$$

Since $v_i > \underline{v}_i$, the solution $\underline{\delta}_i$ for equation 5.5 is less than 1. Taking $\underline{\delta} = \max_i \underline{\delta}_i$ completes the argument. Note that, in deciding whether to deviate, in period t, player i assigns probability 0 to an opponent deviating in the same period. This is a consequence of the definition of Nash equilibrium: Only unilateral deviations are considered.

If payoffs v cannot be generated using pure actions, then we replace the action profile a with a public randomization $a(\omega)$ yielding payoffs with expected value v. The discount factor required to ensure that player i cannot gain by deviating may be somewhat larger in this case, as if player i conforms he does not receive exactly v_i in each period, and his temptation to deviate may be greater in periods where $g_i(a(\omega))$ is relatively low. It will be sufficient to take $\underline{\delta}_i$ such that

$$(1 - \underline{\delta}_i) \max_a g_i(a) + \underline{\delta}_i \underline{v}_i = (1 - \underline{\delta}_i) \min_a g_i(a) + \underline{\delta}_i v_i. \qquad (5.6)$$

To see that equation 5.6 is sufficient, note that for any period-t realization of ω, player i's continuation payoff to conforming from t on is

$$(1 - \delta)g_i(a(\omega)) + \delta v_i,$$

which is at least as large as $(1 - \delta)\min_a g_i(a) + \delta v_i$. By hypothesis, δ is large enough that this latter expression exceeds the continuation payoff from deviating, which is at most $(1 - \delta)\max_a g_i(a) + \delta \underline{v}_i$. ∎

Under the strategies used in the proof of theorem 5.1, a single deviation provokes unrelenting punishment. Now, such punishments may be very costly for the punishers to carry out. For example, in a repeated quantity-setting oligopoly, the minmax strategies require player i's opponents to produce so much output that price falls below player i's average cost, which may be below their own costs as well. Since minmax punishments can be costly, the question arises if player i ought to be deterred from a profitable one-shot deviation by the fear that his opponents will respond with the unrelenting punishment specified above. More formally, the point is that the strategies we used to prove the Nash folk theorems are not subgame perfect. This raises the question of whether the conclusion of the folk theorem applies to the payoffs of perfect equilibrium.

The answer to this question is yes, as shown by the perfect folk theorem. Friedman (1971) proved a weaker result, sometimes called a "Nash-threats" folk theorem.

Theorem 5.2 (Friedman 1971)   Let α^* be a static equilibrium (an equilibrium of the stage game) with payoffs e. Then for any $v \in V$ with $v_i > e_i$ for all players i, there is a $\underline{\delta}$ such that for all $\delta > \underline{\delta}$ there is a subgame-perfect equilibrium of $G(\delta)$ with payoffs v.

Proof   Assume that there is an $\hat{a}$ with $g(\hat{a}) = v$, and consider the following strategy profile: In period 0 each player i plays $\hat{a}_i$. Each player i continues to play $\hat{a}_i$ so long as the realized actions were $\hat{a}$ in all previous periods. If at least one player did not play according to $\hat{a}$, then each player i plays α_i^* for the rest of the game.

This strategy profile is a Nash equilibrium for δ large enough that

$$(1 - \delta) \max_a g_i(a) + \delta e_i < v_i. \tag{5.7}$$

This inequality is satisfied for a range of δ less than 1 because it holds strictly at the $\delta = 1$ limit. To check that the profile is subgame perfect, note that in every subgame off the equilibrium path the strategies are to play α^* forever, which is a Nash equilibrium for any static equilibrium α^*.

If there is no $\hat{a}$ with $g(\hat{a}) = v$, we use public randomizations as in the previous theorem.   ∎

Friedman's result shows that patient, identical Cournot duopolists can "implicitly collude" by each producing half of the monopoly output, with any deviation triggering a switch to the Cournot outcome forever after. This equilibrium is "collusive" in obtaining the monopoly price; the collusion is "implicit" in that it can be enforced without the use of binding contracts. Instead, each firm is deterred from breaking the agreement by the (credible) fear of provoking Cournot competition.

There is ample evidence that firms in some industries have understood the role of repeated play in allowing such collusive outcomes (although other models than the repeated games considered here can be used to capture the effects of repeated play). Some of the agents involved have even recognized the key role of the interval between periods in determining whether the discount factor is large enough to allow collusion to be an equilibrium, and have suggested that the industry take steps to ensure that any defectors from the collusive outcome will be detected quickly. Scherer (1980) quotes the striking example of the American Hardwood Manufacturer's Association, which proclaimed: "Knowledge regarding prices actually made is all that is necessary to keep prices at reasonably stable and normal levels.... By keeping all members fully and quickly informed of what others have done, the work of the Plan results in a certain uniformity of trade practices.... Cooperative competition, not cutthroat competition."

The conclusion of Friedman's theorem is weaker than that of the folk theorem, except in games with a static equilibrium that holds all the players to their minmax values. (This is a fairly special condition, but it does hold in the prisoner's dilemma and in Bertrand competition with perfect substitutes and constant returns to scale.) Thus, Friedman's theorem leaves open the question of whether the requirement of perfect equilibrium restricts the limit set of equilibrium payoffs. The "perfect folk theorems" of Aumann and Shapley (1976), Rubinstein (1979a), and Fudenberg and Maskin (1986a) show that this is not the case: For any feasible, individually rational payoff vector, there is a range of discount factors for which that payoff vector can be obtained in a subgame-perfect equilibrium.

As a first step toward understanding the strategies used in the folk theorems, note that to hold player i's payoff very near to his minmax value in an equilibrium, his opponents' must specify that if player i deviates from the equilibrium path they will "punish" him by playing the minmax strategies m_{-i}^i (or a profile very close to it) for at least one period. (Otherwise, if player i were to play a static best response to his opponents' strategies in every period, his payoff in every period would be bounded away from his minmax value, and so his overall payoff would exceed his minmax value as well.) Thus, the perfect folk theorem requires that there be perfect-equilibrium strategies in which player i's opponents play m_{-i}^i. It is easy to induce player i's opponents to play m_{-i}^i for a finite number of periods when intertemporal preferences are represented by the time-average criterion, as then, even if punishment reduces the punishers' per-period payoff, the overall cost of the punishment is 0. This is the intuition for the following theorem.

Theorem 5.3 (Aumann and Shapley 1976) If players evaluate sequences of stage-game utilities by the time-average criterion, then for any $v \in V$ with $v_i > \underline{v}_i$ for all players i, there is a subgame-perfect equilibrium with payoffs v.

Proof Consider the following strategies: "Begin in the 'cooperative phase.' In this phase, play a public randomization ρ with payoff v, and remain in this phase as long as there are no deviations. If player i deviates, play the minmax strategy $m^i = (m_i^i, m_{-i}^i)$ for N periods, where N is chosen so

$$\max_a g_i(a) + N\underline{v}_i < \min_a g_i(a) + Nv_i,$$

for all players i. After the N periods have elapsed, return to the cooperative phase, regardless of whether there were any deviations from m^i."

Recall that the one-stage deviation principle does not apply in infinite-horizon games with time averaging. Hence, to verify that these strategies are a perfect equilibrium, we must explicitly verify that there is no strategy that improves a player's payoff in any subgame. The condition on N ensures that any gains from deviation in the cooperative phase are removed at the punishment phase, so no sequence of a finite or infinite number of deviations can increase player i's average payoff above v_i. Moreover, even though minmaxing a deviator is costly in terms of per-period payoff, any finite number of such losses are costless with the time-average criterion. Thus, player j's average payoff in a subgame where player i is being punished is v_j, and no player j can gain by deviating in any subgame. Therefore the strategies are subgame perfect. ∎

The strategies in this last proof are not subgame perfect under the overtaking criterion studied by Rubinstein (1979a), since with that criterion players do care about the finite number of periods where they may incur a loss by minmaxing an opponent. To prove the folk theorem for this case, Rubinstein used strategies in which the punishment lengths grow exponentially: The first deviator is punished for N periods, a player who deviates from minmaxing the first deviator is punished for N^2 periods, a player who deviates from punishing a player who deviated from punishing the first deviator is punished for N^3 periods, and so on. Here, N is chosen long enough so that it is better for any player to minmax any opponent for one period and then have play revert to payoffs v per period, than to be minmaxed for N periods, and then have per-period payoffs revert to v.

When the players discount their future payoffs, this kind of scheme will not work: If player i's payoff when minmaxing player j, $g_i(m^j)$, is strictly less than his own minmax value, $\underline{v}_i$, then for any $\delta < 1$ there is a k where punishing player j for N^k periods is not individually rational: The best possible payoff from playing m^j for N^k periods is

$$(1 - \delta^{N^k})g_i(m^j) + \delta^{N^k} \max_a g_i(a),$$

which converges to $g_i(m^j) < \underline{v}_i$ as k goes to infinity.

Thus, to obtain the folk theorem in the limit of discount factors tending to 1, Fudenberg and Maskin (1986a) consider a different type of strategy—one that induces player i's opponents to minmax him not by threatening them with "punishments" if they don't minmax but rather by offering them "rewards" if they do. Abreu (1986, 1988) makes the same observation in his work on the structure of the equilibrium set for fixed discount factors; we discuss his results below. Now, in designing strategy profiles that provide such rewards for punishing a deviator, one must take care not to reward the original deviator as well, for such a reward could undo the effect of the punishments and make deviations attractive. The need to be able to provide rewards for punishing player i without rewarding player i himself leads to the "full-dimension" condition used in the following theorem.

Theorem 5.4 (Fudenberg and Maskin 1986a)   Assume that the dimension of the set V of feasible payoffs equals the number of players. Then, for any $v \in V$ with $v_i > \underline{v}_i$ for all i, there is a discount factor $\underline{\delta} < 1$ such that for all $\delta \in (\underline{\delta}, 1)$ there is a subgame-perfect equilibrium of $G(\delta)$ with payoffs v.

Remarks

(1) Fudenberg and Maskin give an example of a three-player game with $\dim V = 1$ where the folk theorem fails. Abreu and Dutta (1990) weaken the full-dimension condition to $\dim V = I - 1$; Smith (1990) shows that it suffices that the projection of V^* onto the coordinate space of any two players' payoffs is two-dimensional.

(2) Rubinstein's version of the perfect folk theorem supposes that any deviation from a minmax profile was certain to be detected, which requires

either that the minmax profile be in pure actions or that the players' choices of randomizing probabilities, and not just their realized actions, are observed at the end of each period. As is noted above, the restriction to pure minmax strategies can lead to higher minmax values. Indeed, the pure-strategy minmax values can be above the payoffs in any static equilibrium.

Proof

(i) For simplicity, suppose that there is a pure action profile a with $g(a) = v$. The proof for the general case follows essentially the same lines. Assume first that the minmax profile m^i_{-i} against each player i is in pure strategies, so that deviations from this profile are certain to be detected. Case ii below sketches how to modify the proof for the case of mixed minmax profiles.

Choose a v' in the interior of V and an $\varepsilon > 0$ such that, for each i,

$$\underline{v}_i < v'_i < v_i,$$

and the vector

$$v'(i) = (v'_1 + \varepsilon, \ldots, v'_{i-1} + \varepsilon, v'_i, v'_{i+1} + \varepsilon, \ldots, v'_I + \varepsilon)$$

is in V. (The full-dimension assumption ensures that such $v'(i)$ exist for some ε and v'.)

Again, to avoid the details of public randomizations, assume that for each i there is a pure action profile $a(i)$ with $g(a(i)) = v'(i)$. Let $w^j_i = g_i(m^j)$ denote player i's payoff when minmaxing player j. Choose N such that, for all i,

$$\max_a g_i(a) + N\underline{v}_i < \min_a g_i(a) + Nv'_i. \tag{5.8}$$

This is the punishment length such that, for discount factors close to 1, deviating once and then being minmaxed for N periods is worse than getting the lowest payoff once and then N periods of v'_i.

Now consider the following strategy profile:

Play begins in *phase I*. In phase I, play action profile a, where $g(a) = v$. Play remains in phase I so long as in each period either the realized action is a or the realized action differs from a in two or more components. If a single player j deviates from a, then play moves to phase II$_j$.

Phase II$_j$ Play m^j each period. Continue in phase II$_j$ for N periods so long as in each period either the realized action is m^j or the realized action differs from m^j in two or more components. Switch to phase III$_j$ after N successive periods of phase II$_j$. If during phase II$_j$ a single player i's action differs from m^j_i, begin phase II$_i$. (Note that this construction makes sense only if m^j is a pure action profile; otherwise the "realized action" can't be the same as m^j.)

Phase III$_j$ Play $a(j)$, and continue to do so unless in some period a single player i fails to play $a_i(j)$. If a player i does deviate, begin phase II$_i$.

To show that these strategies are subgame perfect, it suffices to check that in every subgame no player can gain by deviating once and then conforming to the strategies thereafter.

In phase I, player i receives at least v_i from conforming, and he receives at most

$$(1 - \delta) \max_a g_i(a) + \delta(1 - \delta^N)\underline{v}_i + \delta^{N+1}v_i'$$

by deviating once. Since v_i' is less than v_i, the deviation will yield less than v_i for δ sufficiently large. Similarly, if player i conforms in phase III$_j$, $j \neq i$, then player i receives $v_i' + \varepsilon$. His payoff to deviating is at most

$$\max_a g_i(a) + \delta(1 - \delta^N)\underline{v}_i + \delta^{N+1}v_i',$$

which is less than $v_i' + \varepsilon$ when δ is sufficiently large.

In phase III$_i$, player i receives v_i' from conforming and at most

$$(1 - \delta) \max_a g_i(a) + \delta(1 - \delta^N)\underline{v}_i + \delta^{N+1}v_i'$$

from deviating once. Inequality 5.8 ensures that deviation is unprofitable for δ sufficiently close to 1.

If player i conforms in phase II$_j$, $j \neq i$, when there are N' periods of phase II$_j$ remaining (including the current period), her payoff is

$$(1 - \delta^{N'})w_i^j + \delta^{N'}(v_i' + \varepsilon).$$

If she deviates, she is minmaxed for the next N periods; the play in phase III$_j$ will then give her v_i' instead of the $v_i' + \varepsilon$ she would get in phase III$_j$ if she conformed now. Once again, the ε differential once phase III is reached outweighs any short-term gains when δ is close to 1. Finally, if player i conforms in phase II$_i$ (i.e., when she is being punished) then when there are $N' \leq N$ periods of punishment remaining player i's payoff is

$$q_i(N') \equiv (1 - \delta^{N'})\underline{v}_i + \delta^{N'}v_i' < v_i.$$

If she deviates once and then conforms, she receives at most $\underline{v}_i$ in the period in which she deviates (because the opponents are playing m_{-i}^i) and her continuation payoff is then $q_i(N) \leq q_i(N' - 1)$.

(ii) The above construction assumes that player i would be detected if she failed to play m_i^j in phase II$_j$. This need not be the case if m_i^j is a mixed strategy. In order to be induced to use a mixed minmax action, player i must receive the same normalized payoff for each action in the action's support. Since these actions may yield different payoffs in the stage game, inducing player i to mix requires that her continuation payoff be lower after some of the pure actions in the support than after others. Now, in the strategies of part i the exact continuation payoffs for player i in phase III_j, $j \neq i$, were irrelevant (the essential requirement was that player i's payoff be higher in phase III_j than in phase III_i). Thus, as Fudenberg and Maskin (1986a) showed, players can be induced to use mixed actions as punishments by specifying that each player i's continuation payoff in phase III_j, $j \neq i$, vary with the actions player i chose in phase II_j in such a way that each action in the support of m_i^j gives player i the same overall payoff.

As an example of the construction involved, consider a two-player game in which player 1's minmax strategy against player 2 is to randomize $\frac{1}{2}$-$\frac{1}{2}$ between U and D and player 1's payoffs to U and D are 2 and 0, respectively, regardless of how player 2 plays. If player 1 plays U for each of the N periods in phase II_2, he receives an average value of $2(1 - \delta^N)$, as opposed to an average value of 0 from playing D and $(1 - \delta^N)$ of playing his minmax strategy. So instead of switching to a fixed payoff vector

$$v'(2) = (v_1'(2), v_2'(2))$$

at the end of phase II_2, as in the proof of case i, we specify that player 1's payoff be $v_i'(2) - 2(1 - \delta^N)$ if he played U each period, $v_1'(2) - 2\delta(1 - \delta^{N-1})$ if he played D at the beginning of phase II_2 and U thereafter, and so on, with the adjustment term chosen so that player 1's average payoff from the start of phase II_2 is $\delta^N v_1'(2)$ for any sequence of phase-II actions that lie in the support of m_1^2. (If player 1 plays an action not in the support of m_1^2, then play switches to phase II_1 as in the proof of i.)   ■

Discussion The various folk theorems show that standard equilibrium concepts do very little to pin down play by patient players. In applying repeated games, economists typically focus on one of the efficient equilibria, usually a symmetric one. This is due in part to a general belief that players may coordinate on efficient equilibria, and in part to the belief that cooperation is particularly likely in repeated games. It is a troubling fact that at this point there is no accepted theoretical justification for assuming efficiency in this setting. The concept called "renegotiation proofness," discussed in section 5.4, has been used by a number of authors to reduce the set of perfect-equilibrium outcomes; some versions of this concept imply that behavior must be inefficient.

5.1.3   Characterization of the Equilibrium Set (technical)

The folk theorem describes the behavior of the equilibrium set as $\delta \to 1$. It is also of interest to determine the set of subgame-perfect equilibria for a fixed δ. (The folk theorem suggests that there will be many such equilibria for large discount factors.) Following Abreu (1986, 1988), we will consider the construction of strategies such that any deviation by player i is "punished" by play switching to the perfect equilibrium in which that player's payoff is lowest. In order for this construction to be well defined, we must first verify that these worst equilibria indeed exist.

Theorem 5.5

(i) (Fudenberg and Levine 1983)   If the stage game has a finite number of pure actions, there exists a worst subgame-perfect equilibrium $\underline{w}(i)$ for each player i.

(ii) (Abreu 1988)   If each player's action space in the stage game is a compact subset of a finite-dimensional Euclidean space, payoffs are continuous for each player i, and there exists a static pure-strategy equilibrium, there is a worst subgame-perfect equilibrium $\underline{w}(i)$ for each player i.

Remarks   From the stationarity of the set of equilibria, $\underline{w}(i)$ is also the worst equilibrium in any subgame. The question of whether worst equilibria exist without the pure-strategy restriction in games with a continuum of actions is still open.

Proof

(i) As in chapter 4, with a finite number of actions and payoffs that are continuous at infinity, the set of subgame-perfect equilibria is compact in the product topology on strategies, and payoffs to strategies are continuous in this topology as well. Thus, there are worst (and best) equilibria for each player.

(ii) Let $y(i)$ be the infimum of player i's payoff in any pure-strategy subgame-perfect equilibrium, and let $s^{i,k}$ be a sequence of pure-strategy subgame-perfect equilibria such that $\lim_{k \to \infty} g_i(s^{i,k}) = y(i)$. Let $a^{i,k}$ be the equilibrium path corresponding to strategies $s^{i,k}$, so that

$$a^{i,k} = \{a^{i,k}(0), a^{i,k}(1), \ldots, a^{i,k}(t), \ldots\}.$$

Since A is compact, so is the set of sequences of pure actions (this is Tychonoff's theorem[10]), and we let $a^{i,\infty}$ be an accumulation point. Note that player i's payoff to $a^{i,\infty}$ is $y(i)$.

Now fix a player i, and consider the following strategy profile: Begin in phase I_i.

Phase I_i   Play the sequence of actions

$$a^{i,\infty} = \{a^{i,\infty}(0), a^{i,\infty}(1), \ldots\}$$

so long as there are no unilateral deviations from this sequence. If player j unilaterally deviates in period t, then begin phase I_j in period $t + 1$. That is, play $a^{j,\infty}(0)$ in period $t + 1$, $a^{j,\infty}(1)$ in period $t + 2$, and so forth.

If all players follow these strategies, player i's payoff is $y(i)$. To check that the strategies are subgame-perfect, note that if they are not there must exist players i and j, action $\hat{a}_j$, $\varepsilon > 0$, and τ such that

---

10. See, e.g., Munkres 1975.

$$(1 - \delta)g_j(\hat{a}_j, a_{-j}^{i,\infty}(\tau)) + \delta\, y(j) > (1 - \delta) \sum_{t=0}^{\infty} \delta^t g_j(a^{i,\infty}(\tau + t)) + 3\varepsilon.$$

Since payoffs are continuous and $a^{i,k} \to a^{i,\infty}$, for k large enough we would have

$$(1 - \delta)g_j(\hat{a}_j, a_{-j}^{i,k}(\tau)) + \delta\, y(j) > (1 - \delta) \sum_{t=0}^{\infty} \delta^t g_j(a^{i,k}(\tau + t)) + \varepsilon. \tag{5.9}$$

Finally, since $s^{i,k}$ is a subgame-perfect equilibrium, it prescribes *some* subgame-perfect equilibrium if profile $s^{i,k}$ is followed until period τ and then player j plays $\hat{a}_j$ instead of $a_j^{i,k}(\tau)$. Let player j's (normalized) continuation payoff in this equilibrium be $z_j(\tau, \hat{a}_j)$. Since $s^{i,k}$ is subgame perfect,

$$(1 - \delta)g_j(\hat{a}_j, a_{-j}^{i,k}(\tau)) + \delta\, z_j(\tau, \hat{a}_j) \le (1 - \delta) \sum_{t=0}^{\infty} \delta^t g_j(a^{i,k}(\tau + t)),$$

which contradicts inequality 5.9 since $y(j) \le z_j(\tau, \hat{a}_j)$.   ∎

Because the players' actions are observed without error, their equilibrium payoffs are not directly affected by the continuation payoffs following actions to which the equilibrium assigns probability 0. Thus, when one is constructing equilibria the magnitudes of such continuation payoffs matter only in that they determine whether or not players can gain by deviating. For this reason, any strategy profile that is "enforced" by some subgame-perfect punishments can be enforced with the harshest punishments available.[11]

Theorem 5.6 (Abreu 1988)

(i) If the stage game is finite, any distribution over infinite histories that can be generated by some subgame-perfect equilibrium σ can be generated with a strategy profile σ^* that specifies that play switches to the worst equilibrium $\underline{w}(i)$ for player i if player i is the first to play an action to which σ assigns probability 0.

(ii) If the stage game has compact finite-dimensional action spaces and continuous payoffs, then any history $\tilde{h}$ that is generated by a pure-strategy subgame-perfect equilibrium s can be generated by a strategy profile $\hat{s}$ that switches to the worst pure-strategy equilibrium $\underline{w}(i)$ for player i if player i unilaterally deviates from the sequence $\tilde{h}$.

Proof

(i) Fix a perfect equilibrium σ, and construct a new profile σ^* as follows: The profile σ^* agrees with σ (i.e., $\sigma^*(h^t) = \sigma(h^t)$) so long as σ gives the history h^t positive probability. If σ gives positive probability to h^τ for all

---

11. For readers familiar with the literature on agency, this is the same as the observation that optimal contracts can "shoot the agent" if the observed signal could not have occurred unless the agent cheated.

$\tau < t$, and player i is the only player to play an action outside the support of $\sigma(h^t)$ at period t, then play switches to the the worst subgame-perfect equilibrium for player i, which is $\underline{w}(i)$. More formally,

$$\sigma^*(h^{t+1}) = \underline{w}(i)(h^0),$$

$$\sigma^*((h^{t+1}, a^{t+1})) \equiv \underline{w}(i)(a^{t+1}),$$

and so on. (As usual, the strategies will ignore simultaneous deviations by two or more players.) Let us verify that σ^* is subgame perfect. In subgames where player i was the first to deviate from the support of σ, σ^* specifies that all players follow profile $\underline{w}(i)$, which is subgame perfect by definition. In all other subgames h^t the actions prescribed by σ^* are the same as those prescribed by σ, and the continuation payoffs are the same as under σ so long as player i plays an action in the support of $\sigma(h^t)$. It remains to check that player i cannot gain by choosing an action $a_i \notin \text{support}(\sigma_i(h^t))$. If he can, then

$$(1 - \delta)g_i(a_i, \sigma_{-i}(h^t)) + \delta u_i(\underline{w}(i)) > u_i(\sigma \mid h^t). \tag{5.10}$$

However, since σ is subgame perfect,

$$u_i(\sigma \mid h^t) \geq (1 - \delta)g_i(a_i, \sigma_{-i}(h^t)) + \delta u_i^\sigma(a_i \mid h^t), \tag{5.11}$$

where the last term on the right-hand side is player i's continuation payoff from period $t + 1$ on under σ if he deviates from σ at h^t by playing a_i. Combining inequalities 5.10 and 5.11 yields the contradiction

$$u_i(\underline{w}(i)) > u_i^\sigma(a_i \mid h^t).$$

(ii) The proof is analogous.   ∎

Finding the worst possible equilibrium for each player is fairly complicated. However, finding the worst strongly symmetric pure-strategy equilibria of a symmetric game is much simpler, particularly if there are strongly symmetric strategies that generate arbitrarily low payoffs. By "strongly symmetric" we mean that, for all histories h^t and all players i and j,

$$s_i(h^t) = s_j(h^t),$$

so that both players play the same way even after asymmetric histories. For example, in the repeated prisoner's dilemma, the profile where both players use the strategy "tit for tat" (that is, play the action the opponent played the previous period) is not strongly symmetric, since the two players' actions are not identical following the history $h^1 = (C, D)$. Note that the profile is symmetric in the weaker sense that if $h_1^t = \tilde{h}_2^t$ and $h_2^t = \tilde{h}_1^t$, then

$$s_1(h_1^t, h_2^t) = s_2(\tilde{h}_1^t, \tilde{h}_2^t),$$

so that permuting the past history permutes the current actions. We use the terms "strongly symmetric" and "symmetric" to distinguish between these two notions of symmetry.

Abreu (1986) shows that the worst strongly symmetric equilibrium is very easy to characterize in symmetric games where the action spaces are intervals of real numbers, payoffs are continuous and bounded above, (a) the payoff to symmetric pure-strategy profiles $\vec{a}$ (i.e., profiles where each player i plays a) is quasiconcave, and decreases without bound as a tends to infinity, and (b) letting $\vec{a}_{-i}$ denote the profile in which all of player i's opponents choose action a, the maximal payoff to deviating from pure-strategy symmetric profile $\vec{a}$,

$$\max_{a_i'} g_i(a_i', \vec{a}_{-i}),$$

is weakly decreasing in a.

Condition b is natural in a symmetric quantity-setting game, where by producing very large outputs firms drive the price to 0 and thus make the best payoff to deviating very small.

We emphasize that in the definition of a strongly symmetric equilibrium, symmetry is required off the equilibrium path as well as on the path, which rules out many asymmetric punishments that could be used to enforce symmetric equilibrium outcomes.

Theorem 5.7 (Abreu 1986)   Consider a symmetric game satisfying conditions a and b. Let e^* and e_* denote the highest and lowest payoff per player in a pure-strategy strongly symmetric equilibrium.

(i) The payoff e_* can be attained in an equilibrium with strongly symmetric strategies of the following form: "Begin in phase A, where the players play an action a_* that satisfies

$$(1 - \delta)g(\vec{a}_*) + \delta e^* = e_*.\tag{5.12}$$

If there are any deviations, continue in phase A. Otherwise, switch to a perfect equilibrium with payoffs e^* (phase B)."

(ii) The payoff e^* can be attained with strategies that play a constant action a^* as long as there are no deviations, and switch to the worst strongly symmetric equilibrium if there are any deviations. (Other feasible payoffs can be attained in a similar way.)

Proof

(i) Fix some strongly symmetric equilibrium $\hat{s}$ with payoff e_* and first-period action a. Since the continuation payoffs under $\hat{s}$ cannot be more than e^*, the first-period payoffs $g(\vec{a})$ must be at least $(-\delta e^* + e_*)/(1 - \delta)$. Thus, under condition a there is an $a_* \geq a$ with $g(\vec{a}_*) = (-\delta e^* + e_*)$. Let s_* denote the strategies constructed in the statement of the theorem. By definition, the strategies s_* are subgame perfect in phase B. In phase A, condition b and $a_* \geq a$ imply that the short-run gain to deviating is no more than that in the first period of $\hat{s}$. Since the punishment for deviating in phase A is the worst possible punishment, the fact that no player preferred to deviate in the first period of $\hat{s}$ implies that no player prefers to deviate in phase A of s_*.

(ii) We leave the proof of part ii to the reader.   ■

Remark   When this theorem applies, the problem of characterizing the best strongly symmetric equilibrium reduces to finding two numbers, representing the actions in the two phases. (One application is given by Lambson (1987).) If the action space is bounded above by some $\bar{a}$, payoffs cannot be made arbitrarily low, and the punishment phase A may have to last for several periods. In this case it is not obvious precisely which actions should be specified in phase A. The obvious extension of the theorem would have the players using $\bar{a}$ for T periods and then, as before, switching to phase B, where the continuation payoff is e^*. The difficulty is that there may not be a T such that the resulting payoffs in phase A, which are $(1 - \delta^T)g(\bar{a}) + \delta^T e^*$, exactly equal e_*, as required by equation 5.12. However, if we assume that public randomizing devices are available, this integer problem can be eliminated. (Remember that for small discount factors the public-randomization assumption can change the set of equilibria.)

Abreu also shows that in general the highest symmetric pure-strategy equilibrium payoff requires "punishments" with payoffs e_* less than the payoffs in any static equilibrium, unless the threat of switching to the static equilibrium forever—i.e., the strategies introduced by Friedman—supports an efficient outcome.

Finally, Abreu shows that under conditions a and b symmetric pure-strategy equilibria support payoffs on the frontier of the equilibrium set if and only if there is a strongly symmetric equilibrium that gives players their minmax values.

Fudenberg and Maskin (1990b) consider stage games with finitely many actions. They observe that when for each player i there is a perfect equilibrium in which player i's payoff is $\underline{v}_i$, the sets of Nash-equilibrium and perfect-equilibrium payoffs coincide and provide conditions in the stage game for which such perfect equilibria exist for all sufficiently large discount factors. (See exercise 5.8.)

5.2   Finitely Repeated Games[†††]

These games represent the case of a fixed known horizon T. The strategy spaces at each $t = 0, 1, \ldots, T$ are as defined above; the utilities are usually taken to be the time average of the per-period payoffs. (Allowing for a discount factor δ close to 1 will not change the conclusions we present.)

The set of equilibria of a finitely repeated game can be very different from that of the corresponding infinitely repeated game, because the scheme of self-reinforcing rewards and punishments used in the folk theorem can unravel backward from the terminal date. The classic example of this is the repeated prisoner's dilemma. As observed in chapter 4, with a fixed finite horizon "always defect" is the only subgame-perfect-equilibrium outcome. In fact, with a bit more work one can show this is the only *Nash* outcome:

Fix a Nash equilibrium σ^*. Both players must cheat in the last period, T, for any history h^T that has positive probability under σ^*, since doing so increases their period-T payoff and since there are no future periods in which they might be punished. Next, we claim that both players must defect in period $T - 1$ for any history h^{T-1} with positive probability: We have already established that both players will cheat in the last period along the equilibrium path, so in particular if player i conforms to the equilibrium strategy in period $T - 1$ his opponent will defect in the last period, and hence player i has no incentive not to defect in period $T - 1$. An induction argument completes the proof. This conclusion, though not pathological, relies on the fact that the static equilibrium gives the players exactly their minmax values, as the following theorem shows.

Theorem 5.8 (Benoit and Krishna 1987) Assume that for each player i there is a static equilibrium $\alpha^*(i)$ with $g_i(\alpha^*(i)) > \underline{v}_i$. Then the set of Nash-equilibrium payoffs of the T-period game with time averaging converges to the set of feasible, individually rational payoffs as $T \to \infty$.

Proof The key idea of the proof is to first construct a "terminal reward phase" in which each player receives strictly more than his minmax value for many periods. To do this, let the "reward cycle" be the sequence of mixed-action profiles $\alpha^*(1), \alpha^*(2), \ldots, \alpha^*(I)$, and let the R-cycle terminal phase be the sequence of profiles of length $R \cdot I$ where the reward cycle is repeated R times. Any R-cycle terminal phase is clearly a Nash-equilibrium path in any subgame of length $R \cdot I$. And since each $\alpha^*(j)$ gives player i at least his minmax value and $\alpha^*(i)$ by assumption gives him strictly more, each player's average payoff in this phase strictly exceeds his minmax level.

Next, fix a feasible, strictly individually rational payoff v, and set R large enough so that each player i prefers payoff v_i followed by the R-cycle terminal phase to getting the largest possible payoff, $\max_a g_i(a)$, in one period and then being minmaxed for $R \cdot I$ periods. Then choose any $\varepsilon > 0$, and choose T so that there is a deterministic cycle of pure actions $\{a(t)\}$ of length $T - R \cdot I$ whose average payoffs are within ε of payoff v.

Finally, we specify the following strategies: Play according to the deterministic cycle $\{a(t)\}$ in each period so long as past play accords with $\{a(t)\}$ and there are more than $R \cdot I$ periods left. If any player unilaterally deviates from this path when there are more than $R \cdot I$ periods left, then minmax that player for the remainder of the game. If play agrees with $\{a(t)\}$ until there are $R \cdot I$ periods left, then follow the R-cycle terminal phase for the remainder of the game regardless of the observed actions in this phase.

These strategies are a Nash equilibrium for any $T > R \cdot I$. For $T > R \cdot I(\max_a g_i(a) - \underline{v}_i)/\varepsilon$, the average payoffs are within 2ε of v. ∎

Benoit and Krishna (1985) give a related result for subgame-perfect equilibria under a stronger condition. (Friedman (1985) and Fraysse and Moreaux (1985) give independent, less complete analyses of special classes of games.) Recall from chapter 4 that if the stage game has a unique equilibrium, backward induction shows that the unique perfect equilibrium of the finitely repeated game is to play the static equilibrium in every period of every subgame. Where there are several static equilibria, it is possible to punish a player for deviating in the next-to-last period by specifying that if he does not deviate the static equilibrium he prefers will occur in the last period, and that deviations lead to the static equilibrium he likes less.

Theorem 5.9 (Benoit and Krishna 1985) Assume that for each player i there are static equilibria $\alpha^*(i)$ and $\hat{\alpha}(i)$ with $g_i(\alpha^*(i)) > g_i(\hat{\alpha}(i))$, and that the dimension of the feasible set equals the number of players. Then, for every feasible payoff $v \in V$ with v_i strictly exceeding player i's pure-strategy minmax level, and for every sufficiently small $\varepsilon > 0$, there is a T such that for all finite horizons $T' > T$ there is a subgame-perfect equilibrium whose payoffs are within ε of v.

Proof Omitted.

As in the infinite-horizon case, the full-dimension condition is needed to allow strategies that reward one player without rewarding another. The question of whether this result can be strengthened to obtain all payoffs above the mixed-strategy minmax levels is still open.

Although the Benoit-Krishna results extend the Nash-equilibrium and perfect-equilibrium folk theorems to a class of finitely repeated games, in games like the prisoner's dilemma the only Nash equilibrium with finite repetitions is to always be "unfriendly." Few "real-world" long-term relationships correspond to the finite-horizon model; however, there have been many experimental studies of games in which the participants are indeed told that the horizon has been set at a fixed finite point, and there is a unique stage-game equilibrium. In such experimental studies of the prisoner's dilemma, subjects do in fact tend to cooperate in many periods, despite what the theory predicts.

One response is that players are known to derive some extra satisfaction from "cooperating" above and beyond the rewards specified in the experimental design. This explanation does not seem implausible, but it is a bit too convenient, and seemingly much too powerful; once we admit the possibility that payoffs are known to be misspecified, it is hard to see how any restrictions on the predicted outcome of the experiment could be obtained.

A second response is to make a smaller change in the model and allow for there to be a *small probability* that the players get extra satisfaction from cooperating, *so long as* their opponent has cooperated with them in the past. This is the basis of the Kreps-Milgrom-Roberts-Wilson (1982) idea of "reputation effects," which we discuss in detail in chapter 9. The ε-equilibrium approach (Radner 1980; Fudenberg and Levine 1983), discussed in section 4.8, is another way of derailing backward induction in the finitely repeated game, although this requires adopting ε-equilibrium as a descriptive model of bounded rationality rather than merely a convenient technical device.

5.3 Repeated Games with Varying Opponents[†††]

Classic repeated games suppose that the same fixed set of players play one another every period. However, results similar to the folk theorem can be obtained in some cases where not all of the players play one another infinitely often. This section discusses several variants of this idea.

5.3.1 Repeated Games with Long-Run and Short-Run Players

The first variant we will consider supposes that some of the players are long-run players, as in standard repeated games, while the roles corresponding to other "players" are filled by a sequence of short-run players, each of whom plays only once.

Example 5.1

Suppose that a single long-run firm faces a sequence of short-run consumers, each of whom plays only once but is informed of all previous play when choosing his actions. Each period, the consumer moves first, and chooses whether or not to purchase a good from the firm. If the consumer does not purchase, then both players receive a payoff of 0. If the consumer decides to purchase, then the firm must decide whether to produce high or low quality. If it produces high quality, both players have a payoff of 1; if it produces low quality, the firm's payoff is 2 and the consumer's payoff is -1. This game is a simplified version of those considered by Dybvig and Spatt (1980), Klein and Leffler (1981), and Shapiro (1982).[12] Simon (1951)

---

12. Dybvig and Spatt (1980) and Shapiro (1982) consider models where the place of the "short-run player" described above is taken by a continuum of long-lived but "small" consumers. Since they assume that the play of any individual consumer cannot be observed, the consumers always act to maximize their current payoff, and the models are equivalent to the case of a sequence of individual, short-run consumers. (See our treatment of open- and closed-loop equilibrium in chapter 3 for a discussion of the assumption that the play of "small" players cannot be observed.)

and Kreps (1986) use a similar game to analyze the employment relation-ship, and to argue that one reason for the existence of "firms" is precisely to provide a long-run player who can be induced to be trustworthy by the prospect of future rewards and punishments.

The following strategies are a subgame-perfect equilibrium of this game when the firm is sufficiently patient: The firm starts out producing high quality every time a consumer purchases, and continues to do so as long as it has never produced low quality in the past. If ever the firm produces low quality, it produces low quality at every subsequent opportunity. The consumers start out purchasing the good from the firm, and continue to do so so long as the firm has never produced low quality. If ever the firm produces low quality, then no consumer ever purchases again. The con-sumer's strategies are optimal because each consumer cares only about that period's payoff, and thus should buy if and only if that period's quality is expected to be high. The firm does incur a short-run cost by producing high quality, but when the firm is patient this cost is offset by the fear that producing low quality will drive away future consumers. Note that this equilibrium suggests why consumers may prefer to deal with a firm that is expected to remain in business for a while, as opposed to a "fly-by-night" firm for whom long-run considerations are unimportant.

Example 5.2

As a second example, consider repeated play of a sequential-move version of the prisoner's dilemma with a single long-run player facing a sequence of short-run opponents. Each period, the short-run player's decision whether to cooperate or to cheat is observed before the long-run player makes his own decision. As in the previous example, if the long-run player's discount factor is close to 1, there is an equilibrium where players always cooperate. One such equilibrium is: "The short-run players cooperate so long as in every past period the long-run player has played the same action as that period's short-run player; if the long-run player has ever failed to match the short-run players cheat. The long-run player matches the play of that period's opponent so long as he has never failed to match in the past, and cheats otherwise."

The key to the cooperative equilibrium in example 5.2 is that since the short-run players move first, they can be provided with an incentive to cooperate without the use of rewards and punishments in future periods. If instead the moves in the stage game are simultaneous, the short-run players will cheat in every period, and so the only equilibrium outcome is for both sides to always cheat. This suggests that the way to extend the folk theorem to these games is to modify the definitions of the feasible payoffs and the minmax values to incorporate the constraint that short-run players always play short-run best responses.

To state this conjecture formally, label the players so that players $i =$ $1, \ldots, \ell$ are long-run players who maximize the normalized discounted sum of their per-period payoffs as in ordinary repeated games, and let players $j = (\ell + 1), \ldots, I$ represent sequences of short-run players who act in each period to maximize that period's payoff. That is, the stage game has I players, and in the repeated game the individuals playing the parts of players $\ell + 1$ through I change each period. (Alternatively, players $\ell + 1$ to I could be long-run players whose discount factor is 0.) Let

$$\text{B}: \mathcal{A}_1 \times \cdots \times \mathcal{A}_\ell \to \mathcal{A}_{\ell+1} \times \cdots \times \cdots \times \mathcal{A}_I$$

be the correspondence that maps any action profile $(\alpha_1, \ldots, \alpha_\ell)$ for the long-run players to the corresponding Nash-equilibrium actions for the short-run players. That is, for each $\alpha \in \text{graph(B)}$ and $i \geq \ell + 1$, α_i is a best response to α_{-i}.

For each long-run player i, define the minmax value $\underline{v}_i$ to be

$$\min_{\alpha \in \text{graph(B)}} \max_{a_i \in A_i} g_i(a_i, \alpha_{-i}). \tag{5.13}$$

(The minimum is attained because the graph of B is compact, and the payoff functions are continuous in the mixed strategies. Note that this definition reduces to the usual one if all the players are long-run.) Let

$$U = \{v = (v_1, \ldots, v_\ell) \in \mathbb{R}^\ell \,|\, \exists \alpha \in \text{graph(B)} \text{ with } g_i(\alpha) = v_i \text{ for } i = 1, \ldots, \ell)\}$$

and set

$$V = \text{convex hull } (U).$$

This is the modified definition of the set of feasible payoffs.

As we remarked, one might suspect that the folk theorem would extend with these modified definitions of feasibility and the minmax levels. However, as shown by Fudenberg, Kreps, and Maskin (1990) this extension obtains only when each player's choice of a mixed action in the stage game is publicly observable. When players observe only their opponents' realized actions, the set of subgame-perfect equilibria can be strictly smaller. The reason for this, as illustrated in exercise 5.9, is that, in order to induce a short-run player to take a particular action along the equilibrium path, some of the long-run players may need to use mixed actions. When the randomizing probabilities are not observable, inducing this randomization will require that the continuation payoffs make the randomizing long-run players indifferent between the pure actions they assign positive probability, which imposes a cost in terms of the efficiency of the possible equilibrium payoffs.

The limit set of equilibria with unobserved randomizing probabilities is the intersection of the feasible, individually rational payoffs with the constraints $v_i \leq \bar{v}_i$, where $\bar{v}_i$ is defined as

$$\bar{v}_i = \max_{\alpha \in \text{graph}(B)} \min_{a_i \in \text{support}(\alpha_i)} g_i(a_i, \alpha_{-i}). \tag{5.14}$$

For a fixed mixed-action profile α, equation 5.14 computes player i's worst payoff among the actions α_i requires him to play with positive probability. Intuitively, if player i is asked to play α_i along the equilibrium path, he must be willing to use every action in α_i.

Theorem 5.10 (Fudenberg, Kreps, and Maskin 1990; Fudenberg and Levine 1990)   Assume that the dimension of V is equal to ℓ, the number of long-run players. Then for every $v \in V$ with $\underline{v}_i < v_i < \bar{v}_i$ for all $i = 1, \ldots, \ell$, there is a $\underline{\delta}$ such that for all $\delta \in (\underline{\delta}, 1)$ there is a subgame-perfect equilibrium with payoffs v.

Proof   Omitted.

5.3.2   Games with Overlapping Generations of Players

Crémer (1986) considered a repeated game in which overlapping generations of players live for T periods, so that at each date t there is one player of age T who is playing his last round, one player of age $T - 1$ who has two rounds still to play, and so on down to the new player who will play T times. Each period, the T players simultaneously choose whether to work or to shirk, and their choices are revealed at the end of each period; players share equally in the resulting output, which is an increasing function of the number who chose to work.[13] The cost of effort exceeds a $1/T$ share of the increases in output, so shirking is a dominant strategy in the stage game, which has the flavor of a T-player prisoner's dilemma. Payoffs in the repeated game are the average of the per-period utilities.

Suppose that the efficient outcome is for all players to work. This outcome cannot occur in any Nash equilibrium, since the age-T player will always shirk. Nevertheless, there can be equilibria where most of the players work. This will be easiest to see if we further specialize the model. Let $T = 10$. Suppose that if k players work the aggregate output is $2k$, and that the disutility of effort is 1. Then if preferences are linear in output and effort, the payoff to working when k opponents work is $2(k + 1)/10 - 1$, and the payoff to shirking is $2k/10$. The efficient outcome is for all players to work, with resulting utility of 1 per player.

Now consider the following strategy profile: "Age-10 players always shirk. So long as no player has ever shirked when his age is less than 10, all players of age less than 10 work. If a player has ever shirked when his age is less than 10, then all players shirk." If all players conform to this profile, each player receives $18/10 - 1 = 4/5$ in the periods he works and

---

13. It is, however, interesting to note that the "cooperative" equilibrium we derive in the next paragraph remains an equilibrium if we suppose that workers observe only the total number of shirkers but not their identities.

9/5 in the period he is of age 10. Clearly, no player can gain by deviating when he is of age 10. If a player of age 9 deviates, he receives 8/5 the period he deviates, and 0 the next period, which is less than $4/5 + 9/5$; younger players lose even more by deviating. Thus, these strategies are a subgame-perfect equilibrium.

Kandori (1989b) and Smith (1989) have generalized this type of construction and provided conditions for the folk theorem to obtain.

5.3.3 Randomly Matched Opponents

Another variant of the repeated-games model supposes that there are *a* many players, each of whom plays infinitely often but against a different opponent each period. More precisely, fix a two-player stage game, and suppose that there are two populations of players of equal size, N. Each period, every player 1 is matched with a player 2. The probability of being matched to a particular player 2 is $1/N$, and matching in each stage is independent.[14]

In the first analyses of this sort of random-matching model, Rosenthal (1979) and Rosenthal and Landau (1979) assumed that when the players in each pair are matched, their information consists of the actions that the two of them played in the previous period. Thus, if the stage game is the prisoner's dilemma, where C is "cooperate" and D is "defect," there are four possible "histories" a pair of players can have, namely (C, C), (D, C), (C, D), and (D, D), and consequently each player has $2^4 = 16$ pure strategies. (Note that players do not have perfect recall!)

With this information structure, the strategy "cooperate if and only if my opponent cooperated last period," or "tit for tat," is feasible. More generally, the action a player chooses in period t can have a direct effect on his opponent's play in period $t + 1$.

If the player expects to face the same opponent in period $t + 1$ and in period $t + 2$, he may anticipate an additional indirect effect of his period-t action on his opponent's play in periods after $t + 1$. For example, if the opponent's strategy is to cooperate only if the history is (C, C), defecting in period t will not only make the opponent defect in period $t + 1$; it will also make the opponent defect in every period thereafter.

Rosenthal (1979) and Rosenthal and Landau (1979) restrict their attention to "Markovian equilibria," where this indirect effect is not present and where each player believes that his action at date t has no effect on the play of his opponent at all dates from $t + 2$ on.[15] Although this belief is incorrect with a single player of each type, it is correct in a model with a continuum

---

14. This kind of model can be used to explain why, e.g., traders may behave honestly even though it is very unlikely that they will ever meet each other again in the future (Greif 1989; Milgrom, North, and Weingast 1989).
15. The Markovian notion here differs from the one defined in chapter 13.

of each kind of player, so that no player ever meets the same opponent twice.[16]

When is all players using "tit for tat" a Markovian equilibrium of the prisoner's dilemma? Each player must be willing to cooperate if the current opponent cooperated last period, and must be willing to defect if the opponent defected last period. Yet, the next period's opponent will not know the past play of the current opponent, and thus cannot distinguish between a "defect" that occurred to punish the current opponent for a past defection and a "defect" that represents a deviation from the strategy "tit for tat." In particular, with the strategy "tit for tat," any defection today will make the next opponent defect. Thus, both players using "tit for tat" is a Markovian equilibrium only if the discount factor is exactly such that the short-run gain to cheating exactly equals the discounted cost of being punished next period: If the discount factor is smaller, then the threat of punishment will not enforce cooperation; if the discount factor is larger, then a player whose opponent defected last period will not be willing to punish him, as doing so will reduce the punisher's future payoff. With payoff as in figure 4.1, this critical value is $\delta = \frac{1}{2}$. More generally, Rosenthal shows that for all but one value of the discount factor, the unique symmetric Markovian equilibrium of the prisoner's dilemma is for all players to cheat in every period. (Exercise 5.7 asks you to check this.)

Kandori (1989b) observes that cooperation *is* an equilibrium outcome for discount factors near 1 if each player observes the *outcome* in his partner's previous match, i.e., the play of both his partner and the partner's opponent. In this case cooperation can be enforced by the strategies "Cooperate in the first period, and continue to cooperate as long as the outcome in each of my matches has been (C, C), and the outcome in my opponent's last match was (C, C); otherwise defect." With these strategies a player whose partner cheated last period can do no better than carry out the prescribed punishment, as the current partner will defect this period, and so the player will be punished next period regardless of how he plays today. Also, a player who deviates will be punished forever, regardless of his future play. Kandori notes that these strategies have the unappealing feature that a deviation by a single player causes the whole "society" to eventually unravel to the all-defect equilibrium. He proposes that researchers should look for equilibria that are "resilient" in the sense that play will eventually return to cooperation in any subgame (that is, after any finite sequence of deviations). Since the motivation for this kind of stability is the idea that there may be some noise in the model that triggers the "punishment

---

16. To avoid technical complications, Rosenthal (1979) and Rosenthal and Landau (1979) suppose that each population of players is finite, so that there is a nonzero chance that a player 1 will be matched with the same player 2 in two successive periods. Thus, player 1's action today may in fact have some influence on the way his next opponent will play, but with many players in each population this influence is small.

scheme," an alternative methodology would be to make the noise explicit. This would transform the prisoner's dilemma into a game with imperfectly observed actions, a topic we discuss in sections 5.5–5.7. Studying random-matching equilibria in games with noise is an open problem in the literature at this time.

Kandori also suggests another type of equilibrium for games in which players observe only the play in their own past contests. In this "contagion" equilibrium, all players initially cooperate, and if a player ever encounters an opponent who plays D, he plays D from then on. With an infinite population of players, so that with probability 1 no player will ever meet his current opponent again (nor will he even meet anyone who has played his current opponent, etc.), players have no long-run loss from playing D, and these strategies are not an equilibrium. However, with a finite population and random matching, playing D today will eventually lead the entire population to play D. Thus, there is the potential for the contagion strategies to be an equilibrium; whether they are or not depends on how fast the contagion spreads, which in turn depends on the number of players. If there are only two players, the contagion strategies are clearly an equilibrium for discount factors close to 1. For fixed stage-game payoffs, Kandori's contagion strategies fail to be an equilibrium, but not because players are tempted to defect in the cooperative phase. Rather, the problem is that players prefer to continue playing C even after meeting an opponent who plays D in order to slow down the spread of the contagion process. Kandori shows that, for any fixed number of players, the contagion strategies are an equilibrium for discount factors close to 1 provided that the payoffs in the stage game are altered to make the payoff to playing C against an opponent who plays D sufficiently negative. In this case, even a very small probability that the next opponent plays D is sufficient to make D the best response.

Ellison (1991) shows that for any number of players and fixed stage-game payoffs there is in fact an equilibrium where all players cooperate. More-over, this equilibrium is partially resilient, in the sense that if a single player cheats once the resulting steady state is for players to continue to cooperate most of the time. (The equilibrium can be made entirely resilient if public randomizing devices are introduced.) Ellison also constructs an approx-imately efficient equilibrium of the random-matching model with noise.

5.4 Pareto Perfection and Renegotiation-Proofness in Repeated Games[†††]

5.4.1 Introduction

Recently many economists have studied the idea of the "renegotiation" of equilibria and, in particular, the consequences of such renegotiation for play in repeated games. The idea is that if equilibrium arises as the result of negotiations between the players, and players have the opportunity to negotiate anew at the beginning of each period, then equilibria that enforce "good" outcomes by the threat that deviations will trigger a "punishment equilibrium" may be suspect, as a player might deviate and then propose abandoning the punishment equilibrium for another equilibrium in which all players are better off. This sort of equilibrium restriction is called "Pareto perfection" because it extends the idea that players will not play a Pareto-dominated equilibrium to dynamic settings by requiring that in any subgame the equilibrium played must not be Pareto dominated given the constraints on the equilibria at future dates.

The restriction is also called "renegotiation-proofness," because the constraint of Pareto optimality in the subgames can be interpreted as the result of the players' "renegotiating" the original agreement. This latter terminology suggests a parallel with the literature on the renegotiation of contracts, which has also developed a notion of "renegotiation-proofness," but the parallel is inexact: If two players agree on a contract, its terms are legally binding unless both players agree to replace the contract with another one; in contrast, the original "negotiations" on an equilibrium are not binding and serve only to coordinate expectations.

Since the process of selecting a Pareto-optimal outcome and the ideas of Pareto perfection and renegotiation-proofness all take as their starting point the premise that in a static game players will always play an equilibrium on the Pareto frontier of the set of equilibrium payoffs, they are subject to the various critiques of that assumption. In particular, consider the game illustrated in figure 5.3, which we discussed in subsection 1.2.4.

We argued in subsection 1.2.4 that even though equilibrium (U, L) Pareto dominates the others, it is not clear that it is the most reasonable prediction of how the game will be played, even if the players can communicate before the game is played. As Aumann (1990) observes, regardless of his own play, player 2 gains if player 1 plays U, and so regardless of how player 2 intends to play he should tell player 1 that he intends to play L. Thus, it is not clear that the players should expect that their opponent believes that their announcements are sincere.

The concepts developed in this section go further and suppose that even if players have deviated from the play prescribed by past "negotiations," future negotiations will have an efficient outcome. For example, in a twice-

	L	R
U	9,9	0,8
D	8,0	7,7

Figure 5.3

repeated version of the game illustrated in figure 5.3, Pareto perfection requires that players play (U, L) in the first period, and that they play (U, L) in the second period even if one or both of them deviated in the first period—the first-period deviation is treated as a "bygone" that has no effect on subsequent play. This is a strong assumption. It can, however, be rationalized if we suppose that, as in subgame perfection, players treat deviations as accidents that are unlikely to be repeated. (Chapter 11 discusses the idea of forward induction, where deviations are interpreted as strategic signals; this yields quite different conclusions about repeated games.)

Despite our reservations about Pareto efficiency in static games, the concept is sufficiently interesting that we want to discuss its dynamic counterpart. There are currently several competing theories of what this dynamic counterpart should be; the folk theorem obtains under some of them but not others, as we explain below. We begin with the case of finitely repeated games, where it is easier to see what the "right" definitions should be.

5.4.2 Pareto Perfection in Finitely Repeated Games

The best-established formal definition of a renegotiation-proof equilibrium concept is the Bernheim-Peleg-Whinston (1987) notion of a Pareto-perfect equilibrium of a finitely repeated game. Pareto perfection combines the ideas of the Pareto optimality of equilibrium with the logic of subgame perfection, resulting in the recursive definition given below.[17]

For any set C in $\mathbb{R}^I$, let Eff(C) be the set of strongly efficient points in C, i.e., the set of $x \in C$ such that there is no $y \in C$ with $y \geq x$ and $y \neq x$.

Definition 5.1 (Bernheim, Peleg, and Whinston 1987) Fix a stage game g, and let G^T be the associated T-fold repetition. Let P^T be the set of payoffs of pure-strategy subgame-perfect equilibria of G^T. Set $Q^1 = P^1$ and $R^1 = \text{Eff}(P^1)$.

For $T > 1$, let $Q^T \subseteq P^T$ be the set of pure-strategy perfect-equilibrium payoffs that can be enforced with continuation payoffs in R^{T-1} in the second period of the game, and set $R^T = \text{Eff}(Q^T)$.

A perfect equilibrium σ of G^T is Pareto perfect if, for every time t and history h^t, the continuation payoffs under σ are in R^{T-t}.

The restriction to pure-strategy equilibria is commonly used in this literature. However, since some games have mixed-strategy equilibria that Pareto dominate all the pure-strategy equilibria, this restriction is not innocuous. Also, recall from subsection 1.2.4 that "negotiation"-type argu-

---

17. Bernheim, Peleg, and Whinston (1987) give a definition for general multi-stage games with observed actions. We specialize to repeated games for notational convenience but the general definition should be clear. Our formulation of the definition is taken from Benoit and Krishna (1988), who restrict their attention to pure-strategy equilibria.

	b_1	b_2	b_3	b_4
a_1	0,0	2,4	0,0	5.5,0
a_2	4,2	0,0	0,0	0,0
a_3	0,0	0,0	3,3	0,0
a_4	0,5.5	0,0	0,0	5,5

Figure 5.4

ments support Pareto-optimal equilibria only in two-player games. Though Bernheim, Peleg, and Whinston extend their concept of coalition-proof equilibrium to that of perfectly coalition-proof equilibrium, most subsequent work has focused on two-player games.

To see the force of renegotiation constraints, consider the example illustrated in figure 5.4, which was used by Benoit and Krishna (1988) and by Bergin and MacLeod (1989). In this game the set R^1 of Pareto-optimal pure-strategy equilibrium payoffs is $\{(4, 2), (3, 3), (2, 4)\}$. Because there are multiple elements in R^1, in the twice-repeated game G^2 we are free to vary the last-period equilibrium with the first-period play. This allows the payoffs (5, 5) to be enforced in the first period if the players are sufficiently patient: Specify that the continuation will be (3, 3) if there are no deviations, and that a deviator will be punished with a continuation payoff of 2. In particular, if (as Benoit and Krishna assume) the discount factor is exactly 1, then R^2 is the single point (8, 8). But now in G^3 there is no way to enforce cooperation in the first period, as the continuation game G^2 has a unique Pareto-perfect payoff! Thus, Pareto perfection requires that one of the static equilibria occur in the first period of the thrice-repeated game, and $Q^3 = R^3 = \{(12, 10), (11, 11), (10, 12)\}$. Given the variation in the payoffs allowed by Q^3, profile (a_4, b_4) can be enforced in the first period of a four-stage game, and so on. Benoit and Krishna show that the sets R^T alternate, having three elements if T is odd and a single element if T is even. Moreover, as $T \to \infty$, the average payoffs per period, R^T/T, converge to the point (4, 4), even though from Benoit and Krishna 1985 the efficient payoffs (5, 5) can be approximated in a subgame-perfect equilibrium when T is large (see section 5.2). Thus, the Pareto-perfect equilibria need not be Pareto efficient in the set of all perfect equilibria, as the restriction to Pareto-perfect continuations reduces the frequency with which players can be induced to play the efficient pair (a_4, b_4).

Note an interesting way in which the set R^T differs from the set P^T of all perfect equilibria: Even with a very long horizon, the play in the first few periods is very sensitive to the exact period length. Thus, the assumption of a precisely known horizon is even more important here than when the renegotiation constraint is not imposed, for then (under the conditions of

Benoit and Krishna (1985)), with a long horizon, play until the "last few periods" need not depend on the exact length of the game.

Benoit and Krishna prove that for general stage games the set of average payoffs R^T/T converges either to a single point, as in the example, or to a subset of the efficient frontier. More precisely, they prove that these properties hold when the recursive definition of R^T is modified to consider at each stage T only the pure-strategy equilibria with continuations in R^{T-1}. (Recall that there is a sense in which this restriction to pure-strategy equilibria conflicts with the criterion of Pareto optimality, as there are stage games in which all pure-strategy equilibria are Pareto dominated by equilibria in mixed strategies.)

Bergin and MacLeod (1989) offer an alternative to renegotiation-proofness for finitely repeated games that they call *recursive efficiency*. Recursive efficiency is defined recursively like Pareto perfection, with Q'^T as the equilibrium payoffs enforceable with continuations in R'^{T-1}; the difference is that the set R'^T of recursively efficient agreements is allowed to be a proper subset of the efficient points of Q'^T.[18]

In the example above, recursive efficiency allows the set R'^1 to be the singleton $(3, 3)$, so that $(5, 5)$ could not be enforced in the first period of G^2. This, in turn, allows R'^2 to be the set $\{(5, 7), (6, 6), (7, 5)\}$, so that in G^3 the outcome (a_4, b_4) with payoffs $(5, 5)$ can be enforced in period 1 by specifying that if there are no deviations the continuation payoffs are $(6, 6)$ while a deviator receives continuation payoff of 5. This contrasts with Pareto perfection, which requires that the continuation payoffs from period 2 on be $(8, 8)$, and thus precludes "cooperation" in the first period.

If the discount factors are exactly 1, this time shift does not affect the players' payoffs in G^3, but if the discount factor is less than 1 the players prefer to have the high payoffs $(5, 5)$ occur in the first period. For example, if $\delta = \frac{1}{2}$ the discounted Pareto-perfect payoffs in G^3 are $(25/4, 25/4)$ and the recursively efficient payoffs are $(29/4, 29/4)$. (If the discount factor is too small, the strategies described are not perfect.)

Bergin and MacLeod justify their alternative definition as follows. Suppose that the players meet before period 1 and agree to play any fixed static equilibrium in the last period, and that this continuation equilibrium then becomes the "social norm." Moreover, all players believe that, regardless of the negotiations in period 2, the social norm for period 3 will be played unless it is unacceptable *once period 3 is reached*. Then in period 2, the suggestion to play the more efficient equilibrium "(a_4, b_4) today, any deviator gets 2 tomorrow" is not credible, even though it is Pareto perfect, as both players would feel free to deviate and then appeal to the "social

18. Recursive efficiency also replaces efficiency with weak efficiency. (For any set C in $\mathbb{R}^I$, Weff(C), the set of weakly efficient points in C is the set of $x \in C$ such that there is no $y \in C$, with $y > x$; i.e., y dominates x in all its components.)

norm" of $(3, 3)$ in the last period. In other words, recursive efficiency gives some weight to the original agreement, while under Pareto perfection the set of original agreements is not considered when choosing agreements for the continuation game. As Bergin and MacLeod put it, Pareto perfection is "a theory for which history is not important," while under recursive efficiency "the agreement in period 1 acts as a default focal point." From another viewpoint, recursive efficiency supposes "less superrationality" at the renegotiation stage, as players cannot renegotiate to the Pareto-perfect equilibrium of the last two periods.

The spirit of the Bergin-MacLeod interpretation can be used to justify a less restrictive notion of recursive efficiency. Bergin and MacLeod allow the set of agreements at time t to be a subset of the efficient agreements relative to the recursively efficient continuations, but do not allow the subset Q'^t chosen to depend on the prior history h^t. Consider a four-period version of the game illustrated in figure 5.4, with discount factor $\frac{1}{2}$. If the players can agree to use an equilibrium in periods 2 and 3 that plays (a_1, b_2) even though it is not Pareto perfect, they might be able to agree to use the recursively efficient equilibrium in the last three periods if there are no deviations in period 1, and to use the Pareto-efficient equilibrium otherwise, thus enforcing outcome (a_4, b_4) in both periods 1 and 2. (See DeMarzo 1988 and Greenberg 1988 for other discussions of equilibrium refinements as social norms.)

5.4.3 Renegotiation-Proofness in Infinitely Repeated Games

Pareto perfection and recursive efficiency for finite-horizon games are both defined using backward recursion from the terminal date. Defining renegotiation or Pareto perfection for infinite-horizon games has proved to be much more difficult, and there are currently many competing definitions. One of the earliest treatments is by Farrell and Maskin (1989), who define "weak renegotiation-proofness" for infinitely repeated games. This concept extends the "bygones are bygones" flavor of Pareto perfection by requiring that the set of renegotiation-proof equilibria at date t be independent not only of the history h^t but also of calendar time t. Weak renegotiation-proofness begins with the point of view that there is an exogenously chosen set of possible equilibrium payoffs Q that is conceivable at any t and h^t, and that each payoff in Q must require only continuation payoffs corresponding to other equilibria in Q. Formally, let $c(\sigma; h^t)$ be the continuation payoffs implied by σ given history h^t, and let $C(\sigma) = \bigcup_{t, h^t} c(\sigma; h^t)$ be the set of all continuation payoffs for strategy profile σ. Then, if $v \in Q$, there must be a perfect equilibrium σ with payoffs v such that $C(\sigma) \subseteq Q$. The set Q is said to be *weakly renegotiation-proof* (WRP) if no equilibrium payoff in Q is Pareto dominated by the payoffs of another equilibrium in Q.

	C	D
C	2,2	−1,3
D	3,−1	0,0

Figure 5.5

This definition assigns a great deal of weight to the exogenous set of "social norms" Q. This allows, for example, any static equilibrium to be weakly renegotiation-proof as a one-point set. However, in the prisoner's dilemma, the "grim" strategies of initial cooperation followed by the static equilibrium forever if someone deviates are not weakly renegotiation-proof, as the payoffs corresponding to the "cooperative phase" of the strategies Pareto dominate those of the punishment phase. That is, once the payoffs of "always cooperate" are included in the set Q of possible "agreements," the players will always renegotiate from the unending punishment back to the cooperative phase. Moreover, the strategies "perfect tit for tat," defined by "play C in the first period, and subsequently play C if last period's outcome was (C, C) or (D, D); play D if last period's outcome was (D, C) or (C, D)," are not WRP either, as in the period immediately following a unilateral deviation it would be more efficient to ignore the deviation and play (C, C). These strategies are, however, subgame perfect for discount factors near 1 with the usual payoffs, i.e., those given in figure 5.5.

Nevertheless, Farrell and Maskin (1989) and van Damme (1989) have shown that cooperation is a weakly renegotiation-proof outcome in the repeated prisoners' dilemma if the discount factor is sufficiently near to 1, and indeed the folk theorem in its renegotiation-proof version holds for this game. In particular, the strategy profile where both players use the following "penance" strategy is WRP and has efficient payoffs: "Begin in the cooperative phase where both players play C. If a single player i deviates to D, switch to the punishment phase for i. In this phase, player i plays C and the other player plays D. Play remains in this phase until the first time player i plays C, at which point play returns to the cooperative phase."

The first step in verifying that this profile is WRP is to check that it is subgame perfect. In the cooperative phase, any deviation triggers a period of punishment, which is not desirable for discount factors near 1 with payoffs as in figure 5.5. When player 1 is being punished, his payoff if he conforms is $-(1 - \delta) + 2\delta > 0$; if he deviates, he obtains $0 - \delta(1 - \delta) + 2\delta^2$, which is less. And when player 1 is being punished, player 2's payoff is $3(1 - \delta) + 2\delta$, which exceeds the payoff of 2 he receives by deviating once and then conforming. So the strategy profile is subgame perfect. Moreover, none of the three continuation payoff vectors involved is Pareto dominated by the others, so the profile is WRP.

The key to obtaining efficient WRP payoffs in the repeated prisoner's dilemma is using the profile (C, D) to punish player 1, which minmaxes player 1 while rewarding player 2. In other games, there can be a tradeoff between rewarding player 2 and punishing player 1, and this can prevent the full set of efficient individually rational payoffs from being WRP. For example, in a repeated Cournot duopoly with costless production and demand $D(p) = 2 - p$, any payoff vector of the form $(x, 1 - x)$, $x \in (0, 1)$, is feasible and individually rational, but regardless of the discount factor the only efficient WRP payoffs give each firm a payoff of at least $\frac{1}{9}$. (See Farrell and Maskin 1989 for the argument.)

Pearce (1988) and Abreu, Pearce, and Stachetti (1989) develop an alternative definition of renegotiation-proofness, to which they unfortunately give the same name. Unlike Farrell and Maskin, Pearce et al. allow some of the equilibria in $C(\sigma)$ to Pareto dominate others—they do not test for "internal" Pareto consistency; instead, they use an external test: They say that σ is renegotiation-proof unless there is a continuation payoff w in $C(\sigma)$ and another subgame-perfect equilibrium σ' such that all the continuation payoffs in $C(\sigma')$ Pareto dominate w. The idea is that the agents cannot renegotiate away from w to an alternative equilibrium that would require payoffs below w in some subgame, for fear that in that subgame the players would renegotiate back to the equilibrium with payoffs w.[19] Unlike WRP, this definition typically rules out infinite repetition of a static equilibrium. For example, in the prisoner's dilemma the infinite repetition of (D, D) is ruled out by the profile where both players play perfect "tit for tat."

Moreover, there can be nontrivial symmetric equilibria σ (equilibria where the continuation payoffs depend nontrivially on the history) that are renegotiation-proof in this sense, so that at some histories h^t all players would gain by "agreeing" to play the strategies $\sigma(\tilde{h}^t)$ corresponding to a different history. For example, the profile where both players play perfect tit for tat can be shown to be renegotiation-proof, even though it is not WRP. Pearce (1988) shows that with this definition of renegotiation-proofness, the folk theorem holds for general games. Abreu, Pearce, and Stachetti (1989) obtain an exact characterization of the symmetric renegotiation-proof equilibria in a class of games that generalizes Cournot competition.

The Farrell-Maskin definition of WRP tests only for "internal Pareto consistency," and the Abreu-Pearce-Stachetti definition tests only for external consistency; both definitions may seem weaker in some ways than Pareto perfection for finitely repeated games. One alternative would be to take a payoff that is Pareto efficient in the set of all payoffs of WRP theories

---

19. Pearce's definition also covers the games with imperfectly observed actions discussed in the next section. If there is a positive probability of any (finite) sequence of observations even if no player cheats, then the "subgames" (observations) that required payoffs below w have positive probability, and this definition seems better founded.

(assuming that the set of all WRP payoffs is closed). Farrell and Maskin propose several alternative definitions. A set of payoffs Q is "strongly renegotiation-proof" if it is weakly renegotiation-proof, and there is no other WRP set with a payoff that strictly Pareto dominates any of the payoffs of Q. The idea is roughly that at any time the players are able to renegotiate to a different WRP set Q' and an initial equilibrium from that theory, so that all the payoffs in Q must be immune to this sort of renegotiation. Unfortunately, such strongly renegotiation-proof payoffs need not exist, as was also observed by Bernheim and Ray (1990). Farrell and Maskin and Bernheim and Ray go on to develop more complicated solution concepts that relax strong renegotiation-proofness enough to guarantee existence.

5.5 Repeated Games with Imperfect Public Information[††]

In the repeated games considered in the last section, each player observed the actions of the others at the end of each period. In many situations of economic interest this assumption is not satisfied, because the information that players receive is only an imperfect signal of the stage-game strategies of their opponents. Although there are many ways in which the assumption of observable actions can be relaxed, economists have focused on games of *public information*: At the end of each period, all players observe a "public outcome," which is correlated with the vector of stage-game actions, and each player's realized payoff depends only on his own action and the public outcome. Thus, the actions of a player's opponents influence his payoff only through their influence on the distribution of outcomes. Games with observable actions are the special case where the public outcome consists of the realized actions themselves.

There are many examples of games in which the public outcome provides only imperfect information. Green and Porter (1984) published the first formal study of these games in the economics literature. Their model, which was intended to explain the occurrence of "price wars," was motivated in part by the work of Stigler (1964). In Stigler's model, each firm observes its own sales but not the prices or quantities of its opponents. The aggregate level of consumer demand is stochastic. Thus, a fall in a firm's sales might be due either to a fall in demand or to an unobserved price cut by an opponent. Since each firm's only information about its opponents' actions is its own level of realized sales, no firm knows what its opponents have observed, and there is no public information about the actions played.[20] In contrast, the Green-Porter model does have public information, which

---

20. Lehrer (1989) and Fudenberg and Levine (1990) study repeated games with imperfect private information.

makes it much easier to analyze. In that model, each firm's payoff depends on its own output and on the publicly observed market price. Firms do not observe one another's outputs, and the market price depends on an unobserved shock to demand as well as on aggregate output. Hence, an unexpectedly low market price could be due either to unexpectedly high output by an opponent or to unexpectedly low demand.

Another example of repeated games with imperfect public information is the partnership models considered by Radner (1986) and others. In these models, each player's payoffs depend on his own effort and on the publicly observed output, each player does not observe his partner's effort, and output is stochastic. Yet another example is a "noisy" prisoner's dilemma where players sometimes inadvertently choose the "wrong" action, so that the observed actions are only an imperfect signal of the intended ones. (Equivalently, each player might sometimes misperceive his opponent's action, with the payoffs a function of the perceived actions and not the intended ones.)

In the standard terminology, the above games are all examples of "repeated moral hazard." The class of games with imperfect public information can be extended to include games of "repeated adverse selection," where player i's stage-game actions a_i are maps from some private information (i.e. "types") to a space of physical actions or announcements, and all that is observed is the realized action. (The function from types to actions is not observed.) An example is Green's (1987) model of repeated insurance, in which the players' endowments are random and independent over time and between players and the stage-game strategies are maps from endowments to "announced" endowment levels. Here the reported endowments are observed, but not the maps from true endowments to reports.

5.5.1 The Model

In the *stage game*, each player $i = 1, \ldots, I$ simultaneously chooses a strategy a_i from a finite set A_i. Each action profile $a \in A = \times_i A_i$ induces a probability distribution over the publicly observed outcomes y, which lie in a finite set Y. Let $\pi_y(a)$ denote the probability of outcome y under a, and let $\pi_{.}(a)$ denote the probability distribution, which we will sometimes view as a row vector. Player i's realized payoff, $r_i(a_i, y)$, is independent of the actions of other players. (Otherwise, player i's payoff could give him *private* information about his opponents' play.) Player i's expected payoff under strategy profile a is

$$g_i(a) = \sum_y \pi_y(a) r_i(a_i, y).$$

The payoffs and distributions over outcomes corresponding to mixed strategies α are defined in the obvious way.

In the repeated game, the *public information* at the beginning of period t is

$$h^t = (y^0, y^1, \ldots, y^{t-1}).$$

Player i also has private information at time t—namely, his own past choices of actions; denote this by z_i^t. A *strategy for player i* is a sequence of maps from player i's time-t information to probability distributions over A_i; $\sigma_i^t(h^t, z_i^t)$ denotes the probability distribution chosen when player i's information is (h^t, z_i^t).

Here are some illustrations of the model:

- In a repeated game with observable actions, the set Y of outcomes is isomorphic to the set A of action profiles: $\pi_y(a) = 1$ if y is equivalent to a, and $\pi_y(a) = 0$ otherwise.
- In the Green-Porter model, $a_i \in [0, \bar{Q}]$ is firm i's output, and the outcome y is the market price. Green and Porter make the additional assumptions that the probability distribution over outcomes depends only on the sum of the firms' outputs and that every price has positive probability under every action profile.
- In the repeated partnership model, a_i is player i's effort level and y is the realized output. In the model of Radner (1986) and Radner et al. (1986), A_i is the set {work, shirk}. Closely related is the repeated principal-agent model of Radner (1981, 1985), where the principal's action is an observed monetary transfer and the agent's effort level is not observed. Here the outcome is the pair (output, transfer).
- In Green's (1987) model of repeated insurance, each period t each player i learns his current endowment θ_i^t, with the θ_i^t distributed i.i.d. according to a known distribution $P_i(\cdot)$. Here a_i is a map from the set Θ_i of all possible types to reports $\hat{\theta}_i \in \Theta_i$. (See chapter 7 for an introduction to static mechanism design.) The public outcome y^t is then the vector $\hat{\theta}^t$ of reports, which reveals neither the players' actual types nor the strategy that they used. (There are only $\prod_{i=1}^I (\#\Theta_i)$ outcomes, but there are $\prod_{i=1}^I (\#\Theta_i)^{\#\Theta_i}$ strategy profiles.) In this case the private information of player i must be extended to include the past values of his types in addition to his past actions. We will not pursue this extension here; see Fudenberg, Levine, and Maskin 1990 for details.
- In a "noisy prisoner's dilemma," the set of outcomes Y is isomorphic to the action space A, but $\pi_y(a) > 0$ even if y does not correspond to a. For example, if both players played $a_i = C$ the distribution on outcomes might be

$$\pi_{(C,C)}(C, C) = (1 - \varepsilon)^2,$$

$$\pi_{(C,D)}(C, C) = \pi_{(D,C)}(C, C) = \varepsilon(1 - \varepsilon),$$

and

$$\pi_{(D, D)}(C, C) = \varepsilon^2$$

for some $0 < \varepsilon < \frac{1}{2}$. This describes a situation where each player has probability ε of making a "mistake," and mistakes are independent. The key assumption here is that the intended actions are not observed, only the realized ones.

5.5.2 Trigger-Price Strategies

In the analysis of their oligopoly model, Green and Porter (1984) focus on equilibria in "trigger-price strategies," which generalize the trigger-strategy equilibria introduced by Friedman (1971). Suppose that the set of outcomes Y are interpreted as prices, so that $Y \subseteq \mathbb{R}$, and each firm's output a_i must lie in the interval $[0, \overline{Q}]$. Payoff functions are assumed to be symmetric and attention is restricted to equilibria where all players choose the same actions in every period—that is, $\sigma_i(h^t) = \sigma_j(h^t)$ for all t and h^t. (Thus, the equilibria are "strongly symmetric" in the sense of subsection 5.1.3.) Trigger-price-strategy profiles are indexed by three parameters, $\hat{a}$, $\hat{y}$, and $\hat{T}$. In these profiles, play can be in one of two possible "phases." In the "cooperative phase," all firms produce the same output, $\hat{a}$. Play remains in the cooperative phase as long as each period's realized price y^t is at least the "trigger price" $\hat{y}$. If $y^t < \hat{y}$, then play switches to a "punishment phase" for $\hat{T}$ periods. In this phase, the players play a static Nash equilibrium a^* in each period, regardless of the realized outcomes; after the $\hat{T}$ periods end, play returns to the cooperative phase.

If we simply take $\hat{a} = a^*$, the strategies prescribe that the static equilibrium a^* be played every period, which is clearly an equilibrium, so trigger-price equilibria exist. More generally, we can characterize the trigger-price equilibria as follows: For fixed $\hat{y}$ and $\hat{a}$, let

$$\lambda(\hat{a}) = \text{Prob}(y^t \geq \hat{y} | \hat{a})$$

be the probability that the outcome is at least the trigger level when players use profile $\hat{a}$. For convenience, normalize the payoff of the static equilibrium a^* to be 0. Then the (normalized) payoff if players conform to the strategies is

$$\hat{v} = (1 - \delta)g(\hat{a}) + \delta \lambda(\hat{a})\hat{v} + \delta(1 - \lambda(\hat{a}))\delta^{\hat{T}}\hat{v}, \tag{5.15}$$

so that

$$\hat{v} = \frac{(1 - \delta)g(\hat{a})}{1 - \delta \lambda(\hat{a}) - \delta^{\hat{T}+1}(1 - \lambda(\hat{a}))}. \tag{5.16}$$

Note that $\hat{v} = g(\hat{a})$ if $\lambda(\hat{a}) = 1$, so that the probability of punishment is 0 so long as all players conform, or if $\hat{T} = 0$, so that "punishments" have length 0. The latter case is possible only if $\hat{a}$ is a static equilibrium, so that no punishment is needed to provide incentives. Even if $\hat{a}$ is not a static equilibrium, it might be that $\lambda(\hat{a}) = 1$, so that there is no punishment unless someone deviates; this is possible, for example, if the actions are perfectly observed. However, under the Green-Porter "full support" assumption that $\pi_y(a) > 0$ for all $y \in Y$ and all $a \in A$, the only trigger-price strategies where punishment never occurs so long as no player deviates also have the property that punishment never occurs after any sequence of outcomes. Since such strategies give players no incentive to look beyond their short-run interest, the only trigger-price equilibria where punishment never occurs are those in which there is repeated play of the static equilibria. Thus, $\lambda(\hat{a})$ will be less than 1 in equilibria that improve on the static equilibrium payoffs, and so there is a cost to imposing strong punishments for deviation. In particular, for fixed $\hat{y}$ and $\hat{a}$, the equilibrium payoffs decrease in the punishment length.

However, very long and even infinite punishments may be optimal, as by increasing the punishment length it may be possible to decrease the trigger price or increase the payoffs in the cooperative phase. The optimal trigger-price equilibria will maximize $\hat{v}$ given by equation 5.16 subject to the incentive constraint that no player gain by deviating in the cooperative phase, which is displayed in equation 5.17:

$$(1 - \delta)g(a_i, \hat{a}_{-i}) + \delta\,\lambda(a_i, \hat{a}_{-i})\hat{v} + \delta(1 - \lambda(a_i, \hat{a}_{-i}))\delta^{\hat{T}}\hat{v}$$

$$\leq (1 - \delta)g(\hat{a}) + \delta\,\lambda(\hat{a})\hat{v} + \delta(1 - \lambda(\hat{a}))\delta^{\hat{T}}\hat{v} \quad \text{for all } a_i. \tag{5.17}$$

(No player can gain by deviating in the punishment phase, since play there is a fixed number of repetitions of a static equilibrium.)

Grouping terms together and substituting for $\hat{v}$ from equation 5.16, we get

$$(1 - \delta)[g(a_i, \hat{a}_{-i}) - g(\hat{a})]$$

$$\leq \frac{\delta[1 - \delta^{\hat{T}}][\lambda(\hat{a}) - \lambda(a_i, \hat{a}_{-i})](1 - \delta)g(\hat{a})}{1 - \delta\,\lambda(\hat{a}) - \delta^{\hat{T}+1}(1 - \lambda(\hat{a}))} \tag{5.18}$$

for all a_i.

The optimal trigger-price equilibrium (from the viewpoint of the firms) is given by the $\hat{a}$, $\hat{T}$, and $\hat{y}$ that maximize equation 5.16 subject to equation 5.18. Porter (1983b) characterizes the optimal trigger-price equilibria with a continuum of output levels and prices, and provides conditions for infinite punishments to be optimal. With a continuum of actions, the best equilibrium is better than the static one, because if the output in the cooperative phase is just a small ε below the static equilibrium levels, payoffs in the cooperative phase are greater than in the static equilibrium, while the incentive to deviate—the left-hand side of equation 5.18—is 0 to first order in ε, so that preventing deviations requires only a probability of punishment that is 0 to first order as well.

In the trigger-price equilibria, there is probability 1 that play eventually enters the punishment phase. This is loosely consistent with the idea of "price wars," but note that in equilibrium all players correctly forecast that their opponents will never deviate. Thus, the "price war" is *not* triggered by the inference that some firm chose high output in the previous period. Rather, all players correctly presume that their opponents chose the "cooperative" output last period, and that price was low because of a demand shock, but the "punishment" occurs anyway as a self-enforcing reaction to a low level of realized demand. (The solution concepts of section 5.4 were introduced in response to the concern that such punishments might not be carried out. Note that if the punishment did not occur when demand was low, players could not trust each other in the cooperative phases.)

The study of trigger-price equilibria leaves open the question of whether there are other equilibria with higher payoffs. By analogy with games with observable actions, one suspects that there may be "punishment equilibria" with payoffs lower than those in the static equilibria, and that one might in some cases be able to do better by using stronger punishments. However, this analogy is inconclusive because the punishments may be carried out even if there are no deviations. This question is one of the motivations of the Abreu-Pearce-Stachetti papers we discuss below.

5.5.3   Public Strategies and Public Equilibria

Though all the players know the public history h^t at date t, each player i also knows z_i^t, the actions he has chosen in the past. We will restrict our attention to equilibria in "public strategies," where players ignore their private information in choosing their actions.

Definition 5.2   Strategy σ_i is a *public strategy* if $\sigma_i^t(h^t, z_i^t) = \sigma_i^t(h^t, \tilde{z}_i^t)$ for all periods t, public histories h^t, and private histories z_i^t and $\tilde{z}_i^t$.

Although not all pure strategies are public strategies, it is easy to see that any payoff to a pure-strategy equilibrium is a payoff of an equilibrium in public strategies. That is, given a pure-strategy equilibrium where players' strategies may depend on their private information, we can find an equivalent equilibrium where the players' strategies depend only on their public information. The idea is that, in a pure-strategy equilibrium, each player perfectly forecasts how each opponent will play in each period— player 1 plays, say, a_1^0 in the first period, and is supposed to play $\sigma_1^1(a_1^0, y^0)$ in the second period—but since player 1's first-period play was deterministic, the conditioning of his second-period play on his first action is redundant—we could replace σ_1^1 by the public strategy $\hat{\sigma}_1^1(y^0) = \sigma_1(a_1^0, y^0)$.

When all players use public strategies, they agree about the subsequent probability distribution of actions and outcomes given any public history h^t. Thus, we can define the continuation payoffs conditional on a public history, and ask whether a profile of public strategies induces a Nash equilibrium from date t on.

Definition 5.3 A profile $\sigma_i = \{\sigma_1, \ldots, \sigma_I\}$ of repeated-game strategies is a perfect public equilibrium if

(i) each σ_i is a public strategy, and

(ii) for each date t and history h^t the strategies yield a Nash equilibrium from that date on.

Note that subgame perfection would not be restrictive in these games, since the only proper subgame is the game starting from date 0: At subsequent dates, the players need not know each other's past moves, and thus the continuation games do not emanate from a single node. However, when players use public strategies, their private information about their own past actions is irrelevant, and so perfect public equilibrium is an obvious extension of the idea of subgame perfection.

A key fact about perfect public equilibria (PPE) is that the payoffs to such equilibria are stationary—that is, the set of possible continuation payoffs of PPE starting in period t with an arbitrary public and private history is same as the set of PPE payoffs starting in period 0. (Exercise: Check this formally.) However, the sets of Nash and sequential equilibria are not, in general, stationary. (Another way of saying this is that the game lacks a "recursive structure.") Loosely speaking, the point is the following: If players 1 and 2 play a mixed strategy in the first period and their actions in the second period depend on their realized first-period action, then the actions to be played in the second period are not common knowledge. Since (in any Nash equilibrium) the first-period strategies are necessarily common knowledge, the strategic possibilities in the first and second periods are different. Exercise 5.10 develops this point further, with an example of a game which has an equilibrium that holds a player to a payoff *below* his minmax level.[21]

5.5.4 Dynamic Programming and Self-Generation

A useful tool for the analysis of perfect public equilibria is the concept of *self-generation*, introduced in Abreu, Pearce, and Stachetti 1986 and developed further in Abreu, Pearce, and Stachetti 1990. Self-generation is a sufficient condition for a set of payoffs to be supportable by perfect public equilibria. It is the multi-player generalization of the principle of optimality of discounted dynamic programming, which gives a sufficient condition for

21. The nonstationarity arises from the possibility that the players may come to have imperfectly correlated forecasts of one another's play. The same sort of imperfectly correlated forecasts arise if players observe private, correlated signals at the start of period, as in the "extensive-form correlated equilibrium" discussed in chapter 8. The set of extensive-form correlated equilibria is stationary, because the imperfect correlation that arises in period 2 from observing the public outcome in period 1 can be reproduced in period 1 with the appropriate distribution over private signals.

a vector of payoffs, one for each state, to be the maximal present values obtainable when commencing play in the corresponding state.

The key difference between self-generation in repeated games and dynamic programming is that in the former the states and the state transition function are exogenous. In repeated games, the physical environment is memoryless—the past has no physical influence on the present and the future. However, each player's strategy can depend on the history—for example, player 1's output today may depend on last period's price, and then the output that player 2 wishes to choose today might depend on last period's price as well. Thus, the control problem faced by each individual player can depend on the history, even though the physical environment does not.

Let us look at the Abreu-Pearce-Stachetti characterization of equilibrium. Recall that Y is the space of publicly observable outcomes, and let w be a function from Y into $\mathbb{R}^I$. The function w is interpreted as being the players' (normalized) continuation payoffs as a function of the realized outcome, but at this point no restrictions are made on the range of W. (Abreu, Pearce, and Stachetti use a model with a continuum of publicly observed outcomes y; we assume a finite number of outcomes for simplicity.)

Definition 5.4 The pair (α, v) is *enforceable* with respect to δ and $W \subseteq \mathbb{R}^I$ if there exists a function $w: Y \to W$ such that, for each player i,

(i) $v_i = (1 - \delta)g_i(\alpha) + \delta \sum_y \pi_y(\alpha)w_i(y)$

and

(ii) α_i solves $\max_{\alpha_i'} \left((1 - \delta)g_i(\alpha_i', \alpha_{-i}) + \sum_y \pi_y(\alpha_i', \alpha_{-i})w_i(y) \right)$.

Condition ii says that playing α_i is an optimal choice if the continuation payoffs are given by $w(\cdot)$; condition i says that when all players play α, the resulting normalized payoffs are v. Clearly, in any period t of any PPE, the actions $\sigma(h^t)$ are enforced by the equilibrium continuation payoffs; otherwise, some player could gain by a one-period deviation.

If, for some v, (α, v) is enforceable with respect to δ and W, we say that α is *enforceable* on W. If, for some α, (α, v) is enforceable with respect to δ and W, we say that v is *generated* by (δ, W). The set of all payoffs v generated by (δ, W) is denoted $B(\delta, W)$.

Let $E(\delta)$ denote the set of all PPE payoffs for a given discount factor. It should be clear that $E(\delta) = B(\delta, E(\delta))$. Given any $v \in B(\delta, E(\delta))$, it is easy to construct a PPE with payoffs v: Choose an α and a w with range in $E(\delta)$ such that w enforces (α, v), and specify that players use α in the first period and a PPE with payoffs $w(y)$ if outcome y occurs. Hence, $B(\delta, E(\delta)) \subseteq E(\delta)$. Conversely, if $v \in E(\delta)$, then no player wishes to deviate from the first-period action profile, and the continuation payoffs must (from the perfectness requirement) be in $E(\delta)$. Hence, $E(\delta) \subseteq B(\delta, E(\delta))$.

Definition 5.5 W is self-generating if $W \subseteq B(\delta, W)$.

In words, W is self-generating if the set of payoffs that can be enforced with continuation payoffs in W includes all of W. A trivial example of a self-generating set is the payoffs of a static equilibrium; static equilibria are the only one-point self-generating sets. At the other extreme, the set $E(\delta)$ of all PPE payoffs is self-generating.

Theorem 5.10[22] (Abreu, Pearce, and Stachetti 1986, 1990)   If W is self-generating, then $W \subseteq E(\delta)$: All payoffs in W are PPE payoffs.

Proof   Fix a $v \in W$. We will exhibit strategies for the repeated game that yield payoff v, and check that the strategies are a PPE. Since W is self-generating, $v \in B(\delta, W)$, so we have an action profile α and a map $w: Y \to W$ that generate payoff v. Set the period-0 strategies to be $\sigma^0 = \alpha^0$, and for each period-0 outcome y^0 set $v^1 = w^0(y^0)$. Since $v^1 \in W \subseteq B(\delta, W)$, there is an action profile $\alpha(v^1)$ and a map $w^1(y^1): Y \to W$ that generates payoff v^1. Set $\sigma^1(y^0) = \alpha^1(w^0(y^0))$, and for each sequence y^0, y^1 set $v^2 = w^1(w^0(y^0))(y^1)$, and so on: The constructed strategies yield payoff v if there are no deviations, and they have been constructed so that there is no history where a player can gain by deviating once and conforming thereafter. Thus, the constructed strategies are a PPE.   ∎

As we remarked above, this argument is essentially that of dynamic programming, applied to a game where the physical situation is memoryless, but the past matters because it influences the opponents' play. Here, the "state" is summarized by the current target payoff v—associated with each payoff vector v we have a first-period action for each player, and a rule that specifies the continuation payoffs as a function of this period's realized outcome.

Example 5.3

To help fix ideas, here is an example of a self-generating set in a game where actions are perfectly observed, namely the prisoner's dilemma with payoffs as in figure 5.5. With observed actions, there are four outcomes y, corresponding to the four action profiles of the stage game, and the probability distribution over outcomes assigns probability 1 to the action profile that was played. Consider the two-point set $W = \{v, \hat{v}\}$, where

$$v = \left[\frac{3 - \delta}{1 + \delta}, \frac{3\delta - 1}{1 + \delta} \right] \tag{5.19}$$

---

22. Abreu, Pearce, and Stachetti consider only pure-strategy equilibria, but the proof extends immediately to all PPE.

and

$$\hat{v} = \left[\frac{3\delta - 1}{1 + \delta}, \frac{3 - \delta}{1 + \delta} \right].$$ (5.20)

We claim that this set is self-generating for $\delta > \frac{1}{3}$.

Given the symmetry of W, it suffices to check that payoff vector v can be enforced with continuation payoffs in W. Let the action profile α corresponding to v be (D, C), and let the continuation payoffs be $w(D, C) = w(C, C) = \hat{v}$ and $w(D, D) = w(C, D) = v$. If both players follow α, the resulting payoffs are

$$(1 - \delta)(3, -1) + \delta\hat{v} = \left[\frac{3(1 - \delta^2) + 3\delta^2 - \delta}{1 + \delta}, -\frac{(1 - \delta^2) + 3\delta - \delta^2}{1 + \delta} \right]$$

$$= v.$$

Since player 1's current action does not influence the continuation payoffs, his average payoff is maximized by playing D, as this maximizes his current payoff. If player 2 plays C as prescribed by α, his payoff is $v_2 = (3\delta - 1)/(1 + \delta)$. If he plays D, he receives payoff 0 today, and continuation payoff v_2, so that C is better than D if $\delta > \frac{1}{3}$.

Abreu, Pearce, and Stachetti (1990) prove that the set of pure-strategy equilibria is compact. There are thus best and worst equilibria. Even though those authors assume a finite number of output levels, this is not immediate, because (in contrast with our finite model) they allow a continuum of prices, so that the number of outcomes is uncountable. Furthermore, and relatedly, Abreu et al. (1986) show that any payoff to a symmetric PPE can be enforced with strategies that threaten to switch to either best or worst equilibria. There is no need for intermediate values. And, more generally, Abreu et al. (1990) show that any pure-strategy PPE payoff can be achieved with continuation values that are extremal points of the equilibrium set. Furthermore, under an additional mild condition, they show that an extremal equilibrium—an equilibrium whose payoffs are on the boundary of the feasible set—*must* have continuation payoffs that are themselves extremal equilibria.

Knowing that it is sufficient to use extremal equilibria as continuation equilibria is particularly useful for characterizing "strongly symmetric" equilibria of symmetric games—that is, equilibria where for every public history all players' actions are identical. (Strong symmetry is discussed in subsection 5.1.3; recall that the trigger-price strategies of subsection 5.5.2 are strongly symmetric.) To characterize strongly symmetric equilibria, only two numbers need be determined: the highest and the lowest strongly symmetric equilibrium payoffs, $\bar{v}$ and $\underline{v}$.

5.6 The Folk Theorem with Imperfect Public Information[††]

Fudenberg, Levine, and Maskin (1990) develop the dynamic-programming approach to equilibrium further and use it to prove a folk theorem for games with imperfect public information.[23] The key question in determining when the folk theorem obtains is: How much information must the public outcome reveal about the players' actions? If players receive no information at all about one another's play, the only equilibrium payoffs will be convex combinations of the payoffs to static equilibria; when actions themselves are observed, the folk theorem obtains under the mild "full-dimensionality" condition.

To begin, consider an extremal payoff v of the feasible set—that is, a point that is not a convex combination of any two other points in V. If there is an equilibrium whose payoff is close to v, it must be possible to enforce a strategy profile a with $g(a)$ close to v. When will this be the case? That is, when will there be some (not necessarily feasible) continuation payoffs that induce the players to play a? The answer is that a is enforceable unless for some player i there is an action a_i' such that

(i) $g_i(a_i', a_{-i}) > g_i(a)$

and

(ii) $\pi_.(a_i', a_{-i}) = \pi_.(a)$.

Condition i implies that player i prefers a_i' to a_i if the expected continuation payoffs are the same, and condition ii ensures that the two actions induce the same distributions of outcomes and thus the same distributions of continuation payoffs. It should be clear that these conditions preclude enforceability; it is also true that when the conditions fail then a is enforceable.

A slightly stronger sufficient condition for enforceability is the following *individual full-rank condition*, which implies that any two distinct mixed strategies for player i lead to different distributions over outcomes.

Definition 5.6 The *individual full-rank condition* is satisfied at profile α if for each player i the vectors $\{\pi_.(a_i', \alpha_{-i})\}_{a_i' \in A_i}$ are linearly independent.

To see why this is called a "full-rank" condition, fix a profile α, let $\Pi_i(\alpha_{-i})$ denote the matrix whose rows are the vectors $\pi_.(a_i', \alpha_{-i})$ corresponding to each a_i', and let $G_i(\alpha_{-i})$ denote the column vector whose elements are $[(1 - \delta)/\delta]g_i(a_i', \alpha_{-i})$. Then player i has the same overall payoff

---

23. Their work extends an earlier result of Fudenberg and Maskin (1986b) on repeated principal-agent games. Matsushima (1989) obtains a partial folk theorem in a model with a continuum of actions on the hypothesis that for each h^t the incentive constraint can be replaced by the corresponding first-order condition.

to each action a_i' under continuation payoffs $w_i(\cdot)$ if and only if, for some constant vector k,

$$\Pi_i(\alpha_{-i}) \circ w_i = -G_i(\alpha_{-i}) + k. \tag{5.21}$$

The individual full-rank condition ensures that matrix $\Pi_i(\alpha_{-i})$ has full row rank, so that equation 5.21 can be solved for any k. (See subsection 7.6.1 for related ideas.) Note that this full-rank condition requires that there be at least as many publicly observed outcomes as there are actions for any player.

However, enforceability of all the extremal actions is not sufficient for a folk theorem in the limit of discount factors tending to 1.[24] The first counterexample was given by Radner, Myerson, and Maskin (1986) in a repeated partnership game like the following.

Example 5.4
Each period, each of two players chooses whether to work or to shirk. Each player's payoff depends on his own effort and on the publicly observed output, which they share equally. The output has only two levels, good and bad, with the probability of good equal to $\frac{9}{16}$ if both players work, $\frac{3}{8}$ if only one of them does, and $\frac{1}{4}$ if both shirk. Note that even if both players choose to work there is a positive probability of bad output. The payoffs are as follows: Working instead of shirking has a utility cost of 1; when output is good, both players receive a payment worth 4 utils; the payment in the bad output state is worth 0. (This will be the case if both players are risk neutral, the output is either 8 or 0, and the players share the output equally.)

The individual full-rank condition is satisfied at the profile where both players work, as the matrix

$$\Pi_i(\text{work}, \text{work}) = \begin{bmatrix} \frac{9}{16} & \frac{7}{16} \\ \frac{3}{8} & \frac{5}{8} \end{bmatrix}$$

is nonsingular. Thus, both players can be induced to work by the appropriate choice of continuation payoffs. And the profile where both players work yields $(\frac{5}{4}, \frac{5}{4})$, which is the highest feasible symmetric payoff. However, regardless of δ, the sum of the equilibrium payoffs is bounded by 2. The intuition for why efficiency cannot even be approximated in this model is that, in order to provide incentives for both players to work, both players' continuation payoffs must be higher after high output than after low. Loosely speaking, this means that both players must be "punished" when a bad outcome is observed. As long as the bad outcome has positive

24. Rubinstein (1979a), Rubinstein and Yaari (1983), and Radner (1986) obtain Nash-threats folk theorems using time-average payoffs in examples of games that do not meet the stronger information conditions we develop below. The literature suggests that individual full rank suffices for the full folk theorem with time-average payoffs, but we are not aware of a formal proof.

probability when both players work, there must be a positive probability of this "mutual punishment," and since mutual punishments are inefficient, the set of equilibrium payoffs is bounded away from efficiency.

Although the bound holds for all Nash equilibria of the game, it is easiest to obtain for the symmetric pure-strategy equilibria. Let v^* be the highest payoff in any pure-strategy symmetric equilibrium. Since the set of these equilibria is stationary, the payoff in the first period of an equilibrium with payoffs v^* must be at least v^*; thus, if $2v^*$ is greater than 2, in any equilibrium with payoffs v^* both players must work in the first period. If v_g is the (symmetric) continuation payoff after the good output and v_b the continuation payoff after bad, incentive compatibility requires that

$$(1 - \delta)[(\tfrac{9}{16}\cdot 4 + \tfrac{7}{16}\cdot 0) - 1] + \delta[\tfrac{9}{16}v_g + \tfrac{7}{16}v_b]$$

$$\geq (1 - \delta)[\tfrac{3}{8}\cdot 4 + \tfrac{5}{8}\cdot 0] + \delta[\tfrac{3}{8}v_g + \tfrac{5}{8}v_b],$$

or

$$v_g - v_b \geq [(1 - \delta)/\delta]\tfrac{4}{3}.$$

Since $v_g \leq v^*$, we conclude that if v^* is close to $\tfrac{5}{4}$ then

$$v^* \leq (1 - \delta)\tfrac{5}{4} + \delta[\tfrac{9}{16}v^* + \tfrac{7}{16}\{v^* - [(1 - \delta)/\delta]\tfrac{4}{3}\}],$$

so $(1 - \delta)v^* \leq (1 - \delta)\tfrac{8}{12}$—a contradiction.

Here, even though the required difference between v_g and v_b goes to 0 as δ goes to 1, the normalized present value of the efficiency loss remains nonnegligible.

In other cases, though, it is possible to provide all players with incentives to take the desired action while incurring a minimal loss of efficiency. In this case one can show that the efficiency loss required to provide incentives becomes negligible as $\delta \to 1$. When can the continuation payoffs be chosen in this way? *A sufficient condition is that the distributions over outcomes induced by different players' deviations be distinct.* This is made precise in the following definition and lemma.

Definition 5.7   The *pairwise full-rank condition* is satisfied at action α for players i and j if the $(|A_i| + |A_j|)$ vectors

$$\{\pi.(a_i', \alpha_{-i})_{a_i' \in A_i}, \pi.(a_j', \alpha_{-j})_{a_j' \in A_j}\}$$

admit only one linear dependency.

This condition implies that the matrix $\Pi_{ij}(\alpha)$ formed by stacking the matrix $\Pi_i(\alpha_{-i})$ on top of the matrix $\Pi_j(\alpha_{-j})$ has maximal rank. This matrix does not have *full* row rank, as it necessarily admits at least one linear dependency. This is easiest to see in the case where all players use their first pure strategy, so the first rows of Π_i and Π_j are identical. More generally, the rows of Π_{ij} satisfy the following equality:

$$\pi_.(\alpha) = \sum_{a_i \in A_i} \alpha_i(a_i)\pi_.(a_i, \alpha_{-i}) = \sum_{a_j \in A_j} \alpha_j(a_j)\pi_.(a_j, \alpha_{-j}). \tag{5.22}$$

If profile α satisfies pairwise full rank, not only can it be enforced, but the continuation payoffs can be chosen to satisfy the additional linear identity $\beta_1 w_1(y) + \beta_2 w_2(y) = k$ for any nonzero β_1 and β_2. That is, player i's continuation payoff can be exchanged for player j's at rate $-\beta_2/\beta_1$, so it is as if utility were transferable between the players. In this case, we can arrange the continuation payoffs so that when player i is punished, player j is rewarded, and conversely. Moreover, when profile α is efficient, the rate of exchange can be taken to be equal to the tangent to the efficient frontier at profile α, which is the key to providing incentives in an efficient way.

Under pairwise full rank, a deviation by player i leads to a distribution over outcomes that is different from that induced by any deviation by player j.

Note that this condition requires that the number of outcomes be at least $|A_i| + |A_j| - 1$. In example 5.4, there are two actions per player, and only two outcomes, so that pairwise full rank cannot be satisfied at any action profile. This is why shirking by player 1 could not be distinguished, even statistically, from shirking by player 2.

Even if the number of outcomes is large enough to permit pairwise full rank to be satisfied, the condition can still fail at some profiles. In particular, regardless of the number of outcomes, pairwise full rank fails at symmetric profiles in games such as the Green-Porter oligopoly or the partnership of example 5.4, where the distribution of outcomes depends only on the sum of the individual player's actions. For example, regardless of the number of outcomes in example 5.4, at a profile where both players work we see that

$$\Pi_{12}(\text{work}, \text{work}) = \begin{bmatrix} \pi_.(\text{work}, \text{work}) \\ \pi_.(\text{shirk}, \text{work}) \\ \pi_.(\text{work}, \text{work}) \\ \pi_.(\text{work}, \text{shirk}) \end{bmatrix},$$

and since $\pi_.(\text{shirk}, \text{work}) = \pi_.(\text{work}, \text{shirk})$ this matrix only has rank 2, instead of the rank 3 that pairwise full rank requires.

However, if there are more than two outcomes, the profile where player 1 works and player 2 shirks does satisfy pairwise full rank for generic probability distributions on outcomes. For example, suppose there are three outcomes, y_1, y_2, and y_3, and that

$$\pi_.(\text{work}, \text{work}) = (\tfrac{1}{2}, \tfrac{3}{8}, \tfrac{1}{8}),$$

$$\pi_.(\text{work}, \text{shirk}) = \pi_.(\text{shirk}, \text{work}) = (\tfrac{1}{4}, \tfrac{1}{2}, \tfrac{1}{4}),$$

and

$$\pi_.(\text{shirk}, \text{shirk}) = (\tfrac{1}{8}, \tfrac{3}{8}, \tfrac{1}{2}).$$

Then

$$\Pi_{12}(\text{work}, \text{shirk}) = \begin{bmatrix} (\tfrac{1}{4}, \tfrac{1}{2}, \tfrac{1}{4}) \\ (\tfrac{1}{8}, \tfrac{3}{8}, \tfrac{1}{2}) \\ (\tfrac{1}{2}, \tfrac{3}{8}, \tfrac{1}{8}) \\ (\tfrac{1}{4}, \tfrac{1}{2}, \tfrac{1}{4}) \end{bmatrix},$$

which has rank 3. Moreover, as observed by Legros (1988), any profile where player 1 works and player 2 shirks with positive probability also satisfies pairwise full rank, since the profile where player 1 shirks and player 2 uses his mixed strategy induces a different distribution than the profile where player 1 works and player 2 shirks. Legros' observation is generalized in the following lemma.

Lemma 5.1   If for each pair of players $i \neq j$ there is an $\alpha^{i,j}$ that satisfies pairwise full rank for players i and j, then there is an open dense set of α's that satisfy pairwise full rank for all pairs of players.

Theorem 5.11 (Fudenberg, Levine, and Maskin 1990)   If (i) the individual full-rank condition is satisfied at every pure strategy a, (ii) for each pair i, j of players there is a profile that satisfies pairwise full rank for i and j, and (iii) the feasible set V has dimension equal to the number of players, then for closed set W in the relative interior of V there is a $\underline{\delta}$ such that, for all $\delta > \underline{\delta}$, $W \subseteq \mathrm{E}(\delta)$.

Outline of Proof   Approximate the set of feasible individually rational payoffs by a smooth convex set W. Condition i implies that the minmax profile against player i and the *best* profile for player i can both be enforced on hyperplanes where player i's payoff is constant. Condition ii implies that almost all profiles can be enforced on hyperplanes where no player's payoff is held constant. Combining these two observations, for any point w on the boundary of W there is an action profile α with $g(\alpha)$ weakly separated from W by the tangent plane H at w, *and* such that profile α can be enforced with continuation payoffs on any linear translate $H + v$ of H. If we choose δ to be close to 1, the required variation in $w(y)$ is small enough that profile α can be enforced with continuation payoffs contained in a translate of H very close to the boundary of W. Intuitively, the "efficiency loss" (relative to W) required to provide incentives becomes negligible, since the smooth set W is approximately (i.e., to first order) linear.

   In a symmetric game, the theorem asserts that there are equilibria with payoffs arbitrarily close to the highest symmetric payoff. This is so even though in such games the highest payoff in a *symmetric equilibrium* may be bounded away from efficiency. The point is that the information revealed at symmetric action profiles can be poor (i.e., fail to satisfy pairwise full rank) even though many nearby almost-symmetric strategy profiles do generate "enough" information.[25]

5.7 Changing the Information Structure with the Time Period[†††]

The folk theorem looks at a set of equilibrium payoffs as $\delta \to 1$, holding $\pi_y(a)$ constant. As we saw, whether the folk theorem holds depends on the amount of information the public outcome y reveals. The interpretation of the result is therefore that almost all feasible, individually rational payoffs are equilibrium payoffs when δ is large in comparison with the information revealed by the outcome. Abreu, Pearce, and Milgrom (1990) show that the folk theorem need not hold if one interprets $\delta \to 1$ as the result of the interval between periods converging to 0, and if the information revealed by y deteriorates as the time interval shrinks. Why might this be the case? In games with observed actions, the public outcome is perfectly informative, and there is no reason to expect the information to change as the time period shrinks. In these games, then, we can interpret $\delta \to 1$ as a situation of either very little time preference or very short time periods. However, if players observe only imperfect signals of one another's actions, it is plausible that the quality of their information depends on the length of each observation period. Thus, one cannot interpret the case of $\delta \cong 1$, with $\pi_y(a)$ fixed, as the study of what would occur if the time period became very short.

Abreu, Pearce, and Milgrom (APM) investigate the effects of changing the time period and the associated information structure in two different examples. We will focus on a variant of their first example, a model of a repeated partnership game. We begin as usual by describing the stage game, which in the APM model is a continuous-time game of length τ. The interpretation is that players lock in their actions at the start of the stage, and at the end of the stage the outcome and the payoffs are revealed. As in example 5.4, each player has two choices: work and shirk. Payoffs are chosen so that shirk is a dominant strategy in the stage game, and so that shirk is the minmax strategy. As in the example, the stage game has the structure of the prisoner's dilemma: "Both shirk" is a Nash equilibrium in dominant strategies, and this equilibrium gives the players their minmax values. Payoffs are normalized so that this minmax payoff is 0, the (ex-

---

25. Fudenberg, Levine, and Maskin go on to develop a Nash-threats folk theorem for games where there are too few outcomes for even the individual full-rank condition to hold but where the information revealed by the outcomes has a "product structure," meaning that $y = (y_1, \ldots, y_I)$, and each player i's action influences only his "own" outcome y_i. This is the case in Green's (1987) model of repeated adverse selection, where the actions are reports of the players' types.

pected) payoffs if both players work are (c, c), and the payoff to shirking when the opponent works is $c + g$. (These are the expected payoffs, where the expectation is taken with respect to the corresponding distribution of output.) The difference between the APM stage game and example 5.4 is that, instead of there being only two outcomes each period (namely high and low output), the outcome is the number of "successes" in the period, which is distributed as a Poisson variable whose intensity is λ if both players work and μ if one of them shirks, with $\lambda > \mu$. Thus, if the time period is short, it is unlikely that there will be more than one success, and the probability of one success in a period of length dt is proportional to dt. This might correspond to a situation where the workers are trying to invent new products.[26]

In the repeated game $G(\tau, r)$, the discount factor δ is $\exp(-r\tau)$, and the public information is simply the number of "successes" each period.

From our earlier discussion of repeated partnership games, we can see that the folk theorem does not apply to the symmetric equilibria of this game, as deviations by the two players are indistinguishable when both use the same strategy. Thus, both players must be punished simultaneously, which causes efficiency losses, and so the set of symmetric equilibria is bounded away from efficiency even as $r \to 0$. However, when r is small, we would expect there to be symmetric equilibria with higher payoffs than the static equilibrium. Even this limited conclusion does not hold in the limit $\tau \to 0$, as the information revealed by the outcomes may "deteriorate" quickly enough to outweigh the effects of a larger discount factor. APM compute the highest symmetric-equilibrium payoffs of the game for small τ, and consider how the payoff of the best symmetric equilibrium varies with τ. To do so they consider a Taylor-series approximation of the game, neglecting terms smaller than τ^2.

This Taylor-series approach is necessary to discover if it is best to increase or decrease τ when τ is small. We will content ourselves with making the simpler point that sending τ to 0, with the corresponding changes in the information structure, has very different effects than sending r to 0, holding the information structure fixed. To this end we simplify the APM model by assuming that there are only two possible outcomes in each period: For $\theta = \lambda, \mu$, there is probability $\exp(-\theta\tau)$ that no events occur, and probability $1 - \exp(-\theta\tau)$ that exactly one event occurs. This simplifies the Poisson distribution by identifying all the events with one or more outcomes; it is a good approximation to the Poisson distribution when periods are short.

Let us consider when the best symmetric pure-strategy equilibrium can have payoff v^* that strictly exceeds 0. (APM show that this maximum is

26. Abreu, Pearce, and Milgrom also consider the case of "bad news," where *low* output is a Poisson event with intensity λ if both players work and μ if one shirks, with $\lambda < \mu$. Here the Poisson event corresponds to "accidents" that are made less likely if both players work.

attained; the following arguments would extend with the addition of a few epsilons if v^* were a supremum rather than a maximum.) Since APM allow public randomizations, it is immediate that an equilibrium with payoffs (v^*, v^*) can be constructed using continuation payoffs which are lotteries between the best continuation payoff of v^* and the worst continuation payoff of 0, as any continuation payoff between these values can be obtained by a public randomization between them. Thus, fix an equilibrium that attains v^*, and let the continuation payoff for each player be the lottery $[(1 - \alpha(0))v^*, \alpha(0) \cdot 0]$ if the first-period outcome is 0, and the lottery $[(1 - \alpha(1))v^*, \alpha(1) \cdot 0]$ if the first-period outcome is 1. One can show that if v^* is greater than 0, the strategies that attain (or closely approximate) v^* must have both players working in the first period. Thus, in order for v^* to exceed 0 there must exist probabilities $\alpha(0)$ and $\alpha(1)$ such that both agents are induced to work, and such that if both players do work the resulting normalized present values are v^*.

Writing out these two equations, we have

$$(1 - e^{-rt})g \le e^{-rt} \cdot (e^{-\mu\tau} - e^{-\lambda\tau}) \cdot [\alpha(0) - \alpha(1)]v^* \tag{5.24}$$

and

$$v^* = (1 - e^{-rt})c + e^{-rt}[(1 - \alpha(0)e^{-\lambda\tau}) - \alpha(1)(1 - e^{-\lambda\tau})]v^*. \tag{5.25}$$

Solving equation 5.25 for v^* yields

$$v^* = \frac{(1 - e^{-rt})c}{1 - e^{-rt}\{1 - \alpha(1) - e^{-\lambda\tau}[\alpha(0) - \alpha(1)]\}}. \tag{5.26}$$

Intuitively, one would expect that in the best equilibrium players would not be "punished" if a success occurs, so that $\alpha(1) = 0$. This can be checked by inspecting the above equations: When $\lambda > \mu$, setting $\alpha(1) = 0$ makes equation 5.24 more likely to be satisfied, and increases the equilibrium payoff by decreasing the probability of switching to the punishment state.

Now we ask when the above system has a solution with $\alpha(1) = 0$ and $\alpha(0) \le 1$. Algebraic manipulation shows that this is possible only if

$$c[e^{-\mu\tau} - e^{-\lambda\tau}] \ge g[e^{rt} - 1 + e^{-\lambda\tau}] \ge ge^{-\lambda\tau}. \tag{5.27}$$

Thus, a necessary condition for $v^* > 0$, regardless of the rate of interest, is that

$$\frac{g}{c} \le \frac{e^{-\mu\tau} - e^{-\lambda\tau}}{e^{-\lambda\tau}} = e^{-(\mu-\lambda)\tau} - 1, \tag{5.28}$$

which says that the likelihood ratio $L(\tau) = e^{-(\mu-\lambda)\tau}$ associated with the event "no successes" should be sufficiently large. However, as τ converges to 0, the likelihood ratio $L(\tau)$ converges to 1: Since it is almost certain that there will be no successes, the information provided by the publicly observed outcome is too poor for there to be an equilibrium that improves on the minmax values.

On the topic of changing the information structure, we should also mention Kandori (1989a), who studies how the set of equilibrium payoffs changes when the public outcome becomes a less informative signal of the players' actions. One probability distribution is a "garbling" of another (Blackwell and Girshik 1954) if it can be obtained from the first one by adding noise. Kandori shows that if the information becomes worse in the sense of garbling, the set of equilibrium payoffs becomes strictly smaller. That the set cannot grow larger is fairly clear, and is obvious in the presence of public randomizations, as the public randomizing device can be used to create a garbling of the original signal; the interesting conclusion is that the set must become strictly smaller.

Exercises

Exercise 5.1* Compute the set of feasible payoffs in the "battle of the sexes" stage game as shown in figure 1.10a. What is the highest feasible symmetric payoff? Let $\delta = \frac{9}{10}$, and find a deterministic strategy profile for the repeated game with payoffs $(\frac{3}{2}, \frac{3}{2})$.

Exercise 5.2* Consider the infinitely repeated play of a finite stage game $(\mathcal{I}, A, g)$. Given $\varepsilon > 0$, show that there exists a $\underline{\delta} > 0$ such that for all $\delta \in [0, \underline{\delta}]$ every Nash equilibrium σ has the property that, at all histories h^t with positive probability under σ, $\sigma(h^t)$ must be within ε of one of the Nash equilibria of the stage game. Give an example to show that the conclusion need not hold for all subgames. Can the equilibrium to be played in period t vary with the history h^t? Why or why not?

Exercise 5.3** Prove that the set of continuation payoff vectors corresponding to all Nash equilibria is the same in every proper subgame of a repeated game. The idea of the proof is to show more strongly that every proper subgame is strategically isomorphic, i.e., there is a one-to-one correspondence between the strategy spaces that preserves the payoffs. The simplest example is the map between the whole game and a subgame: To map a strategy s for the whole game to its equivalent in the subgame starting at h^t, set $\hat{s}(h^t) = s(h^0)$, $\hat{s}(h^t, a^t) = s(a^t)$, $\hat{s}(h^t, a^t, a^{t+1}) = s(a^t, a^{t+1})$, and so on, so that $\hat{s}$ treats period $t + \tau$ just as s treats period τ. Conversely, given a strategy profile s and a subgame h^t, the equivalent strategy for the whole game is $\hat{s}(h^0) = s(h^t)$, $\hat{s}(a^0) = s(h^t, a^0)$, and so on. Use these maps between the subgames to argue that if a profile for the whole game is a Nash equilibrium it must be a Nash equilibrium in the subgame, and conversely.

Exercise 5.4* Consider a finite symmetric repeated game, and assume there is a symmetric mutual minmax profile m^* in pure strategies, i.e., a pure-strategy profile m such that $\max_{a_i} g(a_i, m^*_{-i}) \leq \underline{v}$. Show that, if public randomizations are available, for sufficiently large discount factors the worst strongly symmetric equilibrium payoff e_* can be attained with strategies that have two phases: In phase A, players play m^*. If players conform in phase A, play switches to phase B with a probability specified by the equilibrium strategies; if there are any deviations, play remains in phase A with probability 1. In phase B, play follows strategies that yield the highest equilibrium payoff.

Exercise 5.5* Consider the two-player game illustrated in figure 5.6. In the first period, players 1 and 2 simultaneously choose U1 or D1 (player 1) and L1 or R1 (player 2); these choices are revealed at the end of period 1 with payoffs as in the left-hand matrix. In period 2, players choose U2 or D2 and L2 or D2, with payoff as in the matrix on the right. Each player's objective is to maximize the average of his per-period payoff.

(a) Find the subgame-perfect equilibria of this game, and compute the convex hull of the associated payoffs.

(b) Now suppose that the players can jointly observe the outcome y_1 of a public randomizing device before choosing their first-period actions, where y_1 has a uniform distribution on the unit interval. Find the set of subgame-perfect equilibria, and compare the resulting payoffs against the answer to part a of this exercise.

(c) Suppose that the players jointly observe y_1 at the beginning of period 1 and y_2 at the beginning of period 2, with y_1 and y_2 being independent draws from a uniform distribution on the unit interval. Again, find the subgame-perfect equilibrium payoffs.

(d) Relate your answers to parts a–c to the role of public randomizations in the proof of the Folk Theorem.

Exercise 5.6*** Under the assumptions used by Benoit and Krishna (1985) for pure-strategy equilibria, try to characterize the limit as the horizon $T \to \infty$ of the set of payoffs of all subgame-perfect equilibria of a T-period finitely repeated game.

	L1	R1
U1	2,2	-1,3
D1	3,-1	0,0

	L2	R2
U2	6,4	3,3
D2	3,3	4,6

First-period payoffs   Second-period payoffs

Figure 5.6

Exercise 5.7** Consider a sequence of randomly matched players, with the information structure of Rosenthal (1979), who play the prisoner's dilemma with payoffs as in figure 5.5. Show that unless the discount factor equals $\frac{1}{2}$, the only Markovian equilibrium where all players use the same pure strategy is for all players to always cheat.

Exercise 5.8**

(a) In a repeated game, show that if for each player there is a subgame-perfect equilibrium where that player's payoff is his minmax value, then any payoff of a Nash equilibrium is also the payoff of a subgame-perfect equilibrium.

(b) Suppose that, for each player i and each $j \neq i$, $g_j(m^i) > \underline{v}_j$. Show that the sets of Nash-equilibrium and perfect-equilibrium payoffs are identical for sufficiently large discount factors. Give an economic example where the condition is plausible, and an example where it is not. Show that the sets can differ for small discount factors.

(c) Suppose that the minmax profile is in pure strategies, that the vector where all players simultaneously receive their minmax payoff is in the interior of the feasible set, and that for each player i there is an $\hat{a}_i$ such that $g_i(\hat{a}_i, m^i_{-i}) < \underline{v}_i$. Show that the sets of Nash-equilibrium and perfect-equilibrium payoffs are identical for large enough discount factors. Give an example of a game where the feasible set has full dimension yet the inferiority condition used here does not apply. (Answers are given in Fudenberg and Maskin 1990b.)

Exercise 5.9* Consider infinitely repeated play of the stage game of figure 5.7.

(a) What is the highest perfect equilibrium for player 1 if both sides are long-run players?

(b) If player 1 could publicly commit to always play the same mixed strategy α_1, what α_1 would he choose? What would his payoff be?

(c) Show that when the player 2's are an infinite sequence of short-run players, the highest payoff for player 1 in any Nash equilibrium is 2. To do this, proceed as follows.

• Let $v^*(\delta)$ be the supremum of player 1's payoff in any Nash equilibrium when his discount factor is δ, and suppose $v^*(\delta) > 2$. Let $\varepsilon = (1 - \delta)(v^*(\delta) - 2)/2$, and choose an equilibrium σ where player 1's equilibrium payoff $v(\delta)$ is at least $v^*(\delta) - \varepsilon$. Show that under profile σ player 1's

	L	M	R
U	6,0	−1,−100	0,1
D	2,2	0,3	1,1

Figure 5.7

expected payoff in the first period must be greater than 2. (Hint: Player 1's continuation payoff from the second period on cannot exceed $v^*(\delta)$.)

• Show that under σ player 2 must play L with positive probability in the first period, and thus that player 1 must play D with positive probability in the first period.

• Conclude that $v^*(\delta) - \varepsilon \le v(\delta) \le 2(1 - \delta) + \delta v^*(\delta)$, so that $v^*(\delta) \le 2$.

Exercise 5.10** Consider the following three-person stage game: Players 1 and 2 choose pairs, the first element being Up (U) or Down (D) and the second being Heads (H) or Tails (T). Player 3 chooses Right or Left. Players 1 and 2 receive 0 regardless of what happens. Player 3's payoff is -1 if 1 and 2 both chose Up and he went Right, or if 1 and 2 both chose Down and he went Left; otherwise he gets 0. Notice that the choice of H or T is irrelevant to the payoffs.

Suppose player 3's choices are observable, that whether 1 and 2 play Up or Down is observable, but that the public information y^t about their choice of H or T is only the total number of H chosen by both players. Thus, if y^t is 2 or 0, it reveals the actions of players 1 and 2. If $y^t = 1$, then the actions of players 1 and 2 are common knowledge for players 1 and 2 (since it is common knowledge that they each know their own action), but player 3 does not know which player played H.

(a) Show that player 3's minmax payoff is $-\frac{1}{4}$.

(b) Now consider the repeated version of this game. Construct a Nash equilibrium where players 1 and 2 use the following strategies, e.g., for player 1: "Randomize $\frac{1}{2}$-$\frac{1}{2}$ between H and T in every period. If $y^{t-1} = 1$ and player 2 played H, play D; if $y^{t-1} = 1$ and player 2 played T, then play U. If $y^{t-1} = 0$ or 2, randomize $\frac{1}{2}$-$\frac{1}{2}$ between U and D." Show that player 3's equilibrium payoff is below his minmax value. Explain.

References

Abreu, D. 1986. Extremal equilibria of oligopolistic supergames. *Journal of Economic Theory* 39: 191–228.

Abreu, D. 1988. Towards a theory of discounted repeated games. *Econometrica* 56: 383–396.

Abreu, D., and D. Dutta. 1990. in preparation.

Abreu, D., D. Pearce, and P. Milgrom. 1990. Information and timing in repeated partnerships. *Econometrica*, forthcoming.

Abreu, D., D. Pearce, and E. Stachetti. 1986. Optimal cartel equilibrium with imperfect monitoring. *Journal of Economic Theory* 39: 251–269.

Abreu, D., D. Pearce, and E. Stachetti. 1989. Renegotiation and symmetry in repeated games. Mimeo, Harvard University.

Abreu, D., D. Pearce, and E. Stachetti. 1990. Toward a theory of discounted repeated games with imperfect monitoring. *Econometrica* 58: 1041–1064.

Aumann, R. 1986. Repeated games. In *Issues in Contemporary Microeconomics*, ed. G. Feiwel. Macmillan.

Aumann, R. 1989. Survey of repeated games. In *Essays in Game Theory and Mathematical Economics in Honor of Oskar Morgenstern*. Manheim: Bibliographisches Institut.

Aumann, R. 1990. Communication need not lead to Nash equilibrium. Mimeo.

Aumann, R., and L. Shapley. 1976. Long-term competition—a game theoretic analysis. Mimeo.

Benoit, J. P., and V. Krishna. 1985. Finitely repeated games. *Econometrica* 53: 890–904.

Benoit, J. P., and V. Krishna. 1987. Nash equilibria of finitely repeated games. *International Journal of Game Theory* 16.

Benoit, J. P., and V. Krishna. 1988. Renegotiation in finitely repeated games. Discussion Paper 89-004, Harvard University.

Bergin, J., and B. MacLeod. 1989. Efficiency and renegotiation in repeated games. Mimeo, Queen's University.

Bernheim, B. D., B. Peleg, and M. Whinston. 1987. Coalition-proof Nash equilibria. I: Concepts. *Journal of Economic Theory* 42: 1–12.

Bernheim, B. D., and D. Ray. 1989. Collective dynamic consistency in repeated games. *Games and Economic Behavior* 1: 295–326.

Blackwell, D. and M. Girshik. 1954. *Theory of Games and Statistical Decisions*. Wiley.

Chamberlin, E. 1929. Duopoly: Value where sellers are few. *Quarterly Journal of Economics* 43: 63–100.

Crémer, J. 1986. Cooperation in ongoing organizations. *Quarterly Journal of Economics* 101: 33–49.

DeMarzo, P. M. 1988. Coalitions and sustainable social norms in repeated games. IMSSS Discussion Paper 529, Stanford University.

Dybvig, P., and C. Spatt. 1980. Does it pay to maintain a reputation? Mimeo, Carnegie-Mellon University.

Ellison, G. 1991. Cooperation in random matching games. Mimeo.

Farrell, J., and E. Maskin. 1989. Renegotiation in repeated games. *Games and Economic Behavior* 1: 327–360.

Fisher, I. 1898. Cournot and mathematical economics. *Quarterly Journal of Economics* 12: 1–26.

Forges, F. 1986. An approach to communications equilibrium. *Econometrica* 54: 1375–1386.

Fraysse, J., and M. Moreaux. 1985. Collusive equilibria in oligopolies with long but finite lives. *European Economic Review* 27: 45–55.

Friedman, J. 1971. A noncooperative equilibrium for supergames. *Review of Economic Studies* 38: 1–12.

Friedman, J. 1985. Trigger strategy equilibria in finite horizon supergames. Mimeo.

Fudenberg, D., D. Kreps, and E. Maskin. 1990. Repeated games with long-run and short-run players. *Review of Economic Studies* 57: 555–574.

Fudenberg, D., and D. Levine. 1983. Subgame-perfect equilibria of finite and infinite horizon games. *Journal of Economic Theory* 31: 227–256.

Fudenberg, D., and D. Levine. 1990a. Approximate equilibria in repeated games with imperfect private information. *Journal of Economic Theory*, forthcoming.

Fudenberg, D., and D. Levine. 1990b. Efficiency and observability in games with long-run and short-run players. Mimeo.

Fudenberg, D., D. Levine, and E. Maskin. 1990. The folk theorem in repeated games with imperfect public information. Mimeo, Massachusetts Institute of Technology.

Fudenberg, D., and E. Maskin. 1986a. The folk theorem in repeated games with discounting or with incomplete information. *Econometrica* 54: 533–556.

Fudenberg, D., and E. Maskin. 1986b. Discounted repeated games with one-sided moral hazard. Mimeo.

Fudenberg, D., and E. Maskin. 1990a. On the dispensability of public randomizations in discounted repeated games. *Journal of Economic Theory*. Forthcoming.

Fudenberg, D., and E. Maskin. 1990b. Nash and perfect equilibria of discounted repeated games. *Journal of Economic Theory* 50: 194–206.

Green, E. 1987. Lending and the smoothing of uninsurable income. In *Contractual Arrangements for Intertemporal Trade*, ed. E. Prescott and N. Wallace. University of Minnesota Press.

Green, E., and R. Porter. 1984. Noncooperative collusion under imperfect price information. *Econometrica* 52: 87–100.

Greenberg, J. 1988. The theory of social situations. Mimeo, Haifa University.

Greif, A. 1989. Reputation and coalitions in medieval trade: Evidence from the geniza documents. Mimeo.

Kandori, M. 1989a. Monotonicity of equilibrium payoff sets with respect to obervability in repeated games with imperfect monitoring. Mimeo.

Kandori, M. 1989b. Social norms and community enforcement. Mimeo.

Kandori, M. 1989c. Repeated games played by overlapping generations of players. Mimeo.

Klein, B., and K. Leffler. 1981. The role of market forces in assuring contractual performance. *Journal of Political Economy* 81: 615–641.

Kreps, D. 1986. Corporate culture and economic theory. In *Technological Innovation and Business Strategy*, ed. M. Tsuchiya. Nippon Keizai Shimbunsha Press (in Japanese). Also (in English) in *Rational Perspectives on Political Science*, ed. J. Alt and K. Shepsle. Harvard University Press, 1990.

Kreps, D., P. Milgrom, J. Roberts, and R. Wilson. 1982. Rational cooperation in the finitely repeated prisoner's dilemma. *Journal of Economic Theory* 27: 245–252.

Lambson, V. E. 1987. Optimal penal codes in price-setting supergames with capacity constraints. *Review of Economic Studies* 54: 385–397.

Legros, P. 1988. Sustainability in partnerships. Mimeo, California Institute of Technology.

Lehrer, E. 1988. Two player repeated games with non-observable actions and observable payoffs. Mimeo.

Lehrer, E. 1989. Lower equilibrium payoffs in two-player repeated games with non-obervable actions. *International Journal of Game Theory* 18.

Macaulay, S. 1963. Non-contractual relations in business: A preliminary study. *American Sociological Review* 28: 55–67.

Matsushima, H. 1989. Efficiency in repeated games with imperfect monitoring. *Journal of Economic Theory* 48: 428–442.

Mertens, J.-F. 1987. Repeated games. In *Proceedings of the International Congress of Mathematicians 1986*.

Mertens, J.-F., S. Sorin, and S. Zamir. 1990. Repeated games. Manuscript.

Milgrom, P., D. North, and B. Weingast. 1989. The role of law merchants in the revival of trade: A theoretical analysis. Mimeo.

Munkres, I. 1975. *Topology: A First Course*. Prentice-Hall.

Myerson, R. 1986. Multistage games with communication. *Econometrica* 54: 323–358.

Neyman, J. 1989. Counterexamples with almost common knowledge. Mimeo.

Oye, K., ed. 1986. *Cooperation under Anarchy*. Princeton University Press.

Pearce, D. 1988. Renegotiation-proof equilibria: Collective rationality and intertemporal cooperation. Mimeo, Yale University.

Porter, R. 1983a. A study of cartel stability: The joint economic committee, 1880–1886. *Bell Journal of Economics* 14: 301–314.

Porter, R. 1983b. Optimal cartel trigger-price strategies. *Journal of Economic Theory* 29: 313–338.

Radner, R. 1980. Collusive behavior in non-cooperative epsilon equilibria of oligopolies with long but finite lives. *Journal of Economic Theory* 22: 121–157.

Radner, R. 1981. Monitoring cooperative agreements in a repeated principal-agent relationship. *Econometrica* 49: 1127–1148.

Radner, R. 1985. Repeated principal agent games with discounting. *Econometrica* 53: 1173–1198.

Radner, R. 1986. Repeated partnership games with imperfect monitoring and no discounting. *Review of Economic Studies* 53: 43–58.

Radner, R., R. Myerson, and E. Maskin. 1986. An example of a repeated partnership game with discounting and with uniformly inefficient equilibria. *Review of Economic Studies* 53: 59–70.

Rosenthal, R. 1979. Sequences of games with varying opponents. *Econometrica* 47: 1353–1366.

Rosenthal, R., and H. Landau. 1979. A game-theoretic analysis of bargaining with reputations. *Journal of Mathematical Psychology* 20: 235–255.

Rotemberg, J., and G. Saloner. 1986. A supergame-theoretic model of price wars during booms. *American Economic Review* 76: 390–407.

Rubinstein, A. 1979a. Equilibrium in supergames with the overtaking criterion. *Journal of Economic Theory* 21: 1–9.

Rubinstein, A. 1979b. An optimal conviction policy for offenses that may have been committed by accident. In *Applied Game Theory*, ed. S. Brams, A. Schotter, and G. Schwodiauer. Physica-Verlag.

Rubinstein, A., and M. Yaari. 1983. Repeated insurance contracts and moral hazard. *Journal of Economic Theory* 30: 74–97.

Scherer, F. M. 1980. *Industrial Market Structure and Economic Performance*, second edition. Houghton Mifflin.

Shapiro, C. 1982. Consumer information, product quality, and seller reputation. *Bell Journal of Economics* 13: 20–35.

Simon, H. 1951. A formal theory of the employment relationship. *Econometrica* 19: 293–305.

Smith, L. 1989. Folk theorems in overlapping-generations games. Mimeo.

Smith, L. 1990. Folk theorems: Two-dimensionality is (almost) enough. Mimeo.

Sorin, S. 1986. On repeated games with complete information. *Mathematics of Operations Research* 11: 147–160.

Sorin, S. 1988. Supergames. Mimeo.

Stigler, G. 1964. A theory of oligopoly. *Journal of Political Economy* 72: 44–61.

van Damme, E. 1989. Renegotiation-proof equilibria in repeated prisoner's dilemma. *Journal of Economic Theory* 47: 206–207.

III STATIC GAMES OF INCOMPLETE INFORMATION

6 Bayesian Games and Bayesian Equilibrium

6.1 Incomplete Information[†]

When some players do not know the payoffs of the others, the game is said to have *incomplete information*. Many games of interest have incomplete information to at least some extent; the case of perfect knowledge of payoffs is a simplifying assumption that may be a good approximation in some cases.

As a particularly simple example of a game in which incomplete information matters, consider an industry with two firms: an incumbent (player 1) and a potential entrant (player 2). Player 1 decides whether to build a new plant, and simultaneously player 2 decides whether to enter. Imagine that player 2 is uncertain whether player 1's cost of building is 3 or 0, while player 1 knows her own cost. The payoffs are depicted in figure 6.1. Player 2's payoff depends on whether player 1 builds, but is not directly influenced by player 1's cost. Entering is profitable for player 2 if and only if player 1 does not build. Note also that player 1 has a dominant strategy: "build" if her cost is low and "don't build" if her cost is high.

Let p_1 denote the prior probability player 2 assigns to player 1's cost being high. Because player 1 builds if and only if her cost is low, player 2 enters whenever $p_1 > \frac{1}{2}$ and stays out if $p_1 < \frac{1}{2}$. Thus, we can solve the game in figure 6.1 by the iterated deletion of strictly dominated strategies. Section 6.6 gives a careful analysis of iterated dominance arguments in games of incomplete information.

The analysis of the game becomes more complex when the low cost is only 1.5 instead of 0, as in figure 6.2. In this new game, "don't build" is still a dominant strategy for player 1 when her cost is high. However, when her cost is low, player 1's optimal strategy depends on her prediction of y, the probability that player 2 enters: Building is better than not building if

$$1.5y + 3.5(1 - y) > 2y + 3(1 - y),$$

or

$$y < \tfrac{1}{2}.$$

Thus, player 1 must try to predict player 2's behavior to choose her own action, and player 2 cannot infer player 1's action from his knowledge of player 1's payoffs alone.

Harsanyi (1967–68) proposed that the way to model and understand this situation is to introduce a prior move by nature that determines player 1's "type" (here, her cost). In the transformed game, player 2's *incomplete* information about player 1's cost becomes *imperfect* information about nature's moves, so the transformed game can be analyzed with standard techniques.

	Enter	Don't
Build	0,−1	2,0
Don't Build	2,1	3,0

	Enter	Don't
Build	3,−1	5,0
Don't Build	2,1	3,0

Payoffs if 1's building cost is high

Payoffs if 1's building cost is low

Figure 6.1

	Enter	Don't
Build	0,−1	2,0
Don't Build	2,1	3,0

	Enter	Don't
Build	1.5,−1	3.5,0
Don't Build	2,1	3,0

Payoffs if 1's building cost is high

Payoffs if 1's building cost is low

Figure 6.2

The transformation of incomplete information into imperfect information is illustrated in figure 6.3, which depicts Harsanyi's rendering of the game of figure 6.2. N denotes "nature," who chooses player 1's type. (In the figure, numbers in brackets are probabilities of nature's moves.) The figure incorporates the standard assumption that all players have the same prior beliefs about the probability distribution on nature's moves. (Although this is a standard assumption, it may be more plausible when nature's moves represent public events, such as the weather, than when nature's moves model the determination of the players' payoffs and other private characteristics.) Once this common-prior assumption is imposed, we have a standard game, to which Nash equilibrium can be applied. Harsanyi's *Bayesian equilibrium* (or Bayesian Nash equilibrium) is precisely the Nash equilibrium of the imperfect-information representation of the game.

For instance, in the game of figure 6.2 (or figure 6.3), let x denote player 1's probability of building when her cost is low (player 1 never builds when her cost is high), and let y denote player 2's probability of entry. The optimal strategy for player 2 is $y = 1$ (enter) if $x < 1/[2(1 - p_1)]$, $y = 0$ if $x > 1/[2(1 - p_1)]$, and $y \in [0, 1]$ if $x = 1/[2(1 - p_1)]$. Similarly, the best response for the low-cost player 1 is $x = 1$ (build) if $y < \frac{1}{2}$, $x = 0$ if $y > \frac{1}{2}$, and $x \in [0, 1]$ if $y = \frac{1}{2}$. The search for a Bayesian equilibrium boils down to finding a pair (x, y) such that x is optimal for player 1 with low cost against player 2 and y is optimal for player 2 against player 1 given beliefs p_1 and

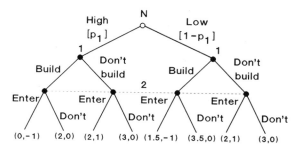

Figure 6.3

player 1's strategy. For instance, $(x = 0, y = 1)$ (player 1 does not build, player 2 enters) is an equilibrium for any p_1, and $(x = 1, y = 0)$ (player 1 builds if her cost is low, and player 2 does not enter) is an equilibrium if and only if $p_1 \leq \frac{1}{2}$.[1]

The remainder of the chapter is organized as follows. Section 6.2 gives a second example of Bayesian equilibrium in a game of incomplete information. Section 6.3 discusses the notion of type, and section 6.4 gives a formal definition of Bayesian equilibrium. Section 6.5 returns to illustrations but emphasizes the details of the characterization of Bayesian equilibria rather than motivation. The details of the analysis are somewhat involved, and many of the examples could be skipped on a first reading. Section 6.6 discusses the iterated deletion of dominated strategies in games of incomplete information. Here the issue arises of whether different "types" of a single player should be viewed as separate individuals, with potentially different beliefs about the strategies of their opponents, or as a single individual with fixed beliefs. Section 6.7 develops an incomplete-information justification of mixed strategies in games of complete information. Section 6.8 presents more technical material on games in which players have a continuum of types.

6.2 Example 6.1: Providing a Public Good under Incomplete Information[†]

The supply of a public good gives rise to the celebrated *free-rider problem*. Each player benefits when the public good is provided, but each would prefer the other players to incur the cost of supplying it. There are numerous variants of the public-good paradigm; we consider one studied experimentally by Palfrey and Rosenthal (1989). There are two players, $i = 1, 2$. Players decide simultaneously whether to contribute to the public good, and contributing is a 0-1 decision. Each player derives a benefit of 1 if at

---

1. In this case, there is also a mixed-strategy equilibrium: $(x = 1/[2(1 - p_1)], y = \frac{1}{2})$.

	Contribute	Don't
Contribute	$1-c_1, 1-c_2$	$1-c_1, 1$
Don't	$1, 1-c_2$	$0, 0$

Figure 6.4

least one of them provides the public good and 0 if none does; player i's cost of contributing is c_i. The payoffs are depicted in figure 6.4.[2]

The benefits of the public good—1 each—are common knowledge, but each player's cost is known only to that player. However, both players believe it is common knowledge that the c_i are drawn independently from the same continuous and strictly increasing cumulative distribution function, $P(\cdot)$, on $[\underline{c}, \bar{c}]$, where $\underline{c} < 1 < \bar{c}$ (so $P(\underline{c}) = 0$ and $P(\bar{c}) = 1$). The cost c_i is player i's "type."

A pure strategy in this game is a function $s_i(c_i)$ from $[\underline{c}, \bar{c}]$ into $\{0, 1\}$, where 1 means "contribute" and 0 means "don't contribute." Player i's payoff is

$$u_i(s_i, s_j, c_i) = \max(s_1, s_2) - c_i s_i.$$

(Note that player i's payoff does not depend $c_j, j \neq i$.)

A Bayesian equilibrium is a pair of strategies $(s_1^*(\cdot), s_2^*(\cdot))$ such that, for each player i and every possible value of c_i, strategy $s_i^*(c_i)$ maximizes $E_{c_j} u_i(s_i, s_j^*(c_j), c_i)$. Let $z_j \equiv \text{Prob}(s_j^*(c_j) = 1)$ be the equilibrium probability that player j contributes. To maximize his expected payoff, player i will contribute if his cost c_i is less than $1 \cdot (1 - z_j)$, which is his benefit from the public good times the probability that player j does not contribute. Thus, $s_i^*(c_i) = 1$ if $c_i < 1 - z_j$, and conversely, $s_i^*(c_i) = 0$ if $c_i > 1 - z_j$.[3] This shows that the types of player i who contribute lie in an interval $[\underline{c}, c_i^*]$: Player i contributes only if his cost is sufficiently low. (We adopt the convention that $[\underline{c}, c_i^*]$ is empty if $c_i^* < \underline{c}$.) Similarly, player j contributes if and only if $c_j \in [\underline{c}, c_j^*]$ for some c_j^*. Such "monotonicity" properties are frequent in economic applications; they will be useful in characterizing Bayesian equilibria in section 6.5, and they will be developed in more detail in chapter 7.

Since $z_j = \text{Prob}(\underline{c} \leq c_j \leq c_j^*) = P(c_j^*)$, the equilibrium cutoff levels c_i^* must satisfy $c_i^* = 1 - P(c_j^*)$. Thus, c_1^* and c_2^* must both satisfy the equation

2. The important feature of this game as a model of the provision of public goods is that if both players choose to contribute they both pay the full cost, as opposed to sharing the cost equally. One in interpretation of this model is that the two players belong to a committee. If either player attends the committee's meeting, the outcome is the one both prefer; if neither attends, the outcome is bad for both. The time to attend the meeting costs c_i utils.

3. Type $c_i = 1 - z_j$ is indifferent between contributing and not contributing, but since $P(\cdot)$ is continuous the probability of this (or any) particular type is 0.

$c^* = 1 - P(1 - P(c^*))$. If there is a unique c^* that solves this equation, then necessarily $c_i^* = c^* = 1 - P(c^*)$. For instance, if P is uniform on $[0, 2]$ ($P(c) \equiv c/2$), then c^* is unique and is equal to $\frac{2}{3}$. (As a check on the analysis, note that if a player does not contribute his expected payoff is $P(c^*) = \frac{1}{3}$, and if a player with cost c^* contributes his payoff is $1 - c^* = \frac{1}{3}$.) A player does not contribute if his cost belongs to $(\frac{2}{3}, 1)$ even though his cost of providing the good is less than his benefit, and even though there is probability $1 - P(c^*) = \frac{2}{3}$ that the good will not be supplied by the other player.

If, instead of $\underline{c} = 0$, we suppose that $\underline{c} \geq 1 - P(1)$, the game has two asymmetric Nash equilibria. In these equilibria, one player never contributes and the other player contributes for all $c \leq 1$. For instance, the equilibrium where player 1 never contributes is $c_1^* = 1 - P(1) < \underline{c}$, and $c_2^* = 1$. The player who never contributes prefers not to, as his minimum cost of $\underline{c}$ exceeds the gain of $1 \cdot (1 - P(1))$ from increased supply of the good; the player who contributes for all $c \leq 1$ is playing optimally in view of the fact that if he does not contribute there is probability 0 of obtaining the public good.

6.3 The Notions of Type and Strategy[†]

In the examples of sections 6.1 and 6.2, a player's "type"—his private information—was simply his cost. More generally, the "type" of a player embodies any private information (more precisely, any information that is not common knowledge to all players) that is relevant to the player's decision making. This may include, in addition to the player's payoff function, his beliefs about other players' payoff functions, his beliefs about what other players believe his beliefs are, and so on.

We have already seen examples where the players' types are identified with their payoff functions. For an example where the type includes more than this, consider disarmament talks between two negotiators. Player 2's objective function is public information; player 1 is uncertain whether player 2 knows player 1's objectives. To model this, suppose that player 1 has two possible types—a "tough" type, who prefers no agreement to making substantial concessions, and a "weak" type, who prefers any agreement to none at all—and that the probability that player 1 is tough is p_1. Furthermore, suppose that player 2 has two types—"informed," who observes player 1's type, and "uninformed," who does not observe player 1's type. The probability that player 2 is informed is p_2, and player 1 does not observe player 2's type.

It is easy to construct more complicated versions of this game where, say, player 1's prior beliefs about player 2 can be either p_2 or p_2', and player 2 does not know which. In practice, though, these sorts of complications make the models difficult to work with, and in most applications a player's beliefs about his opponent are assumed to be completely determined by his own payoff function.

More generally, Harsanyi assumed that the players' types $\{\theta_i\}_{i=1}^I$ are drawn from some objective distribution $p(\theta_1, \ldots, \theta_I)$, where θ_i belongs to some space Θ_i. For simplicity, let us assume that Θ_i has a finite number $\#\Theta_i$ of elements. θ_i is observed by player i only. $p(\theta_{-i} | \theta_i)$ denotes player i's conditional probability about his opponent's types $\theta_{-i} = (\theta_1, \ldots, \theta_{i-1}, \theta_{i+1}, \ldots, \theta_I)$ given his type θ_i. We assume that the marginal $p_i(\theta_i)$ on each type $\theta_i \in \Theta_i$ is strictly positive.

To complete the description of a Bayesian game, we must specify a pure-strategy space S_i (with elements s_i, and mixed strategies $\sigma_i \in \Sigma_i$) and a payoff function $u_i(s_1, \ldots, s_I, \theta_1, \ldots, \theta_I)$ for each player i.[4] As in the previous chapters, the usual interpretation is that all the exogenous data of the game—the strategy spaces, payoff functions, possible types, and prior distributions—are "common knowledge" in an informal sense (i.e., every player knows them, knows that everybody knows them, and so on). In other words, any initial private information that a player may have is included in the description of his type.[5]

As usual, these strategy spaces are abstract objects which may be contingent plans in some extensive-form game, but for the time being it may be easiest to think of the strategy spaces S_i as representing choices of (uncontingent) actions. Paralleling our development of concepts in parts I and II, we will begin by discussing the solution concepts of Nash equilibrium and iterated strict dominance, which are often strong enough for reasonable predictions in static games but which are typically too weak for strong predictions in dynamic games. Chapters 8 and 11 develop "equilibrium refinements" for dynamic games of incomplete information.

Since each player's choice of strategy can depend on his type, we let $\sigma_i(\theta_i)$ denote the (possibly mixed) strategy player i chooses when his type is θ_i. If player i knew the strategies $\{\sigma_j(\cdot)\}_{j \neq i}$ of the other players as a function of their type, player i could use his beliefs $p(\theta_{-i} | \theta_i)$ to compute the expected utility to each choice and thus find his optimal response $\sigma_i(\theta_i)$. (Aumann (1964) pointed out that there are technical (measurability) problems with this way of modeling strategies when there is a continuum of types. We will say more about this at the end of this chapter when we discuss the work of Milgrom and Weber (1986).)

---

4. As in earlier chapters, we allow the payoff functions to be expectations over moves by nature (random variables) not known by any player when the players pick their strategies.
5. For more on this, see Mertens and Zamir 1985 and chapter 3 of Mertens, Sorin, and Zamir 1990.

6.4 Bayesian Equilibrium[†]

Definition 6.1 A *Bayesian equilibrium* in a game of incomplete information with a finite number of types θ_i for each player i, prior distribution p, and pure-strategy spaces S_i is a Nash equilibrium of the "expanded game" in which each player i's space of pure strategies is the set $S_i^{\Theta_i}$ of maps from Θ_i to S_i.[6]

Given a strategy profile $s(\cdot)$, and an $s_i'(\cdot) \in S_i^{\Theta_i}$, let $(s_i'(\cdot), s_{-i}(\cdot))$ denote the profile where player i plays $s_i'(\cdot)$ and the other players follow $s(\cdot)$, and let

$$(s_i'(\theta_i), s_{-i}(\theta_{-i})) = (s_1(\theta_1), \ldots, s_{i-1}(\theta_{i-1}), s_i'(\theta_i), s_{i+1}(\theta_{i+1}), \ldots, s_I(\theta_I))$$

denote the value of this profile at $\theta = (\theta_i, \theta_{-i})$. Then, profile $s(\cdot)$ is a (pure-strategy) Bayesian equilibrium if, for each player i,

$$s_i(\cdot) \in \arg\max_{s_i'(\cdot) \in S_i^{\Theta_i}} \sum_{\theta_i} \sum_{\theta_{-i}} p(\theta_i, \theta_{-i}) u_i(s_i'(\theta_i), s_{-i}(\theta_{-i}), (\theta_i, \theta_{-i})).$$

Because each type has positive probability, this *ex ante* formulation is equivalent to player i maximizing his expected utility conditional on θ_i for each θ_i:

$$s_i(\theta_i) \in \arg\max_{s_i' \in S_i} \sum_{\theta_{-i}} p(\theta_{-i} | \theta_i) u_i(s_i', s_{-i}(\theta_{-i}), (\theta_i, \theta_{-i})).$$

The existence of a Bayesian equilibrium is an immediate consequence of the Nash existence theorem. (Since Bayesian equilibrium, like Nash equilibrium, is essentially a consistency check, players' beliefs about others' beliefs do not enter the definition—all that matters is each player's own beliefs about the distribution of types and his opponents' type-contingent strategies. Beliefs about beliefs, and so on, become relevant when one is considering the likelihood that play resembles a Bayesian equilibrium, and when one is considering equilibrium refinements.)

6.5 Further Examples of Bayesian Equilibria[††]

This section sketches the analyses of several Bayesian games. Although the first example is straightforward, the details of the other examples become somewhat involved, and many readers may wish to skip them. However, we refer to several of them in section 6.7.

Example 6.2: Cournot Competition with Incomplete Information
Consider a duopoly playing Cournot (quantity) competition. Let firm i's profit be quadratic: $u_i = q_i(\theta_i - q_i - q_j)$, where θ_i is the difference between

---

6. The "expanded game" here closely parallels the expanded game used in describing correlated equilibrium in section 2.2.

the intercept of the linear demand curve and firm i's constant unit cost ($i = 1, 2$) and where q_i is the quantity chosen by firm i ($s_i = q_i$). It is common knowledge that, for firm 1, $\theta_1 = 1$ ("firm 2 has complete information about firm 1," or "firm 1 has only one potential type"). Firm 2, however, has private information about its unit cost. Firm 1 believes that $\theta_2 = \frac{3}{4}$ with probability $\frac{1}{2}$ and $\theta_2 = \frac{5}{4}$ with probability $\frac{1}{2}$, and this belief is common knowledge. Thus, firm 2 has two potential types, which we will call the "low-cost type" ($\theta_2 = \frac{5}{4}$) and the "high-cost type" ($\theta_2 = \frac{3}{4}$). The two firms choose their outputs simultaneously.

Let us look for a pure-strategy equilibrium of this game. We denote firm 1's output by q_1, firm 2's output when $\theta_2 = \frac{5}{4}$ by q_2^L, and firm 2's output when $\theta_2 = \frac{3}{4}$ by q_2^H. Firm 2's equilibrium choice $q_2(\theta_2)$ must satisfy

$$q_2(\theta_2) \in \arg\max_{q_2} \{q_2(\theta_2 - q_1 - q_2)\} \Rightarrow q_2(\theta_2) = (\theta_2 - q_1)/2.$$

Firm 1 does not know which type of firm 2 it faces, so its payoff is the expected value over firm 2's types:

$$q_1 \in \arg\max_{q_1} \{\tfrac{1}{2}q_1(1 - q_1 - q_2^H) + \tfrac{1}{2}q_1(1 - q_1 - q_2^L)\}$$

$$\Rightarrow q_1 = \frac{2 - q_2^H - q_2^L}{4}.$$

Plugging in for $q_2(\theta_2)$, we obtain ($q_1 = 1/3, q_2^L = 11/24, q_2^H = 5/24$) as a Bayesian equilibrium. (In fact, this is the unique equilibrium.)

Example 6.3: War of Attrition
Consider an incomplete-information version of the war of attrition discussed in chapter 4. Player i chooses a number s_i in $[0, +\infty)$. Both players choose simultaneously. The payoffs are

$$u_i = \begin{cases} -s_i & \text{if } s_j \geq s_i \\ \theta_i - s_j & \text{if } s_j < s_i. \end{cases}$$

Player i's type, θ_i, is private information, and takes values in $[0, +\infty)$ with cumulative distribution P and density p. Types are independent between the players. θ_i is the prize received by the winner (i.e., the player whose s_i is highest). The game resembles a second-bid auction in that the winner pays the second bid. However, it differs from the second-bid auction in that the loser also pays the second bid.

Let us look for a (pure-strategy) Bayesian equilibrium $(s_1(\cdot), s_2(\cdot))$ of this game. For each θ_i, $s_i(\theta_i)$ must satisfy

$$s_i(\theta_i) \in \arg\max_{s_i} \left\{ -s_i \, \text{Prob}(s_j(\theta_j) \geq s_i) \right.$$

$$\left. + \int_{\{\theta_j | s_j(\theta_j) < s_i\}} (\theta_i - s_j(\theta_j)) p_j(\theta_j) \, d\theta_j \right\}. \tag{6.1}$$

We will look for profiles in which each player's strategy is a strictly increasing and continuous function of his type. In fact, it can be shown that every equilibrium profile satisfies these properties. To see that equilibrium strategies must be nondecreasing, notice that equilibrium requires that type θ_i' prefer $s_i' = s_i(\theta_i')$ to $s_i'' = s_i(\theta_i'')$ and that type θ_i'' prefer s_i'' to s_i'. Thus,

$$\theta_i' \operatorname{Prob}(s_j(\theta_j) < s_i') - s_i' \operatorname{Prob}(s_j(\theta_j) \geq s_i') - \int_{\{\theta_j | s_j(\theta_j) < s_i'\}} s_j(\theta_j) p_j(\theta_j) \, d\theta_j$$

$$\geq \theta_i' \operatorname{Prob}(s_j(\theta_j) < s_i'') - s_i'' \operatorname{Prob}(s_j(\theta_j) \geq s_i'') - \int_{\{\theta_j | s_j(\theta_j) < s_i''\}} s_j(\theta_j) p_j(\theta_j) \, d\theta_j,$$

and

$$\theta_i'' \operatorname{Prob}(s_j(\theta_j) < s_i'') - s_i'' \operatorname{Prob}(s_j(\theta_j) \geq s_i'') - \int_{\{\theta_j | s_j(\theta_j) < s_i''\}} s_j(\theta_j) p_j(\theta_j) \, d\theta_j$$

$$\geq \theta_i'' \operatorname{Prob}(s_j(\theta_j) < s_i') - s_i' \operatorname{Prob}(s_j(\theta_j) \geq s_i') - \int_{\{\theta_j | s_j(\theta_j) < s_i'\}} s_j(\theta_j) p_j(\theta_j) \, d\theta_j.$$

Subtracting the right-hand side of the second inequality from the left-hand side of the first, and subtracting the left-hand side of the second inequality from the right-hand side of the first, yields

$$(\theta_i'' - \theta_i')[\operatorname{Prob}(s_j(\theta_j) \geq s_i') - \operatorname{Prob}(s_j(\theta_j) \geq s_i'')] \geq 0,$$

so $s_i'' \geq s_i'$ if $\theta_i'' \geq \theta_i'$. (This is the monotonicity property mentioned in example 6.1.)

The argument that strategies must be *strictly* increasing and continuous is more involved, and we will only give the intuition. First, if strategies were not strictly increasing, there would be an "atom" at some $s > 0$, i.e., an s such that $\operatorname{Prob}(s_j(\theta_j) = s) > 0$. In this case, player i would assign probability 0 to the interval $[s - \varepsilon, s)$ for ε small, as she does better playing just above s (this argument is a bit loose, but can be made rigorous). Thus, the types of player j that play s would be better off playing $s - \varepsilon$, because this would not reduce the probability of winning and would lead to reduced cost, so there cannot be an atom at s after all. A similar intuition underlies the argument that strategies must be continuous. If they were discontinuous, then there would be an $s' \geq 0$ and an $s'' > s'$ such that $\operatorname{Prob}(s_j(\theta_j) \in [s', s'']) = 0$ while $s_j(\hat{\theta}_j) = s'' + \varepsilon$ for some small $\varepsilon \leq 0$ for some $\hat{\theta}_j$. In this case, player i strictly prefers $s_i = s'$ to any $s_i \in (s', s'')$, as the probability of winning is the same and the expected cost is reduced. But then quitting "at or just beyond" s'' is not optimal for player j with type $\hat{\theta}_j$.

Let us look for a strictly increasing, continuous function s_i with inverse Φ_i—that is, $\Phi_i(s_i)$ is the type that plays s_i. Transforming the variable of integration from θ_j to s_j in equation 6.1 (using the formula for the transformation of the densities[7]) gives

$$s_i(\theta_i) \in \arg\max_{s_i} \left\{ -s_i(1 - P_j(\Phi_j(s_i))) \right.$$

$$\left. + \int_0^{s_i} (\theta_i - s_j)p_j(\Phi_j(s_j))\Phi_j'(s_j)\, ds_j \right\}. \qquad (6.2)$$

The corresponding first-order conditions are that type θ_i cannot increase its payoff by playing $s_i + ds_i$ instead of s_i where $s_i \equiv s_i(\theta_i)$. This change costs ds_i if player j plays above $s_i + ds_i$, which has probability $1 - P_j(\Phi_j(s_i + ds_i))$; thus, the expected incremental cost is $(1 - P_j(\Phi_j(s_i)))\, ds_i$ to the first order in ds_i. The change yields a gain of $\theta_i = \Phi_i(s_i)$ if player j plays in the interval $[s_i, s_i + ds_i)$, which occurs if θ_j is the interval $[\Phi_j(s_i), \Phi_j(s_i + ds_i))$; this has probability $p_j(\Phi_j(s_i))\Phi_j'(s_i)\, ds_i$. Equating the costs and benefits, we obtain the first-order conditions[8]

$$\Phi_i(s_i)p_j(\Phi_j(s_i))\Phi_j'(s_i) = 1 - P_j(\Phi_j(s_i)). \qquad (6.3)$$

7. If random variable x has density $p(x)$ and $f: X \to Y$ is one-to-one, then $y = f(x)$ has density g given by

$$g(y) = \frac{p(f^{-1}(y))}{f'(f^{-1}(y))} = p(f^{-1}(y))(f^{-1})'(y).$$

8. Let us show that the global second-order conditions are satisfied if the first-order conditions are. Let $U_i(s_i, \theta_i)$ denote the maximand in equation 6.2. Note that

$$\frac{\partial^2 U_i}{\partial s_i \partial \theta_i} = p_j(\Phi_j(s_i))\Phi_j'(s_i) > 0.$$

Suppose that there exist a type θ_i and a strategy s_i' such that

$$U_i(s_i', \theta_i) > U_i(s_i, \theta_i),$$

where $s_i = s_i(\theta_i)$. This implies that

$$\int_{s_i}^{s_i'} \frac{\partial U_i}{\partial s}(s, \theta_i)\, ds > 0.$$

Or, using the first-order condition $(\partial U_i/\partial s)(s, \Phi_i(s)) = 0$ for all s,

$$\int_{s_i}^{s_i'} \left(\frac{\partial U_i}{\partial s}(s, \theta_i) - \frac{\partial U_i}{\partial s}(s, \Phi_i(s)) \right) ds > 0$$

or

$$\int_{s_i}^{s_i'} \int_{\Phi_i(s)}^{\theta_i} \frac{\partial^2 U_i}{\partial s \partial \theta}(s, \theta)\, d\theta\, ds > 0.$$

If, for instance, $s_i' > s_i$, then $\Phi_i(s) > \theta_i$ for all $s \in (s_i, s_i']$, and the last inequality cannot hold. And similarly for $s_i' < s_i$. So s_i is globally optimal for type θ_i.

Next, we suppose that $P_1 = P_2 = P$, and look for a symmetric equilibrium. Substituting $\theta = \Phi(s)$ in equation 6.3, and using the fact that $\Phi' = 1/s'$,[9] we have

$$s'(\theta) = \frac{\theta\, p(\theta)}{1 - P(\theta)},$$
                                                        (6.4a)

or

$$s(\theta) = \int_0^\theta \left(\frac{x\, p(x)}{1 - P(x)} \right) dx,$$
                                                        (6.4b)

where the constant of integration is determined by $s(0) = 0$: Types with 0 value for the good are unwilling to fight for it.

We leave it to the reader to check that, for a symmetric exponential distribution $P(\theta) = 1 - \exp(-\theta)$, there exists a symmetric equilibrium: $\Phi(s) = \sqrt{2s}$, which corresponds to $s(\theta) = \theta^2/2$. (Riley (1980) shows that there also exists a continuum of asymmetric equilibria: $\Phi_1(s_1) = K\sqrt{s_1}$ and $\Phi_2(s_2) = (2/K)\sqrt{s_2}$ for $K > 0$.)

Aside We can give this war of attrition the following industrial-organization interpretation: Suppose that there are two firms in the market. Each firm loses 1 per unit of time when they compete. They make a monopoly profit when their opponents have left the market, the present discounted value of which is θ_i. (More realistically, we could allow the duopoly and monopoly profit to be correlated, but this would not change the results very much.) Then, s_i is the length of time firm i intends to stay in the market, if firm j has not exited before.[10,11]

Example 6.4: Double Auction

In a double auction, potential sellers and buyers of a single good move simultaneously, with the sellers submitting asking prices and the buyers submitting bids. An auctioneer then chooses a price p that clears the market: All the sellers who ask less than p sell, all the buyers who bid more

9. This is the inverse-function theorem.
10. See chapter 4 for an introduction to the symmetric-information war of attrition. The incomplete-information war of attrition was introduced in the theoretical biology literature by Bishop, Cannings, and Maynard Smith (1978), and extended by Riley (1980), Kreps and Wilson (1982), Nalebuff (1982), Nalebuff and Riley (1983), and Bliss and Nalebuff (1984). For a characterization of the set of equilibria and a uniqueness result with non stationary flow payoffs and/or large uncertainty over types, see Fudenberg and Tirole 1986.
11. Some readers may wonder whether the concept of Nash equilibrium is sufficiently strong for this dynamic interpretation of the game and whether a stronger equilibrium concept might reduce this multiplicity. In our study of the stationary complete-information war of attrition in chapter 4, we saw that all the Nash equilibria are subgame perfect. Similarly, the multiple equilibria just described satisfy the concept of perfect Bayesian equilibrium we introduce in chapter 8. (They trivially satisfy the concept of subgame perfection introduced in chapter 3, as the only proper subgame is the game itself.)

than p buy, and the total number of units supplied at price p equals the number demanded. (Any buyers or sellers who named exactly p are indifferent, and their allocations are chosen so that the quantity demanded equals the quantity supplied.)

Chatterjee and Samuelson (1983) consider the simplest example of a double auction, in which a single seller and a single buyer may trade 0 or 1 unit of a good. The seller (player 1) has cost c, and the buyer (player 2) has valuation v, where v and c belong to the interval $[0, 1]$. The seller and the buyer simultaneously choose bids b_1, $b_2 \in [0, 1]$. If $b_1 \leq b_2$, the two parties trade at price $t = (b_1 + b_2)/2$.[12] If $b_1 > b_2$, the parties do not trade the good and do not transfer money. The seller's utility is thus $u_1 = (b_1 + b_2)/2 - c$ if $b_1 \leq b_2$, and 0 if $b_1 > b_2$; the buyer's utility is $u_2 = v - (b_1 + b_2)/2$ if $b_1 \leq b_2$, and 0 if $b_1 > b_2$.

Under symmetric information (that is, with v and c common knowledge when the two parties bid), this is known as the *Nash* (1953) *demand game*. If we assume $v > c$ to make things interesting, the symmetric-information game has a continuum of pure-strategy, efficient equilibria in which the two parties bid the same amount: $b_1 = b_2 = t \in [c, v]$. In such equilibria the two traders realize positive surplus. If either tries to be more greedy (the seller asks for more than t or the buyer bids less than t), trade does not occur. There are also inefficient equilibria, in which the parties make nonserious offers: The seller asks for more than v and the buyer bids less than c.[13]

Now consider asymmetric information, where the seller's cost is distributed according to distribution P_1 in $[0, 1]$ and the buyer's valuation has distribution P_2 on the same interval. These distributions are common knowledge. Chatterjee and Samuelson look for a pure-strategy equilibrium $(s_1(\cdot), s_2(\cdot))$ where s_1 and s_2 map $[0, 1]$ into $[0, 1]$. Let $F_1(\cdot)$ and $F_2(\cdot)$ denote the equilibrium cumulative distributions of the seller's and the buyer's bids, respectively. That is, $F_1(b)$ is the probability that the seller has a cost that induces him to bid less than b:

$$F_1(b) = \text{Prob}(s_1(c) \leq b).$$

And similarly for the buyer.

For types who trade with positive probability, equilibrium bids are necessarily increasing in type. Consider for instance two costs, c' and c'', for the seller, and let $b_1' \equiv s_1(c')$ and $b_1'' \equiv s_1(c'')$. Then optimization by the seller requires that

$$\int_{b_1'}^1 \left(\frac{b_1' + b_2}{2} - c' \right) dF_2(b_2) \geq \int_{b_1''}^1 \left(\frac{b_1'' + b_2}{2} - c' \right) dF_2(b_2)$$

12. Chatterjee and Samuelson thus assume that the parties split the gains from the trade. More generally, the price in a two-player double auction is set at $kb_1 + (1 - k)b_2$, where $k \in [0, 1]$.
13. See exercise 1.3 for Nash's suggestion on how to select among equilibria.

and

$$\int_{b_1''}^1 \left(\frac{b_1'' + b_2}{2} - c''\right) dF_2(b_2) \geq \int_{b_1'}^1 \left(\frac{b_1' + b_2}{2} - c''\right) dF_2(b_2).$$

Combining these inequalities gives

$$(c'' - c')[F_2(b_1'') - F_2(b_1')] \geq 0,$$

so that $b_1'' \geq b_1'$ if $c'' > c'$.[14] And similarly for the buyer.

Chatterjee and Samuelson require further that each player's strategy as a function of his type be strictly increasing and continuously differentiable. The maximization problem of the type-c seller is then

$$\underset{b_1}{\text{Max}} \int_{b_1}^1 \left(\frac{b_1 + b_2}{2} - c\right) dF_2(b_2),$$

which implies that either

(i) $\quad \frac{1}{2}[1 - F_2(s_1(c))] - (s_1(c) - c)f_2(s_1(c)) = 0$

or

(ii) $\quad \frac{1}{2}[1 - F_2(s_1(c))] - (s_1(c) - c)f_2(s_1(c)) > 0$ and $s_1(c) = 1$

or

(iii) $\quad \frac{1}{2}[1 - F_2(s_1(c))] - (s_1(c) - c)f_2(s_1(c)) < 0$ and $s_1(c) = 0$.

Since $F_2(1) = 1$ and $F_2(0) = 0$, the boundary constraints $s_1 \in [0, 1]$ do not bind and the relevant condition is i. Note that for cost c above the highest buyer bid $\bar{s}_2$ the seller's optimal bid is any $s_1 > \bar{s}_2$, and all such bids satisfy the seller's first-order condition, since for these bids both $f_2(s_1(c))$ and $1 - F_2(s_1(c))$ equal 0. (Similar remarks apply to the buyer's first-order condition below.) Note that this first-order condition yields the same formula as for a monopoly seller, except that when the seller raises his price by 1 the trading price increases by $\frac{1}{2}$ instead of 1. We have an analogous formula for the buyer:

$$\underset{b_2}{\text{Max}} \int_0^{b_2} \left(v - \frac{b_1 + b_2}{2}\right) dF_1(b_1) \Rightarrow [v - s_2(v)]f_1(s_2(v)) = \frac{1}{2}F_1(s_2(v)).$$

Suppose now, following Chatterjee and Samuelson, that P_1 and P_2 are uniform distributions on $[0, 1]$, and look for linear strategies, so that

$$s_1(c) = \alpha_1 + \beta_1 c$$

14. To make this conclusion rigorous, we observe that $F_2(b_1'') = F_2(b_1') < 1$ is impossible: Type c' of the seller would be better off asking the higher price because this would not affect the probability of trade, and trade would take place at a higher price when it does.

and

$$s_2(v) = \alpha_2 + \beta_2 v.$$

Then

$$F_i(b) = P_i(s_i^{-1}(b)) = s_i^{-1}(b) = (b - \alpha_i)/\beta_i$$

and

$$f_i(b) = 1/\beta_i.$$

Plugging this into the first-order conditions yields

$$2[\alpha_1 + (\beta_1 - 1)c]/\beta_2 = [\beta_2 - (\alpha_1 + \beta_1 c) + \alpha_2]/\beta_2$$

and

$$2[(1 - \beta_2)v - \alpha_2]/\beta_1 = (\alpha_2 + \beta_2 v - \alpha_1)/\beta_1.$$

Since these equations must hold for all c and v, we can identify the constant terms and coefficients of c and v on the two sides of these, obtaining

$$2(\beta_1 - 1) = -\beta_1,$$

$$2(1 - \beta_2) = \beta_2,$$

$$2\alpha_1 = \beta_2 - \alpha_1 + \alpha_2,$$

$$-2\alpha_2 = \alpha_2 - \alpha_1.$$

Solving this system, we have

$$\beta_1 = \beta_2 = \tfrac{2}{3},$$

$$\alpha_1 = \tfrac{1}{4},$$

$$\alpha_2 = \tfrac{1}{12}.$$

With these strategies, player 1's bid of $\tfrac{1}{4} + \tfrac{2}{3}c$ is less than his cost if $c > \tfrac{3}{4}$. However, for costs in this range, $s_1(c)$ also exceeds $\tfrac{3}{4}$, which is player 2's maximum bid, so player 1's strategy never leads him to sell at a price below his cost. Similarly, player 2's bid exceeds his valuation when $v < \tfrac{1}{4}$, but again for such bids trade never takes place.

In equilibrium the parties trade if and only if $\alpha_2 + \beta_2 v \geq \alpha_1 + \beta_1 c$, or $v \geq c + \tfrac{1}{4}$. Comparing with the *ex post* efficient trading pattern (trade if and only if $v \geq c$), we conclude that there is too little trading in equilibrium.

As one would expect from the symmetric-information case, there are other equilibria in this double auction. In particular, both parties making nonserious offers ($b_1 = 1$ and $b_2 = 0$) is an equilibrium. There also exists a continuum of "single-price" equilibria at $b \in [0, 1]$. The seller asks b if $c \leq b$ and 1 if $c > b$, and the buyer offers b if $v \geq b$ and 0 if $v < b$. Because the price is "fixed" at b if trade takes place, no player has any incentive to deviate. More interestingly, Leininger, Linhart, and Radner (1989) show that there exists a one-parameter continuum of differentiable and symmetric (but nonlinear) equilibrium strategies. (There exists a two-parameter continuum of differentiable, asymmetric equilibrium strategies; see Satterthwaite and Williams 1989.) Leininger et al. also show existence of other, discontinuous equilibria.

Example 6.5: First-Price Auction with a Continuum of Types (technical)
In a first-price auction, the bidder who offers the highest price gets the good and pays his bid (in contrast with the second-price auction analyzed in subsection 1.1.3, in which the highest bidder pays the second highest bid); the other bidders do not pay anything. In this example, we study the equilibria of two-bidder, symmetric-uncertainty first-price auctions when the valuations belong to an interval; the next example considers the same game when each valuation belongs to a two-point set. The point of going through two examples of first-price auctions is to illustrate the different techniques used to solve the continuous and discrete cases. (The analysis of the first example is fairly complicated.) There are two bidders, $i = 1, 2$, and one unit of a good for sale. Player i's valuation is θ_i and belongs to $[\underline{\theta}, \overline{\theta}]$, where $\underline{\theta} \geq 0$. Each player knows his own valuation and has beliefs P with positive density p on $[\underline{\theta}, \overline{\theta}]$ about his rival's valuation. The valuations are independent. The seller imposes a reservation price $s_0 > \underline{\theta}$, meaning that bids below s_0 are rejected. Player i's bid is s_i. The utility of player i is $u_i = \theta_i - s_i$ if $s_i > s_j$ and $s_i \geq s_0$; it is $u_i = 0$ if $s_i < s_j$ or $s_i < s_0$. If both players bid the same amount, we assume that each gets the good with probability $\frac{1}{2}$: If $s_i = s_j \geq s_0$, $u_i = (\theta_i - s_i)/2$. Let $s_i(\cdot)$ denote the (pure) equilibrium strategy of player i. (We leave it to the reader to show that s_i is increasing in θ_i, by following the steps of the monotonicity proofs in examples 6.3 and 6.4.)

Bayesian equilibrium strategies can be characterized intuitively as follows.[15] First, note that a player with valuation less than s_0 does not bid (or, rather, bids less than s_0). Second, as in the war of attrition, show that the strategies have no atoms at bids greater than s_0. Next, argue that the strategies have no "gaps." Suppose that player i, whatever his type, does not bid in the interval $[s_i^-, s_i^+]$, where $s_i^- \geq s_0$, but there are types of player i who bid s_i^+ or arbitrarily close to s_i^+. Then player j, whatever his type, ought not to bid $s_j \in (s_i^-, s_i^+)$: Starting from any such s_j, if player j reduces his bid slightly, he does not affect his probability of winning, and he reduces the price he pays when he wins. But then a type of player i who bids s_i^+ or

---

15. The (rigorous) characterization is given in Maskin and Riley 1986a. The style of proof that the distribution of bids has no atoms and that the strategies are strictly increasing is also common in search theory (e.g., Butters 1977) and in wars of attrition (e.g., Fudenberg and Tirole 1986).

arbitrarily close to s_i^+ would be better off bidding just above s_i^-, as he would reduce his probability of winning by an infinitesimal amount (recall that player i has no atom at s_i^+) and would substantially reduce his payment when winning.

In this manner one can show that the strategies are continuous and strictly increasing beyond s_0. It is easy to see that $s_i(\bar{\theta}) = s_j(\bar{\theta}) \equiv \bar{s}$. (If $s_i(\bar{\theta}) > s_j(\bar{\theta})$, then type $\bar{\theta}$ of player i could lower his bid slightly and still win the auction with probability 1.) Let $\theta_i = \Phi_i(s)$ denote the inverse function of $s_i(\cdot)$ on $(s_0, \bar{s}]$. That is, player i bids s when his valuation is $\Phi_i(s)$. The function $\Phi_i(\cdot)$ is differentiable almost everywhere because it is monotonic.

Type θ_i maximizes $(\theta_i - s)P(\Phi_j(s))$ over s. This yields

$$P(\Phi_j(s)) = [\Phi_i(s) - s]p(\Phi_j(s))\Phi_j'(s). \tag{6.5}$$

Equation (6.5) and the symmetric equation obtained by switching i and j yield two first-order differential equations in the functions $\Phi_1(\cdot)$ and $\Phi_2(\cdot)$. Let $G_j(\cdot)$ denote the cumulative distribution of bids, $G_j(s) = P(\Phi_j(s))$, with density $g_j(s) = p(\Phi_j(s))\Phi_j'(s)$. Equation 6.5 can then be rewritten as

$$G_j(s) = [\Phi_i(s) - s]g_j(s). \tag{6.6}$$

Note the analogy with monopoly pricing: A unit increase in price raises revenue by the expected probability of winning, $G_j(s)$, but the bidder's surplus $(\Phi_i(s) - s)$ is lost with probability $g_j(s)$.

We now investigate the boundary conditions for equation 6.5. Recall that $\Phi_i(\bar{s}) = \bar{\theta}$ for all i. Further, $\lim_{s \downarrow s_0} \Phi_i(s) = s_0$ for at least some i. (Suppose both players have atoms at s_0, i.e., that types $\theta_i \in [s_0, s_0 + a_i]$, $a_i > 0$, bid s_0 for $i = 1, 2$. Then type $s_0 + a_i$ of player i could bid slightly more than s_0 and increase its probability of winning by a nonnegligible amount.) So these two boundary conditions might seem to pin down a unique solution to equation 6.5.

Although the solution is indeed unique, the reasoning is a bit more complex than this, because Φ_j' in equation 6.5 is not Lipschitz continuous at s_0 if $\Phi_i(s_0) = s_0$.[16] Integrating equation 6.5 yields

$$\ln \frac{P(\Phi_2(s))}{P(\Phi_1(s))} = \int_s^{\bar{s}} \left(\frac{1}{\Phi_2(x) - x} - \frac{1}{\Phi_1(x) - x} \right) dx. \tag{6.7}$$

Equation 6.5 shows that if $\Phi_1(s) = \Phi_2(s)$ for some $s \in (s_0, \bar{s}]$, then the solution is symmetric: $\Phi_1(s) = \Phi_2(s)$ for all $s \in (s_0, \bar{s}]$ (and, by continuity, for $s = s_0$ as well). Can there exist an asymmetric solution? By the previous reasoning, $\Phi_1(s) \neq \Phi_2(s)$ for all s in $(s_0, \bar{s}]$. Suppose that, without loss of

16. That is, the slope of Φ_j goes to infinity. Standard results on the uniqueness of solutions to differential equations require Lipschitz continuity. The war of attrition of example 6.3 is not Lipschitz continuous at $s = 0$, which is why it is possible for the system represented in equation 6.3 to have multiple solutions.

generality, $\Phi_2(s) > \Phi_1(s)$ for all s in $(s_0, \bar{s}]$. Then equation 6.7 implies that $P(\Phi_2(s))/P(\Phi_1(s))$ is greater than 1 and increases from s to $\bar{s}$. Hence it cannot converge to 1 at $\bar{s}$, a contradiction.

We conclude that any equilibrium is symmetric, which implies that there is no atom at s_0. From equation 6.5, $\Phi_1 = \Phi_2 = \Phi$ satisfies

$$\ln(P(\Phi(s))) = -\int_s^{\bar{s}} \frac{dx}{\Phi(x) - x}. \tag{6.8}$$

To show that there exists a unique equilibrium it suffices to note that there exists a unique $\bar{s}$ such that, if $\Phi(\cdot)$ is given by equation 6.8, then $\Phi(s_0) = s_0$.[17]

We thus conclude that as long as $s_0 > \underline{\theta}$ there is a unique solution; it is symmetric and satisfies $P(\Phi(s)) = [\Phi(s) - s]p(\Phi(s))\Phi'(s)$ and $\Phi(s_0) = s_0$. The equilibrium strategy $s(\cdot)$ is the inverse of the function $\Phi(\cdot)$.

Example 6.6: First-Price Auction with Two Types
As a last example, we analyze equilibrium in the first-price auction (see example 6.5) when each of two bidders has one of two possible valuations, $\underline{\theta}$ and $\bar{\theta}$ (with $\underline{\theta} < \bar{\theta}$). The valuations are independent; let $\bar{p}$ and $\underline{p}$ denote the probability that θ_i equals $\bar{\theta}$ and $\underline{\theta}$, respectively (with $\underline{p} + \bar{p} = 1$). To make things interesting, assume that the seller's reservation price or minimum bid is lower than $\underline{\theta}$. The new technical twist when the support of the distribution of types is discrete rather than continuous is that players must play a mixed strategy in equilibrium.

We look for an equilibrium where, for each player, type $\underline{\theta}$ bids $\underline{\theta}$ and type $\bar{\theta}$ randomizes according to the continuous distribution $F(s)$ on $[\underline{s}, \bar{s}]$. (It can be shown that the equilibrium is unique.) Clearly, $\underline{s} = \underline{\theta}$: If $\underline{s} > \underline{\theta}$, then a player with valuation $\bar{\theta}$ would be better off bidding just above $\underline{\theta}$ rather than bidding $\underline{s}$ (or close to $\underline{s}$), as this would not reduce his probability of winning and would reduce his payment when he wins. In order for player i with type $\bar{\theta}$ to play a mixed strategy with support $[\underline{s}, \bar{s}]$, it must be the case that

$$\forall s \in [\underline{s}, \bar{s}], (\bar{\theta} - s)[\underline{p} + \bar{p}F(s)] = \text{constant}. \tag{6.9}$$

(Type $\bar{\theta}$'s expected payoff is not affected by bids he makes with probability 0. Thus, even though playing $\underline{s}$ with positive probability will result in a lower expected payoff because it risks tying with type $\underline{\theta}$, bid $\underline{s}$ can still belong to the support of type $\bar{\theta}$'s equilibrium strategy.[18]) Because $F(\underline{\theta}) = 0$, the constant is equal to $(\bar{\theta} - \underline{\theta})\underline{p}$. Thus, $F(\cdot)$ is defined by

---

17. The proof is similar to that proving that there cannot be asymmetric equilibria: Consider two highest bids, $\bar{s}^1 > \bar{s}^2$, and let Φ^1 and Φ^2 denote the corresponding solutions. Then $\Phi^2(\bar{s}^2) = \bar{\theta} > \Phi^1(\bar{s}^2)$. For any $s \leq \bar{s}^2$, $P(\Phi^2(s))/P(\Phi^1(s))$ is greater than 1 and is decreasing. Hence, $P(\Phi^2(s))/P(\Phi^1(s))$ cannot converge to 1 when s converges to s_0.
18. Recall that the support of a probability distribution is the smallest closed set that has probability 1.

$$(\bar{\theta} - s)[\underline{p} + \bar{p}F(s)] = (\bar{\theta} - \underline{\theta})\,p. \tag{6.10}$$

Letting $G(s) \equiv \underline{p} + \bar{p}F(s)$ denote the cumulative distribution of bids for $s \geq \underline{\theta}$, we can rewrite equation 6.10 as

$$(\bar{\theta} - s)G(s) = (\bar{\theta} - \underline{\theta})\,p. \tag{6.11}$$

Last, $F(\bar{s}) = 1$ implies that

$$(\bar{\theta} - \bar{s}) = (\bar{\theta} - \underline{\theta})\,p, \text{ or } \bar{s} = \bar{p}\bar{\theta} + \underline{p}\underline{\theta}. \tag{6.12}$$

Since the seller's reservation price is below $\underline{\theta}$, trade always takes place, and the seller's expected profit is equal to the expected social surplus minus the expected utility of the bidders. Expected social surplus is equal to $\underline{p}^2\underline{\theta} + (1 - \underline{p}^2)\bar{\theta}$. Each bidder's net utility is 0 when he has type $\underline{\theta}$ and $\underline{p}(\bar{\theta} - \underline{\theta})$ when he has type $\bar{\theta}$. (Because type $\bar{\theta}$ is indifferent among bids in $(\underline{\theta}, \bar{s}]$, his utility can be computed by assuming he bids just above $\underline{\theta}$, in which case he wins with probability $\underline{p}$.)

It is interesting to note that both expected social surplus and the bidders' utility (and therefore the seller's expected profit) are the same as in the second-price auction studied in chapter 1. This fact, known as the *revenue-equivalence theorem*, would also hold in the continuous case of example 6.5. (We will see in chapter 7 that the first-price and second-price auctions do not maximize the seller's expected revenue in the two-type case; they do maximize revenue under some conditions in the continuum case.)

6.6 Deletion of Strictly Dominated Strategies[††]

6.6.1 Interim vs. *Ex Ante* Dominance

If player i, instead of knowing the type-contingent strategies of his opponents, must try to predict them, then player i must be concerned with how player $j \neq i$ *thinks* player i would play for each possible type player i might have. And player i must also try to estimate player j's beliefs about player i's type, in order to predict the distribution of strategies that player i expects to face.

This brings us to the question of how the players predict their opponents' strategies, which in turn raises the following question: Should different types θ_1 and θ_1' of player 1 be viewed simply as a way of describing different information sets of a *single* player 1, who makes a type-contingent decision at the *ex ante* stage (that is, before he learns his type)? This interpretation seems natural in the Harsanyi formulation, which introduces a move by nature that determines the "type" of a single player 1. Alternatively, should we think of θ_1 and θ_1' as denoting two different "individuals," one of whom is selected by nature to "appear" when the game is played? In the first interpretation, the single *ex ante* player 1 should be thought of as predicting his opponents' play at the *ex ante* stage, so all types of player 1 would make the same prediction about the play of the other players. Under the second interpretation, the "different individuals" corresponding to different θ_1's would each make their predictions at the "interim" stage (i.e., after learning their type), and the different types could make different predictions. (This second interpretation may become more plausible if we imagine that the "types" correspond to aspects of preferences that are genetically determined, for here the "*ex ante*" stage is difficult to interpret literally.)

It is interesting to see that iterated strict dominance is at least as strong in the *ex ante* interpretation as in the interim interpretation and that the *ex ante* interpretation yields strictly stronger predictions in some games. To illustrate this, let us return to the public-good game of example 6.1. Using the interim approach to dominance, we ask which strategies are strictly dominated for player i when his cost is c_i. Not contributing is not dominated for any positive cost level, as it is always better not to contribute if you expect that the opponent will contribute. However, if c_i is greater than the private benefit of the good, which is 1, then contributing is strictly dominated for player i.

If the lowest possible cost, $\underline{c}$, is greater than $1 - P(1)$, the deletion process stops after only one round: For all types in $[\underline{c}, 1]$, neither "contribute" nor "don't contribute" is dominated. In particular, interim dominance does not preclude the situation where, for some c' between $\underline{c}$ and 1, all types in an interval $[\underline{c}, c']$ don't contribute and all types in $(c', 1]$ do contribute—the types in $[\underline{c}, c']$ will not contribute if they expect that their opponent will contribute whenever his cost is less than 1, while the types in $(c', 1]$ should contribute if they expect that no type of their opponent will contribute.

This situation could not arise in a Bayesian equilibrium, because, as we saw, in any Bayesian equilibrium each player's strategy must be a cutoff rule of the form "contribute if and only if $c_i \leq c'$ for some c'." That is, in Bayesian equilibrium, if the type of player i with a given cost level contributes, all player i's types with lower costs must contribute as well.

The conclusion that players' strategies must take the form of cutoff rules also follows from applying strict dominance at the *ex ante* stage. To see this, note that any strategy $s_i(\cdot)$ for player i that has player i contribute with probability $z > 0$ and is *not* a cutoff rule is strictly dominated *ex ante* by the strategy that has player i contribute if and only if $c_i < c'$, where c' is defined by $P(c') = z$. With this cutoff rule, for any strategy $s_j(\cdot)$ of the opponent, player i receives the public good with the same probability as when using s_i, and player i's expected cost of contributing is strictly lower. The point is that if player i is a single individual optimizing against the play of player j, then any beliefs about j's strategy that make it attractive to contribute with cost c' also make it attractive to contribute for all lower costs.

More generally, the reason that more strategies are dominated *ex ante* than *ex post* is that, for a given type-contingent strategy $\hat{\sigma}_1(\cdot)$ of player 1, it is easier to find a $\sigma_1(\cdot)$ satisfying the *ex ante* dominance condition

$$\sum_{\theta_1} p_1(\theta_1) \sum_{\theta_{-1}} p(\theta_{-1}|\theta_1)u_1(\sigma_1(\theta_1), \sigma_{-1}(\theta_{-1}), \theta)$$

$$> \sum_{\theta_1} p_1(\theta_1) \sum_{\theta_{-1}} p(\theta_{-1}|\theta_1)u_1(\hat{\sigma}_1(\theta_1), \sigma_{-1}(\theta_{-1}), \theta)$$

for all $\sigma_{-1}(\cdot)$ than it is to find a s_1 and a θ_1 that satisfy the interim constraints

$$\sum_{\theta_{-1}} p(\theta_{-1}|\theta_1)u_1(s_1, \sigma_{-1}(\theta_{-1}), \theta) > \sum_{\theta_{-1}} p(\theta_{-1}|\theta_1)u_1(\hat{\sigma}_1(\theta_1), \sigma_{-1}(\theta_{-1}), \theta)$$

for all $\sigma_{-1}(\cdot)$. (One way of putting this is that the *ex ante* approach "pools the domination constraints" and allows the use of "wasted slack" on some constraints.) This difference does not arise when the Nash concept is used, as Nash equilibrium supposes that all players make the same predictions about the strategies that will be played, whereas dominance arguments allow two players to make different predictions about the play of a third.

6.6.2 Examples of Iterated Strict Dominance

Now we present two examples of incomplete-information games where iterated dominance does lead to a unique prediction.

The first is the public-good game of example 6.1 when $\underline{c} < 1 - P(1)$ and there exists a unique c^* such that $c^* = 1 - P(1 - P(c^*))$. Here even interim iterated dominance gives a unique prediction.

Recall that at the first round of iteration we concluded that no type with cost over 1 would contribute. (Contributing is strictly dominated for all $c_i \in (c^1, \bar{c}]$, where $c^1 \equiv 1$.) At the second round, not contributing is strictly dominated for all $c_i \in [\underline{c}, c^2)$, where $c^2 \equiv 1 - P(1) = 1 - P(c^1)$. In contrast, the optimal strategy for types $c_i \in [c^2, c^1]$ depends on what types $c_j \in [c^2, c^1]$ do; hence, no strategy for these types can be eliminated in the second round. In the third round, types close to 1 should not contribute, as the cost of contributing is close to the private value of the public good, and there is a probability of at least $P(c^2)$ that the other player contributes. Thus, if $c_i > c^3 \equiv 1 - P(c^2)$, contributing is a strictly dominated strategy for player i, and so on. Iterating the process of deletion of strictly dominated strategies yields, at stage $2k + 1$ ($k = 0, 1, \ldots$), that contributing is a strictly dominated strategy for types greater than $c^{2k+1} \equiv 1 - P(c^{2k})$. And at stage $2k$ ($k = 1, 2, \ldots$), not contributing is a strictly dominated strategy for types lower than $c^{2k} \equiv 1 - P(c^{2k-1})$. The sequences $\{c^{2k+1}\}_{k=0,1,\ldots}$ and $\{c^{2k}\}_{k=1,2,\ldots}$ are strictly decreasing and strictly increasing, respectively. Because they are bounded, they converge to two numbers c^+ and c^-. Because P is continuous, $c^+ = 1 - P(c^-)$ and $c^- = 1 - P(c^+)$. If there is a unique c^* such that $c^* = 1 - P(1 - P(c^*))$, which is the condition for a

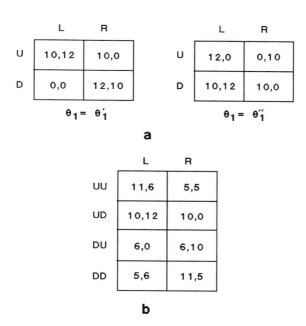

Figure 6.5

unique Nash equilibrium, then $c^+ = c^- = c^*$ and the game is *solvable by (interim) iterated deletion of strictly dominated strategies.*

In our second example, *ex ante* iterated dominance gives a unique prediction, but interim iterated dominance does not.

Consider the game illustrated in figure 6.5. Player 1 has two possible types, θ_1' and θ_1'', each of which has prior probability $\frac{1}{2}$. Figure 6.5a displays the payoff matrices corresponding to player 1's two types; figure 6.5b shows the strategic form for the imperfect-information game where player 1 chooses type-contingent strategies. Here the first component of player 1's strategy is his play when he is of type θ_1', and the second component is his play when he is of type θ_1''; payoffs are obtained by taking the expected value with respect to the prior distribution.

Using interim dominance, neither U nor D can be eliminated for either type of player 1—both types prefer U if player 2 plays L, and both prefer D if player 2 plays R. And interim iterated dominance stops at this point. However, when the two types of player 1 are equally likely, as is assumed in figure 6.5b, the type-contingent strategy DU *is* strictly dominated by UD in figure 6.5b. And once DU is deleted for player 1, L dominates R for player 2. At the next round, UU dominates UD and DD, and the unique outcome surviving *ex ante* iterated dominance is (UU, L). (If the prior probability of θ_1' is 0.9, DU is not dominated by UD.)

6.7   Using Bayesian Equilibria to Justify Mixed Equilibria[†]

6.7.1   Examples

In chapter 1 we saw that simultaneous-move games of *complete* information often admit mixed-strategy equilibria. Some researchers are unhappy with this notion because, they argue, "real-world decision makers do not flip coins." However, as Harsanyi (1973) has shown, mixed-strategy equilibria of complete-information games can usually be interpreted as the limits of pure-strategy equilibria of slightly perturbed games of incomplete information. Indeed, we have already noticed that in a Bayesian game, once the players' type-contingent strategies have been computed, each player behaves as if he were facing mixed strategies by his opponents. (The uncertainty arises through the distribution of types rather than through "coin flips.")

Example 6.7: "Grab the Dollar"
To illustrate the mechanics of this construction, let us consider a one-period variant of the "grab the dollar" game introduced in chapter 4. Each player has two possible actions: invest ("grab") and don't invest. In the complete-information version of the game, a firm gains 1 if it is the only one to invest, loses 1 if both invest, and breaks even if it does not invest. (We can view this game as an extremely crude representation of entry into a natural-monopoly market.) The only symmetric equilibrium is that each firm invests with probability $\frac{1}{2}$. This clearly is an equilibrium, as each firm makes 0 if it does not invest and $\frac{1}{2}(1) + \frac{1}{2}(-1) = 0$ if it does. Now consider the same game with the following type of incomplete information: Each firm has the same payoff structure, except that when it wins it gets $(1 + \theta_i)$, where θ_i is uniformly distributed on $[-\varepsilon, \varepsilon]$. Each firm knows its type, θ_i, but not that of the other firm. Now, it is easily seen that the symmetric pure strategies "$s_i(\theta_i < 0) =$ do not invest, $s_i(\theta_i \geq 0) =$ invest" form a Bayesian equilibrium. From the point of view of each firm, the other firm invests with probability $\frac{1}{2}$. Thus, the firm should invest if and only if $\frac{1}{2}(1 + \theta_i) + \frac{1}{2}(-1) \geq 0$; that is, $\theta_i \geq 0$. Last, note that, when ε converges to 0, the pure-strategy Bayesian equilibrium converges to the mixed-strategy Nash equilibrium of the complete-information game.

Example 6.8: War of Attrition[††]
As another example, consider the symmetric war of attrition. Suppose that, in example 6.3, it is common knowledge that the payoffs are

$$u_i(s_i, s_j) = \begin{cases} -s_i & \text{if } s_j \geq s_i \\ \hat{\theta} - s_j & \text{if } s_j < s_i. \end{cases}$$

This game has asymmetric equilibria (for instance, firm 1 always stays in, and firm 2 always exits in the natural-monopoly interpretation). But there is a single symmetric equilibrium, which is in mixed strategies. Each player uses the distribution function $F(s) = 1 - \exp(-s/\hat{\theta})$, with corresponding density $f(s) = (1/\hat{\theta})\exp(-s/\hat{\theta})$; the hazard rate for this density—i.e., the probability that a player stops between s and $s + ds$ conditional on not stopping before s—is $ds/\hat{\theta}$. That this profile is an equilibrium results from the fact that the expected gain of staying in ds more is $\hat{\theta} \cdot (ds/\hat{\theta})$, which equals the cost of ds. At each instant, conditional on both players still fighting, each player's valuation from that date on (which does not include the sunk cost of having fought to that date) is equal to 0, so the player is indifferent between fighting and quitting.

Can this mixed-strategy equilibrium be purified? That is, does there exist a sequence of continuous distributions of types that weakly converge to a point mass at $\hat{\theta}$, and such that each type plays a pure strategy and the equilibrium distributions of actions converge to the one associated with the mixed-strategy equilibrium of the complete-information game?

Consider a sequence of symmetric densities $p^n(\cdot)$ on $[0, \infty)$, with cumulative distribution functions $P^n(\cdot)$ such that $P^n(0) = 0$ and such that, for all $\varepsilon > 0$,

$$\lim_{n \to \infty} [P^n(\hat{\theta} + \varepsilon) - P^n(\hat{\theta} - \varepsilon)] = 1.$$

Let $s^n(\cdot)$ be the symmetric-equilibrium strategy corresponding to p^n, and let Φ^n be the inverse of s^n.

Integrating equation (6.3) (the first-order condition for maximization) shows that

$$P^n(\Phi^n(s)) = 1 - \exp\left(-\int_0^s db/\Phi^n(b)\right). \tag{6.13}$$

Since $P^n(\hat{\theta} - \varepsilon)$ converges to 0, and $P^n(\hat{\theta} - \varepsilon) = P^n(\Phi^n(s^n(\hat{\theta} - \varepsilon)))$, equation 6.13 implies that $s^n(\hat{\theta} - \varepsilon)$ converges to 0 for all $\varepsilon > 0$. Similarly, one can show that $s^n(\hat{\theta} + \varepsilon)$ converges to infinity. Hence, for any $s > 0$ and $\varepsilon \in (0, \hat{\theta})$,

$$s^n(\hat{\theta} - \varepsilon) < s < s^n(\hat{\theta} + \varepsilon)$$

for n sufficiently large. Rewrite equation 6.13 as

$$P^n(\Phi^n(s)) = 1 - \exp\left(-\int_0^{s^n(\hat{\theta}-\varepsilon)} \frac{db}{\Phi^n(b)}\right)\exp\left(-\int_{s^n(\hat{\theta}-\varepsilon)}^s \frac{db}{\Phi^n(b)}\right)$$

$$= 1 - [1 - P^n(\hat{\theta} - \varepsilon)]\exp\left(-\int_{s^n(\hat{\theta}-\varepsilon)}^s \frac{db}{\Phi^n(b)}\right). \tag{6.14}$$

Since $P^n(\hat{\theta} - \varepsilon)$ and $s^n(\hat{\theta} - \varepsilon)$ converge to 0, for n sufficiently large, $\phi^n(b) \in [\hat{\theta} - \varepsilon, \hat{\theta} + \varepsilon]$ for all $b \in [s^n(\hat{\theta} - \varepsilon), s]$, so that any accumulation point of

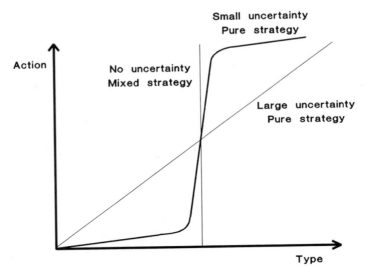

Figure 6.6

$P^n(\Phi^n(s))$ is between $1 - \exp[-s/(\hat{\theta} + \varepsilon)]$ and $1 - \exp[-s/(\hat{\theta} - \varepsilon)]$. Since this is true for all $\varepsilon > 0$, we conclude that

$$P^n(\Phi^n(s)) \rightarrow P(\Phi(s)) = 1 - \exp(-s/\hat{\theta}).$$

Thus, once again, a sequence of pure-strategy equilibria for an incomplete-information game "converges" to a mixed-strategy equilibrium of the complete-information version of the game. Our analysis focused on the convergence of the probability distributions over actions. Figure 6.6 illustrates the convergence in the space of strategies.

Example 6.9: First-Price Auction
As a last example, consider the first-price auction with a continuum of types and with two types (examples 6.5 and 6.6). Differentiating equation 6.11 (corresponding to the two-type case) for $s > \underline{\theta}$ yields

$$G(s) = (\bar{\theta} - s)g(s). \tag{6.15}$$

To compare equation 6.15 with equation 6.6 (which corresponds to the continuum case),[19] consider a sequence of continuous distributions $P^n(\cdot)$ converging to spikes at $\underline{\theta}$ and $\bar{\theta}$ ($\lim_{n \rightarrow \infty} P^n(\theta) = 0$ for $\theta < \underline{\theta}$, $= p$ for $\theta \in [\underline{\theta}, \bar{\theta})$, $= 1$ for $\theta \geq \bar{\theta}$). If $\Phi^n(\cdot)$ denotes the equilibrium strategy for distribution $P^n(\cdot)$, then $\Phi^n(s)$ must converge to $\bar{\theta}$ for $s > \underline{\theta}$, and thus (loosely speaking) equation 6.6 converges to equation 6.15.

---

19. In the continuous example, we imposed a reservation price. Take the reservation price to be equal to $\underline{\theta}$ in the discrete example to make the two games compatible.

6.7.2 Purification Theorem (technical)[††]

Harsanyi (1973) shows that any mixed-strategy equilibrium can "almost always" be obtained as the limit of a pure-strategy equilibrium in a given sequence of slightly perturbed games. Consider a strategic-form game with finite strategy sets S_i and payoff functions u_i. Harsanyi perturbs the payoffs in the following way: Let θ_i^s denote a random variable with range a closed interval ($[-1, 1]$, say) and let $\varepsilon > 0$ denote a positive constant (which will later converge to 0). Player i's perturbed payoff function $\tilde{u}_i$ depends on player i's "type" $\theta_i \equiv \{\theta_i^s\}_{s \in S}$ and on the "scale of perturbation" ε:

$$\tilde{u}_i(s, \theta_i) = u_i(s) + \varepsilon \theta_i^s.$$

Harsanyi assumes that the players' types are statistically independent. Let $P_i(\cdot)$ denote the probability distribution for θ_i. It is assumed that P_i has a density function $p_i(\cdot)$ that is continuously differentiable for all θ_i. Harsanyi first shows that the best reply of any player i is an essentially unique pure strategy. That is, two best replies for player i, $\sigma_i(\cdot)$ and $\tilde{\sigma}_i(\cdot)$, must coincide for almost all θ_i, and furthermore they must be pure strategies for almost all θ_i. This is quite intuitive, since for given strategies of player i's opponents the coincidence of player i's payoffs for two pure strategies must be a rare event if θ_i is continuously distributed. As a consequence, in any equilibrium of a perturbed game, $\sigma_i(\theta_i)$ is a pure strategy for all i and for almost all $\theta \equiv (\theta_1, \ldots, \theta_I)$. Harsanyi shows that an equilibrium exists. He then proves the following result.

Theorem 6.1 (Harsanyi 1973) Fix a set of I players and strategy spaces S_i. For a set of payoffs $\{u_i(s)\}_{i \in \mathcal{I}, s \in S}$ of Lebesgue measure 1, for all independent, twice-differentiable distributions p_i on $\Theta_i = [-1, 1]^{\#S}$, any equilibrium of the payoffs u_i is the limit as $\varepsilon \to 0$ of a sequence of pure-strategy equilibria of the perturbed payoffs $\tilde{u}_i$. More precisely, the probability distributions over strategies induced by the pure-strategy equilibria of the perturbed game converge to the distribution of the equilibrium of the unperturbed game.

Note the order of quantifiers in the statement of the theorem: A single sequence of perturbed games can be used to "purify" *all* the mixed equilibria of the limit game.

Note also the restriction to a set of payoffs of full measure. There are two possible problems that occur for "pathological" payoffs. First, it may be that a given equilibrium can only be approximated by pure-strategy equilibria of a small subset of all perturbed games, and different perturbed games pick out different equilibria. Exercise 6.10 gives an example of this. Second, equilibria in weakly dominated strategies are not limits of equilibria of any perturbed game. In figure 6.7 (taken from Harsanyi 1973), the pure-strategy equilibrium (D, R) is not approachable by any equilibrium

	L	R
U	3,4	2,2
D	1,1	2,1

Figure 6.7

once the game is perturbed. For instance, suppose that the random variables θ_1^{UR} and θ_1^{DR} are symmetrically (e.g., uniformly) distributed on $[-1, 1]$. Then, for any probability that player 2 plays R, the probability that player 1 strictly prefers to play U is at least $\frac{1}{2}$. Hence, player 1's strategy in the perturbed game cannot converge to probability 1 on D. But games like that depicted in figure 6.7 are extraordinary. In the (D, R) equilibrium, players are indifferent between their equilibrium strategy and a dominating strategy—a situation that is unlikely to occur if the entries in figure 6.7 are drawn "randomly."[20]

Our view is that games of complete information are an idealization, as players typically have at least a slight amount of incomplete information about the others' objectives. One consequence of that view, as Harsanyi's argument shows, is that the distinction between pure and mixed strategies may be artificial.

6.8 The Distributional Approach (technical)[†††]

Modeling mixed strategies as maps from types to mixtures over pure strategies has the drawback that it is not well defined in games with a continuum of types, as Aumann (1964) pointed out. Aumann proposed that a mixed strategy be a function σ_i from $[0, 1] \times \Theta_i$ into S_i. The interpretation is that type θ_i chooses among actions s_i on the basis of the outcome x_i of a lottery. Assuming without loss of generality that x_i is drawn from the uniform distribution on $[0, 1]$,[21] the probability that type θ_i of player i plays s_i is equal to the measure of the set of x_i such that $\sigma_i(x_i, \theta_i) = s_i$. There are, of course, an infinity of mixed strategies that describe a given behavior. For instance, the following mixed strategies are "equivalent":

$$\sigma_i(x_i, \theta_i) = s_i \text{ if } x_i \leq \tfrac{1}{3}, \quad \sigma_i(x_i, \theta_i) = s_i' \text{ if } x_i > \tfrac{1}{3},$$

and

---

20. Recall from chapter 3 that the set of all strategic-form payoffs arising from a given extensive form can have measure 0 in the space of all payoffs for that strategic form.

21. To see that we can assume a uniform distribution without loss of generality, consider a mixed strategy $\sigma_i(y_i, \theta_i)$ where y_i is distributed on $[0, 1]$ according to the increasing cumulative distribution function $F_i(y_i)$($F_i(0) = 0$, $F_i(1) = 1$). Define the new strategy $\tilde{\sigma}_i(x_i, \theta_i) \equiv \sigma_i(F_i(x_i), \theta_i)$. This mixed strategy is a function of the random variable x_i, which is uniformly distributed on $[0, 1]$ (as $\text{Prob}(x_i \leq x) = \text{Prob}(F_i^{-1}(x_i) \leq F_i^{-1}(x)) = F_i(F_i^{-1}(x)) = x$).

$$\tilde{\jmath}_i(x_i, \theta_i) = s_i \text{ if } x_i > \tfrac{2}{3}, \quad \tilde{\jmath}_i(x_i, \theta_i) = s_i' \text{ if } x_i \le \tfrac{2}{3}.$$

In other words, the "Aumann fix" is not parsimonious.

In response, Milgrom and Weber (1986) introduced the concept of a "distributional strategy," which refers to the equivalence class of the mixed strategies that yield the same behavior. From the point of view of the other players, what matters is the joint distribution of player i's type and actions. This leads to the definition of a distributional strategy as a joint distribution on $\Theta_i \times S_i$ for which the marginal distribution on Θ_i is the one specified by the prior beliefs.

The equivalence between mixed strategies and distributional strategies is clear. A mixed strategy induces a joint distribution across types and actions. Conversely, a joint distribution can be generated by many mixed strategies.

The reader familiar with the notion of correlated equilibrium introduced in chapter 2 will note the analogy between definitions A and B of correlated equilibrium and the distinction between mixed and distributional strategies. In chapter 2, we noted that we could determine the set of correlated equilibria without considering all possible correlating devices, but instead could restrict attention to joint distributions over strategies. Similarly, we do not need to list all the possible relationships between the randomizing device and the strategy; instead we can focus on the joint distribution of the player's type and action.

Since pure-strategy equilibria need not exist in games of complete information, it is interesting to note that, under certain regularity conditions, pure-strategy equilibria do exist in games with an atomless distribution over types. (Mixed strategies are needed for existence in general incomplete-information games.)

The idea is that the effects of mixing can be duplicated by having each type play a pure strategy. If each player's payoff does not depend on the types of the others, then players care only about the distribution of their opponents' actions, and their payoffs are not affected by the replacement of an opponent's mixed strategy by a pure one that induces the same distribution.

To illustrate this, suppose that type θ_i is uniformly distributed on the interval $[0, 1]$ and that, given the strategies of player i's rivals, all types θ_i in $[0, \tfrac{1}{2}]$ are indifferent between actions s_i and s_i'. These types randomize in such a way that the probability that s_i (respectively, s_i') is chosen given that θ_i belongs to $[0, \tfrac{1}{2}]$ is α (respectively, $1 - \alpha$). Consider the following pure strategy: Types θ_i in $[0, \alpha/2]$ play s_i with probability 1, and types θ_i in $(\alpha/2, \tfrac{1}{2}]$ play s_i' with probability 1. Because types in $[0, \tfrac{1}{2}]$ are indifferent between the two actions, the pure strategy is an equilibrium behavior as long as the rivals do not change their behavior. The rivals' expected payoff function is not affected by the substitution if two conditions hold. The first condition is that θ_i should not enter player j's utility function or, more generally, should be separable from s_i in an appropriate way: Even though the marginal distributions of s_i and θ_i are not affected by the substitution, the distributional strategy is affected, and this matters if there are cross-effects between s_i and θ_i in u_j. The second condition is that the distributions of types should be independent among players. (If this is not the case, the distribution of s_i, conditional on θ_j, may be changed by the substitution.)

Along these lines, we can state a "purification theorem" due to Milgrom and Weber's (1986) extension of a similar result for the single-decision-maker setting of Dvoretzky, Wald, and Wolfowitz (1951). For this purpose (and for the rest of this section), we will assume that the structure of information takes the special form of a commonly observed variable θ_0 (common value) and some piece of private information $\tilde{\theta}_i$ for each player i (private values) such that, conditional on the realization of θ_0, the $\tilde{\theta}_i$ are independent. Let $\theta_i = (\theta_0, \tilde{\theta}_i)$; because θ_0 is commonly observed, $\tilde{\theta}_i$ will be called "player i's type" by abuse of terminology. We assume that $\theta_0 \in \Theta_0$ and $\tilde{\theta}_i \in \Theta_i$.

Definition 6.2 Preferences are *conditionally independent* if each player i's payoff can be written in the form $u_i = u_i(s, \theta_0, \tilde{\theta}_i)$, where $s \equiv (s_1, \ldots, s_I)$, and if, conditional on the realization of θ_0, the players' types $\tilde{\theta}_i$ are independent.

Theorem 6.2 (Milgrom and Weber 1986) Assume that preferences are "conditionally independent," that Θ_0 is finite, that the marginal distributions of types are atomless, that the game has continuous payoffs, and that each S_i is compact. Then every equilibrium point (we will later show that one exists) has a purification.

Remark 1 The assumption of conditionally independent preferences is obviously very strong. One may be able to purify mixed strategies even when it does not hold. When preferences are dependent, one must be able to replicate by a pure strategy not only the distribution on S_i, but the whole distributional strategy on $S_i \times \Theta_i$. That is, the reshuffling of weight we performed earlier must be "local" rather than "global." Unfortunately, we do not quite know what regularity conditions are required to this effect.[22] The issue is that the set of distributional strategies obtained from pure strategies is smaller than the set of distributional strategies obtained from mixed strategies. (Those two sets, however, are close to each other: The former is dense in the latter for the topology of weak convergence of

22. Aumann et al. (1982) allow dependence, but obtain only an approximate purification result. They show that with conditionally atomless distributions, any mixed strategy of a player can be ε-purified (i.e., replaced by a pure strategy that yields all players a payoff within ε of the payoff for the original mixed strategy), no matter what strategies the other players use, for any $\varepsilon > 0$.

probability measures. Hence, for any mixed-strategy equilibrium, there exists a set of nearly pure strategies that form an ε-equilibrium of the game.)

Remark 2 In subsection 6.7.2 we used the term "purification" in a different, although related, sense. We asked to what extent a mixed-strategy equilibrium of a game with complete information (or, more generally, with atoms of types) could be viewed as an approximation of pure-strategy equilibria of nearby games of incomplete information in which each player has a continuum of types.

With a continuum of types and/or a continuum of actions, some regularity conditions must be imposed in order to apply Glicksberg's existence theorem (see subsection 1.3.3). Let η and η_i $(i = 0, \ldots, I)$ denote, respectively, the probability measure over the set $\Theta = \Theta_0 \times \Theta_1 \times \cdots \times \Theta_I$ and the marginal distribution over Θ_i. The following existence result (a slightly stronger version of which can be found in Milgrom and Weber 1986) generalizes one obtained for independent types by Ambruster and Böge (1979).

Theorem 6.3 (Milgrom and Weber 1986) Assume that all S_i are compact; that (continuous information) the measure $\eta(\cdot)$ is absolutely continuous relative to the measure $\hat{\eta}(\cdot) = \eta_0(\cdot) \times \cdots \times \eta_I(\cdot)$;[23] and that (continuous payoffs) either all S_i are finite or, for all i, the function u_i is uniformly continuous on $\Theta \times S$. Then an equilibrium exists.

Exercises

Exercise 6.1** Consider the public-good game of section 6.2. Suppose that there are $I > 2$ players and that the public good is supplied (with benefit 1 for all players) only if at least $K \in \{1, \ldots, I\}$ players contribute. The players' costs of contributing, $\theta_1, \ldots, \theta_I$, are independently drawn from the distribution $P(\cdot)$ on $[\underline{\theta}, \bar{\theta}]$ where $\underline{\theta} < 1 < \bar{\theta}$.

(a) Generalize the Bayesian equilibrium of section 6.2 when $K = 1$.

(b) Suppose $K \geq 2$. Show that there always exists a trivial equilibrium in which nobody contributes. (Assume $\underline{\theta} > 0$.) Derive a more interesting Bayesian equilibrium.

(c) Showoffs: Apply the two concepts of iterated strict dominance to this game.

Exercise 6.2** Two firms simultaneously decide whether to enter a market. Firm i's entry cost is $\theta_i \in [0, +\infty)$. The two firms' entry costs are private

---

23. That is, a null-measure set for $\hat{\eta}$ is also a null-measure set for η. The Radon-Nikodym theorem (Royden 1968) implies that there exists a density f such that, for any subset S of θ, $\eta(S) = \int_S f(\theta) \, d\hat{\eta}(\theta)$. The continuous-information assumption holds for instance when the type spaces are finite or when the types are independently distributed.

information and are independently drawn from the distribution $P(\cdot)$ with strictly positive density $p(\cdot)$. Firm i's payoff is $\Pi^m - \theta_i$ if it is the only one to enter, $\Pi^d - \theta_i$ if both enter, and 0 if it does not enter. Π^m and Π^d are the monopoly and duopoly profits gross of entry costs and are common knowledge. $\Pi^m > \Pi^d > 0$.

(a) Point out the analogies and differences with the public-good game of section 6.2.

(b) Compute a Bayesian equilibrium. Show that it is unique.

(c) Apply *ex ante* and interim strict dominance.

(d) Instead of assuming simultaneous entry, suppose that the two firms are "around" at date 0. They incur cost $r\theta_i$ per unit of time of being in the market, where r is the rate of interest and θ_i is the value of firm i's assets in an alternative use. (Alternatively, $f_i \equiv r\theta_i$ is the fixed cost of production per unit of time.) $r\Pi^m$ and $r\Pi^d$ are the flow monopoly and duopoly payoffs gross of the opportunity or fixed cost. Follow the analysis of example 6.3 to derive the symmetric equilibrium of the war of attrition. (Watch out: There exists some time T after which firms don't drop out.) Show that there is no other equilibrium. For answers, see Fudenberg and Tirole 1986. Compare with the answer to question b.

Exercise 6.3* Consider the first-price auction of example 6.5. There are two bidders with valuations uniformly distributed on $[0, 1]$. The seller's reservation price (or minimum bid) is 0. Find an equilibrium in linear strategies, i.e., where $s_i(\theta_i) = a + c\theta_i$.

Exercise 6.4** This exercise analyzes the first- and second-price auctions with risk-averse bidders and two types per bidder. A bidder with valuation θ has utility $u(\theta - t)$ if he wins and pays transfer t, and utility $u(-t)$ if he loses and pays transfer t; u is increasing and concave. The bidders' valuations are independently drawn from the two-type distribution $\{\underline{\theta}$ with probability $\underline{p}, \bar{\theta}$ with probability $\bar{p}\}$.

(a) Show that in the second-price auction (in which the highest bidder wins but pays the second bid) each player bids his true valuation and so the seller's expected revenue is the same as with risk-neutral bidders.

(b) Now consider the first-price auction. Derive the analogue of equation 6.10 for the case of risk aversion. Show that type $\underline{\theta}$'s distribution of bids, $\tilde{F}$, first-order stochastically dominates the distribution F given by equation 6.10 (i.e., $\tilde{F}(s) \geq F(s)$ for all s). Use the revenue-equivalence theorem under risk neutrality (see example 6.6) to conclude that under risk aversion the seller prefers the first-price to the second-price auction. (The answer is in Maskin and Riley 1985.)

Exercise 6.5** Generalize the analysis of the first- and second-price auctions with two types and risk-neutral bidders to (a) asymmetric distributions (letting $\bar{p}_i \equiv \text{Prob}(\theta_i = \bar{\theta}), \bar{p}_1 \neq \bar{p}_2$) and (b) correlated valuations.

Compare the seller's revenues in the two auctions (do parts a and b separately). (The answers are in Maskin and Riley 1985, 1986b.)

Exercise 6.6** In the examples in the chapter, equilibrium strategies are monotonic in type. Find and informally discuss examples in which such a monotonicity would not necessarily hold. (Hint: Consider the Chatterjee-Samuelson double auction with negatively correlated types. Find other examples. Discuss generally what goes wrong with the usual proof of monotonicity when types are correlated.)

Exercise 6.7* The (present discounted) value of a public good is 1 for all players $i = 1, \ldots, I$. Time is continuous, and the rate of interest is r. Each player's cost c of supplying the public good is distributed according to the cumulative distribution function P on $[0, 1]$. Players' types are independent. The public good is supplied if at least one agent supplies it. The good is supplied at the first time at which at least one player chooses to contribute. Thus, the game is a kind of war of attrition. Look for a symmetric, pure-strategy equilibrium using the following outline:

(a) Argue formally or informally that the date at which a player with cost c supplies the public good, $s(c)$, is increasing in c.

(b) Show that $s(\cdot)$ satisfies

$$s'(c) = \frac{(I - 1)c\, p(c)}{r(1 - c)[1 - P(c)]}.$$

Find a boundary condition. Infer that a player's waiting time to supply the good when there are $I - 1$ other players is $I - 1$ times his waiting time when there are two players. Show that each player's expected utility grows with I.

(Answers can be found in Bliss and Nalebuff 1984.)

Exercise 6.8** Kreps and Wilson (1982) consider the following war of attrition. There are two players, $i = 1, 2$. Time is continuous from 0 to 1. When one player concedes, the game ends. Each player can be either "strong" (with probability p for player 1 and q for player 2) or "weak" (with probabilities $1 - p$ and $1 - q$). A strong player enjoys fighting and therefore never concedes. A weak player loses 1 per unit of time while fighting and makes $a > 0$ per unit of time when his rival has conceded. Thus, a weak player has payoff $a(1 - t) - t$ when it wins at t and payoff $-t$ when it concedes at t. There is no discounting.

(a) Show that from time 0^+ on, the posterior beliefs p_t and q_t of each player about the other must belong to the curve $q = p^{b/a}$.

(b) Show that one of the weak types exits with positive probability at date 0 exactly (that is, a player's cumulative probability distribution of exit times exhibits an atom at $t = 0$). How are the weak types' payoffs affected by a, b, p, and q?

	L	R
U	1,1	1,1
D	1,1	1,1

Figure 6.8

Exercise 6.9** Purify the mixed-strategy equilibrium in the inspection game of example 1.7.

Exercise 6.10** Consider the game illustrated in figure 6.8 (due to Harsanyi). Fix a continuous distribution over perturbations as in Harsanyi's construction (see section 6.7). Is any mixed-strategy equilibrium of the game in figure 6.8 a limit of pure-strategy equilibria of the perturbed games as ε tends to 0? Conclude that this game is "not generic."

Exercise 6.11*** Consider symmetric first-bid and second-bid auctions with correlated information and valuations. There are I bidders. Each bidder i has (unknown) valuation v_i and signal or information θ_i; let $z_i \equiv (\theta_i, v_i)$ and $z \equiv (z_1, \ldots, z_I)$. Player i knows only θ_i. The random variable z is distributed according to distribution $F(z)$ on a rectangular cell, with density $f(z)$. F is invariant under permutations of the bidders (symmetry). Correlation is described by the affiliation property: If $z \vee z'$ and $z \wedge z'$ are the component-wise maximum and minimum of z and z', then

$$f(z \vee z')f(z \wedge z') \geq f(z)f(z') \text{ for all } (z, z').$$

Let $\theta^1 \geq \theta^2 \geq \cdots \geq \theta^I$ denote a reordering in nonincreasing order of the signals. The conditional distribution of θ^2 given $\theta^1 = \gamma$ is denoted $\hat{F}(\cdot|\gamma)$ with density $\hat{f}(\cdot|\gamma)$. Affiliation implies that, for all μ, $\hat{f}(\mu|\gamma)/\hat{F}(\mu|\gamma)$ is nondecreasing in γ (monotone likelihood-ratio property). Let

$$v(\gamma, \mu) \equiv E(v_i|\theta_i = \theta^1 = \gamma \text{ and } \theta^2 = \mu).$$

Look for symmetric, differentiable, and strictly increasing equilibrium bids $s(\theta_i)$.

Note that in the second-price auction $s(\theta) = v(\theta, \theta)$. Show that in the first-price auction

$$s(\theta) = v(\theta, \theta) - \int_{\underline{\theta}}^{\theta} K(\mu) \frac{dv}{d\mu}(\mu, \mu) \Big/ K(\theta),$$

where

$$K(\mu) \equiv \exp\left(\int_{\underline{\theta}}^{\mu} \frac{\hat{f}(\gamma, \gamma)}{\hat{F}(\gamma, \gamma)} d\gamma\right)$$

and $\underline{\theta}$ is the lowest possible signal. (Hint: In the first-price auction, a bidder with type θ maximizes over his bid b:

$$\int_{\underline{\theta}}^{s^{-1}(b)} [v(\theta, \mu) - b] \, d\hat{F}(\mu \mid \theta).$$

For the answer, look in Milgrom and Weber 1982 or Wilson 1990.)

References

Ambruster, W., and W. Böge. 1979. Bayesian game theory. In *Game Theory and Related Topics*, ed. O. Moeschlin and D. Pallascke. North-Holland.

Aumann, R. 1964. Mixed vs. behavior strategies in infinite extensive games. *Annals of Mathematics Studies* 52: 627–630.

Aumann, R., Y. Katznelson, R. Radner, R. Rosenthal, and B. Weiss. 1982. Approximate Purification of Mixed Strategies. *Mathematics of Operations Research* 8: 327–341.

Bishop, D. T., C. Cannings, and J. Maynard Smith. 1978. The war of attrition with random rewards. *Journal of Theoretical Biology* 3: 377–388.

Bliss, C., and B. Nalebuff. 1984. Dragon-slaying and ballroom dancing: The private supply of the public good. *Journal of Public Economics* 25: 1–12.

Butters, G. 1977. Equilibrium distribution of prices and advertising. *Review of Economic Studies* 44: 465–492.

Chatterjee, K., and W. Samuelson. 1983. Bargaining under incomplete information. *Operations Research* 31: 835–851.

Dvoretzky, A., A. Wald, and J. Wolfowitz. 1951. Elimination of randomization in certain statistical decision procedures and zero-sum two-person games. *Annals of Mathematics and Statistics* 22: 1–21.

Fudenberg, D., and J. Tirole. 1986. A theory of exit in duopoly. *Econometrica* 54: 943–960.

Harsanyi, J. 1967–68. Games with incomplete information played by Bayesian players. *Management Science* 14: 159–182, 320–334, 486–502.

Harsanyi, J. 1973. Games with randomly disturbed payoffs: A new rationale for mixed-strategy equilibrium points. *International Journal of Game Theory* 2: 1–23.

Kreps, D., and R. Wilson. 1982. Reputation and imperfect information. *Journal of Economic Theory* 27: 253–279.

Leininger, W., P. Linhart, and R. Radner. 1989. Equilibria of the sealed-bid mechanism for bargaining with incomplete information. *Journal of Economic Theory* 48: 63–106.

Maskin, E., and J. Riley. 1985. Auction theory and private values. *American Economic Review Papers & Proceedings* 75: 150–155.

Maskin, E., and J. Riley. 1986a. Existence and uniqueness of equilibrium in sealed high bid auctions. Discussion paper 407, University of California, Los Angeles.

Maskin, E., and J. Riley. 1986b. Asymmetric auctions. Mimeo, UCLA and Harvard University.

Mertens, J. F., S. Sorin, and S. Zamir. 1990. Repeated games. Manuscript.

Mertens, J. F., and S. Zamir. 1985. Formulation of Bayesian analysis for games with incomplete information. *International Journal of Game Theory* 10: 619–632.

Milgrom, P., and R. Weber. 1982. A Theory of Auctions and Competitive Bidding. *Econometrica* 50: 1089–1122.

Milgrom, P., and R. Weber. 1986. Distributional strategies for games with incomplete information. *Mathematics of Operations Research* 10: 619–631.

Nalebuff, B. 1982. Brinksmanship. Mimeo, Harvard University.

Nalebuff, B., and J. Riley. 1983. Asymmetric equilibria in the war of attrition. Mimeo, University of California, Los Angeles.

Nash, J. 1953. Two-person cooperative games. *Econometrica* 21: 128–140.

Palfrey, T., and H. Rosenthal. 1989. Underestimated probabilities that others free ride: An experimental test. Mimeo, California Institute of Technology and Carnegie-Mellon University.

Riley, J. 1980. Strong evolutionary equilibrium and the war of attrition. *Journal of Theoretical Biology* 82: 383–400.

Royden, H. 1968. *Real Analysis*. Macmillan.

Satterthwaite, M., and S. Williams. 1989. Bilateral trade with the sealed bid k-double auction: existence and efficiency. *Journal of Economic Theory* 48: 107–133.

Wilson, R. 1990. Strategic analysis of auctions. Mimeo, Graduate School of Business, Stanford University.

Bayesian Games and Mechanism Design

This chapter presents a thorough treatment of a special class of games of incomplete information known as *games of (static) mechanism design.* Examples of these games include monopolistic price discrimination, optimal taxation, the design of auctions, and mechanisms for the provision of public goods. In all of these cases, there is a "principal" who would like to condition her actions on some information that is privately known by the other players, called "agents." The principal could simply ask the agents for their information, but they will not report it truthfully unless the principal gives them an incentive to do so, either by monetary payments or with some other instrument that she controls. Since providing these incentives is costly, the principal faces a tradeoff that often results in an inefficient allocation.

The distinguishing characteristic of the mechanism-design approach is that the principal is assumed to choose the mechanism that maximizes her expected utility, as opposed to using a particular mechanism for historical or institutional reasons. This distinction can be illustrated using the topic of auctions: In chapters 1 and 6 we solved for the equilibrium bidding strategies of buyers in two particular mechanisms, the first-price and second-price auctions. When we study auctions in this chapter, we ask which form of auction maximizes the seller's expected revenue. Because of the pervasiveness of the first-price auction, it is very interesting to see that it (and the second-price auction) turns out to be optimal in some situations. Similarly, when we consider models where the principal is the government, we suppose that the government chooses a mechanism that maximizes its utility, which we take to be the total surplus in the economy. Thus, the applications to (e.g.) tax policy may be interpreted as normative as opposed to descriptive models.

Many applications of mechanism design consider games with a single agent. (These single-agent models also apply to situations with a continuum of infinitesimal agents, each of whom interacts with the principal but not with the other agents.) In second-degree price discrimination by a monopolist, the monopolist has incomplete information about the consumer's (the agent's) willingness to pay for her good. The monopolist designs a price schedule that determines the price to be paid by the consumer as a function of the quantity purchased. In the regulation of a natural monopoly under asymmetric information, the government has incomplete information about the regulated firm's (the agent's) cost structure. It designs an incentive scheme that determines the transfer received by the regulated firm as a function of its cost or its price (or both). In the study of optimal taxation, the government would like to raise tax revenue from a consumer (an agent) to finance public goods. The optimal level of tax depends on the consumer's ability to earn money. If the government knew this ability, it could levy an ability-dependent lump-sum tax that would not distort the consumer's labor supply. In the presence of incomplete information about ability, the government can only base the income tax on realized income. The income-tax schedule can be seen as an incentive scheme eliciting information about the consumer's ability.

Mechanism design can also be applied to games with several agents. In the public-good problem, a government must decide whether to supply a public good, but it has incomplete information about how much the good is valued by consumers. The government can then design a scheme determining the provision of the public good as well as transfers to be paid by the consumers as functions of their announced willingnesses to pay for the public good. In the design of auctions, a seller organizes an auction among the buyers for the purchase of a good. The seller, not knowing how much the buyers are willing to pay for the good, set up a mechanism that determines who purchases the good and the sale price. Finally, in problems of bilateral exchange, a mediator designs a trading mechanism between a seller who has private information about the production cost and a buyer who has private information about his willingness to pay for the good.

Mechanism design is typically studied as a three-step game of incomplete information, where the agents' types—e.g., willingness to pay—are private information. In step 1, the principal designs a "mechanism," or "contract," or "incentive scheme." A mechanism is a game in which the agents send costless messages, and an "allocation" that depends on the realized messages. The message game can have simultaneous announcements or a more complex communication process. The allocation is a decision about the level of some observable variable, e.g., the quantity consumed or the amount of public good provided, and a vector of transfers from the principal to the agents (which can be positive or negative). In step 2, the agents simultaneously accept or reject the mechanism. An agent who rejects the mechanism gets some exogenously specified "reservation utility" (usually, but not necessarily, a type-independent number). In step 3, the agents who accept the mechanism play the game specified by the mechanism.

Because a game of mechanism design can have many stages, the distinction between Nash and subgame-perfect equilibria in multi-stage games with complete information (see chapter 3) may suggest that the concept of Bayesian equilibrium is too weak to be useful here. Fortunately, a simple but fundamental result called the "revelation principle" (developed in section 7.2) shows that, to obtain her highest expected payoff, the principal can restrict attention to mechanisms that are accepted by all agents at step 2 and in which at step 3 all agents simultaneously and truthfully reveal their types. This implies in particular that the principal can obtain her highest expected payoff through a static Bayesian game among the agents. This is why we treat mechanism design in part III rather than in part IV of the book. (However, we invoke a mild perfection requirement: We do not allow agents to threaten to reject the principal's mechanism—or to misreport their types—if it is in their interest in steps 2 and 3 to accept the mechanism and to announce truthfully.)

In some situations (mainly situations in which the principal is the government), the "individual-rationality" or "participation" constraints—that the agents must be willing to participate in the principal's mechanism—are not imposed. That is, step 2 of the mechanism-design game is omitted. For instance, a government with coercive powers can choose an income tax that applies to all consumers (unless the possibility of emigration makes the participation constraints binding). Similarly, in some public-good problems, the government may impose decisions that the agents cannot veto. In contrast, the literature has assumed that consumers can refrain from buying from a firm, that bidders are free not to participate in an auction, and that regulated firms (or at least their managers) can refuse to produce (or to work). Whether an individual-rationality constraint should be included in the model depends on the extent of the coercive power of the principal, or, equivalently, on the distribution of property rights.[1]

An important focus of the mechanism-design literature is how the combination of incomplete information and binding individual-rationality constraints can prevent efficient outcomes.[2] Coase (1960) argued that, in the absence of transaction costs and with symmetric information, bargaining among parties concerned by a decision leads to an efficient decision, i.e., to the realization of gains from trade. With some exceptions (see the "efficiency results" in subsection 7.4.3), this is not so under asymmetric information. A constant theme of the mechanism-design literature is that the private information of the agents leads to inefficiency when individual-rationality constraints are binding.

The chapter is organized as follows. Section 7.1 illustrates individual rationality, truthful revelation, and optimal mechanism design in two simple examples. Section 7.2 develops the general framework and derives the revelation principle. Section 7.3 considers the case of a single agent. Besides being of considerable practical interest, this case offers a useful introduction to the more general multi-agent situation. Most of the steps involved in characterizing "implementable" or "incentive-compatible" allocations and in deriving the optimal mechanism for the principal are borrowed from the single-agent framework. Section 7.4 tackles the multi-agent case and characterizes implementable allocations. Subsections 7.4.3–7.4.6 apply this characterization to obtain some efficiency and inefficiency results in public-

---

1. The literature has made reasonable assumptions about the actual distribution of property rights, but little attention has been paid to what determines it in most of the contexts described above.

2. If the principal is a government that does not have a balanced-budget constraint, so that it can give all agents large positive transfers, the individual-rationality constraints will not bind.

and private-goods contexts. Section 7.5 builds on section 7.4 by analyzing the principal's optimal mechanism in two different contexts: auctions, in which a seller tries to extract the maximum expected revenue from buyers, and bilateral exchange, in which a mediator designs a mechanism so as to maximize expected gains from trade between a seller and a buyer. Section 7.6 mentions some additional topics and concludes.

The field of mechanism design is important enough to merit a book of its own.[3] We have not tried to provide a complete review of the field, choosing instead to develop a few main themes. Nevertheless, the material in this chapter could easily take a month to cover. Readers with little interest in mechanism design might choose to skip the chapter entirely, relying on the examples in chapter 6 to illustrate the application of Bayesian equilibrium. Those who are interested in mechanism design but are pressed for time might read through section 7.3, which completes the analysis of mechanism design with a single agent.

7.1 Examples of Mechanism Design[†]

This section contains two examples of mechanism design. To facilitate the exposition, they both involve a seller selling a good—to a single buyer in subsection 7.1.1 and to one of two buyers in subsection 7.1.2. This section is meant to provide motivation for the chapter; it should be skipped by any reader who has already seen some examples.

7.1.1 Nonlinear Pricing

A monopolist produces a good at constant marginal cost c and sells an amount $q \geq 0$ of this good to a consumer. (As is easily checked, nothing would be affected if she sold the good to several consumers who were *ex ante* identical.) The consumer receives utility

$$u_1(q, T, \theta) \equiv \theta V(q) - T,$$

where $\theta V(q)$ is his gross surplus, $V(0) = 0$, $V' > 0$, $V'' < 0$, and T is the transfer from the consumer to the seller. $V(\cdot)$ is common knowledge, but θ is private information to the consumer. The seller knows only that $\theta = \underline{\theta}$ with probability $\underline{p}$ and $\theta = \bar{\theta}$ with probability $\bar{p}$, where $\bar{\theta} > \underline{\theta} > 0$ and $\underline{p} + \bar{p} = 1$. The game proceeds as follows: The seller offers a (possibly nonlinear) tariff $T(q)$ specifying how much the consumer pays if he chooses consumption q. The consumer then either accepts the mechanism, chooses a consumption q, and pays $T(q)$, or else rejects the mechanism. Note that, without loss of generality, we can constrain the seller to offer a tariff such that $T(0) = 0$ and assume that the consumer always accepts the mechanism.

---

3. See, e.g., Green and Laffont 1979 and Laffont 1979.

If the seller knew the true value of θ, she would offer a fixed q and charge $T = \theta V(q)$. Her profit would then be $\theta V(q) - cq$, and it would be maximized at q given by $\theta V'(q) = c$. Because the consumer may have one of two types, the seller will want to offer two different bundles if she does not know θ. Let $(\underline{q}, \underline{T})$ denote the bundle intended for the type-$\underline{\theta}$ consumer, and let $(\overline{q}, \overline{T})$ be the bundle intended for the type-$\overline{\theta}$ consumer.[4] The seller's expected profit is thus

$$Eu_0 = \underline{p}(\underline{T} - c\underline{q}) + \overline{p}(\overline{T} - c\overline{q}).$$

The seller faces two kinds of constraints. The first kind requires that the consumer be willing to purchase. (As we noted above, this is without loss of generality, because the seller can always offer the bundle $(q, T) = (0, 0)$—which corresponds to not purchasing—in her "menu" of bundles.) Such a constraint is called an *individual-rationality* (IR) or *participation* constraint. The "reservation utility" is the level of net utility obtained by the consumer by not purchasing, which is equal to 0 here. Thus, we require that

$(\text{IR}_1) \quad \underline{\theta} V(\underline{q}) - \underline{T} \geq 0$

and

$(\text{IR}_2) \quad \overline{\theta} V(\overline{q}) - \overline{T} \geq 0.$

The second kind of constraint requires that the consumer consume the bundle intended for his type. These are known as *incentive-compatibility* (IC) constraints. Thus, we require that

$(\text{IC}_1) \quad \underline{\theta} V(\underline{q}) - \underline{T} \geq \underline{\theta} V(\overline{q}) - \overline{T}$

and

$(\text{IC}_2) \quad \overline{\theta} V(\overline{q}) - \overline{T} \geq \overline{\theta} V(\underline{q}) - \underline{T}.$

The seller's problem is to choose $\{(\underline{q}, \underline{T}), (\overline{q}, \overline{T})\}$ so as to maximize her expected profit subject to the two IR and the two IC constraints.

A first step in solving this problem is to show that only IR_1 and IC_2 are binding. First, note that if IR_1 and IC_2 are satisfied, then

$$\overline{\theta} V(\overline{q}) - \overline{T} \geq (\overline{\theta} - \underline{\theta})V(\underline{q}) \geq 0,$$

which reflects the fact that type $\overline{\theta}$ receives more surplus from consumption than type $\underline{\theta}$. Hence, IR_2 is satisfied as well. Furthermore, IR_2 will not be binding unless $\underline{q} = 0$, i.e., unless the seller does not sell to the low-type consumer. In contrast, IR_1 must be binding, i.e., $\underline{T} = \underline{\theta} V(\underline{q})$: If the two IR constraints were not binding, the seller could increase $\underline{T}$ and $\overline{T}$ by the same

---

4. The results of section 7.2 imply that the seller will not wish to offer several bundles intended for the same consumer.

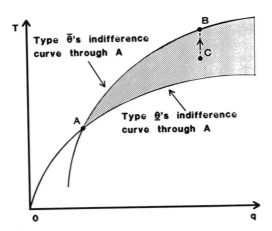

Figure 7.1

small positive amount, which would keep the incentive constraints satisfied, would not violate individual rationality, and would increase revenue.

Next, IC_2 must be binding; that is,

$$\bar{T} = \underline{T} + \bar{\theta}\, V(\bar{q}) - \bar{\theta}\, V(\underline{q}) = \bar{\theta}\, V(\bar{q}) - (\bar{\theta} - \underline{\theta})V(\underline{q}).$$

If IC_2 were not binding, the seller could increase $\bar{T}$ slightly and keep all constraints satisfied. This is illustrated in figure 7.1. Let A denote the low-type consumer's allocation $(\underline{q}, \underline{T})$, and let B denote the high-type consumer's allocation $(\bar{q}, \bar{T})$. Draw the indifference curves of the low type and the high type through A. Note that, because the slope of the indifference curve of type θ is $\theta\, V'(q)$, the high-type consumer's indifference curve is always steeper than that of the low-type consumer at any allocation. The allocation B must belong to the shaded area in figure 7.1, because it must be (weakly) preferred by type $\bar{\theta}$ to A, and A must be (weakly) preferred by type $\underline{\theta}$ to B. (Note that this shows that $\bar{q} \geq \underline{q}$, i.e., a high-demand consumer must consume more than a low-demand one. We will analyze this "monotonicity property" at length in this chapter.) The figure also illustrates the fact that A for type $\underline{\theta}$ and C for type $\bar{\theta}$ cannot be optimal for the seller, who could increase her profit by increasing $\bar{T}$ and replacing C by B for type $\bar{\theta}$. Thus, IC_2 must bind.

Knowing that IR_1 and IC_2 are binding, we ignore IC_1 in the derivation of the seller's optimal nonlinear tariff, and solve the subconstrained program with IR_1 and IC_2 only. If the solution to this subconstrained program turns out to satisfy IC_1 as well (as the previous diagrammatic discussion indicates will be the case), then it is a solution to the overall program.

Maximizing Eu_0 subject to IR_1 and IC_2 binding is equivalent to maximizing

$$\underline{p}(\underline{T} - c\underline{q}) + \bar{p}(\bar{T} - c\bar{q}) = [(\underline{p}\underline{\theta} - \bar{p}(\bar{\theta} - \underline{\theta}))V(\underline{q}) - \underline{p}c\underline{q}]$$
$$+ \bar{p}(\bar{\theta} V(\bar{q}) - c\bar{q}).$$

The first-order conditions are (assuming $\bar{p}\bar{\theta} < \underline{\theta}$ and $V'(0) = 0$)

$$\underline{\theta} V'(\underline{q}) = c \bigg/ \left(1 - \frac{\bar{p}(\bar{\theta} - \underline{\theta})}{\underline{p}\underline{\theta}}\right)$$

and

$$\bar{\theta} V'(\bar{q}) = c.$$

The quantity purchased by the high-demand consumer is socially optimal (the marginal utility of consumption of the good is equal to the marginal cost). If this were not so, the seller could increase or decrease $\bar{q}$ a little and increase or decrease $\bar{T}$ accordingly so as to keep the utility of the $\bar{\theta}$-type constant. The profit from this type would increase because efficiency would increase, and the new tariff would remain incentive compatible because IC_1 was not binding.

In contrast, the quantity purchased by the low-demand consumer is socially suboptimal (recall that $V'' < 0$). This can be easily understood: The seller lowers the consumption of the low-demand consumer to make it less attractive to the high-demand consumer to "cheat" and consume $\underline{q}$. This allows the seller to increase $\bar{T}$ or, equivalently, to reduce the rent of the high-demand consumer, $(\bar{\theta} - \underline{\theta})V(\underline{q})$. Thus, it is optimal for the seller to sacrifice some efficiency for the purpose of rent extraction. Note also that $\bar{q} > \underline{q}$; furthermore, if IR_1 and IC_2 are satisfied with equality, transfers $\underline{T}$ and $\bar{T}$ are determined by $\underline{q}$ and $\bar{q}$.

Last, we check that IC_1 is satisfied by the solution to the subconstrained program; that is,

$$\underline{\theta} V(\underline{q}) - \underline{T} = 0 \geq \underline{\theta} V(\bar{q}) - \bar{T}.$$

We find that

$$\underline{\theta} V(\bar{q}) - \bar{T} = -(\bar{\theta} - \underline{\theta})[V(\bar{q}) - V(\underline{q})] < 0.$$

Since we saw previously that IR_2 will always be satisfied if IR_1 is, we conclude that the solution to the subconstrained program is a solution to the overall program.

This inefficiency associated with incomplete information will be a constant theme of this chapter. The reader may note the resemblance to the analysis of (nondiscriminatory) monopoly pricing, which corresponds to the special case of our model where all consumers have 0-1 demand, i.e., $V(q) = 0$ for $q < 1$ and $V(q) = 1$ for $q \geq 1$. The two-type case we have considered corresponds to a market where demand d as a function of the transfer T is $d(T) = 1$ for $T \leq \underline{\theta}$, $d(T) = \bar{p}$ for $T \in (\underline{\theta}, \bar{\theta}]$, and $d(T) = 0$ for

$T > \bar{\theta}$. (If we supposed a continuum of types, we could have a smooth demand function.) With this step-function demand curve, the seller's optimal tariff is either $T > \bar{\theta}$, with profit 0, or $T = \bar{\theta}$, with profit $\bar{p}(\bar{\theta} - c)$, or $T = \underline{\theta}$, with profit equal to $\underline{\theta} - c$. If $\bar{p}(\bar{\theta} - c) > \max(0, \underline{\theta} - c)$, then it is optimal to set $T = \bar{\theta}$, which "separates" the two types. Type $\bar{\theta}$ consumes one unit at price $\bar{\theta}$ and does not enjoy a rent from private information, and type $\underline{\theta}$ consumes 0. If $\underline{\theta} - c > \max(0, \bar{p}(\bar{\theta} - c))$, it is optimal to "bunch" the two types, i.e., have both of them consume one unit and let type $\bar{\theta}$ enjoy rent $\bar{\theta} - \underline{\theta}$. In section 7.3 and in the appendix to this chapter we will derive conditions under which separation or bunching is optimal in mechanism-design problems.[5]

Looking ahead to section 7.2, we can illustrate the revelation principle in this example. The seller indirectly elicits the consumer's information by having him consume a quantity that varies with his type. Alternatively, the seller could maximize her expected profit by asking the consumer to report his type directly. Let

$$\{(q^*(\theta), T^*(\theta))\}_{\theta \in \{\underline{\theta}, \bar{\theta}\}}$$

denote the solution obtained above. The seller can offer the following direct mechanism to the consumer: "Announce your type. If $\hat{\theta}$ is your announcement, you will consume $q^*(\hat{\theta})$ and pay $T^*(\hat{\theta})$." The incentive constraints IC_1 and IC_2 guarantee that it is optimal for the consumer to announce his type truthfully. Thus, the allocation is the same as under the indirect mechanism.

7.1.2 Auctions

A seller has one unit of a good for sale. There are two potential buyers ($i = 1, 2$) with unit demands, and they are *ex ante* identical. Their valuations, θ_1 and θ_2, take value $\underline{\theta}$ with probability p and $\bar{\theta}$ with probability $\bar{p}$, where $p + \bar{p} = 1$ and θ_1 and θ_2 are independent. Each buyer knows his own valuation, but the seller and the other buyer do not.

One option for the seller is to use the first-price or second-price auctions considered in chapters 1 and 6. But do such auctions maximize the seller's profit? To provide an answer, we solve for the seller's optimal mechanism. As we will see, the familiar auction forms are in fact optimal in some situations.

Suppose that the seller sets up some "message game"—rules for sending and receiving messages—between the buyers, and specifies how the alloca-

5. Section 7.3 solves the more general case of a continuum of types. Approximating the two-type distribution by distributions with a continuum of types, the reader will check that the case of two types does not satisfy the monotone-hazard-rate condition (assumption A10 below), which plays a prominent role throughout the chapter. Bunching (not necessarily of the two "likely types") then occurs.

tion of the good and the transfers will depend on the messages chosen. Let s_1 and s_2 denote the realizations of the two buyers' strategies, σ_1 and σ_2, in this game. The mechanism specifies the probability $x_i(s_1, s_2)$ that the good is transferred to buyer i and the transfer $T_i(s_1, s_2)$ that is paid by buyer i to the seller. For instance, first-price and second-price auctions are mechanisms in which the messages s_i are bids, and bids are made simultaneously. (In both auctions, $x_i(s_1, s_2) = 1$ and $T_j(s_1, s_2) = 0$ if $s_i > s_j$. But when $s_i > s_j$, $T_i = s_i$ in a first-price auction, and $T_i = s_j$ in a second-price auction.)

Let $\{\sigma_1^*(\cdot), \sigma_2^*(\cdot)\}$ denote Bayesian equilibrium strategies in the game or mechanism. Because a buyer is free not to participate in the auction, buyer 1's individual-rationality constraint is that, for each θ_1 and for each s_1 belonging to the support of $\sigma_1^*(\theta_1)$,

(IR) $E_{\theta_2} E_{\sigma_2^*(\theta_2)}[\theta_1 x_1(s_1, s_2) - T_1(s_1, s_2)] \geq 0.$

Similarly, the Bayesian equilibrium or incentive-compatibility condition is that, for each θ_1, each s_1 in the support of $\sigma_1^*(\theta_1)$, and each s_1',

(IC) $E_{\theta_2} E_{\sigma_2^*(\theta_2)}[\theta_1 x_1(s_1, s_2) - T_1(s_1, s_2)]$

$$\geq E_{\theta_2} E_{\sigma_2^*(\theta_2)}[\theta_1 x_1(s_1', s_2) - T_1(s_1', s_2)].$$

There are similar IR and IC constraints for buyer 2.

The optimal auction for the seller would be hard to define, let alone characterize, if one had to consider all possible message spaces. Fortunately, one can restrict one's attention to "direct-revelation games," in which the two buyers simultaneously make (possibly untruthful) announcements of their types $(\hat{\theta}_1, \hat{\theta}_2)$. To see this, define probabilities of consumption and payments by

$$\tilde{x}_i(\hat{\theta}_1, \hat{\theta}_2) \equiv E_{\{\sigma_1^*(\hat{\theta}_1), \sigma_2^*(\hat{\theta}_2)\}}[x_i(s_1, s_2)]$$

and

$$\tilde{T}_i(\hat{\theta}_1, \hat{\theta}_2) \equiv E_{\{\sigma_1^*(\hat{\theta}_1), \sigma_2^*(\hat{\theta}_2)\}}[T_i(s_1, s_2)].$$

IR and IC ensure that the buyers are willing to participate in the direct-revelation game and that a Bayesian equilibrium of this game is for both buyers to announce the truth ($\hat{\theta}_1 = \theta_1, \hat{\theta}_2 = \theta_2$).

We will now solve for the optimal symmetric auction. (As we will see in section 7.5, the optimal auction is indeed symmetric.) Note that IR and IC involve only each buyer's expected probability of getting the good and expected payment to the seller, where the expectations are taken with respect to the other buyer's type. So let $\bar{X}, \underline{X}, \bar{T}$, and $\underline{T}$ denote the expected probabilities of getting the good and the expected payments when the buyer has type $\bar{\theta}$ and $\underline{\theta}$, respectively. The individual-rationality and incentive-compatibility constraints can be written as follows:

(IR$_1$) $\underline{\theta}\underline{X} - \underline{T} \geq 0$

(IR$_2$) $\bar{\theta}\bar{X} - \bar{T} \geq 0$

(IC$_1$) $\underline{\theta}\underline{X} - \underline{T} \geq \underline{\theta}\bar{X} - \bar{T}$

(IC$_2$) $\bar{\theta}\bar{X} - \bar{T} \geq \bar{\theta}\underline{X} - \underline{T}.$

The seller's expected profit per buyer, if his opportunity cost of selling the good is 0, is

$$Eu_0 = (\underline{p}\underline{T} + \bar{p}\bar{T}).$$

We build on the intuition developed in subsection 7.1.1 for the single-buyer case. That is, we guess that the only binding constraints are that the low-valuation type be willing to participate in the mechanism (IR$_1$) and that the high-valuation type not be tempted to claim a low valuation (IC$_2$). As in the single-buyer case, the reader can check that the other two constraints do not bind. IR$_1$ and IC$_2$ determine the expected payments: $\underline{T} = \underline{\theta}\underline{X}$ and $\bar{T} = \bar{\theta}(\bar{X} - \underline{X}) + \underline{\theta}\underline{X}$. Substituting into the seller's expected profit yields

$$Eu_0 = (\underline{\theta} - \bar{p}\bar{\theta})\underline{X} + \bar{p}\bar{\theta}\bar{X}.$$

Until now, we have not imposed constraints on the probabilities $\underline{X}$ and $\bar{X}$. If there were a single buyer, the constraints would obviously be $0 \leq \underline{X}$, $\bar{X} \leq 1$. With two buyers we must take account of the fact that, if one buyer gets the good, the other buyer does not. At the very least, it must be the case that *ex ante* probability of a player's getting the good (i.e., before knowing one's type) does not exceed $\frac{1}{2}$ (by symmetry):

$$\underline{p}\underline{X} + \bar{p}\bar{X} \leq \tfrac{1}{2}. \tag{$*$}$$

As we will see shortly, this constraint does not fully describe the cross-buyer restrictions on probabilities.

First, suppose that $\underline{\theta} \leq \bar{p}\bar{\theta}$. Then Eu_0 is decreasing in $\underline{X}$ and increasing in $\bar{X}$. The seller thus wants to set $\underline{X} = 0$ and $\bar{X}$ "as large as possible." By symmetry, $\bar{X}$ cannot exceed $\underline{p} + \bar{p}/2$ because, when both buyers have valuation $\bar{\theta}$, each receives the good with probability $\frac{1}{2}$ (if any receives it). Hence, $\bar{X} = \underline{p} + \bar{p}/2$. The optimal mechanism is then to not sell if both buyers announce $\underline{\theta}$, to sell to the high type if only one buyer announces $\bar{\theta}$, and to sell with probability $\frac{1}{2}$ to each buyer if both buyers announce $\bar{\theta}$. Notice the strong analogy with the one-buyer case where, if $c = 0$ and $\bar{p}\bar{\theta} > \underline{\theta}$, the buyer buys if and only if $\theta = \bar{\theta}$, and enjoys no informational rent.

Second, suppose that $\underline{\theta} > \bar{p}\bar{\theta}$. Then Eu_0 is strictly increasing in both $\underline{X}$ and $\bar{X}$, and ($*$) must be binding. Substituting $\underline{X}$ using ($*$) in Eu_0 yields

$$Eu_0 = \frac{1}{2\underline{p}}(\underline{\theta} - \bar{p}\bar{\theta}) + \frac{\bar{p}}{\underline{p}}(\bar{\theta} - \underline{\theta})\bar{X}.$$

Hence, again, $\bar{X} = \underline{p} + \bar{p}/2$. And, from $(*)$, $\underline{X} = \underline{p}/2$. If only one buyer announces the high valuation, he receives the good; if both buyers announce the high valuation or both announce the low one, each buyer receives the good with probability $\frac{1}{2}$. This completes the derivation of the optimal mechanism.

A famous result in auction theory (Vickrey 1961) is that, under some assumptions, the first- and second-price auctions yield the seller the optimal expected revenue. We will show in section 7.5 that this is the case if the buyers are symmetric, have independent valuations, and have a continuum of potential valuations (instead of two), and if a technical condition is satisfied.[6] An auction is optimal if it has a (symmetric) equilibrium that yields the same expected transfers, $\underline{T}$ and $\bar{T}$, and the same expected probabilities, $\underline{X}$ and $\bar{X}$, that were obtained above. The symmetric equilibrium of a first-price auction (see chapter 6) and the equilibrium of a second-price auction indeed yield the same $\underline{X}$ and $\bar{X}$ as above if $\underline{\theta} \geq \bar{p}\bar{\theta}$, as the good is sold to the highest-valuation buyer. If $\underline{\theta} < \bar{p}\bar{\theta}$, then the same $\underline{X}$ (i.e., 0) is obtained by adding a "reservation price"—$\bar{\theta}$, say—under which all bids are rejected. However, expected transfers need not be the same as in the optimal mechanism.[7] For instance, in the second-price auction, buyers bid their valuations, and the type-$\bar{\theta}$ buyer obtains rent $\underline{p}(\bar{\theta} - \underline{\theta})$, instead of $\underline{p}(\bar{\theta} - \underline{\theta})/2$, the rent that is optimal when $\underline{\theta} \geq \bar{p}\bar{\theta}$. The optimal revenue can be attained in this case by modifying the second-price auction so that, if one buyer bids $\underline{\theta}$ and the other bids $\bar{\theta}$, the high bidder receives the good at price $\underline{\theta} + (\bar{\theta} - \underline{\theta})/2$. Note that it is still an equilibrium for buyers to bid their valuations: If a high-value buyer bids $\bar{\theta}$, his expected profit is

$$\underline{p}(\bar{\theta} - (\underline{\theta} + (\bar{\theta} - \underline{\theta})/2)) = \underline{p}(\bar{\theta} - \underline{\theta})/2,$$

which equals his expected profit from bidding $\underline{\theta}$.

7.2 Mechanism Design and the Revelation Principle[††]

This section develops the general version of the mechanism-design problem and shows how it can be simplified using the revelation principle.

We suppose that there are $I + 1$ players: a principal (player 0) with no private information, and I agents ($i = 1, \ldots, I$) with types $\theta = (\theta_1, \ldots, \theta_I)$ in

---

6. The technical condition to be satisfied is that the distribution of buyers' types has a monotone hazard rate (see assumption A10). The continuous approximations to a discrete, two-point distribution do not have a monotone hazard rate.

7. This contrasts with the case of a continuum of types. There, incentive compatibility requires $\theta\dot{X}(\theta) - \dot{T}(\theta) = 0$ (see section 7.5) for all θ. Together with the equilibrium condition that the lowest type, $\underline{\theta}$, gets 0 utility, this implies that if $X(\cdot)$ is optimal, so is $T(\cdot)$.

some set Θ. For the time being, we can allow the probability distribution on Θ to be quite general, requiring only that expectations and conditional expectations of the utility functions be well defined.

The object of the mechanism built by the principal is to determine an *allocation* $y = \{x, t\}$. An allocation consists of a vector x, called a *decision*, belonging to a compact, convex, nonempty $\mathscr{X} \subset \mathbb{R}^n$, and a vector of monetary *transfers* $t = (t_1, \ldots, t_I)$ from the principal to each agent (which can be positive or negative).[8] In most applications $\mathscr{X}$ is taken large enough that we are ensured an interior solution; one exception is the auction example mentioned above.

Player i $(i = 0, 1, \ldots, I)$ has a von Neumann-Morgenstern utility $u_i(y, \theta)$. We will assume that u_i $(i = 1, \ldots, I)$ is strictly increasing in t_i, that u_0 is decreasing in each t_i, and that these functions are twice continuously differentiable.

Given a (type-contingent) allocation $\{y(\theta)\}_{\theta \in \Theta}$, agent i $(i = 1, \ldots, I)$ with type θ_i has expected or "interim" utility

$$U_i(\theta_i) \equiv E_{\theta_{-i}}[u_i(y(\theta_i, \theta_{-i}), \theta_i, \theta_{-i}) | \theta_i]$$

and the principal has expected utility

$$E_\theta u_0(y(\theta), \theta).$$

In all applications developed in this chapter, agent i's utility depends on his own transfer t_i and type θ_i, but not on t_{-i} or θ_{-i}. (One situation where u_i depends on θ_{-i} is the common-value auction in which each bidder has private information about the quality of the good for sale.)

The interpretation of x and θ (up to sign adjustments) in the examples mentioned in the introduction is as follows:

Price discrimination x is the consumer's purchase, and t is the price paid to the monopolist; θ indexes the consumer's surplus from consumption.

Regulation x is the firm's cost or price or vector of cost and price, and t is the firm's income; θ is a technological parameter indexing the cost function.

Income tax x is the agent's income, and t is the amount of tax paid by the agent; θ is the agent's ability to earn money.

Public good x is the amount of public good supplied, and t_i is consumer i's monetary contribution to its financing; θ_i indexes consumer i's surplus from the public good.

---

8. In the price-discrimination and auction examples of section 7.1, the agents transferred money to the principal ($t_i = -T_i$).

Auctions $\mathcal{X}$ is the I-dimensional simplex, i.e., $x_i \geq 0$ for all i, and $\sum_{i=1}^{I} x_i \leq 1$. Here, x_i is the probability that consumer i buys the good, and t_i is the amount paid by consumer i; θ_i indexes consumer i's willingness to pay for the good that is auctioned off.

Bargaining x is the quantity sold by a seller to a buyer; t_1 is the transfer to the seller and t_2 is the (negative) transfer to the buyer, such that $t_1 + t_2 = 0$; $\theta_1 = c$ indexes the seller's cost of producing the good, and $\theta_2 = v$ indexes the buyer's willingness to purchase the good.

A *mechanism* or *contract* m defines a message space $\mathcal{M}_i$ for each agent i, and a game form ("step 3" in the introduction) to announce the messages, where $\mu = (\mu_1, \ldots, \mu_I)$ is the vector of all messages sent by the agents in the game form. Because types are private information, y can depend on θ only through the agents' messages; denote this function by $y_m: \mathcal{M} \to Y = \mathcal{X} \times \mathbb{R}^I$.

We can now derive the *revelation principle*, which states that the principal can content herself with "direct" mechanisms, in which the message spaces are the type spaces, all agents accept the mechanism in step 2 regardless of their types, and the agents simultaneously and truthfully announce their types in step 3. This principle has been enunciated by many researchers, including Gibbard (1973), Green and Laffont (1977), Dasgupta et al. (1979), and Myerson (1979).

Note that the game form associated with a mechanism in step 3, together with the acceptance decisions of step 2, defines a larger game among the agents. Without loss of generality, we can include the acceptance decision of the agents into their message $\mu_i(\cdot)$. Consider a Bayesian equilibrium of this larger game. Assume for notational simplicity that this is a pure-strategy equilibrium, which we can thus write $\mu_i^*(\theta_i)$.

Consider the new message space Θ_i for each agent i, so that each agent announces a type $\hat{\theta}_i$ (which may differ from the true value θ_i). Letting $\hat{\theta} \equiv (\hat{\theta}_1, \ldots, \hat{\theta}_I)$, define the new allocation rule $\bar{y}: \Theta \to Y$ by

$$\bar{y}(\hat{\theta}) = y_m(\mu^*(\hat{\theta})),$$

where

$$\mu^*(\hat{\theta}) = (\mu_1^*(\hat{\theta}_1), \ldots, \mu_I^*(\hat{\theta}_I)).$$

It is immediate that truthtelling, $\{\hat{\theta}_i = \theta_i\}$, is a Bayesian equilibrium of the new game, given that $\{\mu_i^*\}$ is a Bayesian equilibrium of the original game[9]: For all i and θ_i,

---

9. We discuss the revelation principle in a Bayesian context, but the same reasoning holds for equilibria in dominant strategies. (See subsection 7.4.2 for the definition of implementation in dominant strategies.)

$$E_{\theta_{-i}}[u_i(\bar{y}(\theta), \theta_i, \theta_{-i}) \,|\, \theta_i]$$

$$= E_{\theta_{-i}}[u_i(y_m(\mu^*(\theta)), \theta_i, \theta_{-i}) \,|\, \theta_i]$$

$$= \sup_{\mu_i \in \mathcal{M}_i} E_{\theta_{-i}}[u_i(y_m(\mu_1^*(\theta_1), \ldots, \mu_i, \ldots, \mu_I^*(\theta_I)), \theta_i, \theta_{-i}) \,|\, \theta_i]$$

$$\geq \sup_{\hat{\theta}_i \in \Theta_i} E_{\theta_{-i}}[u_i(\bar{y}(\theta_1 \ldots, \hat{\theta}_i, \ldots, \theta_I), \theta_i, \theta_{-i}) \,|\, \theta_i],$$

where the first equality results from the definition of the direct-revelation mechanism $\bar{y}$, the second equality is the condition for Bayesian equilibrium in the original mechanism m, and the weak inequality expresses the fact that in the direct-revelation mechanism everything is as if agent i picked an announcement in the subset of messages $\{\mu_i^*(\hat{\theta}_i)\}_{\hat{\theta}_i \in \Theta_i}$ of $\mathcal{M}_i$ (the agent thus has, at most, as many possibilities for deviating as in the original game). When the σ_i^* are random, the same reasoning holds with $\bar{y}(\cdot)$ defined as the appropriate random function of $\hat{\theta}$.

Observation (revelation principle)   Suppose that a mechanism with message spaces $\mathcal{M}_i$ and allocation function $y_m(\cdot)$ has a Bayesian equilibrium

$$\mu^*(\cdot) \equiv \{\mu_i^*(\theta_i)\}_{\substack{i=1,\ldots,I \\ \theta_i \in \Theta_i}}.$$

Then there exists a direct-revelation mechanism (namely, $\bar{y} = y_m \circ \mu^*$) such that the message spaces are the type spaces ($\mathcal{M}_i = \Theta_i$) and such that there exists a Bayesian equilibrium in which all agents accept the mechanism in step 2 and announce their types truthfully in step 3.

Caveat   The direct-revelation game associated with $\bar{y}(\cdot)$ has one equilibrium that yields the same allocation as the original equilibrium. This equilibrium need not be unique. Ma et al. (1988), Mookherjee and Reichelstein (1988), Postlewaite and Schmeidler (1986), and Palfrey and Srivastava (1989) have derived conditions under which a Bayesian allocation can be implemented by a game in the sense of either being achieved by all equilibria of the game or (more strongly) being achieved by the unique equilibrium of the game.[10] The idea, as in Maskin 1977 and in Moore and Repullo 1989, is to use, instead of a direct mechanism, a mechanism where players report information in addition to their type. These additional "nontype" messages turn out to be superfluous in the equilibrium to be implemented, but serve to eliminate other equilibria of the reduced game in which players can only announce their types. The standard methodology is first to derive the principal's optimum, and then, if one is worried about multiple equilibria in the direct-revelation game, to see if the optimal allocation satisfies the sufficient conditions for unique implementation.

10. See also the work of Demski and Sappington (1984) and Ma, Moore, and Turnbull (1988) in more structured environments.

Remark We will be fairly casual about the distinction between a mechanism and an allocation. In a sense, the revelation principle, which we invoke systematically from now on, allows us to merge the two concepts.

7.3   Mechanism Design with a Single Agent[††]

The following methodology, first developed by Mirrlees (1971), was extended and applied to various contexts by Mussa and Rosen (1978), Baron and Myerson (1982), and Maskin and Riley (1984a), among others. The presentation, including the propositions, follows the general analysis of Guesnerie and Laffont (1984).[11]

Because there is a single agent, we omit the subscripts on transfer (t) and type (θ) in this section. We assume that the agent's type lies in an interval $[\underline{\theta}, \overline{\theta}]$. The agent knows θ, and the principal has the prior cumulative distribution function P ($P(\underline{\theta}) = 0, P(\overline{\theta}) = 1$), with differentiable density $p(\theta)$ such that $p(\theta) > 0$ for all θ in $[\underline{\theta}, \overline{\theta}]$. (Differentiability of the density is not necessary, but is assumed for convenience.) The type space is single dimensional,[12] but the decision space may be multidimensional. (Although we consider a multidimensional decision for completeness, the reader can grasp the main ideas from the case of a single-dimensional decision.) A (type-contingent) allocation is a function from the agent's type into an allocation:

$$\theta \to y(\theta) = (x(\theta), t(\theta)).$$

7.3.1   Implementable Decisions and Allocations

Definition 7.1 A decision function $x: \theta \to \mathcal{X}$ is *implementable* if there exists a transfer function $t(\cdot)$ such that the allocation $y(\theta) = (x(\theta), t(\theta))$ for $\theta \in [\underline{\theta}, \overline{\theta}]$ satisfies the incentive-compatibility constraint

(IC) $u_1(y(\theta), \theta) \geq u_1(y(\hat{\theta}), \theta)$ for all $(\theta, \hat{\theta}) \in [\underline{\theta}, \overline{\theta}] \times [\underline{\theta}, \overline{\theta}]$.

We will then say that the allocation $y(\cdot)$ is implementable.

Note that we ignore the individual-rationality constraint (that the agent be willing to participate in step 2) in this definition. Such a constraint, if any, must be reintroduced at the optimization stage.

Remark If $x(\cdot)$ is implementable through transfer $t(\cdot)$, there exists an "indirect" or "fiscal" mechanism $t = T(x)$, in which the agent chooses a decision x, rather than an announcement of his type, that implements the same allocation. Consider the following scheme:

11. See also Laffont 1989, chapter 10.
12. The case of a multi-dimensional type space is considerably harder. See Rochet 1985, Laffont, Maskin, and Rochet 1987, and McAfee and McMillan 1988.

$$T(x) \equiv \begin{cases} t & \text{if } \exists \hat{\theta} \text{ such that } t = t(\hat{\theta}) \text{ and } x = x(\hat{\theta}) \\ & \text{(if there exist several such } \hat{\theta}, \text{ pick one)} \\ -\infty & \text{otherwise.} \end{cases}$$

Choosing an x is *de facto* equivalent to announcing a $\hat{\theta}$.

We restrict our attention to decision profiles $x(\cdot)$ that are piecewise continuously differentiable ("piecewise C^1").[13] We now derive a necessary condition for $x(\cdot)$ to be implementable.

Theorem 7.1 (necessity)   A piecewise C^1 decision function $x(\cdot)$ is implementable only if

$$\sum_{k=1}^{n} \frac{\partial}{\partial \theta} \left(\frac{\partial u_1/\partial x_k}{\partial u_1/\partial t} \right) \frac{dx_k}{d\theta} \geq 0, \tag{7.1}$$

whenever $x = x(\theta)$, $t = t(\theta)$, and x is differentiable at θ.

Proof   Type θ chooses an announcement $\hat{\theta}$ so as to maximize $\Phi(\hat{\theta}, \theta) \equiv u_1(x(\hat{\theta}), t(\hat{\theta}), \theta)$. Because u_1 is C^2 and x is piecewise C^1, any transfer function t that implements x must be piecewise C^1 as well.[14] Maximizing at a point of differentiability yields a first-order condition and a local second-order condition at the optimum $\hat{\theta} = \theta$:

$$\frac{\partial \Phi}{\partial \hat{\theta}}(\theta, \theta) = 0 \text{ (truth telling or IC)} \tag{7.2}$$

and

$$\frac{\partial^2 \Phi}{\partial \hat{\theta}^2}(\theta, \theta) \leq 0. \tag{7.3}$$

(To check that the second derivative in equation 7.3 exists except at a finite number of points, note that $\partial^2 \Phi/\partial \hat{\theta} \partial \theta$ exists except at a finite number of points, because $dx/d\theta$ and $dt/d\theta$ do, and use identity 7.2.)

Differentiating equation 7.2 shows that (except perhaps at a finite number of points)

$$\frac{\partial^2 \Phi}{\partial \hat{\theta}^2}(\theta, \theta) + \frac{\partial^2 \Phi}{\partial \hat{\theta} \partial \theta}(\theta, \theta) = 0. \tag{7.4}$$

Therefore, the local second-order condition can be rewritten as

13. A piecewise-C^1 function admits a derivative except at a finite number of points. And when a derivative does not exist, the function still admits a left and a right derivative. Standard optimal control techniques (Hadley and Kemp 1971) require that functions be piecewise C^1. That the "piecewise" qualifier is needed becomes clear in our analysis of bunching in the appendix to the present chapter.
14. Perform a Taylor expansion of $u_1(x(\hat{\theta}), t(\hat{\theta}), \theta)$ to the left and to the right of $\hat{\theta} = \theta$ at a θ at which $dx/d\theta$ exists and is continuous, and use the fact that $\hat{\theta} = \theta$ is optimal for type θ.

$$\frac{\partial^2 \Phi}{\partial \hat{\theta} \partial \theta}(\theta, \theta) \geq 0, \tag{7.5}$$

or

$$\sum_{k=1}^{n} \frac{\partial}{\partial \theta}\left(\frac{\partial u_1}{\partial x_k}\right)\frac{dx_k}{d\theta} + \frac{\partial}{\partial \theta}\left(\frac{\partial u_1}{\partial t}\right)\frac{dt}{d\theta} \geq 0. \tag{7.6}$$

Rewriting equation 7.2 yields

$$\sum_{k=1}^{n} \frac{\partial u_1}{\partial x_k}\frac{dx_k}{d\theta} + \frac{\partial u_1}{\partial t}\frac{dt}{d\theta} = 0. \tag{7.7}$$

Using equation 7.7 to eliminate $dt/d\theta$ in equation 7.6 yields

$$\sum_{k=1}^{n} \left[\left[\frac{\partial}{\partial \theta}\left(\frac{\partial u_1}{\partial x_k}\right)\frac{\partial u_1}{\partial t} - \frac{\partial}{\partial \theta}\left(\frac{\partial u_1}{\partial t}\right)\frac{\partial u_1}{\partial x_k}\right]\bigg/\frac{\partial u_1}{\partial t}\right]\frac{dx_k}{d\theta} \geq 0, \tag{7.8}$$

which is equivalent to equation 7.1.   ∎

The interpretation of the necessary condition is particularly simple if we make the following assumption:

A1 For all $k \in \{1, \ldots, n\}$, either

$$(CS^+) \quad \frac{\partial}{\partial \theta}\left(\frac{\partial u_1/\partial x_k}{\partial u_1/\partial t}\right) > 0$$

or

$$(CS^-) \quad \frac{\partial}{\partial \theta}\left(\frac{\partial u_1/\partial x_k}{\partial u_1/\partial t}\right) < 0.$$

This is known as the *sorting* (or "constant sign" (CS), or "single crossing," or "Spence-Mirrlees") *condition*.

Note that by changing x_k into $-x_k$ if necessary, one can restrict attention to the case in which all the derivatives are positive if A1 holds. From now on we will assume that CS^+ holds for all k. A1 is very standard and is made in almost all applications of the theory. Note that

$$\frac{\partial u_1/\partial x_k}{\partial u_1/\partial t}$$

is the agent's marginal rate of substitution between decision k and transfer t. The condition asserts that the agent's type affects this marginal rate of substitution in a systematic way.

Suppose for instance that the decision is single dimensional ($n = 1$), and that $\partial u_1/\partial x < 0$, as is the case if x is an output supplied by the agent to the principal. In this case, under the sorting condition, inequality 7.1 is equivalent to the monotonicity of the decision in the agent's type. CS^+ means

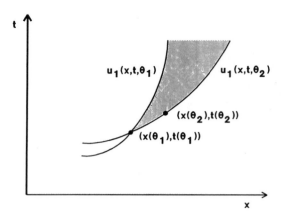

Figure 7.2

that the slope of the agent's indifference curve in the (x, t) space,

$$\left| \frac{\partial u_1 / \partial x}{\partial u_1 / \partial t} \right|,$$

decreases with the agent's type, θ. That is, a high-type (θ_2) agent must be compensated less than a low-type agent (θ_1) for a given increase in the decision x. This situation is depicted in figure 7.2.

In this case the interpretation of inequality CS^+ is straightforward. Let $y(\theta_1) = (x(\theta_1), t(\theta_1))$ and $y(\theta_2) = (x(\theta_2), t(\theta_2))$ denote the allocations of types θ_1 and θ_2. For the allocation to be incentive compatible, it must be the case that $y(\theta_2)$ lies below type θ_1's indifference curve through $y(\theta_1)$ and above type θ_2's indifference curve through $y(\theta_1)$. Hence, $y(\theta_2)$ must belong to the shaded region in figure 7.2. (We put $y(\theta_2)$ on the boundary of the shaded region in the figure because, as we show below, type θ_2's incentive-compatibility constraint is binding in the optimal mechanism.)

Throughout this chapter, we will make substantial use of the following theorem.

Theorem 7.2 (monotonicity)   Assume that the decision space is single dimensional and that CS^+ holds. A necessary condition for $x(\cdot)$ to be implementable is that it be nondecreasing: $\theta_2 > \theta_1 \Rightarrow x(\theta_2) \geq x(\theta_1)$.

Of course, if CS^- held, the necessary condition would be that $x(\cdot)$ be nonincreasing. Note that, whereas theorem 7.1 implies monotonicity at points of differentiability, the proof of theorem 7.2 relies on the simple revealed-preference argument outlined in the discussion of figure 7.2 and has nothing to do with differentiability.

To obtain sufficient conditions for implementability, Guesnerie and Laffont (1984) make the sorting assumption A1 and add the following technical assumption, which guarantees the existence of a solution to a differential equation.

A2 The marginal rates of substitution between decisions and transfer do not increase too fast when the transfer goes to infinity: For all k, there exist K_0 and K_1 such that

$$\left| \frac{\partial u_1 / \partial x_k}{\partial u_1 / \partial t} \right| \leq K_0 + K_1 |t| \text{ uniformly over } x, t, \text{ and } \theta.$$

Assumption A2 is satisfied, for instance, in the case of quasi-linear preferences (for which $\partial u_1 / \partial t = 1$).

It turns out that monotonicity is also sufficient for implementability[15]:

Theorem 7.3 Under assumptions A1 (CS^+) and A2, any piecewise C^1 decision function $x(\cdot)$ satisfying $dx_k / d\theta \geq 0$ for all k is implementable. That is, there exists $t(\cdot)$ such that $(x(\cdot), t(\cdot))$ is incentive compatible.

Proof From the agent's first-order condition (equation 7.7), $t(\cdot)$ must satisfy

$$\frac{dt}{d\theta} = - \sum_{k=1}^{n} \left(\frac{\partial u_1 / \partial x_k}{\partial u_1 / \partial t} \right) \frac{dx_k}{d\theta}. \tag{7.9}$$

Assumption A2 guarantees the existence of a solution to equation 7.9.[16] We are left with showing that $(x(\cdot), t(\cdot))$ is incentive compatible. (By construction, the first-order condition of the agent's maximization with respect to $\hat{\theta}$ is satisfied; so is the local second-order condition in inequality 7.1, from CS^+ and $dx_k / d\theta > 0$. But this is not sufficient, as we must still prove that the global second-order condition for maximization is satisfied.) Suppose that truth telling is not optimal for type θ. That is, there exists $\hat{\theta}$ such that $\Phi(\hat{\theta}, \theta) - \Phi(\theta, \theta) > 0$ (recall that $\Phi(\hat{\theta}, \theta) \equiv u_1(x(\hat{\theta}), t(\hat{\theta}), \theta)$). Then

$$\int_{\theta}^{\hat{\theta}} \frac{\partial \Phi}{\partial a}(a, \theta) \, da > 0,$$

or

15. Thus, the incentive constraints are satisfied globally if they are satisfied locally and if each component of the decision is monotonic. Under some conditions on the utility functions and the distribution of types, the optimal mechanism subject only to the local incentive constraints is monotonic, and thus is optimal subject to the global incentive constraints. We develop this approach below for the case of a one-dimensional decision.

Alternatively, weaker assumptions on the utility functions can be made that, coupled with the hypothesis that all the "downward" incentive constraints are satisfied (that is, $u_1(y(\theta), \theta) \geq u_1(y(\theta'), \theta)$ for all $\theta' \leq \theta$), guarantee that the neglected upward incentive constraints are satisfied and therefore that global incentive compatibility holds. Moreover, one can obtain some properties of the optimal mechanism (without solving explicitly for the optimal mechanism) using the fact that only the downward constraints must be satisfied. This "nonlocal" approach is developed by Moore (1984, 1985, 1988) and by Matthews and Moore (1987).

16. As $|dt/d\theta| \leq (\sup_{\theta,k} |dx_k/d\theta|)(K_0 + K_1 |t|)$. See Hurewicz 1958.

$$\int_{\theta}^{\hat{\theta}} \frac{\partial u_1}{\partial t}(x(a),t(a),\theta) \left(\sum_{k=1}^{n} \frac{\frac{\partial u_1}{\partial x_k}(x(a),t(a),\theta)}{\frac{\partial u_1}{\partial t}(x(a),t(a),\theta)} \frac{dx_k}{da}(a) + \frac{dt}{da}(a) \right) da > 0.$$

$$(7.10)$$

If $\hat{\theta} > \theta$, from the sorting condition CS^+,[17] equation 7.10 implies that

$$\int_{\theta}^{\hat{\theta}} \frac{\partial u_1}{\partial t}(x(a),t(a),\theta) \left(\sum_{k=1}^{n} \frac{\frac{\partial u_1}{\partial x_k}(x(a),t(a),a)}{\frac{\partial u_1}{\partial t}(x(a),t(a),a)} \frac{dx_k}{da}(a) + \frac{dt}{da}(a) \right) da > 0.$$

$$(7.11)$$

But equation 7.9 implies that the integrand in equation 7.11 is equal to 0 for all a, which is a contradiction.

If $\hat{\theta} < \theta$, the same reasoning shows that equation 7.11 cannot hold either.   ∎

An important corollary of theorem 7.3 is that in the case of a single-dimensional decision, under the sorting conditions CS^+ or CS^-, a decision function is implementable if and only if it is monotone (nondecreasing under CS^+, nonincreasing under CS^-).

7.3.2 Optimal Mechanisms

Now that we have characterized the set of implementable allocations, we can determine the optimal one for the principal. To do so, we must re-introduce the individual-rationality constraint for the agent. An imple-mentable allocation that satisfies the individual-rationality constraint is called *feasible*; the principal's problem is to choose the feasible allocation with the highest expected payoff. For simplicity, we assume that the agent's reservation utility (i.e., his expected utility when he rejects the principal's mechanism) is independent of his type.

A3 The reservation utility $\underline{u}$ is independent of type; i.e., the participation constraint is

   (IR) $u_1(x(\theta),t(\theta),\theta) \geq \underline{u}$ for all θ.

Under this assumption, if u_1 increases with the type ($\partial u_1/\partial \theta > 0$), then IR can bind only at $\theta = \underline{\theta}$: Any type $\theta > \underline{\theta}$ can always announce $\hat{\theta} = \underline{\theta}$, which

17. That is,

$$\frac{\frac{\partial u_1}{\partial x_k}(x,t,\theta)}{\frac{\partial u_1}{\partial t}(x,t,\theta)} \leq \frac{\frac{\partial u_1}{\partial x_k}(x,t,a)}{\frac{\partial u_1}{\partial t}(x,t,a)} \text{ for } \theta \leq a.$$

gives him more than type $\underline{\theta}$'s utility, which is at least $\underline{u}$.[18] For notational simplicity, we normalize $\underline{u} = 0$.

We will also make the following assumptions.

A4 Quasi-linear utilities[19]:

$$u_0(x, t, \theta) = V_0(x, \theta) - t,$$

$$u_1(x, t, \theta) = V_1(x, \theta) + t,$$

where V_0 and V_1 are thrice differentiable and concave in x.

A5 $n = 1$: The decision is single dimensional, and CS^+ holds; that is, $\partial^2 V_1 / \partial x \partial \theta \geq 0$.

A6 $\partial V_1 / \partial \theta > 0$.

A7 $\partial^2 V_0 / \partial x \partial \theta \geq 0$ (which is satisfied if V_0 does not depend on θ).

A8 $\partial^3 V_1 / \partial x \partial \theta^2 \leq 0$ and $\partial^3 V_1 / \partial x^2 \partial \theta \geq 0$.

A9 $\mathcal{X}$ is the interval $[0, \bar{x}]$, where $\bar{x} > \arg\max(V_0(x, \bar{\theta}) + V_1(x, \bar{\theta}))$.

We have little information about whether assumption A8 is likely to be satisfied, as it contains third derivatives. This assumption is a sufficient condition (together with the monotone-hazard-rate condition introduced below) for the optimal decision obtained by ignoring the monotonicity constraint to satisfy monotonicity. As can be seen from equation 7.13 below, assumption A8 is not necessary if "uncertainty about θ" is small, i.e., if the hazard rate is large.[20]

The principal maximizes her expected utility subject to the agent's IR and IC constraints:

$$\max_{\{x(\cdot), t(\cdot)\}} \quad E_\theta u_0(x(\theta), t(\theta), \theta)$$

18. There are no general results on mechanism design when the reservation utility is increasing with θ. The issue is that the participation constraint may be binding at points other than $\theta = \underline{\theta}$. For economic examples in which it is binding at $\bar{\theta}$, see Champsaur and Rochet 1989, Laffont and Tirole 1990a, and Lewis and Sappington 1989a; for examples in which it is binding in the middle of the interval, see Lewis and Sappington 1989b. For instance, Champsaur and Rochet (1989) and Laffont and Tirole (1990a) study price discrimination by a firm when the high-demand consumers can purchase an alternative (bypass) product (see also exercise 7.8). Similarly, in a labor market, a high-ability worker might have better outside opportunities, and therefore a higher reservation utility, than a low-ability one. Another common cause of type-dependent reservation utilities and of IR constraints that are binding for a good type is the existence of a prior contract between the principal and the agent (Laffont and Tirole 1990b; Caillaud, Jullien and Picard 1990). Even if, *ex ante*, all types have the same reservation utility before signing a contract with the principal, this contract, once signed, defines a *status quo* allocation in any future contract renegotiation. The *status quo* allocation is then type dependent.
19. Although quasi-linearity is a strong assumption, given quasi-linearity the fact that the coefficients of t are -1 and $+1$ is not an additional restriction, as one can always normalize V_0 and V_1 so that the payoff functions can be written as in assumption A4.
20. The intuition is that, if the uncertainty is small, allocations are close to the symmetric-information allocation, and the study of the symmetric-information case requires only assumptions on second derivatives.

subject to $x \in \mathcal{X}$ and

(IC) $u_1(x(\theta), t(\theta), \theta) \geq u_1(x(\hat{\theta}), t(\hat{\theta}), \theta)$ for all $(\theta, \hat{\theta})$

(IR) $u_1(x(\theta), t(\theta), \theta) \geq \underline{u} = 0$ for all θ.

For the moment, we ignore the constraint $x \in \mathcal{X}$. We will return to it at the end of the analysis; in most applications this constraint does not bind. We noted that assumptions A3 and A6 imply that IR need be satisfied only at $\theta = \underline{\theta}$. Furthermore, because transfers are costly to the principal, it is clear that IR is binding at $\theta = \underline{\theta}$:

(IR') $u_1(x(\underline{\theta}), t(\underline{\theta}), \underline{\theta}) = \underline{u} = 0$.

A useful trick, due to Mirrlees (1971), is to use the indirect utility function. This allows us to eliminate transfers in the above program. Let

$$U_1(\theta) \equiv \max_{\hat{\theta}} u_1(x(\hat{\theta}), t(\hat{\theta}), \theta) = u_1(x(\theta), t(\theta), \theta).$$

The envelope theorem implies that

$$\frac{dU_1}{d\theta} = \frac{\partial u_1}{\partial \theta} = \frac{\partial V_1}{\partial \theta},$$

which implies that

$$U_1(\theta) = \underline{u} + \int_{\underline{\theta}}^{\theta} \frac{\partial V_1}{\partial \tilde{\theta}} (x(\tilde{\theta}), \tilde{\theta}) \, d\tilde{\theta}.$$

Furthermore, $u_0 = V_0 + V_1 - U_1$; that is, the principal's utility is equal to the social surplus minus the agent's utility. The principal's objective function is thus

$$\int_{\underline{\theta}}^{\bar{\theta}} \left[V_0(x(\theta), \theta) + V_1(x(\theta), \theta) - \int_{\underline{\theta}}^{\theta} \frac{\partial V_1}{\partial \tilde{\theta}} (x(\tilde{\theta}), \tilde{\theta}) \, d\tilde{\theta} \right] p(\theta) \, d\theta$$

$$= \int_{\underline{\theta}}^{\bar{\theta}} \left[V_0(x(\theta), \theta) + V_1(x(\theta), \theta) - \frac{1 - P(\theta)}{p(\theta)} \frac{\partial V_1}{\partial \theta} (x(\theta), \theta) \right] p(\theta) \, d\theta$$

after an integration by parts.

Next, we argue that IC is equivalent to the conjunction of the condition that $dU_1/d\theta = \partial V_1/\partial \theta$ and the condition that $x(\cdot)$ be nondecreasing. Theorem 8.2 shows that IC implies these two conditions, and theorem 8.3 shows that the converse holds.

The principal's optimization program is thus

$$\max_{\{x(\cdot)\}} \int_{\underline{\theta}}^{\bar{\theta}} [V_0(x, \theta) + V_1(x, \theta) - \frac{1 - P(\theta)}{p(\theta)} \frac{\partial V_1}{\partial \theta} (x, \theta)] p(\theta) \, d\theta$$

s.t. (monotonicity) $x(\cdot)$ is nondecreasing;

we call this program I. Once the solution $x(\cdot)$ to program I is obtained, we can compute the agent's indirect utility,

$$U_1(\theta) \equiv \int_{\underline{\theta}}^{\theta} \frac{\partial V_1}{\partial \tilde{\theta}}(x(\tilde{\theta}), \tilde{\theta}) \, d\tilde{\theta},$$

and the transfer,

$$t(\theta) \equiv U_1(\theta) - V_1(x(\theta), \theta)).$$

For the moment, let us ignore the monotonicity constraint in program I. The relaxed program is called program II. If the solution to program II turns out to be nondecreasing, then it is also a solution to the full program. Otherwise, one must introduce the monotonicity constraint.

The solution to the relaxed program is given by

$$\frac{\partial V_0}{\partial x} + \frac{\partial V_1}{\partial x} = \frac{1 - P(\theta)}{p(\theta)} \frac{\partial^2 V_1}{\partial x \partial \theta}. \tag{7.12}$$

Let $x^*(\cdot)$ denote a solution to equation 7.12. (From A4 and A8, the relaxed program is concave in x, so the second-order condition is satisfied.)

Interpretation of Equation 7.12   The principal faces a tradeoff between maximizing total surplus $(V_0 + V_1)$ and appropriating the agent's informational rent (U_1). Consider a type θ. By increasing x over the interval $[\theta, \theta + d\theta]$ by δx, the total surplus is increased by

$$\left(\frac{\partial (V_0 + V_1)}{\partial x} \delta x \right) p(\theta) \, d\theta.$$

However, the rent of type $\theta + d\theta$ is increased by

$$\left[\frac{\partial}{\partial x} \left(\frac{\partial V_1}{\partial \theta} \right) \delta x \right] d\theta,$$

as is the rent of types in $[\theta + d\theta, \bar{\theta}]$ (which have weight $1 - P(\theta)$). At the optimum, the increase in total surplus must be equal to the expected increase in the agent's rent. Note that at $\theta = \bar{\theta}$ rent extraction is not a concern, and so $V_0 + V_1$ is maximized; this result is known as "no distortion at the top."

As a trivial illustration, consider monopoly pricing. Let $x \in [0, 1]$ denote the quantity purchased by a buyer with 0-1 demand. Let $V_0(x, \theta) = -cx$ (where c is marginal cost), and $V_1(x, \theta) = \theta x$ (so θ is the buyer's valuation for the good). The maximand in the principal's optimization program is linear in x, and the "bang-bang" solution is $x = 1$ if $\theta \geq \theta^* > c$, where $\theta^* = c + [1 - P(\theta^*)]/p(\theta^*)$, and $x = 0$ otherwise. This is the same solution that is obtained when the monopolist charges a price π and knows that a fraction $1 - P(\pi)$ of consumers have valuation exceeding π: The program

$\max_{\pi}(\pi - c)(1 - P(\pi))$ has solution $\pi = \theta^*$. This is perhaps the simplest example of gains from trade not being realized because the principal trades off efficiency against extraction of the agent's rent.

More generally, a simple revealed-preference argument shows that $x^*(\theta)$ is smaller than the level $\hat{x}(\theta)$ that maximizes total surplus $(V_0 + V_1)$, which is the level that would prevail if the principal knew the agent's type. To see this, note that, by definition,

$$V_0(\hat{x}, \theta) + V_1(\hat{x}, \theta) \geq V_0(x^*, \theta) + V_1(x^*, \theta)$$

and

$$V_0(x^*, \theta) + V_1(x^*, \theta) - \frac{1 - P}{p} \frac{\partial V_1}{\partial \theta}(x^*, \theta)$$

$$\geq V_0(\hat{x}, \theta) + V_1(\hat{x}, \theta) - \frac{1 - P}{p} \frac{\partial V_1}{\partial \theta}(\hat{x}, \theta).$$

Adding up these two inequalities and using the sorting condition $\partial^2 V_1 / \partial x \partial \theta \geq 0$ yields

$$x^*(\theta) \leq \hat{x}(\theta).$$

Inspection of program I suggests the following definition, due to Myerson (1981):

Definition 7.2   The agent's *virtual surplus* is

$$V_1(x, \theta) - \frac{1 - P(\theta)}{p(\theta)} \frac{\partial V_1}{\partial \theta}(x, \theta).$$

Thus, everything is as if the principal maximized total surplus, where the agent's surplus is replaced by his virtual surplus. Note that the principal's virtual surplus is equal to her surplus in the computation of the total virtual surplus. This is due to the fact that the principal has full information about herself.

When is it legitimate to focus on the relaxed program?   One can ignore the monotonicity constraint if and only if the $x^*(\cdot)$ defined by equation 7.12 is nondecreasing. Let us assume for simplicity that the objective function in the relaxed program is strictly concave in x. Totally differentiating equation 7.12 yields

$$\left(\frac{\partial^2 V_0}{\partial x^2} + \frac{\partial^2 V_1}{\partial x^2} - \frac{1 - P(\theta)}{p(\theta)} \frac{\partial^3 V_1}{\partial x^2 \partial \theta} \right) \frac{dx^*}{d\theta}$$

$$= \frac{\partial^2 V_1}{\partial x \partial \theta} \left[\frac{d}{d\theta} \left(\frac{1 - P(\theta)}{p(\theta)} \right) - 1 \right] - \frac{\partial^2 V_0}{\partial x \partial \theta} + \frac{1 - P(\theta)}{p(\theta)} \frac{\partial^3 V_1}{\partial x \partial \theta^2}. \tag{7.13}$$

Thus, under assumptions A5, A7, and A8, and using the second-order condition, $dx^*/d\theta$ is positive if

$$\frac{d}{d\theta}\left(\frac{1-P(\theta)}{p(\theta)}\right) \leq 0.$$

Thus, one can legitimately focus on the relaxed program if the following assumption is satisfied:

A10 (monotone hazard rate) $\dfrac{d}{d\theta}\left(\dfrac{p(\theta)}{1-P(\theta)}\right) \geq 0.$

To see why this is called the monotone-hazard-rate condition, interpret θ as the lifetime of a machine, and let $Q(\theta) = 1 - P(\theta)$ be the reliability function, which gives the probability that the machine lasts at least until time θ. The conditional probability that the machine fails over the interval $[\theta, \theta + d\theta]$, given that it lasts until time θ, is the "hazard rate"

$$\frac{p(\theta)}{Q(\theta)} = \frac{p(\theta)}{1-P(\theta)}.$$

The monotone hazard rate thus indicates that the rate of failure increases as the machine grows older.

Since $1 - P(\theta)$ is decreasing in θ, a sufficient condition for the hazard rate to increase is that the density p is increasing. More generally, the monotone-hazard-rate condition is equivalent to the reliability function Q being log-concave (a function Q is log-concave if $\ln Q$ is concave). One can show that if p is log-concave on $[\underline{\theta}, \overline{\theta}]$, then the reliability function Q is log-concave on $[\underline{\theta}, \overline{\theta}]$ (Bagnoli and Berstrom 1989, theorem 2).[21] Applying these results shows that A10 holds (and thus that one can ignore the monotonicity constraint) if P is uniform, normal, logistic, chi-squared, exponential, Laplace, and, under some restrictions on the parameters, Weibull, gamma, or beta. (See Bagnoli and Bergstrom 1989 for a more complete list of distributions whose reliability function is log-concave.)

Finally, we reintroduce the constraint $x \in \mathcal{X}$, which we have ignored to this point. Because

$$x^*(\overline{\theta}) = \arg\max_{x} \, [V_0(x, \overline{\theta}) + V_1(x, \overline{\theta})],$$

which is less than $\overline{x}$ by assumption A9, and $x^*(\theta) \leq x^*(\overline{\theta})$, the constraint $x(\theta) \leq \overline{x}$ does not bind. If $x^*(\theta) < 0$ for some θ, the optimal allocation will have $x^*(\theta) = 0$ for a range of θ's. In monopoly pricing, for example, the monopolist will choose not to sell to all consumers whose willingness to pay is less than the monopoly price.

---

21. Note that if q is the density of Q, then $q'/q = p'/p$. But a sufficient condition for a strictly monotonic function on an interval $[\underline{\theta}, \overline{\theta}]$ taking value 0 at either $\underline{\theta}$ or $\overline{\theta}$ to be log-concave on this interval is that its derivative is log-concave (Prekova 1973; Bagnoli and Bergstrom 1989, theorem 1). Hence, if $(p'/p)' = (q'/q)' \leq 0$, $(q/Q)' \leq 0$.

Theorem 7.4   Under assumptions A1–A10, the optimal decision $x^*(\theta)$ is given by equation 7.12.

In the appendix to this chapter we analyze what to do if the monotonicity constraint is binding. We also study whether it might be desirable to use stochastic schemes.

7.4   Mechanisms with Several Agents: Feasible Allocations, Budget Balance, and Efficiency[††]

Now we turn to the case of mechanisms with several agents. We will distinguish between the case of a self-interested principal and that of a "benevolent" principal who maximizes the sum of the agents' welfare. Of course, the distinction is relevant only when the principal optimizes over feasible allocations (section 7.5). We will make the following assumptions in the rest of the chapter:

B1   Types are single dimensional. They are drawn from independent distributions P_i on $[\underline{\theta_i}, \overline{\theta_i}]$ with strictly positive and differentiable densities p_i. The distributions are common knowledge.

B2 (private values)   Agent i's preferences depend only on the decision, his own type, and his own transfer: $u_i(x, t_i, \theta_i)$.

B3   Preferences are quasi-linear:

$$u_i(x, t_i, \theta_i) = V_i(x, \theta_i) + t_i, \quad i \in \{1, \dots, I\}$$

and either

$$u_0(x, t, \theta) = V_0(x, \theta) - \sum_{i=1}^{I} t_i \quad \text{(self-interested principal)}$$

or

$$u_0(x, t, \theta) = \sum_{i=0}^{I} V_i(x, \theta) \quad \text{(benevolent principal)},$$

where $V_0(x, \theta) = B_0(x, \theta) - C_0(x)$, $C_0(x)$ is the principal's monetary cost from decision x (for example, for supplying a public good), and $B_0(x, \theta)$ is nonmonetary (representing, for example, side-benefits of the decision in other markets).

We will say that an allocation $y(\cdot)$ is (*ex post*) *efficient* if $x(\theta) \in \mathcal{X}$ for each θ and

(E) $x(\theta)$ maximizes $\sum_{i=0}^{I} V_i(x, \theta)$ over $\mathcal{X}$, for all θ.

The rest of this section proceeds as follows. Subsection 7.4.1 defines budget balance and explains that it may imply that no efficient allocation is implementable. Subsection 7.4.2 discusses the difference between Bayesian implementation and implementation in dominant strategies. Subsection 7.4.3 discusses the case where the reservation utilities of the agents are so low that efficient allocations can be implemented even when budget balance is imposed. Subsection 7.4.4 derives conditions that imply that any allocation that can be implemented under budget balance must be inefficient. Subsection 7.4.5 shows how this inefficiency can disappear with many agents in an exchange economy. Subsection 7.4.6 shows how the inefficiency actually becomes more severe with more agents when the decision is whether to produce a public good.

We should mention that there are other notions of individual rationality, incentive compatibility, budget balance, and efficiency than the ones defined here (see, e.g., Holmström and Myerson 1983). These concepts can be defined *ex ante* (when the agents have not yet received their information), interim (after the agents have received their private information, but before they report), and *ex post* (after announcements are observed, so all types are publicly known). In order not to confuse the reader, we have defined only the notions that we will use.

7.4.1 Feasibility under Budget Balance

In many mechanism problems with several agents, the "principal" is not allowed to be a net source of funds to the agents. Moreover, the principal must raise enough revenue from the transfers to cover her cost. (In some applications, this cost is identically 0.) This leads us to consider mechanisms that meet the additional constraint of budget balance:

(BB) $\sum_{i=1}^{I} t_i(\theta) \leq -C_0(x(\theta))$ for all θ.[22]

As in subsection 7.3.2, we say that an allocation $y = (x, t)$ is *feasible* if x is implementable through t and y is individually rational; y is *feasible under budget balance* if it satisfies BB as well.

One theme of this section will be that efficient allocations are typically not feasible under budget balance when there is incomplete information unless the individual-rationality constraints are very weak. (If budget balance is not required, individual-rationality constraints are irrelevant, as the principal can induce the agents to participate by giving them all very large positive transfers, and efficient allocations are usually feasible.) This kind of inefficiency is different from that in the monopoly-pricing example of

---

22. Either the principal is endowed with limited powers (for instance, a regulator is constitutionally allowed to set up a mechanism, but not to make transfers to the agents), or there is no principal and the analysis aims at characterizing potential outcomes of bargaining among the agents under asymmetric information (see subsections 7.4.4 and 7.5.2).

section 7.1. There, the competitive outcome where price equals the monopolist's cost is both feasible and efficient; the monopolist's optimal mechanism is inefficient because it is designed to maximize the monopolist's profit and not social welfare. In contrast, the inefficiency results of this section pertain to *all* the allocations that are feasible under the budget-balance constraint.

7.4.2 Dominant Strategy vs. Bayesian Mechanisms

This chapter emphasizes Bayesian mechanisms. Another popular concept is that of "dominant-strategy mechanisms," which are mechanisms where each agent's optimal announcement is independent of the announcements of the other agents. (Note that the two solution concepts are equivalent in the single-agent case.) Since the optimal announcement can be taken to be the truth from the revelation principle, the formal definition of a dominant-strategy mechanism is a function $y(\theta)$ such that, for each agent $i = 1, \ldots, I$ and for each θ_i, $\hat{\theta}_i$, and θ_{-i},

(DIC) $u_i(y(\theta_i, \theta_{-i}), \theta_i) \geq u_i(y(\hat{\theta}_i, \theta_{-i}), \theta_i).$

That is, each agent is induced to tell the truth whatever the other agents' reports (or types—this is equivalent). The incentive-compatibility constraint (DIC) for dominant-strategy implementation is much more stringent than the incentive constraint under Bayesian implementation. In the latter, incentive compatibility is required to hold only *on average* over types θ_{-i}, where the expectation is taken over agent i's beliefs about θ_{-i} conditional on his type θ_i. Bayesian incentive compatibility thus pools the incentive constraints of dominant-strategy incentive compatibility. Also, the Bayesian conditions for each player suppose that all other players report truthfully. Thus, the Bayesian incentive-compatibility constraint is

(IC) $\mathrm{E}_{\theta_{-i}} u_i(y(\theta_i, \theta_{-i}), \theta_i) \geq \mathrm{E}_{\theta_{-i}} u_i(y(\hat{\theta}_i, \theta_{-i}), \theta_i).$

When possible, a principal might prefer dominant-strategy implementation of her mechanism, because it is not sensitive to beliefs that players have about each other and it does not require players to compute Bayesian equilibrium strategies. However, focusing on dominant-strategy mechanisms restricts the set of mechanisms considerably. Thus, implementation in dominant strategies is a nice property to have if feasible, but it is not clear how much utility loss a principal should be willing to tolerate in order to have dominant strategies for the agents.

Mookherjee and Reichelstein (1989) identify a class of models in which dominant-strategy implementation involves no welfare loss relative to Bayesian implementation.[23] Suppose that the agents have quasi-linear

23. Their result generalizes a similar observation made by Laffont and Tirole (1987a) in the context of auctions of an incentive contract among firms.

preferences, and that, for $i = 1, \ldots, I$,

$$u_i(x, t, \theta) = V_i(x, \theta_i) + t_i,$$

where t_i is the principal's transfer to agent i. Mookherjee and Reichelstein allow x to be multi-dimensional, but require that V_i depend on x only through a one-dimensional statistic $h_i(x)$:

$$u_i(x, t, \theta) = V_i(h_i(x), \theta_i) + t_i.$$

They further assume that types are drawn independently, that the distribution $P_i(\cdot)$ of player i's type satisfies the monotone-hazard-rate condition ($p_i/(1 - P_i)$ nondecreasing) for each i, and that preferences satisfy the sorting assumption $\partial V_i / \partial \theta_i \partial h_i \geq 0$ and the condition that $\partial^2 V_i / \partial h_i \partial \theta_i$ is decreasing in θ_i. Under these assumptions, they show that an allocation that maximizes the principal's expected utility,

$$E_\theta \left(V_0(x, \theta) - \sum_{i=1}^{I} t_i(\theta) \right),$$

subject to the constraints of Bayesian incentive compatibility (IC) and individual rationality,

(IR) $E_{\theta_{-i}} u_i(y(\theta_i, \theta_{-i}), \theta_i) \geq 0$ for all θ_i,

can be implemented in dominant strategies. That is, one can choose the transfer function in the above program such that DIC, and not only IC, is satisfied.

Though we will mention results about dominant-strategy implementation, we will take the incentive-compatibility and individual-rationality constraints facing the principal to be the *Bayesian* ones, IC and IR, unless we specify otherwise.

7.4.3 Efficiency Theorems

There are two basic results about the implementability of efficient allocations, both of which suppose that the agents' reservation utilities are arbitrarily low.

The Groves Mechanism

An early implementability result was the discovery by Groves (1973) and Clarke (1971) that any efficient provision of public goods can be implemented as long as budget balance is not required. Even more striking is the fact that efficient provision can be implemented in dominant strategies.

The idea is straightforward: Choose agent i's transfer so that agent i's payoff is the same as the total surplus of all parties up to a constant. Because agent i already internalizes his own surplus, it suffices to set the transfer equal to the total surplus minus his surplus. The transfers are thus "externality payments."

To be more specific, let $x^*(\theta)$ maximizing $\sum_{i=0}^{I} V_i(x, \theta_i)$ denote an efficient solution for the type profile θ (by convention, the principal's type space is a singleton, θ_0). Define

$$t_i(\hat{\theta}) \equiv \sum_{\substack{j \in \{0, \ldots, I\} \\ j \neq i}} V_j(x^*(\hat{\theta}_i, \hat{\theta}_{-i}), \hat{\theta}_j) + \tau_i(\hat{\theta}_{-i}), \tag{7.14}$$

where $\tau_i(\cdot)$ is an arbitrary function of $\hat{\theta}_{-i}$.

Consider the mechanism $(x^*(\hat{\theta}), t(\hat{\theta}))$, where $t(\cdot) = \{t_i(\cdot)\}_{i=1}^{I}$. We claim that it is optimal for agent i to announce his true type ($\hat{\theta}_i = \theta_i$) regardless of the other agents' announcements. The proof is simple: Suppose that agent i strictly prefers announcing $\hat{\theta}_i$ to announcing θ_i for some types $\hat{\theta}_{-i}$ of the other agents. Then

$$V_i(x^*(\hat{\theta}_i, \hat{\theta}_{-i}), \theta_i) + \sum_{j \neq i} V_j(x^*(\hat{\theta}_i, \hat{\theta}_{-i}), \hat{\theta}_j)$$

$$> V_i(x^*(\theta_i, \hat{\theta}_{-i}), \theta_i) + \sum_{j \neq i} V_j(x^*(\theta_i, \hat{\theta}_{-i}), \hat{\theta}_j). \tag{7.15}$$

But equation 7.15 contradicts the fact that $x^*(\theta_i, \hat{\theta}_{-i})$ is efficient for type profile $(\theta_i, \hat{\theta}_{-i})$.

A common public-good situation is that in which the principal must decide whether to undertake a project of a fixed size: Should a bridge (of fixed dimensions) be built? The agents' preferences are then $u_i = \theta_i x + t_i$, where x is equal to 0 or 1 and θ_i is agent i's valuation or willingness to pay for the public good. With $c > 0$ denoting the cost of supplying the public good, the efficient supply rule is

$$x^*(\theta) = \begin{cases} 1 & \text{if } \sum_{i=1}^{I} \theta_i \geq c \\ 0 & \text{otherwise.} \end{cases}$$

One Groves mechanism for this example takes the following form:

$$t_i(\hat{\theta}) = \begin{cases} \sum_{j \neq i} \hat{\theta}_j - c & \text{if } \sum_{j=1}^{I} \hat{\theta}_j \geq c \\ 0 & \text{otherwise} \end{cases} \tag{7.16}$$

and

$$x^*(\hat{\theta}) = \begin{cases} 1 & \text{if } \sum_{j=1}^{I} \hat{\theta}_j \geq c \\ 0 & \text{otherwise.} \end{cases}$$

Agent i's payment is independent of his announcement except in the region of $\hat{\theta}_i$, where agent i's announcement changes the public-good level from 0 to 1 or from 1 to 0, i.e., the region in which agent i is "pivotal."

Note that, although the public-good interpretation is natural, there is nothing in the mechanism that requires focusing on public goods. Transactions of private goods are also amenable to this analysis as long as IR and BB are not imposed.

There is a large literature on Groves mechanisms (see in particular Green and Laffont 1979). Green and Laffont (1977) show that, up to the irrelevant transfers $\tau_i(\cdot)$, the Groves transfers given by equation 7.14 are the only transfers that make truthful revelation a dominant strategy when no restriction is put on the domains Θ_i of agents' types.[24]

Another result of Green and Laffont is that, in general, no member of the Groves class satisfies BB. This brings us to the second classic mechanism in this literature: that of d'Aspremont and Gerard-Varet (1979), hereafter cited as AGV. (Arrow (1979) derived this mechanism independently.) The question is whether incentive compatibility is consistent with efficiency *and* budget balance. The Green-Laffont results taken together show that the answer is no for implementation in dominant strategies. It turns out that the answer is yes when Bayesian implementation (IC) is considered.

The AGV Mechanism

The AGV-Arrow mechanism is, in a sense, an extension of the Groves mechanism. Instead of being paid the surpluses of the other agents on the basis of their reports, each agent is paid the expected value of the other agents' surpluses conditional on his own report. Then each agent again internalizes the social surplus and has no incentive to distort the decision by manipulating his announcement. More precisely, assume that agent i receives the transfer

$$t_i(\hat{\theta}) = E_{\theta_{-i}}\left(\sum_{j\neq i} V_j(x^*(\hat{\theta}_i, \theta_{-i}), \theta_j)\right) + \tau_i(\hat{\theta}_{-i}). \tag{7.17}$$

The function $\tau_i(\cdot)$ will be determined shortly so as to ensure BB. Let us first note that (x^*, t) is incentive compatible. For this to be so, $\hat{\theta}_i = \theta_i$ must maximize

$$E_{\theta_{-i}}\left(V_i(x^*(\hat{\theta}_i, \theta_{-i}), \theta_i) + \sum_{j\neq i} V_j(x^*(\hat{\theta}_i, \theta_{-i}), \theta_j)\right).$$

But (by the same reasoning as for the Groves mechanism), $\hat{\theta}_i = \theta_i$ maximizes the term inside the expectation operator for all θ_{-i} and, *a fortiori*, maximizes the expectation.[25]

Assume to begin with that $C_0(x)$ is identically 0, so that the decision does not impose a monetary cost on the principal; we explain below how to handle general cost functions. Budget balance then requires that

24. When the domain of θ_i is not smoothly connected (e.g., made of two disjoint closed intervals), there are dominant-strategy mechanisms that are not Groves mechanisms (i.e., for which no $\tau_i(\cdot)$ functions can be found to fit with equation 7.14); see Holmström 1979.
25. Note that equation 7.17 does not yield a dominant-strategy mechanism. If player i's rivals do not announce truthfully, the realized distribution of $\hat{\theta}_{-i}$ differs from the prior distribution, and it may be in the interest of player i to lie as well.

$$\sum_{i=1}^{I} t_i(\hat{\theta}) = 0.$$

Let

$$\mathscr{E}_i(\hat{\theta}_i) \equiv E_{\theta_{-i}}\left(\sum_{j\neq i} V_j(x^*(\hat{\theta}_i, \theta_{-i}), \theta_j)\right)$$

denote the "expected externality" for agent i when he announces $\hat{\theta}_i$. $\mathscr{E}_i(\hat{\theta}_i)$ is the first part of the transfer to agent i; because $\tau_i(\cdot)$ is supposed not to depend on $\hat{\theta}_i$, $\mathscr{E}_i(\hat{\theta}_i)$ must be paid by the other agents. One can, for instance, have them share the payment, i.e., allocate $\mathscr{E}_i(\hat{\theta}_i)/(I-1)$ to each $\tau_j(\cdot), j \neq i$. Thus, the following functions ensure budget balance[26]:

$$\tau_i(\hat{\theta}_{-i}) = -\sum_{j\neq i} \mathscr{E}_j(\hat{\theta}_j)\Big/(I-1)$$

$$= -\frac{1}{I-1} \sum_{j\neq i} E_{\theta_{-j}}\left(\sum_{k\neq j} V_k(x^*(\hat{\theta}_j, \theta_{-j}), \theta_k)\right). \qquad (7.18)$$

Now suppose that the principal incurs a cost $C_0(x)$ from any decision $x \neq 0$, so that budget balance requires

$$\sum_{i=1}^{I} t_i(\hat{\theta}) \leq -C_0(x(\hat{\theta})).$$

To implement the efficient decision under this constraint, we consider the "fictional problem" where the agents' utility functions are

$$\tilde{V}_i(x, \theta_i) \equiv V_i(x, \theta_i) - C_0(x)/I$$

and the principal's cost is $\tilde{C}_0(x) \equiv 0$. We then compute the transfers $\tilde{t}_i(\cdot)$ for this fictional problem, and set

$$t_i(\cdot) = \tilde{t}_i(\cdot) - C_0(x^*(\cdot))/I.$$

We claim that these transfers implement the efficient decision with budget balance in the original problem. Budget balance is trivial: Since $\sum_{i=1}^{I} \tilde{t}_i(\hat{\theta}) = 0$ for all $\hat{\theta}$,

$$\sum_{i=1}^{I} t_i(\hat{\theta}) = -C_0(x^*(\hat{\theta})).$$

As for incentive compatibility, note that

$$\tilde{V}_i(x^*(\hat{\theta}), \theta_i) + \tilde{t}_i(\hat{\theta}) = V_i(x^*(\hat{\theta}), \theta_i) + t_i(\hat{\theta})$$

26. Crémer and Riordan (1985) show that one can strengthen the AGV result by having a dominant strategy for $I-1$ agents. The first mover maximizes his expected payoff by announcing his true type, and it is a dominant strategy for the $I-1$ "Stackelberg followers" to announce truthfully also.

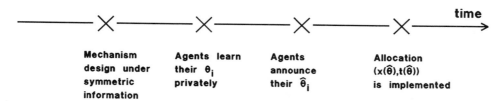

Figure 7.3

for every $\hat{\theta}$ and θ_i; thus, if reporting truthfully is an equilibrium in the fictional problem under transfers $\tilde{t}$, it is an equilibrium in the original problem under transfers t.

An implication of the AGV result is the following: Suppose that I agents meet and agree on a mechanism $(x(\cdot), t(\cdot))$ *ex ante*, i.e., before they receive their private information. The timing is then as shown in figure 7.3. We claim that if $x^*(\cdot)$ is the efficient-decision rule, the contract signed among the I agents *ex ante* will yield $x^*(\theta)$ for each realization of θ, even if the agents can refuse to sign the contract (i.e., $\underline{u}$ is not unboundedly low). Clearly, if $x^*(\cdot)$ is implementable, any mechanism $(x^*(\cdot), t(\cdot))$ that implements it is optimal, as it maximizes the "pie" to be divided among the agents, and agents have an incentive to maximize the pie and distribute it among themselves, possibly with *ex ante* compensatory transfers (recall that the contract is signed under symmetric information, so one expects *ex ante* gains from trade to be realized). But to implement $x^*(\cdot)$, it suffices to give *ex post* transfers $t_i(\cdot)$ specified by the AGV formulas 7.17 and 7.18.

7.4.4 Inefficiency Theorems

We saw in subsection 7.4.3 that, with quasi-linear preferences, efficiency is attainable if budget balance is not required or agents can be forced to participate. In contrast, inefficiency "tends to occur" when the principal must provide the agents with an exogenous reservation utility level and budget balance is required. There are two basic inefficiency results, one due to Laffont and Maskin (1979) and the other to Myerson and Satterthwaite (1983).[27] In this subsection, we first develop the Myerson-Satterthwaite analysis and then sketch that of Laffont and Maskin.

Myerson and Satterthwaite consider a two-agent trading game. The seller can supply one unit of a good at cost c drawn from distribution $P_1(\cdot)$ with differentiable, strictly positive density $p_1(\cdot)$ on $[\underline{c}, \overline{c}]$. The buyer has unit demand and valuation v drawn from distribution $P_2(\cdot)$ on $[\underline{v}, \overline{v}]$ with

27. As Eric Maskin pointed out to us, the inefficiency result can with hindsight be seen in Mirrlees' (1971) treatment of the optimal-taxation problem. Suppose that the social planner's objective is, as proposed by Rawls, to maximize the minimum utility in society. One can consider the minimum utility as a reservation utility and look for incentive-compatible, balanced-budget allocations. Mirrlees shows that there exists no such allocation that is efficient.

differentiable, strictly positive density $p_2(\cdot)$. Let $x(c, v) \in [0, 1]$ denote the probability of trade and $t(c, v)$ denote the transfer from the buyer to the seller (so $t_1 \equiv t$, $t_2 \equiv -t$, and $t_1 + t_2 = 0$). We do not specify how the two players end up with the type-contingent allocation $\{x(\cdot), t(\cdot)\}$. For instance, they might bargain as in the Chatterjee-Samuelson model described in chapter 6, or they could use a more complex sequential-bargaining process such as the ones described in chapter 10, or they might respond to a mechanism designed by a principal (see subsection 7.5.2 for more on this). Rather, the question is whether efficiency is consistent with equilibrium strategies (i.e., IC), individual rationality, and budget balance in general games.

Let

$$X_1(c) \equiv E_v[x(c, v)]$$

and

$$X_2(v) \equiv E_c[x(c, v)]$$

denote, respectively, the seller's and the buyer's probabilities of trading as a function of their type; let

$$T_1(c) \equiv E_v[t(c, v)]$$

and

$$T_2(v) \equiv -E_c[t(c, v)]$$

denote their expected transfers; and let

$$U_1(c) \equiv T_1(c) - c\, X_1(c)$$

and

$$U_2(v) \equiv v\, X_2(v) + T_2(v)$$

denote their expected utilities when they have types c and v, respectively. Because the sorting condition (see subsection 7.3.1) is satisfied, we conclude from theorem 7.2 that X_1 and X_2 must be monotonic if incentive compatibility obtains: X_1 is nonincreasing and X_2 is nondecreasing. Also, from section 7.3, we have

$$U_1(c) = U_1(\bar{c}) + \int_c^{\bar{c}} X_1(\gamma)\, d\gamma \tag{7.19}$$

and

$$U_2(v) = U_2(\underline{v}) + \int_{\underline{v}}^v X_2(v)\, dv. \tag{7.20}$$

Substituting for $U_1(c)$ and $U_2(v)$ and adding up equations 7.19 and 7.20 yields

$$T_1(c) + T_2(v) = c\,X_1(c) - v\,X_2(v) + U_1(\bar{c}) + U_2(\underline{v})$$

$$+ \int_c^{\bar{c}} X_1(\gamma)\,d\gamma + \int_{\underline{v}}^v X_2(v)\,dv. \tag{7.21}$$

But budget balance $(t_1(c,v) + t_2(c,v) = 0)$ in particular implies that

$$E_c\,T_1(c) + E_v\,T_2(v) = 0.$$

Or, using equation 7.21,

$$0 = \int_{\underline{c}}^{\bar{c}} \left(c\,X_1(c) + \int_c^{\bar{c}} X_1(\gamma)\,d\gamma \right) p_1(c)\,dc + U_1(\bar{c})$$

$$+ \int_{\underline{v}}^{\bar{v}} \left(\int_{\underline{v}}^v X_2(v)\,dv - v\,X_2(v) \right) p_2(v)\,dv + U_2(\underline{v}). \tag{7.22}$$

Integrating by parts in equation 7.22 yields

$$U_1(\bar{c}) + U_2(\underline{v}) = - \int_{\underline{c}}^{\bar{c}} \left(c + \frac{P_1(c)}{p_1(c)} \right) X_1(c) p_1(c)\,dc$$

$$+ \int_{\underline{v}}^{\bar{v}} \left(v - \frac{1 - P_2(v)}{p_2(v)} \right) X_2(v)\,p_2(v)\,dv, \tag{7.23}$$

so that, by replacing X_1 and X_2 by their definitions, we get

$$U_1(\bar{c}) + U_2(\underline{v})$$

$$= \int_{\underline{c}}^{\bar{c}} \int_{\underline{v}}^{\bar{v}} \left[\left(v - \frac{1 - P_2(v)}{p_2(v)} \right) - \left(c + \frac{P_1(c)}{p_1(c)} \right) \right] x(c,v) p_1(c) p_2(v)\,dc\,dv. \tag{7.24}$$

Since individual rationality is equivalent to $U_1(\bar{c}) \geq 0$ and $U_2(\underline{v}) \geq 0$, a necessary condition for $x(\cdot)$ to be implementable is that the right-hand side of equation 7.24 be nonnegative.

Now, efficiency requires that $x(\cdot) = x^*(\cdot)$, where $x^*(c,v) = 1$ if $v \geq c$ and $= 0$ otherwise. One can check that equation 7.24 is not satisfied for $x(\cdot) = x^*(\cdot)$ if $\bar{c} > \underline{v}$ and $\underline{c} < \bar{v}$, which establishes the following result.

Theorem 7.5 (Myerson and Satterthwaite 1983)   Suppose that the seller's cost and the buyer's valuation have differentiable, strictly positive densities on $[\underline{c}, \bar{c}]$ and $[\underline{v}, \bar{v}]$, that there is a positive probability of gains from trade $(\underline{c} < \bar{v})$, and that there is a positive probability of no gains from trade $(\bar{c} > \underline{v})$. Then there is no efficient trading outcome that satisfies individual rationality, incentive compatibility, and budget balance.[28]

28. The hypothesis that the distributions are represented by strictly positive densities is important. To see this, consider the following discrete example: $v = \underline{v}$ with probability p and $= \bar{v}$ with probability $\bar{p}$; $c = \underline{c}$ with probability q and $= \bar{c}$ with probability $\bar{q}$, where $\underline{p} + \bar{p} = q + \bar{q} = 1$, $\underline{c} < \underline{v} < \bar{c} < \bar{v}$ and $\underline{v} - \underline{c} > \bar{p}(\bar{v} - \underline{c})$. And consider the following bargaining scheme, in which the seller makes a "take it or leave it" offer, which the buyer accepts

Equation 7.24 exhibits the now familiar virtual surpluses:

$$\left(v - \frac{1 - P_2(v)}{p_2(v)} \right) x$$

for the buyer and

$$-\left(c + \frac{P_1(c)}{p_1(c)} \right) x$$

for the seller.[29,30] Furthermore, as in section 7.3, incentive costs must be taken into account when evaluating gains from trade. This explains the inefficiency result. For example, take two types c and v such that $v = c + \varepsilon$ where ε is "small" (two such types exist as long as $\bar{v} > \bar{c} > \underline{v}$). While the buyer's valuation exceeds the seller's cost, the buyer's virtual valuation is lower than the seller's virtual cost, so that there are no "implementable gains from trade."

Note that the inefficiency result is as tight as possible. When it is common knowledge that there are gains from trade ($\underline{v} \geq \bar{c}$), there exist efficient mechanisms that satisfy IR, IC, and BB: "$x(\hat{c}, \hat{v}) = 1$ and $t(\hat{c}, \hat{v}) = t$ for all $(\hat{c}, \hat{v})$ where $\bar{c} \leq t \leq \underline{v}$."

Cramton, Gibbons, and Klemperer (1987) extend the work of Myerson and Satterthwaite (1983) by allowing arbitrary ownership patterns and more than two agents. In the seller-buyer example, the initial ownership pattern is ($\alpha_1 = 1, \alpha_2 = 0$), where α_i is player i's share of the good; the bargaining is about transforming the ownership structure to ($\alpha'_1 = 0$, $\alpha'_2 = 1$). More generally, suppose that there are I agents who initially hold shares ($\alpha_1, \ldots, \alpha_I$) of a good, with $\sum_{i=1}^I \alpha_i = 1$. Suppose that the final shares are ($\alpha'_1, \ldots, \alpha'_I$) with $\sum_{i=1}^I \alpha'_i = 1$, and that agent i's surplus is $V_i(\alpha_i, \theta_i) = \alpha_i \theta_i$, where the θ_i are independently drawn from some symmetric distribution $P(\cdot)$ on $[\underline{\theta}, \bar{\theta}]$. Cramton et al. show that if the initial shares are fairly evenly distributed (close to ($1/I, \ldots, 1/I$)), there exist efficient mechanisms that satisfy IC, IR, and BB.

Laffont and Maskin (1979, section 6) obtain an inefficiency result in a framework more general than that of Myerson and Satterthwaite: The decision variable x need not be binary but can take values in $\mathbb{R}^n$. Agents have quasi-linear utilities $u_i = V_i(x, \theta_i) + t_i$. Laffont and Maskin assume (i) that the efficient solution $x^*(\theta)$ that maximizes $\sum_{i=1}^I V_i(x, \theta_i)$ is continuously differentiable in θ and (ii) that the optimal expected transfers $t_i(\theta_i)$ are differentiable. Assumption i, although it does not allow for the discontin-

---

or rejects. Clearly, the $\bar{c}$-seller offers price $\bar{v}$ and sells if and only if $v = \bar{v}$. The $\underline{c}$-seller offers price $\underline{v}$ as $\underline{v} - \underline{c} > \bar{p}(\bar{v} - \underline{c})$, and always sells. The bargaining outcome, which satisfies IR, BB, and IC, is efficient.

29. $v - (1 - P_2)/p_2$ and $c + P_1/p_1$ can be called virtual valuation and virtual cost, respectively.
30. The relevant hazard rate for the seller is P_1/p_1 rather than $(1 - P_1)/p_1$. This comes from the fact that the seller dislikes, rather than likes, higher decisions.

uous x^* considered by Myerson and Satterthwaite (which can only be approximated by continuously differentiable x), is natural and of little concern. Assumption ii, which involves endogenous variables, seems more controversial. However, in most applications, incentive compatibility requires that t_i be monotonic: Decisions concerning the agents—e.g., the expected probability of trade, $X_i(\theta_i)$—can be shown to be monotonic, and t_i must be monotonic if X_i is (for instance, purchasing more at a lower price would not be incentive compatible). But a monotonic function is differentiable almost everywhere, and continuouly differentiable functions can be approximated by almost-everywhere-differentiable ones. Hence, Laffont and Maskin's assumption on differentiable transfers is satisfied in many applications of interest.[31]

7.4.5   Efficiency Limit Theorems[†††]

The Myerson-Satterthwaite result shows that a buyer and a seller are unable to exhaust gains from trade if they have incomplete information about each other and there is positive probability that there are no gains from trade. This strengthens our earlier observation that the Coase theorem may not extend to asymmetric-information bargaining. One would want to know whether inefficiency remains substantial when there are many buyers and many sellers. In particular, one would expect that, with a large number of traders, any one trader would be unable to have much influence on his terms of trade by misrepresenting his preferences, and therefore allocations that approximate Walrasian equilibria or Pareto optima could be implemented despite asymmetric information.

Confirming this intuition with a continuum of buyers and sellers is straightforward. Suppose for instance that each seller has one unit of the good for sale and has opportunity or production cost c drawn (independent from those of the other sellers and buyers) from the distribution P_1 on $[\underline{c}, \overline{c}]$; similarly, suppose that the buyers have unit demands and their valuations are drawn independently from the distribution P_2 on $[\underline{v}, \overline{v}]$.

---

31. There have been other extensions of the inefficiency result. Spier (1989) considers bargaining between a plaintiff and a defendant where both have private information about the outcome of the case if they go to court (so, unlike Myerson and Satterthwaite, this is a model of "common values"; that is, each agent cares directly about the other agent's information). Going to court involves a judicial cost for both parties. Settling out of court is the Pareto-superior outcome, as it avoids the judicial costs. Thus, it is common knowledge that there are gains from trade (i.e., gains from agreeing). Yet, Spier shows that if the judicial costs are small (but positive), efficiency is inconsistent with IR, IC, and BB. (The intuition is that efficiency requires that the probability of going to court be equal to 0 and, therefore, that all types of defendants pay and all types of plaintiffs receive the same monetary transfer. But if the judicial costs are small, it pays the plaintiff or the defendant to go to court when they have information that is very favorable to them.) Ledyard and Palfrey (1989) consider the case of a public-good mechanism which is designed by the principal. They show that the principal may choose a mechanism that does not maximize the sum of the agents' willingness to pay for the public good even if she faces no IR constraint, as long as the agents' private information is the marginal utility of income ($u_i \equiv x - t/\theta_i$) and the principal cares about income distribution (i.e., about $\sum_{i=1}^{I} u_i$).

With $\bar{c} > \underline{v}$ (not everyone ought to trade), the market-clearing price π is given by $P_1(\pi) = 1 - P_2(\pi)$ (if sellers and buyers are in equal numbers). Let $x_1 \in [0, 1]$ and $x_2 \in [0, 1]$ denote a seller's probability of selling and a buyer's probability of purchasing, respectively. A social planner can obtain the efficient outcome by offering the Walrasian mechanism: "$x_1(\hat{c}) = 1$ and $t_1(\hat{c}) = \pi$ if $\hat{c} \leq \pi$, and $x_1(\hat{c}) = t_1(\hat{c}) = 0$ otherwise; $x_2(\hat{v}) = 1$ and $t_2(\hat{v}) = -\pi$ if $\hat{v} \geq \pi$, and $x_2(\hat{v}) = t_2(\hat{v}) = 0$ otherwise."

With a large but finite number of traders, one cannot generally obtain efficient outcomes under IR. Indeed, Hurwicz (1972) shows that, in general economies, any mechanism that asks the traders to announce their preference orderings, that is efficient, and that satisfies the individual-rationality constraint that the traders prefer their assigned consumption bundle to their initial endowment vector must violate incentive compatibility for some preference orderings. Hurwicz's informational assumption is that traders know one another's preferences (this is called a "Nash environment" in the literature, to distinguish it from "Bayesian environments," where preferences are private information). Roberts and Postlewaite (1976) pursued this line of research and showed that, under some regularity conditions, a trader's gain in utility from distorting his announcement of preferences is bounded above by a number that tends to zero as the number of traders tends to infinity.

Wilson (1985) and Gresik and Satterthwaite (1989) (see also Cramton et al. 1987) perform a similar analysis in a Bayesian context. Suppose with Wilson that there are I_1 sellers, $i = 1, \ldots, I_1$, and I_2 buyers, $i = 1, \ldots, I_2$; that the sellers' costs and the buyers' valuations are drawn independently from distributions on $[\underline{c}, \bar{c}]$ and $[\underline{v}, \bar{v}]$; and that $\underline{c} \leq \underline{v} < \bar{c} \leq \bar{v}$, so that $v > c$ and $v < c$ have positive probability.

Wilson studies "double auctions," in which the sellers and the buyers make bids $\{\hat{c}_i\}_{i=1,\ldots,I_1}$ and $\{\hat{v}_i\}_{i=1,\ldots,I_2}$, respectively (bids are similar to announcements of costs or valuations). Without loss of generality, we can reorder the bids so that

$$\hat{c}_{I_1} \geq \hat{c}_{I_1-1} \geq \cdots \geq \hat{c}_1$$

and

$$\hat{v}_1 \geq \hat{v}_2 \geq \cdots \geq \hat{v}_{I_2}.$$

Then the number of units traded in a double auction is the largest k such that $\hat{v}_k \geq \hat{c}_k$, and those who trade are sellers 1 through k and buyers 1 through k. The transfer price π is an arbitrary price in $[\hat{c}_k, \hat{v}_k]$ (for instance, $(\hat{v}_k + \hat{c}_k)/2$). The other sellers and buyers do not trade and do not give or receive transfers. Note that if each player's bid equals his type, a double auction maximizes social surplus. Of course, traders have an incentive to misrepresent their preferences, and the equilibrium need not be efficient. Yet Wilson shows that, under some assumptions (existence of an equilibrium in symmetric strategies that are differentiable functions of private information and have uniformly bounded derivatives), a double auction (a very simple mechanism indeed) yields efficiency in the limit when I_1 and I_2 tend to infinity. Gresik and Satterthwaite (1989) provide results on the rate of convergence to Walrasian equilibria.

7.4.6   Strong Inefficiency Limit Theorems[†††]

The efficiency limit theorems for private goods mentioned in the previous subsection are in stark contrast with limit results by Rob (1989) and Mailath and Postlewaite (1990) for public goods when each agent has veto power. In the private-good case with a large number of traders, a trader has little influence on the price at which he trades, so he has little incentive to manipulate the announcement of his preferences to trade off more favorable prices against a lower probability of trading. The reverse holds for public goods with a large number of traders. A trader has a low probability of being pivotal, i.e., of influencing the decision of whether to produce the public good. Hence, the probability of "trade"—the probability of the public good being supplied—cannot be affected, but under some conditions (to be described) each agent can manipulate his "terms of trade"—the amount of his contribution toward the provision of the public good.

Consider a fixed-sized public-good project with I agents. Agent i ($i = 1, \ldots, I$) has utility $u_i = \theta_i x + t_i$ where $x = 1$ if the public good is supplied and $x = 0$ otherwise (t_i is likely to be negative). Let the parameters θ_i be independently drawn from distributions P_i with positive density p_i on $[\underline{\theta}_i, \bar{\theta}_i]$. Assume further that the cost of realizing the project is a function $C(I)$ of the number of agents.

Let us look for mechanisms $m = \{x, t\}$ that satisfy the following properties:

$x(\hat{\theta}) \in [0, 1]$ for all $\hat{\theta}$,

(IC) $E_{\theta_{-i}}[x(\theta_i, \theta_{-i})\theta_i + t_i(\theta_i, \theta_{-i})] \geq E_{\theta_{-i}}[x(\hat{\theta}_i, \theta_{-i})\theta_i + t_i(\hat{\theta}_i, \theta_{-i})]$

$$\text{for all } (i, \theta_i, \hat{\theta}_i),$$

(IR) $E_{\theta_{-i}}[x(\theta_i, \theta_{-i})\theta_i + t_i(\theta_i, \theta_{-i})] \geq 0$ for all (i, θ_i),

(BB) $\sum_{i=1}^{I} t_i(\theta) + x(\theta)C(I) \leq 0$ for all θ.

The Rob-Mailath-Postlewaite result is that, in the limit with a large number of traders, IC, IR, and BB imply that *no* gains from trade are realized if $C(I)$ is proportional to I and $C(I)/I > \underline{\theta}_i$ for all i.

Actually, an apparently stronger result will be proved, by replacing **BB** (which is an *ex post* concept) by *ex ante* budget balance:

(EABB) $E_\theta \left(\sum_{i=1}^I t_i(\theta) + x(\theta)C(I) \right) \le 0.$

Of course, BB implies EABB.[32]

Let

$$U_i(\theta_i) \equiv E_{\theta_{-i}}[x(\theta_i, \theta_{-i})\theta_i + t_i(\theta_i, \theta_{-i})]$$

denote agent i's expected utility when he has type θ_i, and let

$$X_i(\theta_i) \equiv E_{\theta_{-i}}[x(\theta_i, \theta_{-i})]$$

denote the probability that the good is supplied. The analysis of section 7.3 implies that

$$U_i(\theta_i) = U_i(\underline{\theta}_i) + \int_{\underline{\theta}_i}^{\theta_i} X_i(\tilde{\theta}_i) \, d\tilde{\theta}_i. \tag{7.25}$$

The expected total surplus, W, which is equal to the expectation of the sum of the budget surplus and the agents' utilities, is then

$$W = E_\theta \left(\sum_i [-t_i(\theta)] - C(I)x(\theta) + \sum_i U_i(\theta_i) \right)$$

$$\cdot = E_\theta \left(\sum_i [-t_i(\theta)] - C(I)x(\theta) + \sum_i U_i(\underline{\theta}_i) \right)$$

$$+ \sum_i E_{\theta_i} \left[\left(\frac{1 - P_i(\theta_i)}{p_i(\theta_i)} \right) X_i(\theta_i) \right], \tag{7.26}$$

where

$$\int_{\underline{\theta}_i}^{\bar{\theta}_i} \int_{\underline{\theta}_i}^{\theta_i} X_i(\tilde{\theta}_i) \, d\tilde{\theta}_i p_i(\theta_i) \, d\theta_i$$

has been integrated by parts. Now,

$$E_\theta \left(\sum_i [-t_i(\theta)] - C(I)x(\theta) \right) \ge 0$$

by *ex ante* budget balance, and $U_i(\underline{\theta}_i) \ge 0$ for all i by individual rationality. Because

$$0 \le E_\theta \left[\sum_i U_i(\theta_i) \right] = E_\theta \left[\sum_i (t_i(\theta) + \theta_i x(\theta)) \right] \le E_\theta \left[\sum_i \theta_i x(\theta) - C(I)x(\theta) \right]$$

(using the budget constraint), an integration by parts yields

32. But Mailath and Postlewaite show that if EABB, IC, and IR are satisfied, one can choose the transfers $t_i(\cdot)$ such that BB, IC, and IR are satisfied as well.

$$E_\theta\left\{\left[\sum_i\left(\theta_i - \frac{1 - P_i(\theta_i)}{p_i(\theta_i)} - \frac{C(I)}{I}\right)\right]x(\theta)\right\} \geq 0. \tag{7.27}$$

We will use the following lemma:

Lemma 7.1 The expectation of the virtual valuation is equal to the lower bound of the interval.

Proof Integrating by parts,

$$\int_{\underline{\theta}_i}^{\bar{\theta}_i}\left(\theta_i - \frac{1 - P_i(\theta_i)}{p_i(\theta_i)}\right)p_i(\theta_i)\,d\theta_i$$

$$= \int_{\underline{\theta}_i}^{\bar{\theta}_i}\theta_i p_i(\theta_i)\,d\theta_i - \{[1 - P_i(\theta_i)]\theta_i\}_{\underline{\theta}_i}^{\bar{\theta}_i} - \int_{\underline{\theta}_i}^{\bar{\theta}_i}\theta_i p_i(\theta_i)\,d\theta_i$$

$$= \underline{\theta}_i. \qquad\blacksquare$$

We next assume that the per-capita cost of supplying the public good is constant, $C(I)/I = c$, and that $c > \underline{\theta}_i$ for all i. Let us further assume (for simplicity) that all θ_i are drawn from the same distribution $P(\cdot)$ on $[\underline{\theta}, \bar{\theta}]$, so that $c > \underline{\theta}$.

Note that the left-hand side of equation 7.27 is maximized by $x(\theta) = 1$ if

$$\sum_i\left(\theta_i - \frac{1 - P(\theta_i)}{p(\theta_i)} - c\right) \geq 0$$

and by $x(\theta) = 0$ otherwise.

With a continuum of agents, the realized distribution of types in the population of agents coincides with the prior distribution. As $\underline{\theta} < c$, and (from the lemma) the expected total virtual surplus is equal to $E_\theta[(\underline{\theta} - c)x(\theta)]$, in order for expected surplus to be nonnegative, x must be equal to 0 with probability 1. With a large but finite number of agents, the law of large numbers suggests that the same result holds approximately. Technical work is needed to make this intuition precise, but Rob (1989) and Mailath and Postlewaite (1990) show that, as I tends to ∞, the probability that the public project is implemented tends to 0 if $c > \underline{\theta}$, and IR, IC, and BB are required.[33]

Thus, with a large number of agents it becomes very hard to reach agreement. The inefficiency involved can be large. For, suppose that $P(c)$ is very small, so that, with probability close to 1, each agent's valuation for the public good exceeds the per-capita cost of supplying the public good.

---

33. A similar result is obtained by Roberts (1976) for dominant-strategy mechanisms rather than Bayesian ones. In contrast, Green and Laffont (1979) show that, in the absence of the IR constraint, efficiency can be obtained in the limit when the number of agents becomes large, with dominant strategies and budget balance.

Then there are gains from trade with probability close to 1, yet gains from trade are realized with probability close to 0.

The intuition for this result is straightforward. The probability of being pivotal (changing x through a change in $\hat{\theta}_i$) is very small with many agents, and is 0 with a continuum of them.[34] Thus, agent i's objective is simply to maximize his expected transfer, i.e., minimize his expected contribution to the public good. This expected contribution cannot exceed $\underline{\theta}$, because that would violate individual rationality for type $\underline{\theta}$, and agent i can always report type $\underline{\theta}$. But, if the expected contribution is at most $\underline{\theta}$, the cost of realizing the project cannot be covered, contradicting budget balance.

To avoid inefficiencies, one needs subsidies from an external source (such as a "government"). Mailath and Postlewaite show that the per-capita subsidy to implement the efficient provision of the public good (i.e., $x = 1$ if and only if $\sum_i \theta_i \geq cI$) is asymptotically equal to $c - \underline{\theta}$, as one would suspect.

7.5 Mechanism Design with Several Agents: Optimization[††]

In section 7.4 we looked at general properties of the implementable allocations. We now look at the *optimal* choice of a mechanism for two allocation problems. In the first, the auction example, a self-interested principal sells a good to one of several buyers with private information about their willingness to pay for the good. In the second, the bilateral-trade example, a seller and a buyer with private information about their cost and valuation may trade an object. In both cases we will assume that the mechanism is designed by an uninformed party to maximize her objective function, which will allow us to abstract from issues arising from information leakages through contract design (see subsection 7.6.3). In the auction example this corresponds to the assumption that the seller has no private information and maximizes her expected revenue. In the bilateral-trade example the interpretation is more difficult. There, it will be assumed that a benevolent third party maximizes the expected gains from trade between the buyer and the seller; as we will discuss, the existence of this third party is mysterious and the main point of the analysis is to supply an upper bound on the efficiency of bilateral exchange under asymmetric information.

7.5.1 Auctions

Suppose that a seller (the principal) has $\hat{x}$ units of a good for sale. There are I potential buyers (agents): $i = 1, \ldots, I$. All parties have quasi-linear preferences:

---

34. A similar idea underlies the "paradox of voting"—with a large number of voters, the probability of affecting the outcome of an election is infinitesimal. Palfrey and Rosenthal 1985 is a recent paper on this topic.

$$u_i = V_i(x_i, \theta_i) + t_i \text{ for } i = 0, 1, \ldots, I$$

where $x_i \in [0, \hat{x}]$ is the amount consumed by party i and t_i is his (or her) income (in this section, $t_0 = -\sum_{i=1}^{I} t_i$). We assume that V_i is increasing in x_i, and that the sorting condition holds:

$$\frac{\partial^2 V_i}{\partial x_i \partial \theta_i} \geq 0,$$

that is, the marginal utility of the good increases in θ_i.

The seller's parameter θ_0 is common knowledge. In contrast, the buyers' types θ_i are independently drawn from cumulative distributions $P_i(\cdot)$ with strictly positive densities $p_i(\cdot)$ on $[\underline{\theta}, \overline{\theta}]$.

The seller attempts to maximize her expected utility. From the revelation principle, she can restrict attention to direct revelation mechanisms $\{x(\cdot), t(\cdot)\}$. Thus, she maximizes her expected (net) revenue:

$$R = E_\theta \left[V_0 \left(\hat{x} - \sum_{i=1}^{I} x_i(\theta), \theta_0 \right) - \sum_{i=1}^{I} t_i(\theta) \right]$$

subject to

(IC) $\quad E_{\theta_{-i}}[V_i(x_i(\theta_i, \theta_{-i}), \theta_i) + t_i(\theta_i, \theta_{-i})]$

$$\geq E_{\theta_{-i}}[V_i(x_i(\hat{\theta}_i, \theta_{-i}), \theta_i) + t_i(\hat{\theta}_i, \theta_{-i})]$$

$$\text{for all } (i, \theta_i, \hat{\theta}_i),$$

(IR) $\quad E_{\theta_{-i}}[V_i(x_i(\theta_i, \theta_{-i}), \theta_i) + t_i(\theta_i, \theta_{-i})] \geq 0 \text{ for all } (i, \theta_i),$

and

$$x_i(\theta) \geq 0 \text{ and } \sum_{i=1}^{I} x_i(\theta) \leq \hat{x} \text{ for all } \theta.$$

Let

$$U_i(\theta_i) \equiv E_{\theta_{-i}}[V_i(x_i(\theta_i, \theta_{-i}), \theta_i) + t_i(\theta_i, \theta_{-i})]$$

denote buyer i's expected utility when he has type θ_i. The seller's objective function can be rewritten as a function of the buyers' expected utilities by substituting for the transfers:

$$R = E_\theta \left[V_0 \left(\hat{x} - \sum_{i=1}^{I} x_i(\theta), \theta_0 \right) + \sum_{i=1}^{I} V_i(x_i(\theta), \theta_i) \right] - \sum_{i=1}^{I} E_{\theta_i} U_i(\theta_i). \quad (7.28)$$

But, from the envelope theorem,

$$\frac{dU_i}{d\theta_i} = E_{\theta_{-i}} \left(\frac{\partial V_i}{\partial \theta_i}(x_i(\theta_i, \theta_{-i}), \theta_i) \right) \quad (7.29)$$

or

$$U_i(\theta_i) = U_i(\underline{\theta}) + \int_{\underline{\theta}}^{\theta_i} E_{\theta_{-i}} \left(\frac{\partial V_i}{\partial \theta_i}(x_i(\tilde{\theta}_i, \theta_{-i}), \tilde{\theta}_i) \right) d\tilde{\theta}_i. \tag{7.30}$$

At the optimum, $U_i(\underline{\theta}) = 0$, as the seller does not want to leave unnecessary rents to the buyers. Substituting equation 7.30 into equation 7.28 and integrating by parts yields

$$R \equiv E_\theta \left[V_0 \left(\hat{x} - \sum_{i=1}^{I} x_i(\theta), \theta_0 \right) \right.$$

$$\left. + \sum_{i=1}^{I} \left(V_i(x_i(\theta), \theta_i) - \frac{1 - P_i(\theta_i)}{p_i(\theta_i)} \frac{\partial V_i}{\partial \theta_i}(x_i(\theta), \theta_i) \right) \right]. \tag{7.31}$$

The optimal auction defines an allocation $x_i(\cdot)$ of the good so as to maximize R subject to the agents' incentive compatibility. Rather than give a comprehensive study of incentive compatibility, we content ourselves with a full treatment of a special case. Assume that

$$V_i(x_i, \theta_i) = \theta_i x_i, \quad i = 0, 1, \dots, I$$

and

$$\hat{x} = 1.$$

We know from theorem 7.2 that incentive compatibility for agent i is equivalent to equation 7.30 plus the condition that $X_i(\theta_i) \equiv E_{\theta_{-i}} x_i(\theta_i, \theta_{-i})$ be nondecreasing.

Hence, the optimal auction solves

$$\text{Max } E_\theta \left[\sum_{i=1}^{I} \left(\theta_i - \frac{1 - P_i(\theta_i)}{p_i(\theta_i)} \right) x_i(\theta) + \theta_0 \left(1 - \sum_{i=1}^{I} x_i(\theta) \right) \right] \tag{7.32}$$

subject to

$$\sum_{i=1}^{I} x_i(\theta) \le 1, x_i(\theta) \ge 0 \text{ for all } \theta \tag{7.33}$$

and

$$X_i(\cdot) \text{ nondecreasing.} \tag{7.34}$$

The expected transfers associated with the optimal auction are obtained by computing $U_i(\theta_i)$ and using the definition of U_i:

$$T_i(\theta_i) = E_{\theta_{-i}} t_i(\theta_i, \theta_{-i}) = -\theta_i X_i(\theta_i) + \int_{\underline{\theta}}^{\theta_i} X_i(\tilde{\theta}_i) d\tilde{\theta}_i. \tag{7.35}$$

Note that maximizing equation 7.35 determines only *expected* transfers $T_i(\cdot)$, so there is a lot of leeway in defining the *ex post* transfers $t_i(\cdot)$. We will see that this leeway translates into a multiplicity of ways of implementing the optimal auction.

Let

$$J_i(\theta_i) \equiv \theta_i - \frac{1 - P_i(\theta_i)}{p_i(\theta_i)}$$

denote the virtual valuation of buyer i, and let $J_0(\theta_0) \equiv \theta_0$ denote the seller's valuation. We first maximize expression 7.32, ignoring the incentive-compatibility constraint 7.34. This yields

$$x_i(\theta) = 1 \text{ iff } J_i(\theta_i) = \max_{j \in \{0,...,I\}} J_j(\theta_j).$$

(We ignore the cases in which the maximum is reached for at least two players. Such cases have probability 0.)

If $J_i(\cdot)$ is nondecreasing for all i (which is true in particular if the monotone-hazard-rate condition holds—see section 7.3), then, if $x_i(\theta_i, \theta_{-i}) = 1$,

$$x_i(\theta_i', \theta_{-i}) = 1 \text{ for all } \theta_i' > \theta_i.$$

Hence, $X_i(\cdot)$ is nondecreasing, and the ignored incentive-compatibility constraint is automatically satisfied. If $J_i(\cdot)$ decreases over some interval, one must proceed along the lines of the analysis of bunching in the appendix to this chapter. (See Myerson 1981 for details.) In the following, we will assume that $J_i(\cdot)$ is nondecreasing.

We now examine the implications of this analysis.

First, note that the relevant comparison concerns the parties' *virtual valuations*, and not their valuations. The seller's virtual valuation is equal to her true valuation θ_0, because the seller has full information about herself and therefore needs not introduce the incentive cost of revelation of information.

Second, all auctions that yield the same decision $x_i(\cdot)$ and give zero surplus to type $\underline{\theta}$ of each buyer yield the same revenue. We will shortly give an implication of this fact, known as the *revenue equivalence theorem*.

Third, the analysis yields a number of standard results in the symmetric case $(P_i(\cdot) = P(\cdot))$. In this case, the good goes to the highest-valuation buyer if it is sold at all. The good is sold if and only if

$$\max_{i \in \{1,...,I\}} \theta_i \geq \theta^*,$$

where $\theta^* > \theta_0$ is defined by

$$\theta^* - \frac{1 - P(\theta^*)}{p(\theta^*)} \equiv \theta_0.^{35}$$

---

35. Again, this result generalizes the monopoly-pricing paradigm. Note that if $\theta_0 < \max_{i \in \{1,...,I\}} \theta_i < \theta^*$, gains from trade are not realized. The seller distorts the auction in her favor.

Furthermore, all auctions that give the good to the highest bidder (i.e., $X_i(\theta_i) \equiv [P(\theta_i)]^{I-1}$ if $\theta_i \geq \theta^*$, $\equiv 0$ otherwise) and yield zero surplus to a bidder with valuation θ^* (or, equivalently, to valuation $\underline{\theta}$ from equation 7.30) yield the same revenue to the seller.

In particular, a first-price auction (see chapter 6) and a second-price auction (see chapter 1), each with minimum or reservation price θ^*, yield the same revenue and are optimal (Vickrey 1961; Myerson 1981; Riley and Samuelson 1981). Although the first- and second-price auctions yield the same x_i and T_i, they yield different t_i: When bidder i wins, his payment depends only on his bid and therefore only on θ_i in a first-price auction, and depends only on the second bid ($\max_{j \neq i; j \in \{1,\ldots,I\}} \theta_j$) in a second-price auction. This illustrates the leeway one has in building the *ex post* transfers t_i to implement an optimal auction. Note also that the two-type example in section 7.1 shows that neither the first- nor the second-price auction is optimal when the distribution of types is discrete. The problem with both of these auctions is that the high-valuation type receives an unnecessarily high rent. Starting from the second-price auction, for example, the seller can increase her revenue while still inducing buyers to bid their valuations if she specifies that when one buyer bids $\bar{\theta}$ and the other bids $\underline{\theta}$ the high bidder receives the good at price $\underline{\theta} + (\bar{\theta} - \underline{\theta})/2$.

In the asymmetric case, the auction does not necessarily allocate the good to the bidder with the highest willingness to pay (Myerson 1981; McAfee and McMillan 1987b). In particular, suppose that there are two bidders ($i = 1, 2$) and that, for all θ,

$$\frac{1 - P_1(\theta)}{p_1(\theta)} \geq \frac{1 - P_2(\theta)}{p_2(\theta)}.$$

That is, bidder 1 is "on average" more eager to buy than bidder 2. Then the auction should be biased in favor of bidder 2. There exist θ_1 and θ_2 such that $\theta_1 > \theta_2$ but $x_2(\theta_1, \theta_2) = 1$, while there exist no θ_1 and θ_2 such that $\theta_2 > \theta_1$ and $x_1(\theta_1, \theta_2) = 1$.

7.5.2 Efficient Bargaining Processes[†††]

Consider now a single buyer and a single seller. The seller has one unit for sale and has private information about his cost c of supplying the unit. The buyer has unit demand and has private information about his willingness to pay or valuation v for the unit. Thus, $\theta_1 \equiv c$, $\theta_2 \equiv v$, and $\theta \equiv (c, v)$. c and v are independently drawn from cumulative distributions $P_1(\cdot)$ and $P_2(\cdot)$ on $[\underline{c}, \bar{c}]$ and $[\underline{v}, \bar{v}]$, with strictly positive densities $p_1(\cdot)$ and $p_2(\cdot)$. The two parties are risk neutral.

A balanced-budget mechanism is a probability $x(c, v) \in [0, 1]$ that the traders exchange the good given that their types are c and v and a payment $w(c, v)$ (or, equivalently, given that the parties are risk neutral, an expected payment $w(c,v)$) from the buyer to the seller (in our previous notation, $t_1(\theta) = w(c,v) = -t_2(\theta)$). Let

$$X_1(c) \equiv \mathrm{E}_v\, x(c,v); \quad X_2(v) \equiv \mathrm{E}_c\, x(c,v);$$

$$W_1(c) \equiv \mathrm{E}_v\, w(c,v); \quad W_2(v) \equiv \mathrm{E}_c\, w(c,v);$$

$$U_1(c) \equiv -c\, X_1(c) + W_1(c); \quad U_2(v) \equiv v\, X_2(v) - W_2(v).$$

The mechanism is individually rational if $U_1(c) \geq 0$ for all c and $U_2(v) \geq 0$ for all v. It is incentive compatible if

$$U_1(c) \geq -c\, X_1(\hat{c}) + W_1(\hat{c}) \text{ for all } (c, \hat{c})$$

and

$$U_2(v) \geq v\, X_2(\hat{v}) - W_2(\hat{v}) \text{ for all } (v, \hat{v}).$$

Consider a benevolent principal trying to maximize expected social surplus $\mathrm{E}_{\{c,v\}}[(v-c)x(c,v)]$, and suppose that she is able to design a (balanced-budget) mechanism to which the seller and the buyer must comply as long as it is individually rational and incentive compatible. The role of the principal here is difficult to interpret. She might stand for a government, but then it is not clear why the mechanism must satisfy the individual-rationality constraints, since governments have coercive powers. Another potential interpretation is that the parties appeal to a mediator (the principal) to design an efficient mechanism. This interpretation also is often questionable. If the parties appeal to the mediator once they have received their private information (at the "interim stage"), the bargaining over whether to have a mediator and over which objective function to give to the mediator is likely to reveal information about the cost and the valuation; the IR and IC constraints are then misspecified in that the mechanism is played under posterior beliefs that differ from the prior beliefs $P_1(\cdot)$ and $P_2(\cdot)$. If the two parties decide to use a mediator *before* they receive their private information (at the "*ex ante*" stage), they may be able to commit themselves to use the mechanism once they learn their valuations, and so the interim IR constraint may not be relevant. Such commitments can sometimes be accomplished by contractually specified damages for "opting out" or "breach of contract."[36] If parties can commit to use the mechanism, they typically prefer to do so, as binding interim IR constraints generally create inefficiency, whereas in the absence of these constraints AGV mechanisms can be built that implement the *ex post* efficient outcome (i.e., $x = 1$ if $v \geq c$, $= 0$ if $v < c$).

---

36. However, these commitment options may be limited: If we interpret the "seller" as a worker who is providing labor to a firm, workers are not allowed to agree to fines for quitting; however, firm might still be able to commit. This suggests a hybrid model with only one player subject to an individual-rationality constraint.

Because of these reservations about interpretations where a principal designs the mechanism, the best interpretation of the model may be as a characterization of utilities that can be achieved by equilibria of non-cooperative bargaining games. Suppose that the seller and the buyer bargain over whether to trade and over the price. The bargaining process can be a simultaneous sealed-bid auction (à la Chatterjee and Samuelson—see chapter 6) or a more complex, sequential bargaining game (see chapter 10). It has been known for a while as part of the profession's folklore that any (Bayesian) equilibrium of a bargaining process gives rise to an allocation that can be interpreted as a mechanism that satisfies IC and IR, as long as the two traders have identical time preferences.[37] This is a straightforward application of the revelation principle: Suppose that bargaining starts at date 0 and that both traders discount the future at interest rate $r > 0$ (we allow for either discrete-time bargaining—at dates $t = 0, 1, 2, \ldots$—or continuous-time bargaining). Let agreement to trade between the seller with cost c and the buyer with valuation v be reached at time $\tau(c, v)$ at price $z(c, v)$ (we assume that τ and z are deterministic; the reasoning extends straightforwardly to stochastic τ and z). $\tau = +\infty$ corresponds to the case in which agreement is never reached. One can then define

$$x(c, v) \equiv e^{-r\tau(c, v)} \in [0, 1],$$

$$w(c, v) \equiv e^{-r\tau(c, v)}z(c, v),$$

$$U_1(c) \equiv \mathrm{E}_v[w(c, v) - c\,x(c, v)],$$

$$U_2(v) \equiv \mathrm{E}_c[v\,x(c, v) - w(c, v)].$$

Note that delay in reaching agreement ($\tau > 0$) amounts to a probability that exchange does not take place ($x < 1$) in the mechanism reinterpretation.

Observe that the mechanism $\{x(\cdot, \cdot), w(\cdot, \cdot)\}$ satisfies IR, IC, and BB. It is individually rational because each trader can always refuse to trade (by making outrageous demands, and rejecting all offers), and thus get 0. By definition of a Bayesian equilibrium, it satisfies incentive compatibility: A type θ_i of player i cannot adopt the strategy of type $\hat{\theta}_i$ of the same player and obtain a higher expected payoff. Budget balance follows from the absence of a third party.

Viewed from this perspective, the program of computing the highest expected social surplus that can be obtained through individually rational, incentive-compatible, balanced-budget mechanisms can be interpreted as deriving an upper bound on the efficiency of unmediated bilateral bargaining.

---

37. If the traders have different rates of time preference, then having the more patient trader make loans to the less patient one allows the attainment of utility levels that are not feasible in the static problem.

Remark In the same spirit, one can derive the set of allocations that can be implemented by a mediator. The question is then whether any element in this set may arise as *an* equilibrium of *some* unmediated bargaining game. This line of research will be discussed in chapter 10.

Let us now derive the mechanism that maximizes expected gains from trade,

$$E_{c,v}[(v - c)x(c, v)],\tag{7.36}$$

subject to IR, IC, and BB. We saw in subsection 7.4.4 that IR, IC, and BB imply

$$E_{c,v}\{[J_2(v) - J_1(c)]x(c, v)\} \geq 0,\tag{7.37}$$

where

$$J_1(c) \equiv c + \frac{P_1(c)}{p_1(c)}$$

and

$$J_2(v) = v - \frac{1 - P_2(v)}{p_2(v)}.$$

Conversely, if the function $x(\cdot, \cdot)$ maximizes expression 7.36 subject to inequality 7.37, there exists a transfer function $t(\cdot, \cdot)$ that satisfies BB (by definition), satisfies IR, and satisfies IC as long as $X_1(c) = E_v x(c, v)$ is nonincreasing and $X_2(v) = E_c x(c, v)$ is nondecreasing. With $\mu \geq 0$ denoting the multiplier of equation 7.37, the Lagrangian for the above program is

$$\mathscr{L} = E_{c,v}(\{(v - c) + \mu[J_2(v) - J_1(c)]\}x(c, v)).\tag{7.38}$$

The first-order condition is thus

$$x(c, v) = \begin{cases} 1 \text{ if } v + \mu J_2(v) \geq c + \mu J_1(c) \\ 0 \text{ otherwise.} \end{cases}\tag{7.39}$$

Thus, trade occurs if and only if

$$v - \left(\frac{\mu}{1 + \mu}\right)\frac{1 - P_2(v)}{p_2(v)} \geq c + \left(\frac{\mu}{1 + \mu}\right)\frac{P_1(c)}{p_1(c)}.\tag{7.40}$$

Equation 7.40 does not quite yet define the solution, as the coefficient $\alpha \equiv \mu/(1 + \mu) \in [0, 1)$ must still be specified. To this purpose, it suffices to note that equation 7.37 must be satisfied with equality if $\bar{c} > \underline{v}$.[38] (Ideally, one would want the trading rule to come as close as possible to the first-best

---

38. If the inequality is strict in equation 7.37, $\mu = 0$ and equation 7.40 is the first-best rule. But we know from subsection 7.4.4 that, as long as $\bar{c} > \underline{v}$, efficient trade is inconsistent with IR, IC, and BB.

trading rule (trade if and only if $v \geq c$); i.e., one would want μ (or α) to be as small as possible. Equation 7.37 has been relaxed as much as is consistent with IR, IC, and BB by imposing $U_1(\bar{c}) = U_2(\underline{v}) = 0$.)

Note again that if the monotone-hazard-rate conditions hold ($p_2/(1 - P_2)$ nondecreasing, p_1/P_1 nonincreasing), equation 7.40 yields monotonic $X_1(\cdot)$ and $X_2(\cdot)$, so the optimal trading rule has indeed been obtained.

Myerson and Satterthwaite apply equation 7.40 to the case of uniform densities on $[0, 1]$ ($P_1(c) = c$ and $P_2(v) = v$ for $(c, v) \in [0, 1]^2$). Equation 7.40 then yields

$$v - c \geq \frac{\alpha}{1 + \alpha}. \tag{7.41}$$

Substituting into equation 7.37 yields

$$\int_0^{1-(\alpha/(1+\alpha))} \left(\int_{c+(\alpha/(1+\alpha))}^1 [(2v - 1) - 2c]\, dv \right) dc = 0, \tag{7.42}$$

which has solution $\alpha/(1 + \alpha) = \frac{1}{4}$. In the optimal trading rule, trade occurs if and only if the buyer's valuation exceeds the seller's cost by at least one-fourth. Thus, in the uniform case, the linear equilibrium of the Chatterjee-Samuelson double auction exhibited in chapter 6 yields the optimal amount of trade constrained by IR, IC, and BB![39]

7.6 Further Topics in Mechanism Design[†††]

The bare-bones analysis of this chapter has ignored many of the recent extensions of the mechanism-design paradigm. In this concluding section, we give the flavor of a few of these extensions.

7.6.1 Correlated Types

Section 7.5 assumed that the agents' types were independent. Maskin and Riley (1980), Crémer and McLean (1985, 1988), McAfee, McMillan, and Reny (1989), Johnson, Pratt, and Zeckhauser (1990), and d'Aspremont, Crémer, and Gérard-Varet (1990a,b) have shown in various environments that, when preferences are quasi-linear (risk neutrality) and the agents' types are correlated, the principal can implement the same allocation she would implement if she knew the agents' types.[40] Thus, IC is not binding under risk neutrality and correlated types.

---

39. This result is not robust. Satterthwaite and Williams (1989) show that optimal trading allocations cannot be implemented by double auctions for "generic" pairs of prior distributions.

40. Recall from subsection 7.4.3 that correlation is not needed when the principal wants to maximize the sum of the agents' utilities. The result here is interesting when there is a conflict between the objectives of the principal and the agents.

To get some intuition about why this is so, suppose that the agents' types are perfectly correlated. Then each knows the others' types. Let the principal organize a "shoot them all" mechanism: The principal asks the agents to announce the vector of the I types simultaneously. If all announcements coincide, the principal implements the optimal full-information allocation corresponding to the announced types (which may or may not satisfy IR constraints, depending on the case); if they do not coincide, the principal "shoots all agents": $t_i = -\infty$ for all i. Clearly, if all other agents announce the true vector of types, it is in the interest of the remaining agent to announce the true vector of types as well. Hence, the principal can costlessly obtain the agents' information and *de facto* has full information.[41]

This idea generalizes to the case of (even small) imperfect correlation of the agents' types. One can use the fact that an agent's information yields the best predictor of the other agents' information[42] to "shoot the agent stochastically" if he misreports his type. Because the agents and the principal are risk neutral, using transfers that depend not only on the agent's type but also on the other agents' types and therefore impose risk on the agent creates no social loss in terms of risk bearing.

The papers in the literature make a full-rank assumption. Assume that there are a *finite* number of types per agent. Let $p(\theta_{-i}|\theta_i)$ denote the probability of types θ_{-i} for players other than i conditional on player i's having type θ_i. Let $p_i^{\theta_i}$ denote the vector of

$$\{p(\theta_{-i}|\theta_i)\}_{\theta_{-i} \in \Theta_{-i}}.$$

The full-rank condition is satisfied if, for each i, the vectors

$$\{p_i^{\theta_i}\}_{\theta_i \in \Theta_i}$$

are linearly independent. That is, there do not exist an agent i, a type θ_i,

41. Note that there are many other equilibria in the "shoot them all" mechanism. For instance, all agents could announce the same incorrect vector of types. This multiplicity is precisely what gave rise to a large literature on unique Nash implementation, starting with Maskin 1977 (see Moore 1990 for a survey and a list of references). Some authors, including Maskin and Riley (1980), have also looked at equilibrium uniqueness in the imperfect-correlation case (see also the more general literature mentioned in section 7.2). Crémer and McLean (1985, 1988) obtain results on dominant-strategy as well as Bayesian implementation.

42. One formalization of the notion that "an agent's information yields the best predictor of the other agents' information" is obtained by considering the "proper scoring rules" familiar in the statistics literature: Suppose that agent i is asked to reveal his type $\hat{\theta}_i$, and is given transfer $\tau_i(\hat{\theta}) = \ln p(\hat{\theta}_{-i}|\hat{\theta}_i)$ when the other agents announce $\hat{\theta}_{-i}$. Suppose in a first step that no decision x is at stake, so that agent i aims at maximizing his expected transfer. It is easily checked that, if the other agents announce truthfully, it is in the interest of agent i to announce his type truthfully, and strictly so if the vectors of conditional probabilities differ.

When there is a payoff-relevant decision x, such as allocating a good among bidders or supplying a public good, agent i's payoff function depends on the decision as well as his report, and the above proper scoring rule τ_i may no longer induce truthful revelation. However, one can "scale up" τ_i by multiplying by a large positive constant K. Then, any misreport of type implies substantial losses in the transfer $K\tau_i$, which swamps any effect on V_i of misreporting the type. Johnson et al. (1990) use such inflated proper scoring rules (to which they add further terms to meet other constraints such as budget balance).

and a vector of positive numbers $\rho_i(\theta_i')$ such that

$$p_i^{\theta_i} = \sum_{\theta_i' \neq \theta_i} \rho_i(\theta_i') p_i^{\theta_i'}.$$

In words, the full-rank condition means that the vectors of agent i's conditional probabilities about the other agents' types can be told apart.

Crémer and McLean (1985) show that the principal can implement any decision rule $x^*(\cdot)$ and agents' utilities $U_i^*(\cdot)$ under risk neutrality and full rank, even if the principal does not know θ. We illustrate their construction in the case of two agents and two types per agent. Agent i can have type $\underline{\theta}_i$ or $\bar{\theta}_i$. Let q_{11} and q_{12} denote the conditional probabilities that $\theta_2 = \underline{\theta}_2$ and $\theta_2 = \bar{\theta}_2$ when $\theta_1 = \underline{\theta}_1$; the conditional probabilities when $\theta_1 = \bar{\theta}_1$ are q_{21} and q_{22}. The full-rank condition for player 1 is $q_{11}q_{22} \neq q_{21}q_{12}$. Let t_{11} and t_{12} denote the transfers to agent 1 when he announces $\underline{\theta}_1$ and agent 2 announces $\underline{\theta}_2$ and $\bar{\theta}_2$, respectively. And similarly for t_{21} and t_{22}. The decisions and utilities are indexed in the same way. To yield the desired utilities, the transfers must satisfy, for some constants A_1 and A_2 determined by the data of the problem,[43]

$$q_{11}t_{11} + q_{12}t_{12} = A_1 \tag{7.43}$$

and

$$q_{21}t_{21} + q_{22}t_{22} = A_2. \tag{7.44}$$

The transfers must also ensure incentive compatibility for player 1 with type $\underline{\theta}_1$ or $\bar{\theta}_1$. That is,

$$q_{11}(t_{11} - t_{21}) + q_{12}(t_{12} - t_{22}) \geq A_3 \tag{7.45}$$

and

$$q_{21}(t_{21} - t_{11}) + q_{22}(t_{22} - t_{12}) \geq A_4, \tag{7.46}$$

where A_3 and A_4 are constants determined by the data.[44]

Substituting equations 7.43 and 7.44 into equations 7.45 and 7.46 yields

$$(q_{11}q_{22} - q_{21}q_{12})t_{11} \geq A_5 \equiv A_1 q_{22} + (A_4 - A_2)q_{12} \tag{7.47}$$

---

43. Where

$$A_1 \equiv q_{11}(U_{11}^* - V_1(x_{11}^*, \underline{\theta}_1)) + q_{12}(U_{12}^* - V_1(x_{12}^*, \underline{\theta}_1))$$

and

$$A_2 \equiv q_{21}(U_{21}^* - V_1(x_{21}^*, \bar{\theta}_1)) + q_{22}(U_{22}^* - V_1(x_{22}^*, \bar{\theta}_1)).$$

44. The reader will check that

$$A_3 \equiv q_{11}(V_1(x_{21}^*, \underline{\theta}_1) - V_1(x_{11}^*, \underline{\theta}_1)) + q_{12}(V_1(x_{22}^*, \underline{\theta}_1) - V_1(x_{12}^*, \underline{\theta}_1)).$$

and

$$A_4 \equiv q_{21}(V_1(x_{11}^*, \bar{\theta}_1) - V_1(x_{21}^*, \bar{\theta}_1)) + q_{22}(V_1(x_{12}^*, \bar{\theta}_1) - V_1(x_{22}^*, \bar{\theta}_1)).$$

and

$$(q_{11}q_{22} - q_{21}q_{12})t_{21} \leq A_6 \equiv -A_2 q_{12} - (A_3 - A_1)q_{22}. \tag{7.48}$$

Transfers satisfying equations 7.47, 7.48, 7.43, and 7.44 yield the desired allocation for the principal, and such transfers always exist under the full-rank condition. Note, however, that, as types become less correlated, $(q_{11} - q_{21})$ and $(q_{12} - q_{22})$ both converge to 0, so $q_{11}q_{22} - q_{21}q_{12}$ converges to 0, and so the transfers required to satisfy inequalities 7.47 and 7.48 become very large. Transfers for agent 2 can be constructed in a similar manner (given the full-rank condition for player 2). More generally, with an arbitrary number of types (and players), Farkas's lemma (which gives conditions for a system of linear inequalities and equalities to have a solution—see, e.g., section 22 of Rockafellar 1970) and the full-rank condition can be used to prove the existence of appropriate transfers.[45]

Of course, the result that the principal can use any arbitrarily small amount of correlation to achieve the full-information outcome while she usually suffers from the asymmetry of information under independent distributions of types is extreme. The point is that the credibility of risk neutrality is stretched by the very large transfers required for small correlations.

7.6.2 Risk Aversion

Most of the literature on mechanism design has focused on the case of quasi-linear preferences. We saw in sections 7.4 and 7.5 that in this case optimal mechanism design with several agents is a simple extension of mechanism design with a single agent. With risk-averse agents, one still makes heavy use of the single-agent framework and its optimal-control techniques, but things become harder.

To illustrate the issues, consider the problem of designing an optimal auction for one unit of a good when the buyers are risk averse, have the same preferences, and have types that are independently drawn from the same distribution $P(\cdot)$ on $[\underline{\theta}, \overline{\theta}]$ (the theory was developed by Maskin and Riley (1984) and Matthews (1983)). To allow for the case in which the agents have utility functions that are not separable in income and consumption, one must consider two transfers, $t_i(\hat{\theta})$ and $\tilde{t}_i(\hat{\theta})$, according to whether the agent wins or loses in the auction (for simplicity, assume that these transfers are deterministic). Let $u(t_i(\hat{\theta}), \theta_i)$ and $w(\tilde{t}_i(\hat{\theta}))$ denote the utilities of agent i when he wins and when he loses the auction. Let

$$t_i(\hat{\theta}_i) \equiv E_{\theta_{-i}} t_i(\hat{\theta}_i, \theta_{-i})$$

and

---

45. With a continuum of types, an agent can approximate the true conditional probability distribution arbitrarily closely by lying. One must then solve "Fredholm equations" (see McAfee et al. 1989, Caillaud et al. 1986, and Melumad and Reichelstein 1989).

$$\tilde{t}_i(\hat{\theta}_i) \equiv \mathrm{E}_{\theta_{-i}}\tilde{t}_i(\hat{\theta}_i, \theta_{-i}).$$

Eliminating the dependence of transfers on the other agents' announcements reduces the agent's risk and raises his utility. Doing so,[46] assuming a symmetric auction and eliminating subscripts under t_i and $\tilde{t}_i$, yields utility function for an agent of type θ_i:

$$U(\theta_i) = \max_{\hat{\theta}_i} \{X(\hat{\theta}_i)u(t(\hat{\theta}_i), \theta_i) + [1 - X(\hat{\theta}_i)]w(\tilde{t}(\hat{\theta}_i))\}, \qquad (7.49)$$

where $X(\hat{\theta}_i) \equiv \mathrm{E}_{\theta_{-i}}x(\hat{\theta}_i, \theta_{-i})$ is the probability that the agent wins the auction. Let

$$U(\theta_i) = X(\theta_i)u(t(\theta_i), \theta_i) + [1 - X(\theta_i)]w(\tilde{t}(\theta_i)). \qquad (7.50)$$

The envelope theorem implies that

$$\frac{dU}{d\theta_i} = X(\theta_i)\frac{\partial u}{\partial \theta_i}(t(\theta_i), \theta_i). \qquad (7.51)$$

The principal maximizes her expected revenue per buyer,

$$\mathrm{Max} \int_{\underline{\theta}}^{\bar{\theta}} \{X(\theta_i)t(\theta_i) + [1 - X(\theta_i)]\tilde{t}(\theta_i)\}p(\theta_i)\,d\theta_i, \qquad (7.52)$$

subject to equation 7.50, equation 7.51, (IR) $U(\underline{\theta}) \geq 0$, and "consistency."

The "consistency" constraint arises from the fact that, if equation 7.52 is maximized subject to only equation 7.50, equation 7.51, and IR, nothing guarantees that, given $X(\cdot)$, one can find a decision function $x(\cdot) \in [0, 1]$ such that

$$X(\theta_i) = \mathrm{E}_{\theta_{-i}}[x(\theta_i, \theta_{-i})] \text{ for all } (i, \theta_i). \qquad (7.53)$$

In other words, analyzing isolated single-buyer problems ignores the constraint that there is a single unit of the good to be distributed among all buyers. The consistency constraint means that one must restrict attention to probabilities $X(\cdot)$ such that there exist a function $x(\cdot)$ satisfying equation 7.53.

In the case of an auction, there is fortunately a characterization of consistent $X(\cdot)$ that preserves the simple structure of an optimal-control problem. (This characterization is due to Maskin and Riley (1984) and Matthews (1983) and finds its most general formulation in Matthews 1984.) Namely, if $X(\cdot)$ is nondecreasing and satisfies

---

46. Further analysis is needed to prove that it is indeed optimal to eliminate this dependence. Assumptions on preferences must be made so that the agent's incentive-compatibility constraint is not relaxed through the use of a random scheme. (Even if these conditions on preferences are not met and optimal auctions involve random transfers, the optimal randomness has in general little to do with that created by the uncertainty about θ_{-i}.)

$$\int_{\underline{\theta}}^{\bar{\theta}} [P(\tilde{\theta})^{I-1} - X(\tilde{\theta})]p(\tilde{\theta})\,d\tilde{\theta} \geq 0 \text{ for all } \theta \in [\underline{\theta}, \bar{\theta}], \tag{7.54}$$

then it is consistent.

It is easy to see that equation 7.54 is a necessary condition for consistency: The probability that a buyer with a valuation in $[\theta, \bar{\theta}]$ wins,

$$I \int_{\theta}^{\bar{\theta}} X(\tilde{\theta})p(\tilde{\theta})\,d\tilde{\theta},$$

cannot exceed the total probability that at least one buyer has valuation in $[\theta, \bar{\theta}]$,

$$1 - P(\theta)^I.$$

As

$$\frac{1 - P(\theta)^I}{I} = \int_{\theta}^{\bar{\theta}} P(\tilde{\theta})^{I-1} p(\tilde{\theta})\,d\tilde{\theta},$$

this yields equation 7.54. The difficult part of the characterization is to prove that equation 7.54 is sufficient for consistency.

7.6.3 Informed Principal

In this chapter we have assumed that the agents perfectly know the principal's preferences. It may be that the principal (the mechanism designer) also has private information. For instance, she may have information about the cost of supplying a public good, about her private cost of departing with the object in an auction, or about her willingness to pay for a good purchased from the agent.

Once the principal has private information, it must be recognized that the very proposal of a mechanism by the principal will reveal information about her type, as Myerson (1983) pointed out. Whereas Myerson analyzes this situation from a cooperative-game viewpoint, Maskin and Tirole (1989, 1990) keep the three-stage structure described in the introduction and used throughout the chapter and apply noncooperative game theory. (They use perfect Bayesian equilibrium rather than Bayesian equilibrium — see the next chapter. The concept mainly adds the extra requirement that, after observing the principal's contract offer, the agents update their beliefs about her type using Bayes' rule.)

One must distinguish between two situations. In the "private values" case, the principal's type does not enter the agents' preferences (but the agents' types are allowed to enter the principal's preferences). With y denoting the allocation and θ_0 the principal's type, the principal's utility is $u_0(y, \theta, \theta_0)$ and agent i's utility is $u_i(y, \theta)$. In contrast, if θ_0 affects some agent's utility, we have "common values." The difference between private and common values is that in the former case the agents care about the principal's type only to the extent that it affects the principal's behavior in the implementation of the mechanism, whereas in the latter case the agents care about her type *per se*. The three examples given at the beginning of this subsection exhibit private values. In contrast, if, in an auction, the seller's cost of departing with the good is correlated with an unknown-to-the-buyers quality of the good, we have common values.

A simple observation is that under *private values* the principal can guarantee herself the expected payoff she would obtain if the agents knew her type: It suffices that the principal offer the mechanism that is optimal for her when the agents know her type. Because the principal is not a player in the third stage (implementation of the mechanism), nothing is altered by the asymmetry of information about θ_0. The issue is then whether the principal can do better when her type is unknown to the agents than when it is common knowledge. Clearly, to do better the principal must participate in the third stage—for instance, by announcing her private information at the same time that the agents announce theirs. By delaying revelation of her information until after the proposal of the contract, the principal may be able to pool the agents' (IR or IC) constraints across her types. Indeed, Maskin and Tirole (1990) show that any equilibrium of the mechanism-design game can be computed as a Walrasian equilibrium of a fictitious economy. In this economy, the traders are the different types of principals, in proportions equal to those of the prior beliefs about θ_0, the goods traded are the slack variables on the agents' (IC and IR) constraints, and the traders have zero initial endowments of the goods.[47]

When preferences are quasi-linear, it turns out that the multipliers associated with the agents' IR and IC constraints do not depend on θ_0 when the agents know θ_0. Hence, the different types of principal do not gain by pooling when they offer a mechanism, as they do not gain by pooling constraints and trading slack. This implies that the unique equilibrium is the same as when the agents know θ_0. Thus, the single-agent theory of section 7.3 and the multi-agent theory of section 7.5 remain valid when the principal has private information, values are private, and preferences are quasi-linear.

In contrast, the analysis of this chapter must be amended when preferences are not quasi-linear. Generically, the multipliers of the agents' constraints do not coincide for different types of principal, and these types gain by trading slack on the constraints. In equilibrium, the principal does not reveal any of her information in the first step (contract proposal) and waits until the third step (contract implementation) to do so. And she does strictly better than when the agents know her type.

---

47. The paper considers a single agent, but the ideas extend to multiple agents, as this chapter would suggest.

The case of *common values* is more complex. For one thing, the principal may no longer be able to guarantee herself the same payoff as when agents know θ_0. The point is that the optimal mechanism when the agents know θ_0 need no longer be accepted by the agents if they draw the wrong inference about θ_0, as their utilities are directly affected by θ_0. Maskin and Tirole (1989) consider the restrictive case in which there is a single agent and this agent has no private information (and generalize their results to bilateral asymmetric information only in the case of quasi-linear preferences). The mechanism-design game is then similar to the standard signaling game we describe in section 8.2, except that the "sender" (the principal) has a large strategy space (the space of all contracts). The set of equilibria can be fully characterized, has a unique element for a subset of the agents' prior beliefs about θ_0, and has a continuum of elements for the complementary subset of beliefs.

7.6.4 Dynamic Mechanism Design

The static analysis of this chapter can be used to characterize repeated mechanism design as long as the principal and the agents can commit intertemporally (see, e.g., Baron and Besanko 1984a). Consider a multi-period problem, with periods $\tau = 0, 1, \dots, T$. Suppose for instance that there is a single agent, with preferences

$$\sum_{\tau=0}^{T} \delta^\tau u_1(y_\tau, \theta)$$

where $y_\tau = (x_\tau, t_\tau)$ is the allocation at date τ and δ is the discount factor. The principal has preferences

$$\sum_{\tau=0}^{T} \delta^\tau u_0(y_\tau, \theta).$$

Note that we assume that the agent's type is invariant.[48]

Let $y^*(\theta)$ denote the optimal allocation for the principal subject to the agent's IR and IC constraints in a one-period context (see section 7.3). We claim that the allocation $y_\tau(\theta) = y^*(\theta)$ for all τ is optimal (i.e., the optimal allocation is the $(T + 1)$ replica of the static one). To see this, suppose that the principal could do better than replicate the optimal static allocation. That is, assume that there exists an allocation $\{y_\tau(\cdot)\}_{\tau=0,\dots,T}$ that satisfies the agent's multi-period IR and IC constraints,

(multi-period IR) $\displaystyle\sum_{\tau=0}^{T} \delta^\tau u_1(y_\tau(\theta), \theta) \geq \sum_{\tau=0}^{T} \delta^\tau \underline{u}_1(\theta)$ for all θ

(where $\underline{u}_1(\theta)$ is the invariant per-period reservation utility of type θ), and

---

48. See Baron and Besanko 1984a for the case of a type that changes over time.

(multi-period IC) $\sum_{\tau=0}^{T} \delta^{\tau} u_1(y_\tau(\theta), \theta) \geq \sum_{\tau=0}^{T} \delta^{\tau} u_1(y_\tau(\hat{\theta}), \theta)$ for all $(\theta, \hat{\theta})$,

and that yields more expected utility to the principal than y^* repeated $T + 1$ times:

$$E_\theta \left(\sum_{\tau=0}^{T} \delta^{\tau} u_0(y_\tau(\theta), \theta) \right) > (1 + \delta + \cdots + \delta^T)(E_\theta[u_0(y^*(\theta), \theta)]). \qquad (7.55)$$

Now consider the random static mechanism that, for an announcement $\hat{\theta}$, gives the agent allocation $y_0(\hat{\theta})$ with probability $1/(1 + \cdots + \delta^T)$, $y_1(\hat{\theta})$ with probability $\delta/(1 + \cdots + \delta^T)$, ..., $y_T(\hat{\theta})$ with probability $\delta^T/(1 + \cdots + \delta^T)$. Dividing (multiperiod IR), (multiperiod IC), and equation 7.55 by $(1 + \cdots + \delta^T)$, this random allocation satisfies the (static) IR and IC constraints and yields more expected utility than $y^*(\cdot)$, a contradiction. Hence, the optimal static allocation remains optimal in a dynamic context with commitment.[49]

To implement the dynamic optimum, the principal asks the agent to reveal his type $\hat{\theta}$ at date 0, and then implements allocation $y^*(\hat{\theta})$ repeatedly until the end of the horizon. Note that it is important that the principal can commit. Otherwise we face the time-consistency problem studied in chapter 3. We saw in section 7.3 that (if $\underline{u}_1(\theta) = \underline{u}$ and u_1 is increasing in θ), except "at the bottom" ($\theta = \underline{\theta}$), the agent enjoys a rent associated with his private information ($u_1(y^*(\theta), \theta) > \underline{u}$). At the end of period 0, the principal has learned the agent's type and would want to put the agent at his IR level at dates $\tau = 1, \ldots, T$. That is, the principal would want to renege on her commitment to keep the same allocation over time once she had learned the agent's type.

Actually, the ability to commit to a long-term contract that any of the parties (principal or agent) can have enforced by a court if she or he wants to is not sufficient for the optimal static mechanism repeated $T + 1$ times to be feasible, as was demonstrated by Dewatripont (1989). To see this, recall from section 7.3 that $y^*(\cdot)$ involves (under the assumptions made there) a distortion except "at the top" (at $\theta = \bar{\theta}$). The principal trades off efficiency and rent extraction. Now, if at the end of period 0 the principal knows the agent's type to be θ, it is common knowledge that the two parties can improve upon $y^*(\cdot)$ at dates $1, \ldots, T$ to their mutual benefit. They will then renegotiate the initial contract. Thus, the commitment assumption underlying the result that the dynamic allocation is the replicated static one must be taken to mean that the parties not only sign an enforceable long-term contract at date 0 but also can commit never to renegotiate the contract in the future, even if it is in their interest to do so. When the parties cannot

---

49. The above notation implicitly assumes that $y^*(\cdot)$ is deterministic, but the same reasoning clearly holds when the optimal static allocation is random.

commit not to renegotiate, dynamic mechanism design does not boil down to a static one, and the dynamic equilibrium notions developed in chapter 8 must be employed. Hart and Tirole (1988) and Laffont and Tirole (1990b) show that, in the quasi-linear case, the dynamics of the equilibrium alloca- tion $y_\tau(\cdot)$ coincides with the Coasian dynamics of the durable-good models analyzed in chapter 10.[50]

Besides these two paradigms, "full commitment" and "commitment and renegotiation," economists have considered a third one, called "noncom- mitment." Suppose that the parties are unable to sign long-term contracts, either for transactional reasons or for legal ones (as is sometimes the case when the principal is a government). One can then consider the repeated version of the three-step game of section 7.3. In each period τ, the principal offers a mechanism $y_\tau(\cdot)$ that applies only to that period.[51] A main issue in such a situation is the "ratchet effect." Suppose for instance that the agent reveals his type in period 0. The continuation game from date 1 on is then a symmetric-information one, and, in the unique subgame equilibrium of this continuation game, the principal offers in each period an allocation that puts the agent at his IR level. Thus, revealing one's type is very costly in a dynamic setting without commitment, and the different types of agent will have a tendency to "pool." We will not give an analysis of the ratchet problem, which requires the tools of dynamic games of incomplete informa- tion developed in chapter 8.

7.6.5 Common Agency

In some situations an agent may serve several principals. For example, a distributor may carry the products of several manufacturers, a firm may be regulated by several government agencies, and a consumer may buy from several producers. Martimort (1990) and Stole (1990a) have developed a theory of common agency.[52]

Suppose that there are two principals, A and B. Principal i, $i = A, B$, is interested in decision $x_i \in \mathbb{R}$, and has utility

$$u_i = V_i(x_i, \theta) - t_i.$$

The agent has utility

$$u_1 = V_1(x_A, x_B, \theta) + t_A + t_B.$$

50. The issue of contract renegotiation under asymmetric information also arises in moral- hazard models of the principal-agent relationship. Once the agent has chosen his effort, this effort, if private information, becomes a type for the agent. (See Fudenberg and Tirole 1990.)
51. See Freixas et al. 1985 and Laffont and Tirole 1987b, 1988. See Baron and Besanko 1987 for an approach using a different solution concept.
52. An early example of common agency is found in Baron 1985. Other examples are found in Gal-Or 1989, where the two principals' decisions do not interact in the agent's utility function ($\partial^2 V_1/\partial x_A \, \partial x_B = 0$), and in Laffont and Tirole 1990c, where the decisions are perfect comple- ments ($\partial^2 V_1/\partial x_A \, \partial x_B = +\infty$).

A Nash equilibrium in contracts is a pair

$$\{t_A(x_A), t_B(x_B)\},$$

or

$$\{(t_A(\hat{\theta}_A), x_A(\hat{\theta}_A)), (t_B(\hat{\theta}_B), x_B(\hat{\theta}_B))\}$$

where $\hat{\theta}_i$ is the agent's announcement of type to principal i, such that each principal, given the other principal's contract and the agent's optimal reaction to contract offers, maximizes her expected payoff. Note that principal i observes only the report $\hat{\theta}_i$ (or equivalently, the decision x_i) meant for her.

A natural generalization of equation 7.12 to a common-agency differentiable equilibrium (if one exists—see below) is, for all $i = A, B$,

$$
\frac{\partial V_i}{\partial x_i} + \frac{\partial V_1}{\partial x_i}
$$

$$
= \frac{1 - P(\theta)}{p(\theta)} \left(\frac{\partial^2 V_1}{\partial x_i \partial \theta} + \frac{\partial^2 V_1}{\partial x_j \partial \theta} x_j'(\theta) \frac{\frac{\partial^2 V_1}{\partial x_i \partial x_j}}{\frac{\partial^2 V_1}{\partial x_j \partial \theta} + \frac{\partial^2 V_1}{\partial x_i \partial x_j} x_i'(\theta)} \right). \tag{7.56}
$$

Equation 7.56 coincides with equation 7.12 except for the second (interaction) term on the right-hand side. When principal i induces an increase dx_i in $x_i(\theta)$, she changes the marginal utility of decision x_j. The resulting change in decision x_j is

$$
dx_j = dx_i x_j'(\theta) \frac{\partial^2 V_1}{\partial x_i \partial x_j} \Big/ \left(\frac{\partial^2 V_1}{\partial x_j \partial \theta} + \frac{\partial^2 V_1}{\partial x_i \partial x_j} x_i'(\theta) \right).
$$

(To obtain this, differentiate the first-order condition for x_j totally with respect to x_i and $\hat{\theta}_j$ to get an expression for $\partial \hat{\theta}_j / \partial x_i$ and note that $dx_j = x_j'(\theta)(\partial \hat{\theta}_j / \partial x_i) dx_i$.) The change dx_i thus has both a direct $((\partial^2 V_1 / \partial x_i \partial \theta) dx_i)$ and an indirect $((\partial^2 V_1 / \partial x_j \partial \theta) dx_j)$ effect on the rate of growth of the agent's rent, which yields equation 7.56.

Contract complements $(\partial^2 V_1 / \partial x_i \partial x_j > 0)$ lead to a double rent extraction, with the reduction in x_i by principal i making a reduction in x_j more desirable by principal j. The distortion in the decisions thus exceeds that under cooperative contracting by the principals (i.e., that of the single-principal case). In contrast, in a symmetric equilibrium, the decisions lie between the cooperative-contracting decisions and the full-information (or first-best) ones for contract substitutes $(\partial^2 V_1 / \partial x_i \partial x_j < 0)$.

The analysis focuses on finding sufficient conditions for implementability. In the single-principal case, and under the sorting condition, monotonicity is sufficient for local- and global-second-order conditions to be satisfied (theorem 7.3). With two principals, if the agent does not announce his type truthfully to principal i, he may also lie to principal j, and perhaps in a different way. That is, misreporting of θ occurs in a two-dimensional space instead of a single-dimensional one. Martimort and Stole derive sufficient conditions for implementation, and are thus able to prove the existence of a differentiable equilibrium. There is a unique symmetric differentiable equilibrium for contract substitutes and quadratic payoff functions. There is a continuum of symmetric equilibria for contract complements, but the one involving the smallest distortions Pareto dominates the others for the principals and the agent.[53]

Appendix[†††]

What to Do if the Monotonicity Constraint Is Binding

When $x^*(\cdot)$ given by equation 7.12 is not nondecreasing everywhere, one must analyze the full program. There are then two subsets of $[\underline{\theta}, \overline{\theta}]$, both composed of a set of disconnected intervals. In the first subset, the monotonicity constraint is not binding and thus $x(\theta) = x^*(\theta)$. Note that this subset is never empty, because for θ close to $\overline{\theta}$, $p/(1 - P)$ is necessarily increasing.[54] In particular, the "no distortion at the top" result is a general result and does not depend on the monotone-hazard-rate assumption.

In the second subset, the monotonicity constraint is binding and therefore $x(\cdot)$ is constant on each interval in this subset.

We first derive a characterization of the bunching levels, i.e., of decisions x that are chosen by more than one θ. We then sketch an algorithm to obtain the bunching regions. Consider an interval $[\theta_1, \theta_2]$ over which there is "bunching" so that $x(\theta) = \hat{x}$ for all $\theta \in [\theta_1, \theta_2]$, but such that the monotonicity constraint is not binding just outside the interval.

Maximize the principal's expected payoff, and replace the monotonicity constraint by

$$\frac{dx}{d\theta} = \gamma(\theta) \tag{7.57}$$

and

$$\gamma(\theta) \geq 0. \tag{7.58}$$

If $v(\theta)$ and $\lambda(\theta)$ denote the shadow prices of equations 7.57 and 7.58, the

---

53. Another difference with the single-principal case is the treatment of the agent's IR constraint. This treatment depends on whether the agent can accept zero, one, or two contracts (as is the case for a consumer), or whether he can accept zero or two contracts only (as is the case for a regulated firm). For instance, in the second case, the individual transfers for the lowest type, $t_A(\underline{\theta})$ and $t_B(\underline{\theta})$, are not uniquely defined (but their sum is).
54. Recall that we assumed that p is continuous and strictly positive on the whole interval.

Hamiltonian for program I is then

$$H = \left(V_0 + V_1 - \frac{1 - P}{p} \frac{\partial V_1}{\partial \theta} \right) p + v\gamma + \lambda\gamma,$$

where x is taken as a state variable and γ as a control variable. The Pontryagin conditions are

$$\frac{\partial H}{\partial \gamma} = 0 = v + \lambda \tag{7.59}$$

and

$$\frac{dv}{d\theta} = -\frac{\partial H}{\partial x} = -\left(\frac{\partial V_0}{\partial x} + \frac{\partial V_1}{\partial x} - \frac{1 - P}{p} \frac{\partial^2 V_1}{\partial x \partial \theta} \right) p. \tag{7.60}$$

Now we exploit the assumption that the monotonicity constraint is not binding at the two boundaries of the interval. Thus, $v(\theta_1) = v(\theta_2) = 0$, and equation 7.60 can be rewritten as

$$\int_{\theta_1}^{\theta_2} \left(\frac{\partial V_0}{\partial x} + \frac{\partial V_1}{\partial x} - \frac{1 - P}{p} \frac{\partial^2 V_1}{\partial x \partial \theta} \right) p \, d\theta = 0. \tag{7.61}$$

That is, the average distortion of the total virtual surplus is equal to 0 over the interval. Together, equation 7.61 and the condition $x^*(\theta_1) = x^*(\theta_2)$ (which results from the boundary conditions $x(\theta_1) = x^*(\theta_1)$ and $x(\theta_2) = x^*(\theta_2)$ and the fact that $x(\theta_1) = x(\theta_2)$) yield two equations with two unknowns. Figure 7.4 depicts the case in which A10 is not satisfied.

Using this characterization of the bunching regions, we now determine where such regions are located. From our assumptions, x^* is continuously differentiable. Let us assume that the curve x^* has a finite number of interior peaks on $[\underline{\theta}, \bar{\theta}]$.

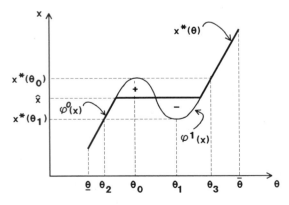

Figure 7.4

If there is no interior peak, x^* is nondecreasing (recall that $x^*(\bar{\theta}) \geq x^*(\theta)$ for all θ) and is therefore the solution to program I. If there is a single interior peak θ_0, then there is also a single interior through θ_1 (see figure 7.4). The inverse image of the interval $[x^*(\theta_1), x^*(\theta_0)]$ is composed of two intervals, $[\theta_2, \theta_0]$ and $[\theta_1, \theta_3]$, over which $x^*(\cdot)$ is increasing (if there is no $\theta_2 < \theta_0$ such that $x^*(\theta_2) = x^*(\theta_1)$, let $\theta_2 \equiv \underline{\theta}$), and one interval, $[\theta_0, \theta_1]$, over which $x^*(\cdot)$ is decreasing. Let $\varphi^0(x)$ and $\varphi^1(x)$ denote the inverse functions of x over the intervals $[\theta_2, \theta_0]$ and $[\theta_1, \theta_3]$. Last, for each $x \in [x^*(\theta_1), x^*(\theta_0)]$, define

$$\Delta(x) \equiv \int_{\varphi^0(x)}^{\varphi^1(x)} \left(\frac{\partial V_0}{\partial x}(x, \theta) + \frac{\partial V_1}{\partial x}(x, \theta) - \frac{1 - P(\theta)}{p(\theta)} \frac{\partial^2 V_1}{\partial x \partial \theta}(x, \theta) \right) d\theta.$$

Note that at $x = x^*(\theta_0)$, $\varphi^0(x) = \theta_0$ and $\varphi^1(x) = \theta_3$ and $\Delta(x) < 0$ as $x > x^*(\theta)$ for all $\theta \in (\theta_0, \theta_3)$ and the objective function

$$V_0 + V_1 - \frac{1 - P}{p} \frac{\partial^2 V_1}{\partial x \partial \theta}$$

is strictly concave in x. Similarly, at $x = x^*(\theta_1)$, $\varphi^0(x) = \theta_2$ and $\varphi^1(x) = \theta_1$, and if $\theta_2 > \underline{\theta}$, $\Delta(x) > 0$ as $x < x^*(\theta)$ for all $\theta \in (\theta_2, \theta_1)$. Furthermore, with x optimal at $\varphi^0(x)$ and $\varphi^1(x)$,

$$\Delta'(x) = \int_{\varphi^0(x)}^{\varphi^1(x)} \left(\frac{\partial^2 V_0}{\partial x^2}(x, \theta) + \frac{\partial^2 V_1}{\partial x^2}(x, \theta) - \frac{1 - P(\theta)}{p(\theta)} \frac{\partial^3 V_1}{\partial x^2 \partial \theta} \right) d\theta < 0.$$

If $\theta_2 > \underline{\theta}$, then the intermediate-value theorem shows that there exists a (unique) $\hat{x} \in [x^*(\theta_1), x^*(\theta_0)]$ such that $\Delta(\hat{x}) = 0$. From our previous characterization, the bunching interval is $[\varphi^0(\hat{x}), \varphi^1(\hat{x})]$, so the solution is $x^*(\theta)$ for $\theta \notin [\varphi^0(\hat{x}), \varphi^1(\hat{x})]$ (see the bold curve in figure 7.4).[55]

Now suppose there are two interior peaks. Intuitively, if we can independently design two bunching levels $\hat{x}_1$ and $\hat{x}_2$ as in figure 7.5a, such that $\hat{x}_1 \leq \hat{x}_2$ and $\hat{x}_1, \hat{x}_2$ and the associated boundaries of the two bunching intervals satisfy the property that the average distortion over each bunching interval is equal to 0, we have the solution (represented by the bold curve in figure 7.5a). If treating the two bunching regions separately yields $\hat{x}_1 > \hat{x}_2$, the resulting decision schedule is not monotonic and therefore not incentive compatible (see the broken segments in figure 7.5b). We must then merge the two into a single bunching interval at some level $\hat{x}_3$ such that the average distortion over the interval $[\theta_5, \theta_6]$ in figure 7.5b is equal to 0.

---

55. If $\theta_2 = \underline{\theta}$, there may or may not exist such an $\hat{x}$. More precisely, if $\Delta(x^*(\underline{\theta})) \geq 0$, there exists such an $\hat{x}$ and the answer is as above. If $\Delta(x^*(\underline{\theta})) < 0$, then the bunching interval is $[\underline{\theta}, \theta_4]$, where $\theta_4 \in [\theta_1, \theta_3]$ and where

$$\int_{\underline{\theta}}^{\theta_4} \left(\frac{\partial V_0}{\partial x} + \frac{\partial V_1}{\partial x} - \frac{1 - P}{p} \frac{\partial^2 V_1}{\partial x \partial \theta} \right) d\theta = 0.$$

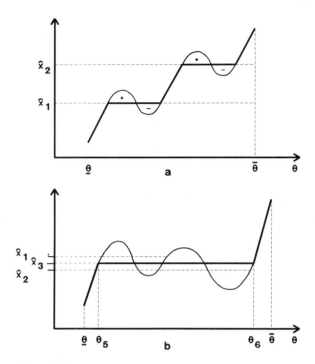

Figure 7.5

We leave it to the reader to construct an algorithm to obtain the solution with more than two peaks.

Assumption A9 implies that the constraint $x(\theta) \leq \bar{x}$ is never binding. First, monotonicity implies that $x(\theta) \leq x(\bar{\theta})$ for all θ. Second, we saw that there is no distortion at the top, so that $x(\bar{\theta}) = x^*(\bar{\theta})$. But $x^*(\bar{\theta}) \leq \bar{x}$ from A9.

When Is It Legitimate to Focus on Deterministic Mechanisms?

We restricted our attention to mechanisms in which the decision x and the transfer t are deterministic functions of the announced type $\hat{\theta}$. More generally, one can allow x and t to have random values $\tilde{x}(\hat{\theta})$ and $\tilde{t}(\hat{\theta})$. It is clear that with quasi-linear utilities there is no gain to be had from introducing random transfers, as the principal and the agent care only about the expectation $t(\hat{\theta}) \equiv \mathscr{E}\,\tilde{t}(\hat{\theta})$. (In this discussion, the expectations are with respect to the random variable underlying the stochastic allocation, and not with respect to type. To distinguish between the two, we denote the new expectations by $\mathscr{E}(\cdot)$.) Thus, only random decisions need be considered.

In many applications, the functions V_0 and V_1 are concave in x, which we assume in the following discussion. Then, V_0 and V_1 can be increased by replacing the random variable $\tilde{x}$ by its expectation $x(\hat{\theta}) \equiv \mathscr{E} \tilde{x}(\hat{\theta})$. Increasing V_0 benefits the principal directly; increasing V_1 helps her indirectly by allowing her to reduce the agent's income. Thus, if there is any benefit to introducing randomness in the decision, it must be the case that the randomness relaxes the incentive constraint. Recall that the incentive constraint can be expressed by the speed at which the agent's rent or utility increases with his type (together with the condition that the decision be monotonic in the agent's type, if the sorting condition holds). For a random scheme, the envelope theorem yields

$$\dot{U}_1(\theta) = \mathscr{E}\left[\frac{\partial V_1}{\partial \theta}(\tilde{x}(\theta), \theta) \right].$$

Suppose, for instance, that u_1 increases with θ. Then, to minimize the slope of the $U_1(\cdot)$ function, the principal wants to minimize $\mathscr{E}[\partial V_1 / \partial \theta]$. If $\partial V_1 / \partial \theta$ is convex in x ($\partial^3 V_1 / \partial \theta \partial x^2 \geq 0$), which is part of assumption A8), Jensen's inequality implies that

$$\mathscr{E}\left[\frac{\partial V_1}{\partial \theta}(\tilde{x}(\theta), \theta) \right] \geq \frac{\partial V_1}{\partial \theta}(\mathscr{E}(\tilde{x}(\theta)), \theta) = \frac{\partial V_1}{\partial \theta}(x(\theta), \theta).$$

That is, $\dot{U}_1(\theta)$ can be reduced by using the deterministic decision $x(\theta)$ instead of the random decision $\tilde{x}(\theta)$. Because random schemes reduce V_0 and V_1, and raise $\dot{U}_1$, they yield less utility to the principal:

$$E_\theta\left[\mathscr{E} V_0(\tilde{x}(\theta), \theta) + \mathscr{E} V_1(\tilde{x}(\theta), \theta) - \int_{\underline{\theta}}^\theta \mathscr{E} \frac{\partial V_1}{\partial \eta}(\tilde{x}(\eta), \eta) \, d\eta \right]$$

$$\leq E_\theta\left[V_0(\mathscr{E}(\tilde{x}(\theta)), \theta) + V_1(\mathscr{E}(\tilde{x}(\theta)), \theta) - \int_{\underline{\theta}}^\theta \frac{\partial V_1}{\partial \eta}(\mathscr{E}(\tilde{x}(\eta)), \eta) \, d\eta \right].$$

Turning things around, transforming a deterministic decision $x(\theta)$ into a random one $\tilde{x}(\theta)$ with the same mean for each θ reduces the principal's welfare. We thus conclude that if the agent's incentive-compatibility constraint for the deterministic allocation is fully characterized by the equation $\dot{U}_1(\theta) = \partial V_1(x(\theta), \theta)/\partial \theta$, as it is under the assumptions of theorem 7.4, the principal cannot gain by using a random mechanism.

In contrast, if $\partial V_1 / \partial \theta$ is strictly concave in x (that is, $\partial^3 V_1 / \partial \theta \partial x^2 < 0$), the principal can reduce the agent's rent by using stochastic decisions. The principal must then trade off the costs (the reduction in efficiency, i.e., in $V_0 + V_1$) and the benefits (the reduction in the agent's rent U_1) of random schemes. For more on random mechanisms, see Maskin 1981.[56]

---

56. Maskin and Riley (1984b) give a sufficient condition for random incentive schemes not to be optimal in the case of non-quasi-linear utilities.

Exercises

Exercise 7.1** Maskin and Riley (1984b) and Matthews (1983) show that with risk-averse bidders an optimal auction may require payments from the bidders even when they lose. The idea is that the seller can exploit the difference in the marginal utilities of income when a bidder wins or loses. Suppose that there are two bidders, $i = 1, 2$. Each bidder can have one of two valuations: $\underline{\theta}$ (with probability $\underline{p}$) or $\bar{\theta}$ (with probability $\bar{p}$), where $0 \le \underline{\theta} < \bar{\theta}$. Let W and L denote the transfers to the seller when the buyer wins or loses and the bidder has announced $\underline{\theta}$ (define $\bar{W}$ and $\bar{L}$ similarly for type $\bar{\theta}$). A bidder has utility $u(\theta - W)$ when winning and paying W, and $u(-L)$ when losing and paying L. Solve for the optimal symmetric auction and show that, at the optimum, $\bar{L} < 0$ (and the $\bar{\theta}$ type is perfectly insured), and $\underline{L} > 0$. (Hint: Proceed as in section 7.1 in your selection of IR and IC constraints. Letting $\bar{X}$ and $\underline{X}$ denote the probabilities of getting the good when $\bar{\theta}$ and $\underline{\theta}$, note that $\frac{1}{2} \ge \underline{p}\underline{X} + \bar{p}\bar{X}$ and $\bar{X} \le \underline{p} + \bar{p}/2$. Solve and check that the ignored constraints are satisfied.)

Exercise 7.2** Consider the problem of inducing firms to reveal information about their cost of reducing their pollution.

(a) Take the case of a single firm. The damage created by an amount of pollution x is $D(x)$. The production cost for the firm is $C(x, \theta)$, where θ is a private-information parameter and C is decreasing in x. Assume that $C_\theta > 0$, $C_{x\theta} < 0$, $C_{xx} \ge 0$, $D'' \ge 0$, and $C_{\theta xx} \ge 0$. Show that if the government has coercive power, it can obtain the socially optimal amount of pollution $x^*(\theta)$ by giving the firm a transfer equal to a constant minus the damage cost $D(x)$. How does this scheme link with the Groves scheme in section 7.3?

(b) Still in the single-firm context, suppose that the firm can refuse to participate (it has property rights and is free to pollute if it wants to). Can the first-best outcome still be implemented if the government cares about the sum of consumer surplus and producer profit? Next, suppose that the government faces a shadow cost of public funds $\lambda > 0$, so that its objective function is

$$W = -D(x) - (1 + \lambda)t + (t - C(x, \theta))$$

(up to a constant). Derive the optimal incentive scheme (Note: The IR level may be type dependent. Perform the analysis as if it were type independent and check *ex post* that everything is fine.)

(c) Assume there are I firms, with production costs $C_i(x_i, \theta_i)$, and that total damage is $D(x_1, \dots, x_I)$. Coming back to question a's assumption that the government faces no individual rationality constraint, derive a d'Aspremont–Gérard-Varet scheme for this model.

Exercise 7.3** Prove theorem 7.2 (monotonicity in the single-dimensional case under CS^+ or CS^-) without making the differentiability assumptions on $y(\cdot)$. (Hint: Use the methodology introduced in chapter 13 to prove that reaction curves are monotonic in separable sequential games.)

Exercise 7.4** A buyer and a seller sign a contract for the delivery of one unit of a good. The buyer has valuation v and designs the contract (is the principal). The seller will receive an outside offer equal to $\theta + \varepsilon$ for the single unit he produces. A contract specifies an unconditional payment t by the buyer to the seller, and a liquidated damage ℓ to be paid by the seller to the buyer if the seller breaches the contract and accepts the outside offer instead. The seller knows θ when signing the contract and learns ε after signing the contract and before deciding whom to serve. The independent random variables θ and ε are drawn from distributions $P(\cdot)$ and $\tilde{P}(\cdot)$, respectively. The expectation of ε is equal to 0, and both parties are risk neutral. The buyer screens the seller's type θ by offering a menu of contracts $\{t(\hat{\theta}), \ell(\hat{\theta})\}$.

(a) What are the analogues of the variables t and x in the text for this model?

(b) Show that the seller's *net* utility (given that he will accept the outside offer if he does not sign the contract), $U_1(\theta)$, satisfies $\dot{U}_1(\theta) = -\tilde{P}(\ell(\theta) - \theta)$.

(c) Show that the buyer sets liquidated damages under the real damage v:

$$\ell(\theta) = v - \frac{P(\theta)}{p(\theta)}$$

(where $p(\theta) \equiv P'(\theta)$). Interpret. (For answers, see Stole 1990b.)

Exercise 7.5* Consider the following insurance model with adverse selection. The insuree can have a low probability of accident ($\underline{\theta}$) or a high one ($\bar{\theta} > \underline{\theta}$), with probabilities $\underline{p}$ and $\bar{p}$, respectively. The insuree knows his probability of accident, but the insurance company (which is a monopolist and which offers a menu of contracts) does not. The insuree has objective function

$$u_1(W_1, W_2, \theta) = (1 - \theta)U(W_1) + \theta\, U(W_2),$$

where W_1 and W_2 are his net incomes in states of nature 1 (no accident) and 2 (accident) and U is his von Neumann-Morgenstern utility function ($U' > 0$, $U'' < 0$). With W_0 denoting the insuree's initial wealth and D the (monetary) damage in case of accident, the (risk-neutral) insurer's expected utility is

$$u_0(W_1, W_2, \theta) = (1 - \theta)(W_0 - W_1) + \theta(W_0 - D - W_2).$$

(a) Give a diagrammatic description of the optimal (two-contract) menu for the insurance monopolist. In particular, draw the status-quo (no contract) point in the (W_1, W_2) space, and indifference curves corresponding to the two types for both the insuree and the insurer. Show that the binding IC constraint is that the high-risk insuree would not want to mimic the low-risk insuree, and that the high-risk insuree gets full insurance $(\bar{W}_1 = \bar{W}_2)$. Argue informally that the high-risk insuree may or may not get a rent (depending on the probability $\bar{p}$), and that he gets a rent if some insurance is given to the low-risk insuree.

(b) Perform the same analysis as in question a, but use algebra. Hints: Use question a to guess which constraints are binding (ignore the others and check them later); let Γ denote the inverse function of the insuree's von Neumann-Morgenstern utility function U. Describe a menu as $(\underline{U}_1, \underline{U}_2)$, $(\bar{U}_1, \bar{U}_2)$, where $\underline{U}_1$ is the low type's number of utils in state 1, etc. Notice that the monopoly's objective is concave in these utility levels and the constraints are linear. Solve the monopoly's program. (For the answer, see Stiglitz 1977.)

Exercise 7.6** A labor-managed firm under contract with the government makes profit $\pi = \theta f(x) - K + t$, where $\theta f(x)$ is output $(f' > 0, f'' < 0)$, K is a known fixed cost, t is a subsidy (possibly negative) from the government, and x is the number of workers. The goverment observes x and t, but not θ and π (which are private information to the firm). The firm's objective function is profit per worker: $u_1 = \pi/x$.

(a) Show that an increasing function $x(\theta)$ is implementable if and only if the marginal productivity of labor exceeds its average productivity $(f' > f/x)$.

(b) Suppose that the government has objective function $u_0 = \theta f(x) - K - wx$, where w is the opportunity wage of workers, and has prior density $p(\theta)$ over $[\underline{\theta}, \bar{\theta}]$. Using question a, show that if $f' < f/x$, the optimal policy for the government is to "bunch" all types at a single contract (t, x). (This exercise is from Guesnerie and Laffont 1984.)

Exercise 7.7** An entrepreneur has a project that yields revenue R with probability θ and 0 with probability $1 - \theta$. A debt contract specifies a reimbursement t to the lender if the project is successful, and an amount of collateral $C \geq 0$ to be paid to the lender if the project fails. The value of the collateral is βC for the lender, where $0 \leq \beta < 1$. The project involves a fixed nonmonetary cost b for the entrepreneur (the opportunity cost of his time). The entrepreneur's expected utility, u_1, is 0 if he does not borrow (the project is not realized) and $\theta(R - t) - (1 - \theta)C - b$ if he borrows. The amount borrowed is fixed and is equal to 1. Assume $\theta R > b + 1$ for any θ. The lender's utility is $u_0 = \theta t + (1 - \theta)\beta C - 1$ if he lends, and 0 otherwise.

The entrepreneur has private information about θ, which takes value $\underline{\theta}$ with probability $\underline{p}$ and $\bar{\theta}$ with probability $\bar{p}$ ($\underline{p} + \bar{p} = 1$). Suppose first that there is a single creditor, who offers a debt contract to the entrepreneur.

(a) Show that if the lender knew θ he would not require any collateral.

(b) Suppose that the lender does not know θ. Proceeding by analogy with the price-discrimination example of section 7.1, what do you think are the binding IR and IC constraints? Assuming that the creditor wants to lend whatever θ, show that choosing $\underline{C} > 0$ *tightens* the IC constraint and that the lender offers the *pooling* contract $\{t = R - b/\underline{\theta}, C = 0\}$. Explain intuitively the difference with the price-discrimination example. (Hint: Think of the sorting condition and of which type's allocation ought to be distorted.)

(c) Suppose now that there is a *competitive* credit market (many lenders). Argue that the relevant IC constraint is not the same as in question b. Show that if a zero-profit, separating equilibrium exists, the levels of collateral for types $\underline{\theta}$ and $\bar{\theta}$ are

$$\underline{C} = 0$$

and

$$\bar{C} = \frac{\bar{\theta} - \underline{\theta}}{\bar{\theta}(1 - \underline{\theta}) - \beta(1 - \bar{\theta})\underline{\theta}}$$

(assuming that the entrepreneur's initial wealth is at least $\bar{C}$; otherwise credit rationing may occur). (This exercise is from Besanko and Thakor 1987. See also Bester 1985.)

Exercise 7.8* A monopolist faces a single consumer. The consumer has utility $u_1 = \theta q - t$, where q is consumption and t is the transfer to the monopolist. The monopolist has cost $cq^2/2$ and offers a sales contract to the consumer. The consumer has reservation utility 0.

(a) Compute the transfer and the consumption under full information about θ.

(b) Suppose from now on that the monopolist has incomplete information about θ, which takes the value $\underline{\theta}$ with probability $\underline{p}$ and $\bar{\theta}$ with probability $\bar{p}$. Assume that $\underline{\theta} > \bar{p}\bar{\theta}$. The monopolist's utility is

$$\underline{p}\left(\underline{t} - c\frac{\underline{q}^2}{2}\right) + \bar{p}\left(\bar{t} - c\frac{\bar{q}^2}{2}\right).$$

Compute the optimal nonlinear tariff. Show that the equilibrium utility of type $\bar{\theta}$ is $\bar{S} = (\bar{\theta} - \underline{\theta})(\underline{\theta} - \bar{p}\bar{\theta})/c\underline{p}$.

(c) Suppose now that the consumer can purchase at the fixed cost f an alternative (bypass) technology that allows him to produce any amount q of the same good at cost $\tilde{c}q^2/2$. Suppose for simplicity that the consumer can consume only the monopolist's good or the alternative good (but not a mix of both), and that

$$\frac{\bar{\theta}^2}{2\tilde{c}} - f > \bar{S} > 0 > \frac{\underline{\theta}^2}{2\tilde{c}} - f.$$

Is the tariff derived in question b still optimal for the monopolist? Discuss what may be optimal for the monopolist—in particular, why it may be optimal to have $c\bar{q} > \bar{\theta}$. For example, consider what happens when f decreases from $\bar{\theta}^2/2\tilde{c} - \bar{S}$.

Exercise 7.9** A regulated firm has cost $C = (\theta - e)q + f$, where q is output, e is effort, and f is a known fixed cost. The regulator observes C and q. The firm's utility is $u_1 = t - \psi(e)$, where t is a net transfer from the regulator ($\psi(0) = 0, \psi' > 0, \psi'' > 0$). The firm has reservation utility 0. The technology parameter θ takes values $\underline{\theta}$ with probability $\underline{p}$ and $\bar{\theta}$ with probability $\bar{p}$. The social welfare function is

$$u_0 = S(q) - R(q) - (1 + \lambda)(t + C - R(q)) + u_1,$$

where $S(q)$ is gross consumer surplus, $R(q) \equiv P(q)q = S'(q)q$ is the firm's revenue from selling quanity q, and $\lambda > 0$ is the shadow cost of public funds.

(a) Determine optimal quantities and effort when the regulator has perfect information. Show that price is determined by a Ramsey-type formula. (The Lerner index—price minus marginal cost over price—is equal to a fraction of the inverse of the elasticity of demand.)

(b) Suppose the regulator does not observe the components of C. Argue intuitively that the regulator will base the allocation on marginal cost $c = (C - f)/q$. Infer from this that (for a given marginal cost) the price is given by the same Ramsey formula as in question a, but that the marginal cost changes. Show that the firm chooses effect $\underline{e}$ when $\theta = \underline{\theta}$ and $\bar{e}$ when $\theta = \bar{\theta}$, where

$$\psi'(\underline{e}) = \underline{q},$$

$$\psi'(\bar{e}) = \bar{q} - \frac{\lambda \underline{p}}{\bar{p}(1 + \lambda)}\Phi'(\bar{e}),$$

and

$$\Phi(e) \equiv \psi(e) - \psi(e - (\bar{\theta} - \underline{\theta})).$$

(This part of the exercise is from Laffont and Tirole 1986.)

(c) A regulator is responsible for two public utilities ($i = 1, 2$) located in separate geographic areas. Each utility produces a fixed amount of output (normalized at $q = 1$) and has a cost function $C_i = \alpha + \beta_i - e_i$, where α can be interpreted as some shock common to both firms, and β_i is an idiosyncratic shock with β_i independent of β_j (so $\theta_i = \alpha + \beta_i$). Social welfare is

$$u_0 = \sum_{i=1}^{2} [S - (1 + \lambda)(C_i + t_i) + u_i],$$

where t_i is the net transfer paid by the regulator to firm i, $u_i = t_i - \psi(e_i)$ is firm i's rent and S is the social surplus associated with a firm's production. Suppose first that $\beta_i = 0$ is common knowledge, but α is known only to the firms (common shock). Show that by offering the contract

$$t_i = -(C_i - C_j) + \psi(e^*),$$

where $\psi'(e^*) = 1$, to each firm, the regulator does as well as under full information. Explain. Suppose, second, that both α and β_i are random (common and idiosyncratic shock). Show that the regulator's lack of information about α has no welfare consequence as long as the two firms do not collude.

Exercise 7.10

(a)* A seller owns one unit of a good, which she values at c. (The value c can be thought of as the quality of the good.) A buyer may buy the unit from the seller. The seller's valuation is equal to $\underline{c}$ or $\overline{c}$ with equal probabilities (where $\underline{c} < \overline{c}$) and is private information to the seller. The buyer's valuation for the good is $\overline{v}$ if $c = \overline{c}$ and $\underline{v}$ if $c = \underline{c}$, where $\overline{v} > \overline{c}$ and $\underline{v} > \underline{c}$. The buyer thus has no private information. Assume that $(\overline{v} + \underline{v})/2 < \overline{c}$ (which implies that $\overline{c} > \underline{v}$). Show that efficiency is inconsistent with the seller's and the buyer's individual rationality and incentive compatibility. Give two reasons why the Myerson-Satterthwaite inefficiency result (theorem 7.5) cannot be applied here. With the quality interpretation in mind, suppose there are a continuum of sellers and a continuum of buyers. Buyers are homogeneous and have the same valuation for the good (which is either $\overline{v}$ or $\underline{v}$, depending on the quality of the seller's good). Each seller has probability $\frac{1}{2}$ of having a high-quality item (and therefore valuation $\overline{v}$). Qualities are "independent" across sellers. (Ignore technical subtleties concerning a continuum of independent variables.) Show that the inefficiency result carries over. (This is Akerlof's (1970) "lemons problem.")

(b)** Replicate the exchange economy of question a, but in such a way that each seller has private information that is relevant to a single buyer instead of to all buyers. Suppose that there are many duplexes (a continuum of them). In each duplex, there is one inhabitant on the first floor, who owns a smoke alarm, and one inhabitant on the second floor, who owns none. Second-floor residents have higher valuations (v) than first-floor residents (c) for the smoke alarm, but both valuations depend on whether the first-floor resident smokes ($\overline{v}$ and $\overline{c}$) or not ($\underline{v}$ and $\underline{c}$). Half of the first-floor residents smoke and half do not. Construct an efficient, individually rational, and incentive-compatible trading mechanism. (Hint: Construct a mechanism in which trade, if it occurs, occurs at a fixed price.) (Gul and Postlewaite (1988) give general limit efficiency results when an agent's private information is relevant only to a fixed number—one, in the above example—of agents, independent of the total number of agents in the economy.)

Exercise 7.11   A firm's profit is $x = \theta + e$, where e is the (single) manager's effort and θ is a productivity parameter known only to the manager.

θ takes value $\underline{\theta}$ with probability $\underline{p}$ and $\overline{\theta}$ with probability $\overline{p}$. The manager's objective is $u_1 = t - g(e)$, and the shareholders' utility function is $u_0 = x - t - Kq$, where q is the probability of audit and K the cost of auditing. The shareholders offer a contract $\{x(\hat{\theta}), t(\hat{\theta}), q(\hat{\theta})\}$, where $\hat{\theta}$ is the firm's announcement of its productivity parameter. If it announces $\hat{\theta}$, the firm is required to attain profit level, $x(\hat{\theta})$. After production takes place, the shareholders audit with probability $q(\hat{\theta})$. The audit yields a signal $\tilde{\theta} \in \{\underline{\theta}, \overline{\theta}\}$. The probability that the signal is truthful ($\tilde{\theta} = \theta$) is $r \in [\frac{1}{2}, 1]$. If $\tilde{\theta} = \hat{\theta}$, the manager receives $t(\hat{\theta})$. If $\tilde{\theta} \neq \hat{\theta}$, the manager, who is protected by limited liability, receives 0 (if you have time, show that this "maximal punishment" is indeed optimal).

Show that $q(\overline{\theta}) = 0$. Show that (for "K not too big") auditing always occurs for r close to 1 and $\hat{\theta} = \underline{\theta}$, and that when K varies there are three regimes (including one in which the first-best effort is attained). Indicate how $x(\underline{\theta})$ changes with r and K. Explain. (For more on auditing, see Baron and Besanko 1984b and Kofman and Lawarrée 1989.)

References

Akerlof, G. 1970. The market for lemons. *Quarterly Journal of Economics* 89: 488–500.

Arrow, K. 1979. The property rights doctrine and demand revelation under incomplete information. In *Economies and Human Welfare*. Academic Press.

Bagnoli, M., and T. Bergstrom. 1989. Log-concave probability and its applications. Discussion paper 89-23, University of Michigan.

Baron, D. 1985. Noncooperative regulation of a nonlocalized externality. *Rand Journal of Economics* 16: 553–568.

Baron, D., and D. Besanko. 1984a. Regulation and information in a continuing relationship. *Information Economics and Policy* 1: 447–470.

Baron, D., and D. Besanko. 1984b. Regulation, asymmetric information and auditing. *Rand Journal of Economics* 15: 447–470.

Baron, D., and D. Besanko. 1987. Commitment and fairness in a continuing relationship. *Review of Economic Studies* 54: 413–436.

Baron, D., and R. Myerson. 1982. Regulating a monopolist with unknown costs. *Econometrica* 50: 911–930.

Besanko, D., and A. Thakor. 1987. Collateral and rationing: Sorting equilibria in monopolistic and competitive credit markets. *International Economic Review* 28: 671–689.

Bester, H. 1985. Screening vs. rationing in credit markets with imperfect information. *American Economic Review* 75: 850–855.

Caillaud, B., R. Guesnerie, and P. Rey. 1986. Noisy observation in adverse selection models. Mimeo, EHESS, Paris.

Caillaud, B., B. Jullien, and P. Picard. 1990. Publicly announced contracts, private renegotiation and precommitment effects. Mimeo, CEPREMAP, Paris.

Champsaur, P., and J.-C. Rochet. 1989. Multiproduct duopolists. *Econometrica* 57: 533–558.

Clarke, E. 1971. Multipart pricing of public goods. *Public Choice* 8: 19–33.

Coase, R. 1960. The problem of social cost. *Journal of Law and Economics* 3: 1–44.

Cramton, P., R. Gibbons, and P. Klemperer. 1987. Dissolving a partnership efficiently. *Econometrica* 55: 615–632.

Crémer, J., and R. McLean. 1985. Optimal selling strategies under uncertainty for a discriminating monopolist when demands are interdependent. *Econometrica* 53: 345–361.

Crémer, J., and R. McLean. 1988. Full extraction of the surplus in Bayesian and dominant strategy auctions. *Econometrica* 56: 1247–1258.

Crémer, J., and M. Riordan. 1985. A sequential solution to the public goods problem. *Econometrica* 53: 77–84.

Dasgupta, P., P. Hammond, and E. Maskin. 1979. The implementation of social choice rules. *Review of Economic Studies* 46: 185–216.

d'Aspremont, C., J. Crémer, and L. A. Gérard-Varet. 1990a. On the existence of Bayesian and non-Bayesian revelation mechanisms. *Journal of Economic Theory*, forthcoming.

d'Aspremont, C., J. Crémer, and L. A. Gérard-Varet. 1990b. Bayesian implementation of non Pareto-optimal social choice functions. Mimeo, CORE.

d'Aspremont, C., and L. A. Gerard-Varet. 1979. Incentives and incomplete information. *Journal of Public Economics* 11: 25–45.

Demski, J., and D. Sappington. 1984. Optimal incentive contracts with multiple agents. *Journal of Economic Theory* 33: 152–171.

Dewatripont, M. 1989. Renegotiation and information revelation over time: The case of optimal labor contracts. *Quarterly Journal of Economics* 104: 589–620.

Freixas, X., R. Guesnerie, and J. Tirole. 1985. Planning under incomplete information and the ratchet effect. *Review of Economic Studies* 52: 173–192.

Fudenberg, D., and J. Tirole. 1990. Moral hazard and renegotiation in agency contracts. *Econometrica* 58: 1279–1320.

Gal-Or, E. 1989. A common agency with incomplete information. Mimeo, University of Pittsburgh.

Gibbard, A. 1973. Manipulation for voting schemes. *Econometrica* 41: 587–601.

Green, J., and J.-J. Laffont. 1977. Characterization of satisfactory mechanisms for the revelation of preferences for public goods. *Econometrica* 45: 427–438.

Green, J., and J.-J. Laffont. 1979. *Incentives in Public Decision Making*. North-Holland.

Gresik, T., and M. Satterthwaite. 1989. The rate at which a simple market converges to efficiency as the number of traders increases: An asymptotic result for optimal trading mechanisms. *Journal of Economic Theory* 48: 304–332.

Groves, T. 1973. Incentives in teams. *Econometrica* 41: 617–631.

Guesnerie, R., and J.-J. Laffont. 1984. A complete solution to a class of principal-agent problems with an application to the control of a self-managed firm. *Journal of Public Economics* 25: 329–369.

Gul, F., and A. Postlewaite. 1988. Asymptotic efficiency in large exchange economies with asymmetric information. Mimeo, Stanford Graduate School of Business and University of Pennsylvania.

Hadley, G., and M. Kemp. 1971. *Variational Methods in Economics*. North-Holland.

Harris, M., and A. Raviv. 1981. Allocation mechanisms and the design of auctions. *Econometrica* 49: 1477–1500.

Hart, O., and J. Tirole. 1988. Contract renegotiation and Coasian dynamics. *Review of Economic Studies* 55: 509–540.

Holmström, B. 1979. Groves' scheme on restricted domains. *Econometrica* 47: 1137–1144.

Holmström, B., and R. Myerson. 1983. Efficient and durable decision rules with incomplete information. *Econometrica* 51: 1799–1820.

Hurewicz, W. 1958. *Lectures on Ordinary Differential Equations*. MIT Press.

Hurwicz, L. 1972. On informationally decentralized systems. In *Decision and Organization*, ed. M. McGuire and R. Radner. North-Holland.

Johnson, S., J. Pratt, and R. Zeckhauser. 1990. Efficiency despite mutually payoff-relevant private information: The finite case. *Econometrica* 58: 873–900.

Kofman, F., and J. Lawarrée. 1989. Collusion in hierarchical agency. Mimeo, University of California, Berkeley.

Laffont, J.-J. (ed.) 1979. *Aggregation and Revelation of Preferences*. North-Holland.

Laffont, J.-J. (1989). *The Economics of Uncertainty and Information*. MIT Press.

Laffont, J.-J., and E. Maskin. 1979. A differentiable approach to expected utility maximizing mechanisms. In *Aggregation and Revelation of Preferences*, ed. J.-J. Laffont. North-Holland.

Laffont, J.-J., and E. Maskin. 1980. A differential approach to dominant strategy mechanisms. *Econometrica* 48: 1507–1520.

Laffont, J.-J., and E. Maskin. 1982. The theory of incentives: An overview. In *Advances in Economic Theory*, ed. W. Hildenbrand. Cambridge University Press.

Laffont, J.-J., E. Maskin, and J.-C. Rochet. 1987. Optimal non-linear pricing with two-dimensional characteristics. In *Information, Incentives, and Economic Mechanisms*, ed. T. Groves, R. Radner, and S. Reiter. University of Minnesota Press.

Laffont, J.-J., and J. Tirole. 1986. Using cost observation to regulate firms. *Journal of Political Economy* 94: 614–641.

Laffont, J.-J., and J. Tirole. 1987a. Auctioning incentive contracts. *Journal of Political Economy* 95: 921–937.

Laffont, J.-J., and J. Tirole. 1987b. Comparative statics of the optimal dynamic incentives contract. *European Economic Review* 31: 901–926.

Laffont, J.-J., and J. Tirole. 1988. The dynamics of incentive contracts. *Econometrica* 56: 1153–1176.

Laffont, J.-J., and J. Tirole. 1990a. Optimal bypass and creamskimming. *American Economic Review* 80: 1042–1061.

Laffont, J.-J., and J. Tirole. 1990b. Adverse selection and renegotiation in procurement. *Review of Economic Studies* 57: 597–626.

Laffont, J.-J., and J. Tirole. 1990c. Privatization and incentives. Mimeo, Université de Toulouse.

Ledyard, J., and T. Palfrey. 1989. Interim efficient public good provision and cost allocation with limited side payments. Mimeo, California Institute of Technology.

Lewis, T., and D. Sappington. 1989a. Inflexible rules in incentive problems. *American Economic Review* 79: 69–84.

Lewis, T., and D. Sappington. 1989b. Countervailing incentives in agency problems. *Journal of Economic Theory* 49: 294–313.

Ma, C., J. Moore, and S. Turnbull. 1988. Stopping agents from "cheating." *Journal of Economic Theory* 46: 355–372.

Mailath, G., and A. Postlewaite. 1990. Asymmetric-information bargaining problems with many agents. *Review of Economic Studies* 57: 351–368.

Martimort, D. 1990. Multiple principals and asymmetric information. Mimeo, Université de Toulouse.

Maskin, E. 1977. Nash equilibrium and welfare optimality. Mimeo.

Maskin, E. 1981. Randomization in incentive schemes. Mimeo, Harvard University.

Maskin, E., and J. Riley. 1980. Auction design with correlated values. Mimeo, University of California, Los Angeles.

Maskin, E., and J. Riley. 1984a. Monopoly with incomplete information. *Rand Journal of Economics* 15: 171–196.

Maskin, E., and J. Riley. 1984b. Optimal auctions with risk averse buyers. *Econometrica* 52: 1473–1518.

Maskin, E., and J. Tirole. 1989. The principal-agent relationship with an informed principal. II: Common values. *Econometrica*, forthcoming.

Maskin, E., and J. Tirole. 1990. The principal-agent relationship with an informed principal. I: Private Values. *Econometrica* 58: 379–410.

Matthews, S. 1983. Selling to risk-averse buyers with unobservable tastes. *Journal of Economic Theory* 30: 370–400.

Matthews, S. 1984. On the implementability of reduced form auctions. *Econometrica* 52: 1619–1522.

Matthews, S., and J . Moore. 1987. Monopoly provision of quality and warranties: An exploration in the theory of multidimensional screening. *Econometrica* 52: 441–468.

McAfee, P., and J. McMillan. 1987a. Auctions and bidding. *Journal of Economic Literature* 25: 699–738.

McAfee, P., and J. McMillan. 1987b. Government procurement and international trade: Implications of auction theory. Mimeo, University of Western Ontario.

McAfee, P., and J. McMillan. 1988. Multidimensional incentive compatibility and mechanism design. *Journal of Economic Theory* 46: 335–354.

McAfee, P., J. McMillan, and P. Reny. 1989. Extracting the surplus in the common-value auction. Mimeo, University of Western Ontario.

Melumad, N., and S. Reichelstein. 1989. Value of communication in agencies. *Journal of Economic Theory* 47: 334–368.

Milgrom, P. 1987. Auction theory. In *Advances in Economic Theory, Fifth World Congress,* ed. T. Bewley. Cambridge University Press.

Mirrlees, J. 1971. An exploration in the theory of optimum income taxation. *Review of Economic Studies* 38: 175–208.

Mookherjee, D., and S. Reichelstein. 1988. Implementation via augmented revelation mechanisms. Discussion paper 985, Graduate School of Business, Stanford University.

Mookherjee, D., and S. Reichelstein. 1989. Dominant strategy implementation of Bayesian incentive compatible allocation rules. Mimeo, Stanford University.

Moore, J. 1984. Global incentive constraints in auction design. *Econometrica* 52: 1523–1535.

Moore, J. 1985. Optimal labour contracts when workers have a variety of privately observed reservation wages. *Review of Economic Studies* 52: 37–67.

Moore, J. 1988. Contracting between two parties with private information. *Review of Economic Studies* 55: 49–70.

Moore, J. 1990. Implementation, contracts, and renegotiation in environments with symmetric information. In *Advances in Economic Theory, Sixth World Congress,* ed. J.-J. Laffont. Cambridge University Press.

Moore, J., and R. Repullo. 1989. Subgame perfect implementation. *Econometrica* 56: 1191–1220.

Mussa, M., and S. Rosen. 1978. Monopoly and product quality. *Journal of Economic Theory* 18: 301–317.

Myerson, R. 1979. Incentive compatibility and the bargaining problem, *Econometrica* 47: 61–73.

Myerson, R. 1981. Optimal auction design. *Mathematics of Operations Research* 6: 58–73.

Myerson, R. 1983. Mechanism design by an informed principal. *Econometrica* 51: 1767–1797.

Myerson, R., and M. Satterthwaite 1983. Efficient mechanisms for bilateral trading. *Journal of Economic Theory* 28: 265–281.

Palfrey, T., and H. Rosenthal. 1985. Voter participation and strategic uncertainty. *American Political Science Review* 79: 62–78.

Palfrey, T., and S. Srivastava. 1989. Implementation with incomplete information in exchange economies. *Econometrica* 56: 115–134.

Postlewaite, A., and D. Schmeidler. 1986. Implementation in differential information economies. *Journal of Economic Theory* 39: 14–33.

Prekova, A. 1973. On logarithmic concave measures and functions. *Acta Sci. Math. (Szeged)* 34: 335–343.

Riley, J., and W. Samuelson. 1981. Optimal auctions. *American Economic Review* 71: 381–392.

Rob, R. 1989. Pollution claims settlements with private information. *Journal of Economic Theory* 47: 307–333.

Roberts, J. 1976. The incentives for correct revelation of preferences and the number of consumers. *Journal of Public Economics* 6: 359–374.

Roberts, J., and A. Postlewaite. 1976. The incentives for price-taking behavior in large economies. *Econometrica* 44: 115–128.

Rochet, J.-C. 1985. The taxation principle and multi-time Hamilton-Jacobi equations. *Journal of Mathematical Economics* 14: 113–128.

Rockafellar, T. 1970. *Convex Analysis.* Princeton University Press.

Satterthwaite, M., and S. Williams. 1989. Bilateral trade with the sealed-bid k-double auction: Existence and efficiency. *Journal of Economic Theory* 48: 107–133.

Spier, K. 1989. Efficient mechanisms for pretrial bargaining. Mimeo, Harvard University.

Stiglitz, J. 1977. Monopoly, nonlinear pricing, and imperfect information: The insurance market. *Review of Economic Studies* 44: 407–430.

Stole, L. 1990a. Mechanism design under common agency. Mimeo, Massachusetts Institute of Technology.

Stole, L. 1990b. The economics of liquidated damage clauses in contractual environments with private information. Mimeo, Massachusetts Institute of Technology.

Vickrey, W. 1961. Counterspeculation, auctions and competitive sealed tenders. *Journal of Finance* 16: 8–37.

Wilson, R. 1985. Incentive efficiency of double auctions. *Econometrica* 53: 1101–1117.

IV DYNAMIC GAMES OF INCOMPLETE INFORMATION

8 Equilibrium Refinements: Perfect Bayesian Equilibrium, Sequential Equilibrium, and Trembling-Hand Perfection

8.1 Introduction[†]

The concept of subgame perfection, introduced in chapter 3, has no bite in games of incomplete information, even if the players observe one another's actions at the end of each period: Since the players do not know the others' types, the start of a period does not form a well-defined subgame until the players' posterior beliefs are specified, and so we cannot test whether the continuation strategies are a Nash equilibrium.[1]

The complications that incomplete information causes are easiest to see in "signaling games"—leader-follower games in which only the leader has private information. The leader moves first; the follower observes the leader's action, but not the leader's type, before choosing his own action. One example is Spence's (1974) famous model of the job market. In that model, the leader is a worker who knows her productivity and must choose a level of education; the follower, a firm (or a number of firms), observes the worker's education level but not her productivity and then decides what wage to offer her. The spirit of subgame perfection in this model is that, for any education level the worker chooses, the continuation play—that is, the offered wage—should be "reasonable" in the sense of being consistent with equilibrium play in the continuation game. Now, the reasonable wage to offer will typically depend on the firm's beliefs about the worker's productivity, which in turn can depend on the worker's observed level of education. If this level is one to which the equilibrium assigns positive probability, the posterior distribution of the worker's productivity can be computed using Bayes' rule. However, Bayes' rule does not determine the posterior distribution over productivity after the observation of an education level to which the equilibrium assigns probability 0, and the reasonable wage will depend on which posterior distribution is specified. Thus, in order to extend subgame perfection to these games, we will need to specify how players update their beliefs about their opponents' types after an observation that has prior probability 0.

This chapter starts by developing two solution concepts that extend subgame perfection to games of incomplete information: "perfect Bayesian equilibrium" and Kreps and Wilson's (1982a) "sequential equilibrium." Perfect Bayesian equilibrium results from combining the ideas of subgame perfection, Bayesian equilibrium, and Bayesian inference: Strategies are required to yield a Bayesian equilibrium in every "continuation game" given the posterior beliefs of the players, and the beliefs are required to be updated in accordance with Bayes' law whenever it is applicable. Sequential equilibrium is similar, but it imposes more restrictions on the way players update their beliefs. In the signaling games described above,

---

1. Formally the only proper subgame of a game of incomplete information is the whole game, so any Nash equilibrium is subgame perfect.

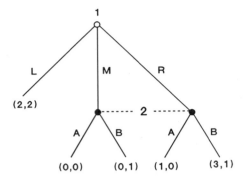

Figure 8.1

these two concepts are identical, and they place very weak restrictions on beliefs following probability-0 events: Any posterior beliefs that assign probability 1 to the support of the prior distribution of types are allowed. In more complex games, the two concepts can impose more restrictions on the allowed beliefs, and thus more restrictions on which equilibria are reasonable.

Since incomplete information is modeled as a kind of imperfect information (Harsanyi 1967–68), it should not be surprising that similar issues concerning out-of-equilibrium beliefs arise in games where the information is complete but imperfect. What matters is that some players' actions can convey information to other players; this information can be anything that one player has observed and the other has not, including the players' own past actions. The simple example illustrated in figure 8.1 is due to Selten (1975). In this game, player 1 has three actions: L, M, and R. If he plays L, the game ends, with payoffs (2, 2). If he plays M or R, then player 2 must choose between A and B, and when making this choice player 2 does not know whether player 1 chose M or R. (This is why the dashed lines connect the nodes following M and R—in the terminology of section 3.3, they belong to the same information set for player 2.) If player 1 chooses M and player 2 chooses A, the payoffs are (0, 0); the payoffs to (M, B) are (0, 1), the payoffs to (R, A) are (1, 0), and the payoffs to (R, B) are (3, 1).

This game has two pure-strategy Nash equilibria, (L, A) and (R, B), both of which are subgame perfect. (To see that (L, A) is subgame perfect, note that since player 2 does not know player 1's action when choosing between A and B, we cannot test whether playing A is part of a "Nash equilibrium" from that point on. Any mixed-strategy profile where player 1 plays L and player 2 plays A with probability at least $\frac{1}{2}$ is a Nash equilibrium as well.) Note, though, that for any specification of player 2's "beliefs" about the relative probability of M and R when player 1 deviates and does not play L, player 2's optimal action is to play B, so that playing A in this game is analogous to offering the worker a wage that is not reasonable for any posterior beliefs about her productivity.

The simple version of perfect Bayesian equilibrium we develop in this chapter is limited to multi-stage games with observed actions and incomplete information, which we will simply call "multi-stage games" in this chapter. In contrast, sequential equilibrium is defined for general games and does rule out the equilibrium (L, A) in figure 8.1. This equilibrium is also ruled out by Selten's (1975) concept of "trembling-hand perfection," which historically preceded Kreps and Wilson's sequential equilibrium, and is quite closely related to it, as both papers develop refinements by considering perturbed games in which players "tremble" and play suboptimal actions with vanishingly small probability.

We reverse the historical order and discuss sequential equilibrium before trembling-hand perfection, because sequential equilibrium emphasizes the formation of the players' beliefs, and we find this approach easier to explain. Another way our development may be idiosyncratic is that, while most of the literature on refinements considers general extensive forms, we will begin by studying multi-stage games with observed actions, where the only relevant private information is each player's knowledge of his own type. This is the kind of private information most frequently encountered in the economics literature, and it is prominent in the applications of chapters 9 and 10.[2,3]

Section 8.2 introduces the concept of perfect Bayesian equilibrium (PBE). Subsection 8.2.1 begins with the special case of PBE in signaling games. Subsection 8.2.2 gives applications to a game of predatory pricing (inspired by Kreps and Wilson (1982b) and Milgrom and Roberts (1982b)) and to Spence's model of job-market signaling; readers familiar with these or with similar examples will want to skip this subsection. Subsection 8.2.3 extends PBE to multi-stage games, and applies it to a repeated version of the public-good game of example 6.1.

PBE imposes more restrictions on beliefs than Bayes' rule alone, as it imposes some restrictions on beliefs after probability-0 events. Specifically, when initial beliefs are that types are independent, PBE requires that the posterior beliefs be that types are independent, that any two players have the same beliefs about the type of a third, and that if player i deviates and player j does not then beliefs about player j are updated in accordance with Bayes' rule.

2. This is not to say that no other games are economically relevant. For example, situations of moral hazard do not correspond to multi-stage games with observed actions.
3. Many early models of uncertainty and information were dynamic in nature and made some implicit use of perfect Bayesian equilibrium. Examples include the disarmament game of Aumann and Machler (1966), the market games of Akerlof (1970) and Spence (1974), and Ortega-Reichert's (1967) analysis of repeated first-bid auctions. The first formal application of the idea was that of Milgrom and Roberts (1982a), which was followed by those of Kreps and Wilson (1982b) and Milgrom and Roberts (1982b).

Section 8.3 develops the concept of sequential equilibrium, which is defined for general extensive-form games and which places even more restrictions on beliefs after probability-0 events than PBE does. In a sequential equilibrium, players' beliefs are as if there were a small probability of a "tremble" or mistake at each information set, with the trembles at each information set being statistically independent of trembles at the others and with the probability of each tremble depending only on the information available at that information set. We discuss the desiderata that Kreps and Wilson used to motivate their concept, and compare it with PBE in the special case of multi-stage games. As we explain, sequential equilibrium is stronger than PBE unless the game has at most two periods or each player has at most two types. To illustrate the sequential-equilibrium definition, we then develop an "extended" version of PBE that is equivalent to sequential equilibrium in general multi-stage games with observed actions.

Section 8.4 describes related (and historically prior) refinements based on the strategic form. The focus on the strategic form does not imply a neglect of extensive-form considerations such as perfection. Selten's original (1975) idea was to introduce small trembles so that all pure strategies have positive probability, and to require that players optimize (subject to the constraint that they tremble with small probability) against their opponents' garbled strategies. Because all outcomes have positive probability, the issue of perfection—that in a Nash equilibrium a player can costlessly play a crazy strategy in some unreached subgame—does not arise. A "trembling-hand perfect equilibrium" is a limit of Nash equilibria with trembles as the trembles tend to 0. The sets of trembling-hand perfect and sequential equilibria coincide for almost all games. Section 8.4 then describes a refinement of trembling-hand perfect equilibrium due to Myerson (1978). A "proper equilibrium" requires that a player tremble less on strategies that are worse responses. Chapter 11 discusses stronger refinements related to the idea of "forward induction."

Finally, before introducing more equilibrium refinements, we should note that, since the concepts of this chapter strengthen subgame perfection, they are subject to the reservations we expressed in chapter 3, as well as to other reservations we will develop along the way. In particular, all these refinements suppose that all players expect an opponent to continue to play according to the equilibrium strategies even after that opponent deviates from the equilibrium path.

8.2 Perfect Bayesian Equilibrium in Multi-Stage Games of Incomplete Information[†]

8.2.1 The Basic Signaling Game

Signaling games are the simplest kind of game in which the issues of updating and perfection both arise. In these games, there are two players.

Player 1 is the leader (also called the sender, because he sends a signal), and player 2 is the follower (or receiver). Player 1 has private information about his type θ in Θ and chooses action a_1 in A_1. (We delete the subscript on player 1's type, as this will not lead to confusion.) Player 2, whose type is common knowledge for simplicity, observes a_1 and chooses a_2 in A_2. The spaces of mixed actions are $\mathscr{A}_1$ and $\mathscr{A}_2$ with elements α_1 and α_2. Player i's payoff is denoted $u_i(\alpha_1, \alpha_2, \theta)$. Before the game begins, it is common knowledge that player 2 has prior beliefs p about player 1's type. A strategy for player 1 prescribes a probability distribution $\sigma_1(\cdot|\theta)$ over actions a_1 for each type θ. A strategy for player 2 prescribes a probability distribution $\sigma_2(\cdot|a_1)$ over actions a_2 for each action a_1. Type θ's payoff to strategy $\sigma_1(\cdot|\theta)$ when player 2 plays $\sigma_2(\cdot|a_1)$ is

$$u_1(\sigma_1, \sigma_2, \theta) = \sum_{a_1} \sum_{a_2} \sigma_1(a_1|\theta)\sigma_2(a_2|a_1)u_1(a_1, a_2, \theta).$$

Players 2's (*ex ante*) payoff to strategy $\sigma_2(\cdot|a_1)$ when player 1 plays $\sigma_1(\cdot|\theta)$ is

$$\sum_{\theta} p(\theta)\left(\sum_{a_1} \sum_{a_2} \sigma_1(a_1|\theta)\sigma_2(a_2|a_1)u_2(a_1, a_2, \theta)\right).$$

Player 2, who observes player 1's move before choosing her own action, should update her beliefs about θ and base her choice of a_2 on the posterior distribution $\mu(\cdot|a_1)$ over Θ. How is this posterior formed? In a Bayesian equilibrium, player 1's action can depend on his type. Let $\sigma_1^*(\cdot|\theta)$ denote this strategy. Knowing σ_1^* and observing a_1, player 2 can use Bayes' rule to update $p(\cdot)$ into $\mu(\cdot|a_1)$. The natural extension of the subgame-perfect equilibrium to the signaling game is the perfect Bayesian equilibrium, which requires that player 2 maximize her payoff conditional on a_1 for each a_1, where the conditional payoff to strategy $\sigma_2(\cdot|a_1)$ is

$$\sum_{\theta} \mu(\theta|a_1)u_2(a_1, \sigma_2(\cdot|a_1), \theta) = \sum_{\theta} \sum_{a_2} \mu(\theta|a_1)\sigma_2(a_2|a_1)u_2(a_1, a_2, \theta).$$

Definition 8.1 A *perfect Bayesian equilibrium* (PBE) of a signaling game is a strategy profile σ^* and posterior beliefs $\mu(\cdot|a_1)$ such that:

(P$_1$) $\forall \theta, \sigma_1^*(\cdot|\theta) \in \arg \max_{\alpha_1} u_1(\alpha_1, \alpha_2, \theta)$,

(P$_2$) $\forall a_1, \sigma_2^*(\cdot|a_1) \in \arg \max_{\alpha_2} \sum_{\theta} \mu(\theta|a_1)u_2(a_1, \alpha_2, \theta)$,

and

(B) $\mu(\theta|a_1) = p(\theta)\sigma_1^*(a_1|\theta) \bigg/ \sum_{\theta' \in \Theta} p(\theta')\sigma_1^*(a_1|\theta')$

  if $\sum_{\theta' \in \Theta} p(\theta')\sigma_1^*(a_1|\theta') > 0$, and $\mu(\cdot \,|\, a_1)$ is any probability distribution on Θ

$$\text{if } \sum_{\theta' \in \Theta} p(\theta')\sigma_1^*(a_1 \,|\, \theta') = 0.$$

P_1 and P_2 are the perfection conditions. P_1 says that player 1 takes into account the effect of a_1 on player 2's action[4]; P_2 states that player 2 reacts optimally to player 1's action given her posterior beliefs about θ. B corresponds to the application of Bayes' rule. Note that if a_1 is not part of player 1's optimal strategy for some type, observing a_1 is a probability-0 event, and Bayes' rule does not pin down posterior beliefs. *Any* posterior beliefs $\mu(\cdot \,|\, a_1)$ are then admissible, and so any action a_2 can be played that is a best response for some beliefs. (This means that the only actions excluded are those which are dominated given that a_1 is played.) Indeed, the purpose of the refinements of the perfect Bayesian equilibrium concept is to put some restrictions on these posterior beliefs. As we will see in section 8.3, the concept of PBE defined here is equivalent to sequential equilibrium for the class of signaling games.

Thus, a PBE is simply a set of strategies and beliefs such that, at any stage of the game, strategies are optimal given the beliefs, and the beliefs are obtained from equilibrium strategies and observed actions using Bayes' rule.

Note the link between strategies and beliefs: The beliefs are consistent with the strategies, which are optimal given the beliefs. Because of this circularity, PBE cannot be determined by backward induction when there is incomplete information, even if players move one at a time. (Recall that, with perfect information, perfect equilibria can be determined by backward induction.)

8.2.2 Examples of Signaling Games

To help build intuition, we will analyze two examples of signaling games in a fair bit of detail. Readers already familiar with the ideas of separating and pooling equilibrium in, say, the Milgrom-Roberts limit-pricing model should probably skip to subsection 8.2.3.

Example 8.1: Two-Period Reputation Game
The following is a much-simplified version of the Kreps-Wilson (1982b)–Milgrom-Roberts (1982b) reputation model. There are two firms ($i = 1, 2$). In period 1, both firms are in the market. Only firm 1 (the "incumbent") takes an action a_1. The action space has two elements: "prey" and "accommodate." Firm 2 (the "entrant") has profit D_2 if firm 1 accommodates and

---

4. Recall that a mixed strategy is a best response if all actions in its support maximize the player's payoff, so condition P_1 is equivalent to

$$a_1 \in \text{support } \sigma_1^*(\cdot \,|\, \theta) \Leftrightarrow a_1 \in \arg\max_{\tilde{a}_1} u_1(\tilde{a}_1, \sigma_2^*(\cdot \,|\, \tilde{a}_1), \theta).$$

P_2 if firm 1 preys, with $D_2 > 0 > P_2$. Firm 1 has one of two potential types: "sane" and "crazy." A sane firm 1 makes D_1 when it accommodates and P_1 when it preys, where $D_1 > P_1$. Thus, a sane firm prefers to accommodate rather than to prey. However, it would prefer to be a monopoly, in which case it would make $M_1 > D_1$ per period. When crazy, firm 1 enjoys predation and thus preys (its utility function is such that it is always worth preying). Let p (respectively, $1 - p$) denote the prior probability that firm 1 is sane (respectively, crazy).

In period 2, only firm 2 chooses an action a_2. This action can take two values: "stay" and "exit." If firm 2 stays, it obtains payoff D_2 if firm 1 is actually sane and P_2 if it is crazy; if firm 2 exits, it obtains payoff 0. The idea is that, unless it is crazy, firm 1 will not prey in the second period, because there is no point to building or keeping a reputation at the end. (This assumption can be derived more formally from the description of the second-period competition.) The sane firm gets D_1 if firm 2 stays and M_1 if firm 2 exits. We let δ denote the discount factor between the two periods.

We presumed that the crazy type always preys. The interesting thing to study is thus the sane type's behavior. From a static point of view, it would want to accommodate in the first period; however, by preying it might convince firm 2 that it is of the crazy type, and thus induce exit (as $P_2 < 0$) and increase its second-period profit.

Let us first start with a taxonomy of potential perfect Bayesian equilibria. A *separating equilibrium* is an equilibrium in which the two types of firm 1 choose two different actions in the first period. Here, this means that the sane type chooses to accommodate. Note that in a separating equilibrium firm 2 has complete information in the second period:

$$\mu(\theta = \text{sane} \,|\, a_1 = \text{accommodate}) = 1$$

and

$$\mu(\theta = \text{crazy} \,|\, a_1 = \text{prey}) = 1.$$

A *pooling equilibrium* is an equilibrium in which firm 1's two types choose the same action in the first period. Here, this means that the sane type preys. In a pooling equilibrium firm 2 does not update its beliefs when observing the equilibrium action:

$$\mu(\theta = \text{sane} \,|\, a_1 = \text{prey}) = p.$$

There can also be *hybrid* or *semi-separating equilibria*. In the reputation game, the sane type may randomize between preying and accommodating, i.e., between pooling and separating. The posterior beliefs are then

$$\mu(\theta = \text{sane} \,|\, a_1 = \text{prey}) \in (0, p)$$

and

$\mu(\theta = \text{sane} \mid a_1 = \text{accommodate}) = 1.$

When do separating equilibria exist? In these equilibria, the sane type accommodates, thus revealing its type, and its payoff is $D_1(1 + \delta)$. (Firm 2 stays in because it expects $D_2 > 0$ in the second period.) If the sane type preyed, it would convince firm 2 that it was crazy and would obtain $P_1 + \delta M_1$. Thus, a necessary condition for the existence of a separating equilibrium is

$$\delta(M_1 - D_1) \leq (D_1 - P_1). \tag{8.1}$$

Conversely, suppose that equation 8.1 is satisfied, and consider the following strategies and beliefs: The sane incumbent accommodates, and the entrant (correctly) infers that the incumbent is sane when observing accommodation and therefore stays; the crazy incumbent preys and the entrant (correctly) infers that the incumbent is crazy when observing predation and therefore exits. Clearly, these strategies and beliefs form a PBE, so equation 8.1 is sufficient as well as necessary for the existence of a separating equilibrium.

In a pooling equilibrium, both types of incumbent prey, so the entrant's posterior beliefs are the same as its prior when predation is observed. Since predation is costly for the sane incumbent, it will prey only if doing so induces a positive probability of exit. Thus, a necessary condition for a pooling equilibrium is that the entrant's expected second-period payoff if it stays in is negative; that is,

$$pD_2 + (1 - p)P_2 \leq 0. \tag{8.2}$$

Conversely, assume that equation 8.2 holds, and consider the following strategies and beliefs: Both types prey; the entrant has posterior beliefs $\mu(\text{sane} \mid a_1 = \text{prey}) = p$ and $\mu(\text{sane} \mid a_1 = \text{accommodate}) = 1$, and stays in if and only if accommodation is observed. The sane type's equilibrium profit is $P_1 + \delta M_1$; it would receive $D_1(1 + \delta)$ from accommodation. Thus, if equation 8.1 is violated, the proposed strategies and beliefs form a pooling PBE. (Note that if equation 8.2 is satisfied with equality, there exists a *continuum* of such equilibria.[5])

We leave it to the reader to check that if both equation 8.1 and equation 8.2 are violated, the unique equilibrium is a hybrid PBE (with the entrant randomizing when observing predation and the sane incumbent randomizing between preying and accommodating[6]).

---

5. When $pD_2 + (1 - p)P_2 = 0$, any probability $x \geq \bar{x}$ that the entrant exits induces the sane incumbent to prey, where $\delta \bar{x}(M_1 - D_1) = D_1 - P_1$, so $0 < \bar{x} < 1$.

6. In this equilibrium, the entrant exits with the probability $\bar{x}$ defined in the previous footnote, and the sane incumbent preys with probability $\bar{y}$ such that $\tilde{p} = p\bar{y}/(p\bar{y} + 1 - p)$, where $\tilde{p}D_2 + (1 - \tilde{p})P_2 = 0$.

Remark The (generic) uniqueness of the PBE in this model is due to the fact that the "strong" type (the crazy incumbent) is assumed to always prey. Thus, predation is not a probability-0 event and, furthermore, if the sane type accommodates with positive probability, then accommodation reveals that player 1 is sane. The next example illustrates a more complex and a more common structure, for which refinements of the PBE concept are required if one insists on uniqueness of equilibrium.

Example 8.2: Spence's Education Game

Spence (1974) developed the following model of the choice of education level: Player 1 (a worker) chooses a level of education $a_1 \geq 0$. His private cost of investing a_1 units in education is a_1/θ, where θ is his type or "ability." The worker's productivity in a firm is equal to θ (to simplify, it is not affected by education). Player 2's (the firm's) objective is to minimize the quadratic difference of the wage a_2 offered to player 1 and player 1's productivity, so player 2 offers the expected productivity $a_2(a_1) = E(\theta|a_1)$ in equilibrium. (Alternatively, we could suppose that there are several firms who make simultaneous wage offers.) Player 1's objective function is $a_2 - a_1/\theta$.

Player 1 has two possible types, θ' and θ'', with $0 < \theta' < \theta''$; the probabilities of these types are p' and p'', respectively. Player 1 knows θ, but player 2 does not.

Let σ_1' and σ_1'' denote the equilibrium strategies of types θ' and θ''. Note that if $a_1' \in$ support σ_1' and $a_1'' \in$ support σ_1'', then $a_1' \leq a_1''$.[7] For, from equilibrium behavior,

$$a_2(a_1') - a_1'/\theta' \geq a_2(a_1'') - a_1''/\theta' \tag{8.3}$$

and

$$a_2(a_1'') - a_1''/\theta'' \geq a_2(a_1') - a_1'/\theta'' \tag{8.4}$$

Adding up these two inequalities yields $(1/\theta' - 1/\theta'')(a_1'' - a_1') \geq 0$, or $a_1' \leq a_1''$.

As in the reputation game of example 8.1, we can distinguish among separating, pooling, and hybrid equilibria.

In a *separating equilibrium*, the low-productivity worker reveals his type and therefore receives a wage equal to θ'. He therefore must choose $a_1' = 0$; if he did not, he would strictly gain by choosing $a_1' = 0$, because he would save on the education cost and would receive a wage which is necessarily a convex combination of θ' and θ'' and therefore is at least equal to θ'. Let $a_1'' > 0$ denote the equilibrium action of type θ'' (note that in a separating equilibrium type θ'' cannot play a mixed strategy, because all his equilibrium actions yield the same wage θ'' and therefore type θ'' prefers the

---

7. This monotonicity property is a special case of the general result in theorem 7.2.

one with the lowest education level). In order for $(a_1' = 0, a_1'')$ to be part of a separating equilibrium, it must be the case that type θ' does not prefer a_1'' to a_1':

$$\theta' \geq \theta'' - a_1''/\theta'$$

or

$$a_1'' \geq \theta'(\theta'' - \theta'). \tag{8.5}$$

Similarly, type θ'' cannot prefer a_1' to a_1'':

$$a_1'' \leq \theta''(\theta'' - \theta'). \tag{8.6}$$

Hence, $\theta'(\theta'' - \theta') \leq a_1'' \leq \theta''(\theta'' - \theta')$.

Conversely, suppose that a_1'' belongs to this interval. Consider the beliefs

$$\{\mu(\theta'|a_1) = 1 \text{ if } a_1 \neq a_1'', \mu(\theta'|a_1'') = 0\}.$$

Clearly, the two types prefer $a_1 = 0$ to any $a_1 \notin \{0, a_1''\}$, because any such a_1 yields the low wage θ' anyway. Because θ' prefers 0 to a_1'' (equation 8.5) and θ'' prefers a_1'' to 0 (equation 8.6), we have a continuum of separating equilibria. This continuum illustrates how the leeway in specifying off-the-equilibrium-path beliefs leads to a multiplicity of equilibria. We used the "pessimistic" beliefs under which any action other than a_1'' convinces player 2 that player 1 is the low type θ'. However, the separating equilibria can be supported by less extreme posterior beliefs. In particular, we can specify that $\mu(\theta'|a_1) = 0$ for all $a_1 \geq a_1''$, so that the posterior beliefs are monotonic in a_1, and we can use beliefs $\mu(\theta'|a_1)$ that are continuous in a_1.[8]

In a *pooling equilibrium*, both types choose the same action: $\tilde{a}_1 = a_1' = a_1''$. The wage is then $a_2(\tilde{a}_1) = p'\theta' + p''\theta''$. The easiest way to support $\tilde{a}_1$ as a pooling outcome is to assign pessimistic beliefs $\mu(\theta'|a_1) = 1$ to any action $a_1 \neq \tilde{a}_1$, as this minimizes both types' temptation to deviate. Therefore, $\tilde{a}_1$ is a pooling-equilibrium education level if and only if, for each θ,

$$\theta' \leq p'\theta' + p''\theta'' - \tilde{a}_1/\theta.$$

Since $\theta' < \theta''$, type θ' is the most tempted to deviate to $a_1 = 0$, to minimize education costs, and the binding constraint is

8. It is interesting to note that, of this continuum of separating equilibria, all but the "least-cost" one, where $a_1'' = \theta'(\theta'' - \theta') \equiv a_1^*$, can be eliminated by the following argument: No matter what education level player 1 chooses, player 2 should never choose any wage outside the interval $[\theta', \theta'']$. If player 1 realizes this, then type θ' will never choose any $a_1 > a_1^*$. If player 2 realizes that this is so, then she should respond to $a_1 > a_1^*$ with wage θ''; in that case, type θ'' will never choose $a_1 > a_1^*$. (This argument can be viewed either as an extension of the concept of iterated conditional dominance defined in chapter 4 or as an implication of iterated weak dominance.) However, with three types, θ', θ'', and θ''', $0 < \theta' < \theta'' < \theta'''$, this argument has little force. If as before we let a_1^* denote the education level where type θ' is indifferent between $(0, \theta')$ and (a_1^*, θ''), then even when the out-of-equilibrium wages are restricted to the interval $[\theta'', \theta''']$ type θ' may still be willing to choose $a_1 > a_1^*$.

$$\tilde{a}_1 \le p''\theta'(\theta'' - \theta'), \tag{8.7}$$

so there is also a continuum of pooling equilibria. We leave it to the reader to derive the set of hybrid equilibria.

8.2.3   Multi-Stage Games with Observed Actions and Incomplete Information[††]

We now consider a more general class of games we call "multi-stage games with observed actions and incomplete information." Each player i has a type θ_i in a finite set Θ_i. Letting $\theta \equiv (\theta_1, \ldots, \theta_I)$, we assume for the time being that types are independent, so that the prior distribution p is the product of marginals; that is,

$$p(\theta) = \prod_{i=1}^{I} p_i(\theta_i),$$

where $p_i(\theta_i)$ is the probability that player i's type is θ_i. At the beginning of the game, each player learns his type but is given no information about his opponents' types.

As in the multi-stage games of chapters 4, 5, and 13, these games are played in periods $t = 0, 1, 2, \ldots, T$, with the property that, at each period t, all players simultaneously choose an action that is revealed at the end of the period. (Recall that the set of feasible actions can be dependent on time and history, so that games with sequential moves such as the signaling game are included.) Players never receive additional observations of θ. For notational simplicity, we assume that each player's action set at each date is type-independent. Let $a_i^t \in A_i(h^t)$ denote player i's date-t action, $a^t = (a_1^t, \ldots, a_I^t)$ the vector of date-t actions, and let $h^t = (a^0, \ldots, a^{t-1})$ denote history at the beginning of date t. A behavior strategy σ_i maps the set of possible histories and types into the action spaces: $\sigma_i(a_i | h^t, \theta_i)$ is the probability of a_i given h^t and θ_i. Player i's payoff is $u_i(h^{T+1}, \theta)$.

To extend the spirit of subgame perfection to these games, we would like to require that the strategies yield a Bayesian Nash equilibrium, not only for the whole game, but also for the "continuation games" starting in each period t after every possible history h^t. Of course, these continuation games are not "proper subgames" because they do not stem from a singleton information set. Thus, to make the continuation games into true games we must specify the players' beliefs at the start of each continuation game. We denote player i's conditional probability that his opponents' types are θ_{-i} by $\mu_i(\theta_{-i} | \theta_i, h^t)$, and assume that it is defined for all players i, dates t, histories h^t, and types θ_i.

What restrictions should be imposed on player i's beliefs? Economic applications of incomplete-information games with independent types have typically made the following assumptions either explicitly or implicitly:

B(i)   Posterior beliefs are independent, and all types of player i have the same beliefs: For all θ, t, and h^t,

$$\mu_i(\theta_{-i} | \theta_i, h^t) = \prod_{j \neq i} \mu_i(\theta_j | h^t).$$

B(i) requires that even unexpected observations do not make player i believe that his opponents' types are correlated.

B(ii)   Bayes' rule is used to update beliefs from $\mu_i(\theta_j | h^t)$ to $\mu_i(\theta_j | h^{t+1})$ whenever possible: For all i, j, h^t, and $a_j^t \in A_j(h^t)$, if there exists $\hat{\theta}_j$ with $\mu_i(\hat{\theta}_j | h^t) > 0$ and $\sigma_j(a_j^t | h^t, \hat{\theta}_j) > 0$ (that is, player i assigns a_j^t positive probability given h^t), then, for all θ_j,

$$\mu_i(\theta_j | (h^t, a^t)) = \frac{\mu_i(\theta_j | h^t)\sigma_j(a_j^t | h^t, \theta_j)}{\sum_{\tilde{\theta}_j} \mu_i(\tilde{\theta}_j | h^t)\sigma_j(a_j^t | h^t, \tilde{\theta}_j)}.$$

B(ii) is stronger than simply using Bayes' rule in the usual fashion, as it applies to updating from period t to period $t + 1$ when the history h^t at period t has probability 0, and to beliefs about player j when h^t has positive probability and some player $k \neq j$ chooses a probability-0 action at date t. The motivation for this requirement is that if $\mu_i(\cdot | h^t)$ represents player i's beliefs given h^t, and nothing "surprising" occurs at t, then player i should use Bayes' rule to form his beliefs in period $t + 1$.

Note that B(ii) does not restrict the way beliefs about player j are updated if player j's period-t action had conditional probability 0.

The next condition says that even if player j does deviate at period t, the updating process should not be influenced by the actions of other players.

B(iii)   For all h^t, i, j, θ_j, a^t, and $\hat{a}^t$,

$$\mu_i(\theta_j | (h^t, a^t)) = \mu_i(\theta_j | (h^t, \hat{a}^t)) \text{ if } a_j^t = \hat{a}_j^t.$$

This condition might be called "no signaling what you don't know," since players $k \neq j$ have no information about j's type not already known to player i.

Finally, most applications assume further that when types are independent players i and j should have the same beliefs about the type of a third player k. This restriction is defended as being in the spirit of equilibrium analysis, since equilibrium supposes that players have the same beliefs about each other's strategies.

B(iv)   For all h^t, θ_k, and $i \neq j \neq k$,

$$\mu_i(\theta_k | h^t) = \mu_j(\theta_k | h^t) = \mu(\theta_k | h^t).$$

This condition implies that the posterior beliefs are consistent with a common joint distribution on Θ given h^t with

$$\mu(\theta_{-i} | h^t)\mu(\theta_i | h^t) = \mu(\theta | h^t).$$

Section 8.3 gives an example in which this restriction reduces the set of equilibrium outcomes. Although it is a standard assumption, we find it the least compelling of the four.

With a strategy σ and beliefs μ satisfying B(i)–B(iv), the natural way to extend subgame-perfect equilibrium is to require that for any t and h^t the strategies from h^t on are a Bayesian equilibrium of the continuation game. Formally, given probability distribution q and history h^t, let $u_i(\sigma|h^t, \theta_i, q)$ be type θ_i's expected payoff under profile σ conditional on reaching h^t. The relevant condition is then as follows:

(P)   For each player i, type θ_i, player i's alternative strategy σ_i', and history h^t,

$$u_i(\sigma|h^t, \theta_i, \mu(\cdot|h^t)) \geq u_i((\sigma_i', \sigma_{-i})|h^t, \theta_i, \mu(\cdot|h^t)).$$

Definition 8.2   A *perfect Bayesian equilibrium* is a (σ, μ) that satisfies P and B(i)–B(iv).

We now give an example of an application of the PBE concept. Other simple examples can be found in sections 9.1 and 10.1.

Example 8.3: The Repeated Public-Good Game

To illustrate the concept of PBE, we analyze the twice-repeated version of the public-good game studied in section 6.2. There are two players, $i = 1, 2$. In each period, $t = 0, 1$, players decide simultaneously whether to contribute to the period-t public good, and contributing is a 0-1 decision. In a given period, each player derives a benefit of 1 if at least one of them provides the public good and 0 if none does; player i's cost of contributing in a period is c_i and is the same in both periods. Per-period payoffs are depicted in figure 6.4. We assume that payoffs are discounted so that a player's objective function is the sum of his first-period payoff plus δ times his second-period payoff, where $0 < \delta < 1$. Though the benefits of the public good—1 each—are common knowledge, each player's cost is known only to that player. However, both players believe that the c_i are drawn independently from the same continuous and strictly increasing cumulative distribution function $P(\cdot)$ on $[0, \bar{c}]$, where $\bar{c} > 1$.

From chapter 6, we know that if there is a unique solution to the equation $c^* = 1 - P(1 - P(c^*))$ then the single-period version of the game has a unique Bayesian equilibrium, and c^* is (also) given by the equation $c^* = 1 - P(c^*)$ (the cost of contributing is equal to the probability that one's opponent won't contribute). Types $c_i \leq c^*$ contribute and the other types do not contribute.

In the repeated version of the game, the action space for each player is $\{0, 1\}$ in each period. A strategy for player i is a pair consisting of $\sigma_i^0(1|c_i)$ (player i's probability of contributing in the first period when his cost is c_i) and $\sigma_i^1(1|h^1, c_i)$ (the probability that player i contributes in the second period when his cost is c_i and when the history is $h^1 \in \{00, 01, 10, 11\}$).

Exercise 8.1 asks you to show that, in any PBE, there exists a cutoff cost $\hat{c}_i$ for each player i such that player i contributes in the first period if and only if $c_i \leq \hat{c}_i$, and also to show that $0 < \hat{c}_i < 1$. We now look for a *symmetric* PBE, where $\hat{c}_1 = \hat{c}_2 = \hat{c}$. We start by solving the second-period Bayesian equilibrium given the posterior beliefs, which are determined by the equilibrium strategies and the first-period outcome.

Neither player contributed.   Both players have learned that their opponent's cost exceeds $\hat{c}$. Posterior cumulative beliefs are thus the truncated beliefs

$$P(c_i \mid 00) = \frac{P(c_i) - P(\hat{c})}{1 - P(\hat{c})}$$

for $c_i \in [\hat{c}, \overline{c}]$, and

$$P(c_i \mid 00) = 0$$

for $c_i \leq \hat{c}$. In a symmetric second-period equilibrium, each player i contributes if and only if $\hat{c} \leq c_i \leq \check{c}$ (from section 6.2, we know that a Bayesian equilibrium in period 2 involves a cutoff rule for each player). The cutoff cost $\check{c}$ is equal to the probability,

$$\frac{1 - P(\check{c})}{1 - P(\hat{c})},$$

that the opponent does not contribute. Note that $\hat{c} < \check{c} < 1$. We will later use the result that, because type $\hat{c}$ contributes in period 2 if no one has contributed in period 1, his second-period utility is $v^{00}(\hat{c}) = 1 - \hat{c}$.

Both players contributed.   The posterior cumulative distribution is then

$$P(c_i \mid 11) = \frac{P(c_i)}{P(\hat{c})}$$

for $c_i \in [0, \hat{c}]$, and

$$P(c_i \mid 11) = 1$$

for $c_i \in [\hat{c}, \overline{c}]$. In a symmetric second-period equilibrium, each player i contributes if and only if $c_i \leq \tilde{c}$, where $0 < \tilde{c} < \hat{c}$. Each player's cutoff cost is equal to the conditional probability that his opponent does not contribute:

$$\tilde{c} = \frac{P(\hat{c}) - P(\tilde{c})}{P(\hat{c})}. \tag{8.8}$$

Note in particular that type $\hat{c}$ does not contribute, so that his second-period utility is $v^{11}(\hat{c}) = P(\tilde{c})/P(\hat{c})$.

Only one player contributed. Suppose player i contributed in period 0 and player j did not. Hence, $c_i \le \hat{c}$ and $c_j \ge \hat{c}$. One equilibrium in period 1 has player i contribute (recall that $\hat{c} < 1$) and player j not contribute, and this is the equilibrium we specify. (For some distributions—e.g., $P(\cdot)$ uniform on $[0, 2]$—this equilibrium is unique.[9]) The second-period utilities of type $\hat{c}$ are thus $v^{10}(\hat{c}) = 1 - \hat{c}$ and $v^{01}(\hat{c}) = 1$.

Let us now derive the first-period equilibrium. Type $\hat{c}$ must be indifferent between contributing and not contributing, or

$$1 - \hat{c} + \delta\{P(\hat{c})v^{11}(\hat{c}) + [1 - P(\hat{c})]v^{10}(\hat{c})\}$$
$$= P(\hat{c}) + \delta\{P(\hat{c})v^{01}(\hat{c}) + [1 - P(\hat{c})]v^{00}(\hat{c})\}. \tag{8.9}$$

Using the formulas of the second-period utilities and equation 8.8, we obtain

$$1 - P(\hat{c}) = \hat{c} + \delta P(\hat{c})\tilde{c}. \tag{8.10}$$

Equations 8.8 and 8.10 define $\hat{c}$. Equation 8.10 has a straightforward interpretation: By contributing in period 1, type $\hat{c}$ spends $\hat{c}$ but provides the public good when it would not have been provided otherwise (which has probability $1 - P(\hat{c})$). Moreover, he reveals that his type is at most $\hat{c}$ instead of signaling a type above $\hat{c}$ by not contributing. This makes no difference when the opponent's type is above $\hat{c}$, because type $\hat{c}$ will contribute in the second period whether or not he contributes today. Contributing does change type $\hat{c}$'s second-period payoff when the opponent has type under $\hat{c}$: Whereas not contributing in the first period would induce the opponent to contribute in the second period, contributing makes the opponent more reluctant to contribute in that he will contribute in the second period only if his cost is lower than $\tilde{c}$. Because a player's second-period payoff when not contributing is independent of his cost, and because type $\tilde{c}$ is indifferent between contributing and not contributing if both have contributed in the first period, type $\hat{c}$ gains $1 - (1 - \tilde{c}) = \tilde{c}$ by signaling a high cost in the first period when the opponent's cost is less than $\hat{c}$.

---

9. Note that it is a dominant strategy for type $c_j > 1$ not to contribute. It is therefore optimal for type $c_i = \varepsilon$ (very small) to contribute. Recalling that strategies in the second period are necessarily cutoff rules, let $\tilde{c}_i \le \hat{c}$ denote the cutoff cost for player i in period 1, and $\check{c}_j \ge \hat{c}$ that for player j. They are given by

$$\tilde{c}_i = \frac{1 - P(\check{c}_j)}{1 - P(\hat{c})}$$

and

$$\check{c}_j = \frac{P(\hat{c}) - P(\tilde{c}_i)}{P(\hat{c})}.$$

For P uniform on $[0, 2]$, $\tilde{c}_i$ would thus be given by $\tilde{c}_i = -\hat{c}/(1 - \hat{c})^2$, which is impossible because $\tilde{c}_i$ must be a positive number.

Because this occurs with probability $P(\hat{c})$, the expected second-period gain from not contributing in the first period is $P(\hat{c})\tilde{c}$.

Equation 8.10 implies that $\hat{c} < c^*$: In this equilibrium (which is the unique symmetric equilibrium under some assumptions), there is less contribution in the first period of the two-period game than in the one-period game. This follows from the fact that each player gains by developing a reputation for not being willing to supply the public good.

8.3 Extensive-Form Refinements[††]

8.3.1 Review of Game Trees

We defined extensive-form games in chapter 3. In the next two subsections and in section 8.4, we will consider games of perfect recall with a finite number of players $(i = 1, \dots, I)$ and a finite number of decision nodes $(x \in X)$. Let $h(x)$ denote the information set containing node x. (We follow standard notation; we hope that this will not create any confusion with the related notion of history.) The player playing at node x is denoted $i(x)$; terminal nodes are denoted by z. The mixed or behavior strategy of player $i = i(x)$ at node x is $\sigma_i(\cdot \,|\, x)$ or $\sigma_i(\cdot \,|\, h(x))$. (We will sometimes delete the conditioning of σ_i on the information set if player i moves at a single information set, as there cannot be any ambiguity.) Let Σ denote the set of

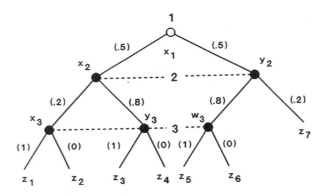

$$P^\sigma(x_1) = 1$$
$$P^\sigma(x_2) = P^\sigma(y_2) = .5$$
$$P^\sigma(x_3) = .1; \quad P^\sigma(y_3) = P^\sigma(w_3) = .4; \quad P^\sigma(h) = .9 \quad \text{where} \quad h = \{x_3, y_3, w_3\}$$
$$P^\sigma(z_1) = P^\sigma(z_7) = .1; \quad P^\sigma(z_2) = P^\sigma(z_4) = P^\sigma(z_6) = 0;$$
$$P^\sigma(z_3) = P^\sigma(z_5) = .4$$
$$\mu(x_3) = 1/9, \quad \mu(y_3) = \mu(w_3) = 4/9$$

Figure 8.2

all strategy profiles $\sigma = (\sigma_1, \ldots, \sigma_I)$, and let p denote the exogenous probability distribution over nature's moves. For example, in a game of incomplete information, nature's move is a choice of type for each player. As in our development so far, nature's moves are not considered to be given by a "strategy"; thus, when we perturb the game with "trembles" nature's moves will not be affected.

With σ given, $P^\sigma(x)$ and $P^\sigma(h)$ denote the respective probabilities that node x and information set h are reached. (These probabilities depend on the prior p, but we omit the superscript p because in a given extensive form the prior is fixed.) A system of beliefs μ specifies beliefs at each information set h: $\mu(x)$ denotes the probability player $i(x)$ assigns to node x conditional on reaching information set $h(x)$. In figure 8.2, which illustrates these concepts, the strategy profile σ is depicted on the tree.

Payoffs are determined by the terminal node of the game, and player i's payoff if z is reached is denoted $u_i(z)$. (Recall that z is a complete description of everything that happens before the terminal node is reached, including nature's choice of the players' private information.) Let $u_{i(h)}(\sigma \mid h, \mu(h))$ be the expected utility of player $i(h)$ given that information set h is reached, that the player's beliefs are given by $\mu(h)$, and that the strategies are σ.

An *assessment* (σ, μ) specifies a strategy profile σ and a system of beliefs μ. The set of all possible assessments is denoted by Ψ.

8.3.2 Sequential Equilibrium

We now describe how Kreps and Wilson (1982a) extend condition P into condition S (S for sequential rationality) and extend and refine condition B into condition C (C for consistency) for general finite games of perfect recall.

We noted in section 8.1 that the requirement that the players' strategies form a Nash equilibrium in each (proper) subgame is too weak, as there are few (proper) subgames in games of incomplete or imperfect information. We saw that in the imperfect-information game of figure 8.1 the only subgame is the whole game, and the Nash equilibrium (L, A) is subgame perfect. This equilibrium is nevertheless implausible, because whatever beliefs player 2 forms about which of M and R was player 1's move, he ought to play B if given the opportunity to move.

The appropriate generalization of condition P is that, given the system of beliefs, no player can gain by deviating at any information set:

(S)   An assessment (σ, μ) is *sequentially rational* if, for any information set h and alternative strategy $\sigma'_{i(h)}$,

$$u_{i(h)}(\sigma \mid h, \mu(h)) \geq u_{i(h)}((\sigma'_{i(h)}, \sigma_{-i(h)}) \mid h, \mu(h)).$$

Note that players believe that their opponents will adhere to the equilibrium profile σ at every information set (including ones that should not be reached if all players adhere to σ). Condition S is equivalent to condition P for multi-stage games.

What conditions one should put on beliefs at an information set off the equilibrium path is a more difficult and controversial question. Kreps and Wilson introduce the notion of consistency. We first define consistency, and then discuss the desiderata that led Kreps and Wilson to offer this definition; we later explore what consistency implies for multi-stage games.

Let Σ^0 denote the set of all completely mixed (behavioral) strategies, i.e., profiles σ such that $\sigma_i(a_i | h) > 0$ for all h and $a_i \in A(h)$. If $\sigma \in \Sigma^0$, then $P^\sigma(x) > 0$ for all nodes x, so that Bayes' rule pins down beliefs at each information set: $\mu(x) = P^\sigma(x)/P^\sigma(h(x))$. Let Ψ^0 denote the set of all assessments (σ, μ) such that $\sigma \in \Sigma^0$ and μ is (uniquely) defined from σ by Bayes' rule.

(C)   An assessment (σ, μ) is *consistent* if

$$(\sigma, \mu) = \lim_{n \to +\infty} (\sigma^n, \mu^n)$$

for some sequence (σ^n, μ^n) in Ψ^0.

Note that the strategies σ need not be totally mixed; however, they and the beliefs can be regarded as limits of totally mixed strategies and associated beliefs. Note also that condition C implies condition B for multi-stage games.

Since the probability distribution over nature's moves is not represented by a strategy, the definition of consistency does not apply "trembles" to nature's moves. Subsection 8.3.3 explains how allowing trembles by nature would change the properties of the equilibrium concept.

Definition 8.3   A *sequential equilibrium* is an assessment (σ, μ) that satisfies conditions S and C.

We now discuss the considerations that led Kreps and Wilson to propose the definition of consistency. Consider figure 8.3 (taken from their paper). Player 1 assigns probabilities $\frac{1}{3}$ and $\frac{2}{3}$ to nodes x and $x' \in h(x)$, respectively. His strategy is to play U. What should player 2 believe if player 1 deviates and plays D? Since player 1 cannot distinguish x and x', it seems natural to require that player 1 be "as likely" to deviate at both nodes. This idea leads to the requirement that player 2 put weights $\frac{1}{3}$ and $\frac{2}{3}$ on nodes y and y', respectively. However, any $\mu(y)$ is compatible with Bayes' rule, because event D has probability 0 in the equilibrium. Consistency yields the "right beliefs" in this game. Consider an arbitrary sequence ε^n converging to 0, and interpret ε^n as the probability that player 1 "trembles" and plays down. For this sequence,

$$\mu^n(y) = \frac{\mu^n(x)\varepsilon^n}{\mu^n(x)\varepsilon^n + \mu^n(x')\varepsilon^n} = \frac{1}{3}.$$

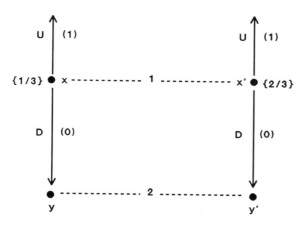

Figure 8.3

Thus, trembles ensure that the players' beliefs respect the information structure. This example also motivates Kreps and Wilson's definition of "structural consistency." (In subsection 8.3.4 we will exploit this example in a different way.)

An assessment (σ, μ) is *structurally consistent* if, for each information set h, there exists a strategy profile $\sigma_h \in \Sigma$ such that $P^{\sigma_h}(h) > 0$ and $\mu(x) = P^{\sigma_h}(x)/P^{\sigma_h}(h)$ for all x in h. That is, for each information set, the player on the move at the information set can find a strategy profile (not necessarily the same as σ) that would yield exactly the specified beliefs at the information set. The significance of structural consistency is the following: Suppose a player unexpectedly finds himself on the move at some information set h. What beliefs should he hold concerning the nodes in h? If he can find an alternative strategy profile σ_h that would reach h with positive probability, he could use this σ_h as a conjecture of the way the game had been played and then use Bayes' rule to form his beliefs at h. If the original equilibrium assessment (σ, μ) is structurally consistent, every player can, for every one of his off-the-equilibrium-path information sets, find such an alternative hypothesis to guide the formation of his beliefs.

Kreps and Wilson assert without proof that consistency implies structural consistency. Kreps and Ramey (1987) use figure 8.4 to show that this is incorrect. In figure 8.4, any assessment

$$\{\sigma_1(R_1) = \sigma_2(R_2) = 1, \sigma_3(R_3) \in (0,1); \mu(x_2) = 0, \mu(y_2) = 1, \mu(x_3) = 0,$$

$$\mu(y_3) = \mu(w_3) = \tfrac{1}{2}\}$$

is consistent, because it is the limit of the assessment derived from the totally mixed assessments

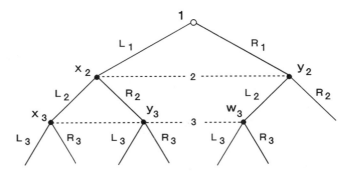

Figure 8.4

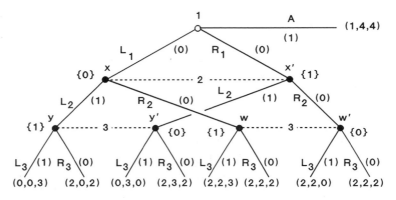

Figure 8.5

$$\{\sigma_1^n(R_1) = \sigma_2^n(R_2) = 1 - 1/n, \sigma_3^n(R_3) = \sigma_3(R_3); \mu^n(x_2) = 1/n,$$

$$\mu^n(y_2) = (n-1)/n, \mu^n(x_3) = 1/(2n-1),$$

$$\mu^n(y_3) = \mu^n(w_3) = (n-1)/(2n-1)\}.$$

This assessment is not structurally consistent, as no strategy giving positive weight to reaching nodes y_3 and w_3 gives weight 0 to reaching node x_3.[10,11]

Figure 8.5 shows how consistency reduces the set of equilibria by imposing common beliefs after deviations from equilibrium behavior. In this game, player 1 gets 2 by playing either L_1 or R_1 as long as at least one of

10. If y_3 is ever reached, then L_1 is sometimes played; if w_3 is ever reached, then L_2 is sometimes played. Therefore, the combination L_1 and L_2 is sometimes played, and therefore x_3 is reached with positive probability.
11. In response to this example, one might be tempted to add structural consistency to the definition of sequential equilibrium. However, Kreps and Ramey provide an example in which the unique sequential equilibrium does not satisfy structural consistency. They also show that any consistent assessment is the convex combination of structurally consistent assessments. In the example, the consistent beliefs are the convex combination of the structurally consistent beliefs $\mu(y_2) = \mu(w_3) = 1$ and $\mu(x_2) = \mu(y_3) = 1$.

the other players "cooperates" by playing right, so player 1 should play A only if both players 2 and 3 are likely to play left. Player 2's action does not affect player 3's payoff, and vice-versa. Consider the assessment (σ, μ), depicted in the figure, where $\sigma_1(A) = 1$, $\sigma_2(L_2) = 1$, $\sigma_3(L_3) = 1$ for each of player 3's two information sets, and $\mu(x') = \mu(y) = \mu(w) = 1$. This assessment is sequentially rational: If players 2 and 3 play left, player 1 should play A; if $\mu(x') = 1$, then player 2 gets 3 from L_2 and 2 from R_2; and if $\mu(y) = \mu(w) = 1$, then player 3 gets payoff 3 from L_3 and 2 from R_3. The assessment (σ, μ) is structurally consistent and obeys Bayes' rule where possible, but it is not consistent, as players 2 and 3 have different beliefs about the relative likelihood of player 1 playing L_1 and R_1, and this is not possible if the beliefs of player 2 and player 3 both are the limit of μ^n derived from the *same* sequence of totally mixed σ^n.

Moreover, there is no consistent assessment where player 1 plays A, since when players 2 and 3 have the same beliefs about player 1's move then at least one of them will play right: If $\mu(x) > \frac{1}{3}$ player 2 plays R_2, and if $\mu(y) < \frac{2}{3}$ player 3 plays R_3.

Although (σ, μ) is not consistent, it satisfies the weaker condition that for each player i there is a sequence $\sigma^n(i) \to \sigma$ of totally mixed strategy profiles such that, at each information set h, $\mu(x)$ is the limit of the beliefs $\mu^n(i)$ computed using Bayes' rule from $\sigma^n(i)$. Why should all players have the same theory to explain deviations that, after all, are either probability-0 events or very unlikely, depending on one's methodological point of view? The standard defense is that this requirement is in the spirit of equilibrium analysis, since equilibrium supposes that all players have common beliefs about the others' strategies. Although this restriction is usually imposed, we are not sure that we find it convincing.

8.3.3 Properties of Sequential Equilibrium (technical)

Existence
For any finite extensive-form game there exists at least one sequential equilibrium. Existence will be proved indirectly in section 8.4: Any trembling-hand perfect equilibrium is sequential, and because trembling-hand perfect equilibria exist in finite games, so do sequential equilibria.

Upper Hemi-Continuity
Like the Nash-equilibrium correspondence, the sequential-equilibrium correspondence is upper hemi-continuous with respect to payoffs. More precisely, fix an extensive form and prior beliefs p. For any sequence of utility functions u^n (defining a game) converging to some u, if the assessment (σ^n, μ^n) is a sequential equilibrium of game u^n for all n and converges to an assessment (σ, μ), then (σ, μ) is a sequential equilibrium of game u.

The proof of this is simple. We must show that (σ, μ) satisfies conditions S and C. The proof that it satisfies S follows the same lines as the proof that the Nash correspondence is upper hemi-continuous. That it satisfies C results from the fact that for each n there exists a sequence of assessments $(\sigma^{m,n}, \mu^{m,n})$ in Ψ^0 converging as m tends to infinity to (σ^n, μ^n), which in turn converges to (σ, μ). This upper-hemi-continuity property distinguishes sequential equilibrium from trembling-hand perfect equilibrium (see section 8.4).

What about upper hemi-continuity with respect to prior beliefs p? Consider a sequence p^n on a *fixed* set of initial nodes that converges to a distribution p. It is straightforward to check that the proof of upper hemi-continuity in the previous paragraph carries over as long as p assigns strictly positive probability to all of nature's moves.[12] However, if p assigns probability 0 to some of nature's moves, upper hemi-continuity with respect to beliefs may not hold. This lack of upper hemi-continuity can be illustrated in Spence's signaling model (example 8.2). In the (least-cost) separating equilibrium the high-productivity worker invests $\theta'(\theta'' - \theta')$ in education even if the probability of a low-productivity worker is very small. But when the latter probability is equal to 0, the high-productivity worker does not invest in education in the unique (subgame-) perfect equilibrium.

Note that if we modify the definition of consistency by requiring nature to tremble as well as the players, the separating equilibrium is still a sequential equilibrium when the prior probability of a low-productivity type is 0. More generally, with this definition the set of sequential equilibria is upper hemi-continuous with respect to prior beliefs on a fixed set of nature's moves. However, with the modified definition, the set of sequential equilibria can change when a probability-0 move by nature is added. That is, the set of sequential equilibria would depend not only on the prior beliefs but also on the set of nature's moves that are "conceivable." (A similar observation applies to the set of perfect Bayesian equilibria.)

Structure of Equilibria

Theorem 8.1 (Kreps and Wilson 1982a)   For generic (i.e., generic end-point payoffs of) finite extensive-form games of perfect recall, the set of sequential-equilibrium probability distributions on terminal nodes is finite.

That is, for a fixed extensive form and fixed prior beliefs, the closure of the set of payoffs u such that the associated game has an infinite number of sequential-equilibrium outcomes has Lebesgue measure 0. The set of sequential-equilibrium assessments is in general infinite because of the leeway in specifying beliefs off the equilibrium path.

The set of sequential equilibrium strategies is in general infinite as well, because, when a player is indifferent between two actions at an off-path

---

12. To extend the proof, note that, although the sequence of beliefs $\mu^{m,n}$ is Bayes consistent with p^n and $\sigma^{m,n}$, it need not be consistent with p and $\sigma^{m,n}$. Thus, one replaces $\mu^{m,n}$ by $\bar{\mu}^{m,n}$, which is Bayes consistent with p and $\sigma^{m,n}$.

information set, many different randomizing probabilities at that information set can be specified. This point is developed in detail in the appendix to the present chapter.

Addition of "Irrelevant Moves or Strategies"

Several authors have shown that the set of sequential equilibria can change when an apparently irrelevant move or strategy is added. We will return to this in more detail in chapter 11; we content ourselves with an example here.

Sequential equilibrium and the related concepts defined in section 8.4 have been criticized by Kohlberg and Mertens (1986) for allowing "strategically neutral" changes in the game tree to affect the equilibrium. Compare, for instance, figures 8.6a and 8.6b. Figure 8.6b is the same as figure 8.6a except that an apparently irrelevant move NA ("not across") has been added. Whereas A is a sequential-equilibrium outcome in figure 8.6a,[13] A is not a sequential-equilibrium outcome in figure 8.6b. In the "simultaneous-move" subgame following NA, the only Nash equilibrium

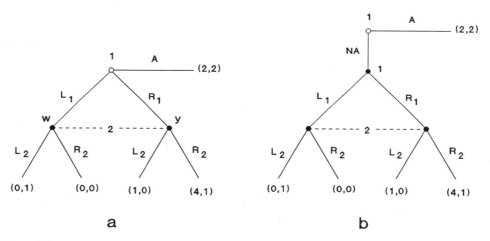

Figure 8.6

13. Consider the assessment $\{\sigma_1(A) = 1, \sigma_2(L_2) = 1; \mu(w) = 1\}$. This assessment satisfies condition S. To see that it satisfies condition C, consider the trembles

$$\sigma_1^n(A) = 1 - \frac{1}{n} - \frac{1}{n^2},$$

$$\sigma_1^n(L_1) = \frac{1}{n},$$

$$\sigma_2^n(L_2) = 1 - \frac{1}{n}.$$

Clearly, $\mu^n(w)$ converges to 1.

is (R_1, R_2), as L_1 is strictly dominated by R_1 for player 1. Hence, the only sequential-equilibrium payoffs are $(4, 1)$. This example also illustrates that the deletion of a strictly dominated strategy affects the set of sequential-equilibrium payoffs: If L_1 is deleted in figure 8.6a, the unique sequential-equilibrium payoffs are $(4, 1)$.

Chapter 11 has more discussion of when two similar trees should be expected to have the same solution. For now, let us note that if we take seriously the idea that players make "mistakes" at each information set, as the definition of sequential equilibrium might suggest, then it is not clear that the two figures are equivalent. In figure 8.6b, if player 1 makes the "mistake" of not playing A, he is still able to ensure that R_1 is more likely than L_1; in figure 8.6a he might play either action by "mistake" when intending to play A.

Correlated Sequential Equilibrium

Just as Nash equilibrium can be generalized to allow preplay observation of correlated signals, sequential equilibrium can be generalized to allow correlated strategies. There are three ways of doing so in multi-stage games (Forges 1986; Myerson 1986). First, one can allow players to receive information in the preplay phase ("at date -1") only. Second, one can allow the players to receive information slowly over time ("at each date"). Third, one can have players send private messages (inputs) at the beginning of each period to a "mediator" or a "machine," which then conveys private (but possibly correlated) messages (outputs) to the players. What differentiates the third possibility from the second is that the messages sent to the players can be contingent on their information. To show that each possibility allows more equilibria than the previous one (it clearly allows at least as many), consider the two examples shown in figures 8.7 and 8.8. Figure 8.7 illustrates the possibility that delaying the observation of "sunspots" may increase the equilibrium set. Payoffs $(3, 3)$ can be obtained by having the players coordinate on (L_1, L_2) or (R_1, R_2) with equal probabilities after the occurrence of a sunspot at the beginning of period 1. If the realization of the sunspot leading to the coordination on (R_1, R_2) were known at date 0, player 1 would play r_1 and payoffs $(3, 3)$ could no longer be obtained.

In figure 8.8, player 2 would like to predict the state of nature. Suppose that player 1, a dummy player in figure 8.8, learns the state of nature and can communicate it before player 2 picks an action. Payoffs $(0, 1)$ are now attainable, and thus communication increases the equilibrium payoff set.

These examples raise the question of the interpretation of an "extensive form." Should the complete rules of the game, including the "observation of sunspots" and "cheap talk," be explicitly described in the extensive form, or should any equilibrium concept allow for correlated moves and communi-

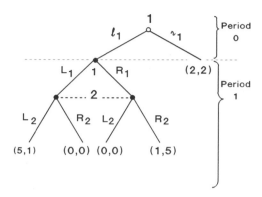

Figure 8.7

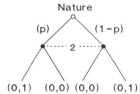

Figure 8.8

cation even if those possibilities are not explicitly described in the extensive form?

In the spirit of the revelation principle (see also sections 2.2 and 7.2), Forges and Myerson show that the set of equilibrium payoffs has a canonical representation. Any sequential equilibrium can be obtained by having, at each period, each player announce his information privately and truthfully to a mediator, who, having observed all messages, privately sends a recommended action or mixed strategy to each player, who then obeys the recommendation. For finite games, the sets of equilibrium strategies with *ex ante* correlation, correlation at each period, and correlation and communication at each period are convex polyhedrons, because in each case the equilibrium strategies are defined by a set of linear inequalities (the incentive-compatibility conditions).

8.3.4 Sequential Equilibrium Compared with Perfect Bayesian Equilibrium

The trembles underlying the definition of consistency give all paths positive probability so Bayes' rule pins down beliefs everywhere. There are two types of objections that can be made to the use of trembles. First, checking that an assessment in a finite game is consistent is a tedious process that is rarely carried out in applications. Furthermore, many applications involve an infinite number of actions or types; extending the formal definition of consistency to infinite games does not seem to require a conceptual innovation, but would face some technical difficulties. Second, and more important, one would like to know more about what consistency implies for behavior. In order to explain what the sequential-equilibrium restriction entails, we will compare it in some detail with the PBE concept of subsection 8.2.3.

Theorem 8.2 (Fudenberg and Tirole 1991)   Consider a multi-stage game of incomplete information with independent types. If either each player has at most two possible types ($\# \Theta_i \le 2$ for each i) or there are two periods, condition B is equivalent to condition C and therefore the sets of PBE and sequential equilibria coincide.

With more than two types per player and/or more than two periods, condition B is no longer sufficient to guarantee consistency, as figure 8.9 shows. This figure depicts a situation where player 1 has three possible types, θ_1', θ_1'', and θ_1^*, but where at time t Bayesian inference from the previous play has led to the conclusion that player 1 must be type θ_1^*. The equilibrium strategies at this point, which are given in parentheses in the figure, are for type θ_1' to play a_1', type θ_1'' to play a_1'', and type θ_1^* to play a_1^*. (For clarity, we do not depict type θ_1^*'s probability-0 actions.) Since the first two types have probability 0, player 2 expects to see player 1 play a_1^*. What should he believe if he sees one of the other two actions? The beliefs in figure 8.9 (given in braces) are that if player 2 sees a_1' he concludes that he is facing type θ_1'', while a_1'' is taken as a signal that player 1 is of type θ_1'. Since the definition of PBE places no constraints on the beliefs about a player who has just deviated (except that these beliefs are common to all players and that they do not depend on actions chosen by players other than the deviating one), the situation in figure 8.9 is compatible with PBE.

However, the situation of figure 8.9 cannot be part of a sequential equilibrium. To see this, imagine that there were trembles σ^n that converged to the given strategies σ and such that the associated beliefs μ^n converged to the given beliefs μ. Let the probability that μ^n assigns to type θ_1' at period t be ε'^n, and let the probability of type θ_1'' be ε''^n. Since μ^n converges to μ,

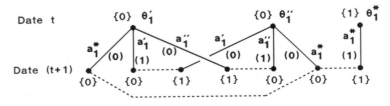

Figure 8.9

both ε''' and ε''' converge to 0, and $\sigma^n(a_1'|\theta_1'')$ and $\sigma^n(a_1''|\theta_1')$ converge to 0 as well. Since

$$\mu^n(\theta_1''|a_1') = \frac{\mu^n(\theta_1'')\sigma^n(a_1'|\theta_1'')}{\sum_{\theta_1}\mu^n(\theta_1)\sigma^n(a_1'|\theta_1)},$$

in order to have $\mu^n(\theta_1''|a_1')$ converge to 1 it must be that $\varepsilon'''/\varepsilon'''$ converges to 0: In order for the beliefs following a_1' to be concentrated on type θ_1'' when θ_1' plays the action with probability 1 while θ_1'' assigns it probability 0, the prior beliefs must be that θ_1'' is infinitely more likely than θ_1'. On its own, this requirement is compatible with sequential equilibrium. However, considering the beliefs that follow a_1'' leads to the conclusion that $\varepsilon'''/\varepsilon'''$ converges to 0, i.e., that θ_1' is infinitely more likely than θ_1'', and these two conditions are jointly incompatible with the beliefs being consistent. This restriction, though implied by sequential equilibrium, is stronger and different in spirit than the restrictions described by Kreps and Wilson to motivate the consistency requirement.

For PBE to imply consistency, the definition of beliefs must be extended to capture the relative probabilities of probability-0 types, and restrictions must be imposed on the way these relative probabilities are updated. Requiring that players assess relative probabilities of probability-0 states of nature is strong, but easy to formalize. Formally, one wants the posterior beliefs about each player to form a "system of relative beliefs," or a "conditional probability system."[14] That is, the players have beliefs $\mu^*(\theta_i|(\theta_i, \theta_i'), h^t)$ that player i is of type θ_i conditional on player i's being type θ_i or θ_i' and on history h^t, even if being type θ_i or θ_i' has probability 0 conditional on h^t. Note that a system of relative beliefs generates a system of absolute beliefs by the formula $\mu(\theta_i|h^t) \equiv \mu^*(\theta_i|\Theta_i, h^t)$.[15] A pair (σ, μ^*) is called a *generalized assessment*.

We now extend Bayesian conditions B(i)–B(iv) to require that *Bayes' rule and the no-signaling condition hold for relative beliefs and not only for absolute beliefs* (to simplify the statements, we assume right away that beliefs are common):

14. A conditional probability system (Myerson 1986) on a finite space Ω is a collection of functions $v(\cdot|\cdot)$ from $2^\Omega \times 2^\Omega$ to $[0, 1]$ such that for each $A \in 2^\Omega$, $v(\cdot|A)$ is a probability distribution on A, and such that for $A \subseteq B \subseteq C \in 2^\Omega$ with $B \neq \varnothing$, $v(A|B)v(B|C) = v(A|C)$.
15. More precisely,

$$\mu^*(\theta_i|\Theta_i, h^t) \equiv 1 \Big/ \sum_{\substack{\theta_i' \in \Theta_i \\ \theta_i' \neq \theta_i}} \left[\left(\frac{1}{\mu^*(\theta_i|(\theta_i, \theta_i'), h^t)} - 1 \right) + 1 \right],$$

if, for all $\theta_i' \neq \theta_i$, $\mu^*(\theta_i|(\theta_i, \theta_i'), h^t) > 0$,

$\equiv 0$ otherwise.

(B*) A generalized assessment $(\sigma, \mu*)$ satisfies condition B* if

(i) Bayes' rule is used to update relative beliefs $\mu*(\theta_i|(\theta_i, \theta_i'), h^t)$ into $\mu*(\theta_i|(\theta_i, \theta_i'), (h^t, a^t))$ whenever possible: If a_i^t has positive probability conditional on (θ_i, θ_i') and h^t,

$$\mu*(\theta_i|(\theta_i, \theta_i'), (h^t, a^t)) = \frac{\mu*(\theta_i|(\theta_i, \theta_i'), h^t)\sigma_i(a_i^t|h^t, \theta_i)}{\displaystyle\sum_{\tilde{\theta}_i = \theta_i, \theta_i'} \mu*(\tilde{\theta}_i|(\theta_i, \theta_i'), h^t)\sigma_i(a_i^t|h^t, \tilde{\theta}_i)};$$

(ii) the posterior beliefs are independent:

$$\mu(\theta|h^t) = \prod_i \mu(\theta_i|h^t);$$

(iii) the relative beliefs about player i at date $t + 1$ depend only on h^t and on player i's period-t action:

$$\mu*(\theta_i|(\theta_i, \theta_i'), (h^t, a^t)) = \mu*(\theta_i|(\theta_i, \theta_i'), (h^t, \tilde{a}^t)) \text{ if } a_i^t = \tilde{a}_i^t.$$

Note that these conditions are the same as B(i)–B(iii) except that they apply to relative probabilities. Indeed, conditions B and B* coincide when $\mu(\cdot|h^t)$ has full support for all h^t. In particular, in a two-period game (such as the signaling game) all types have positive probability in period 0, so condition B* does not refine condition B (it does refine condition B for beliefs formed at the end of period 1, but those beliefs are irrelevant, as the game is over). In the case of at most two types per player, at most one type has probability 0 after any history and the issue of relative beliefs does not arise (absolute beliefs are also relative beliefs), so again condition B* coincides with condition B.

Condition B*(i) implies that if θ_i is infinitely more likely than θ_i' given h^t, and $\sigma_i(a_i^t|h^t, \theta_i) > 0$, then after observing a_i^t in period t, θ_i is still infinitely more likely than θ_i'. Similarly, if two types are "as likely" (in the sense that none is infinitely more likely than the other) given h^t, and both play action a_i^t with positive probability, the two types remain "as likely." Combined, these two implications rule out the beliefs in figure 8.9.

Definition 8.4 A *perfect extended Bayesian equilibrium* (PEBE) of a multistage game of incomplete information with independent types is a generalized assessment satisfying conditions P and B*.

Theorem 8.3 (Fudenberg and Tirole 1991) For multi-stage games of incomplete information with independent types, condition B* implies condition C, and any assessment satisfying C can be extended to a generalized assessment satisfying B*. Therefore, the sets of PEBE and sequential equilibria coincide.

The idea of the proof of these two results (that condition B for two types or two periods, or more generally that condition B* implies condition

C) is as follows: Suppose one has built trembles up to date t that yield strictly positive beliefs at the beginning of date t and converge to $\mu(\cdot\,|\,h^t)$. One then constructs trembles on probability-0 actions so as to obtain the posterior beliefs $\mu(\cdot\,|\,(h^t, a^t))$ in the limit. The no-signaling condition guarantees that these trembles can be built independently among players, and, with more than two types, condition B* guarantees that appropriate trembles exist that vindicate the relative beliefs. One then subtracts trembles on probability-0 actions from the (strictly positive) probabilities on equilibrium actions to ensure that, along the sequence of trembles, the probabilities of each player's actions add up to 1.

Correlated Types[†††]

When types are correlated, it is convenient to transform the game into one with independent types and then map the resulting equilibrium strategies and beliefs to strategies and beliefs in the original game. Myerson (1985) shows that any Bayesian game can be transformed into one with independent types. Suppose that the prior distribution $\rho(\theta) = \rho(\theta_1, \ldots, \theta_I)$ has full support on Θ. And let $\hat{\rho}$ be the product of independent uniform marginal distributions $\hat{\rho}_i$ on Θ_i:

$$\hat{\rho}(\theta) \equiv 1 \left/ \left(\prod_{i=1}^{I} (\#\Theta_i) \right) \right. \text{ for all } \theta \text{ in } \Theta.$$

Define the fictitious von Neumann-Morgenstern payoff functions

$$\hat{u}_i(h^{T+1}, \theta_i, \theta_{-i}) \equiv \rho(\theta_{-i}\,|\,\theta_i) u_i(h^{T+1}, \theta_i, \theta_{-i}) \text{ for all } (h^{T+1}, \theta_i, \theta_{-i}).$$

Let (by the familiar abuse of notation) $u_i(\sigma, \theta)$ and $\hat{u}_i(\sigma, \theta)$ denote the utilities for strategy profile σ and types θ. With E_ρ and $\mathrm{E}_{\hat{\rho}}$ denoting the expectation operators with respect to distributions ρ and $\hat{\rho}$, $\mathrm{E}_\rho(u_i\,|\,\theta_i)$ and $\mathrm{E}_{\hat{\rho}}(\hat{u}_i\,|\,\theta_i)$ represent the same preferences for player i with type θ_i. The Bayesian equilibria of the game (u, ρ) with correlated types and the game $(\hat{u}, \hat{\rho})$ with independent types are therefore the same.

More generally, in a multi-stage game with incomplete information, it is straightforward to check that an assessment $(\hat{\sigma}, \hat{\mu})$ is a sequential equilibrium of the transformed game $(\hat{u}, \hat{\rho})$ if and only if the assessment (σ, μ) defined by $\sigma = \hat{\sigma}$ and

$$\mu(\theta_{-i}\,|\,\theta_i, h^t) \equiv \frac{\rho(\theta_{-i}\,|\,\theta_i)\hat{\mu}(\theta_{-i}\,|\,h^t)}{\sum_{\theta'_{-i}} \rho(\theta'_{-i}\,|\,\theta_i)\hat{\mu}(\theta'_{-i}\,|\,h^t)}$$

is a sequential equilibrium of the original game (u, ρ).

Imposing condition B or B* on the transformed game yields restrictions on beliefs for the original game. In particular, in a game with correlated types, a player's action conveys information about other players' types only to the extent that it conveys information about his own type. The date-

$(t + 1)$ beliefs about θ_{-i} conditional on θ_i depend on the history h^t, the actions a^t_{-i}, and the conditional beliefs $\mu(\theta_{-i}|\theta_i, h^t)$ at date t, but not on player i's action a^t_i.

General Extensive-Form Games[†††]

Necessary conditions for sequential equilibrium in general extensive-form games can be given in terms of "no signaling what you don't know." The trembles associated with sequential equilibrium give rise to a conditional probability system on all terminal nodes. Conversely, assume that there is a conditional probability system on all terminal nodes that is compatible with the strategy profile σ. Let $\mu(x|x, y)$ denote the relative beliefs generated by the conditional probability system when nodes x and y belong to the same information set ($\mu(x|x, y)$ is the probability of terminal nodes that are successors of x conditional on the terminal node's being a successor of either x or y); similarly, let $\mu(\mathfrak{s}(x, a)|x)$ denote the probability of the direct successor $\mathfrak{s}(x, a)$ of node x through action $a \in A(h(x))$. The no-signaling conditions are simply (1) for any information set h and any node $x \in h$ and action $a \in A(h)$,

$$\mu(\mathfrak{s}(x, a)|x) = \sigma(a|h)$$

and (2) for any information set h, nodes x and y in h and action a,

$$\mu(\mathfrak{s}(x, a)|\mathfrak{s}(x, a), \mathfrak{s}(y, a)) = \mu(x|x, y).$$

That is, the player on move at h cannot distinguish among the nodes in h and therefore cannot signal information he does not have. Fudenberg and Tirole (1991) claimed that these conditions implied consistency, but Battigalli (1991) shows that this claim is false. The no-signaling conditions 1 and 2 are very weak, but the existence of a conditional probability system on all terminal nodes is quite strong: It allows the (common across players) comparison of probabilities of nodes at any two information sets. In contrast, for multi-stage games with incomplete information it suffices to be able to compare the likelihoods of types of a player in any given period.

8.4   Strategic-Form Refinements[††]

This section reviews two strategic-form refinements of Nash equilibrium. The concept of sequential equilibrium is closely related to that of trembling-hand perfect equilibrium (henceforth "perfect equilibrium") of Selten (1975). Perfect equilibrium requires that the strategies be the limit of totally mixed strategies and that, subject to the requirement that it must put at least a minimum weight (must tremble) on each pure strategy on the converging sequence, each player's strategy is (constrained) optimal against his opponents' (which include trembles themselves). The distinction with sequential equilibrium is thus that strategies must be in equilibrium along the converging subsequence and not only in the limit. This distinction turns out to make only a minor difference, as the sets of sequential and perfect equilibria coincide "for almost all games." We will also review Myerson's (1978) concept of proper equilibrium, which refines perfect equilibrium by requiring that, along the converging sequence of perturbed strategies, players are less likely to make "mistakes" that are more costly.

8.4.1 Trembling-Hand Perfect Equilibrium

Now we consider the concept of trembling-hand perfection in the *strategic form* and in the *agent-strategic form*. (Selten called the latter the "agent normal form.") As we will see, perfection in the strategic form does not imply subgame perfection. Selten introduced the agent-strategic form in order to rule out subgame-imperfect equilibria.

There are three equivalent definitions of trembling-hand perfection in the strategic form:

Definition 8.5A An "ε-constrained equilibrium" of a strategic-form game is a totally mixed strategy profile σ^ε such that, for each player i, σ_i^ε solves $\max_{\sigma_i} u_i(\sigma_i, \sigma_{-i}^\varepsilon)$ subject to $\sigma_i(s_i) \geq \varepsilon(s_i)$ for all s_i, for some $\{\varepsilon(s_i)\}_{s_i \in S_i, i \in \mathcal{I}}$ where $0 < \varepsilon(s_i) < \varepsilon$. A *perfect equilibrium* is any limit of ε-constrained equilibria σ^ε as ε tends to 0.

According to definition 8.5A, a perfect equilibrium is a limit of Nash equilibria of some sequence of constrained games. The standard sort of closed-graph argument shows that any perfect equilibrium is a Nash equilibrium of the game without the constraints.

For given $\{\varepsilon(s_i)\}$, a constrained equilibrium exists for the usual reasons. (The only difference with the proof of existence of a Nash equilibrium in mixed strategies (see section 1.3) is that each mixed strategy must belong to a subset of a simplex, as opposed to the simplex itself, but this difference is irrelevant because the subset is compact, convex and, for ε small, non-empty.) Thus, for any sequence of constraints $\{\varepsilon(s_i)\}$ there is a corresponding sequence of constrained equilibria. Because strategy spaces are compact, this sequence has a convergent subsequence, so a perfect equilibrium exists.

To see how trembles help refine the set of Nash equilibria, consider the game illustrated in figure 8.10, which Selten used to motivate subgame perfection. The Nash equilibrium $\{R_1, L_2\}$ is not the limit of constrained equilibria: If player 1 plays L_1 with positive probability, player 2 puts as much weight as possible on R_2.

The idea behind definition 8.5A is that players may tremble (make mistakes) and that their constrained strategies should be optimal when the trembles of their rivals are taken into account. Selten's second definition does not explicitly introduce minimum trembles, but requires that the

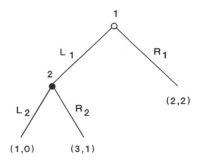

Figure 8.10

profile σ be a limit of a sequence of totally mixed profiles σ^n and that σ_i be a best response to the opponents' perturbed strategies σ^n_{-i}:

Definition 8.5B Strategy profile σ of a strategic form is a *perfect equilibrium* if there exists a sequence of totally mixed strategy profiles $\sigma^n \to \sigma$ such that, for all i, $u_i(\sigma_i, \sigma^n_{-i}) \geq u_i(s_i, \sigma^n_{-i})$ for all $s_i \in S_i$.

Let us emphasize that in definition 8.5B strategy σ_i is a best response to *some* sequence σ^n_{-i} and not necessarily to all sequences converging to σ_{-i}. Likewise, in definition 8.5A, it suffices that σ is the limit of ε-constrained equilibria for *some* sequence of constraints, as opposed to all such sequences. The uniform versions of these definitions—requiring in definition 8.5B that σ_i be a best response to any sequence $\sigma^n_{-i} \to \sigma_{-i}$—yield the concept of "truly perfect equilibrium," which is much more demanding. For some games, truly perfect equilibria do not exist (see chapter 11).

The third definition of perfect equilibrium, due to Myerson (1978), does not quite refer to a conventional optimization:

Definition 8.5C Strategy profile σ^ε of a strategic form is an ε-perfect equilibrium[16] if it is completely mixed, and, for all i and any s_i, if there exists s_i' with $u_i(s_i, \sigma^\varepsilon_{-i}) < u_i(s_i', \sigma^\varepsilon_{-i})$, then $\sigma_i^\varepsilon(s_i) < \varepsilon$. A *perfect equilibrium* σ is any limit of ε-perfect strategy profiles σ^ε for some sequence ε of positive numbers that converges to 0.

That is, player i is not required to optimize against his rivals' strategies subject to an explicit constraint on minimum weights, but must put less than ε weight on strategies that are not best responses.

Theorem 8.4 The three definitions of perfect equilibrium (8.5A–8.5C) are equivalent.

Proof We show that definition A implies definition C, which implies definition B, which in turn implies definition A. First, by construction, the

---

16. Here the ε does *not* refer to ε-optimization, as in the ε-perfect equilibrium discussed in section 4.8.

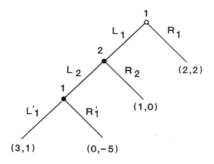

a. Extensive Form

	L$_2$	R$_2$
R$_1$	2,2	2,2
L$_1$,L$_1'$	3,1	1,0
L$_1$,R$_1'$	0,-5	1,0

b. Strategic Form

Figure 8.11

sequence σ^ε defined in definition A is an ε-perfect equilibrium, so that σ^ε satisfies definition C if it satisfies definition A. Second, suppose that σ satisfies definition C. Then there is a sequence $\sigma^\varepsilon \to \sigma$ and a constant $d > 0$ with $\sigma_i^\varepsilon(s_i) > d$ for every s_i in the support of σ_i. Thus, every s_i in the support of σ_i must be a best response to σ_{-i}^ε, so that definition B is satisfied. Third, suppose that σ satisfies definition B, and let $\sigma^n \to \sigma$ be the hypothesized totally mixed strategy profiles. For s_i not in the support of σ_i define $\varepsilon^n(s_i) \equiv \sigma_i^n(s_i)$, and for s_i in the support of σ_i let $\varepsilon^n(s_i) \equiv 1/n$. Then consider the program $\{\text{Max}_{\sigma_i} u_i(\sigma_i, \sigma_{-i}^\varepsilon)$ subject to $\sigma_i(s_i) \geq \varepsilon^n(s_i)$ for all $s_i \in S_i\}$. Because σ_i is a best response to σ_{-i}^n by assumption, one of the corresponding ε-constrained equilibria, σ^ε, has $\sigma_i^\varepsilon(s_i) = \varepsilon^n(s_i)$ for $s_i \notin$ support (σ_i), and $\sigma_i^\varepsilon(s_i) = \sigma_i(s_i)$ for $s_i \in$ support (σ_i).[17] $\varepsilon^n \equiv \max\{\varepsilon^n(s_i)\}$ tends to 0 as n tends to ∞. Hence, definition A is satisfied. ∎

As Selten noted, perfection in the strategic form is not totally satisfactory. Consider figure 8.11. The only subgame-perfect equilibrium is $\{L_1, L_2, L_1'\}$. But the subgame-imperfect Nash equilibrium $\{R_1, R_2, R_1'\}$ is the limit of equilibria with trembles. To see why, consider the corresponding (reduced) strategic form (displayed in figure 8.11b), and let player 1 play (L_1, L_1') with

17. There can be other ε-constrained equilibria, because some $s_i \notin$ support (σ_i) could also be best responses to σ_{-i}^n.

probability ε^2 and (L_1, R_1') with probability ε. Then player 2 should put as much weight as possible on R_2, because player 1's probability of "playing" R_1' conditional on having "played" L_1 is $\varepsilon/(\varepsilon + \varepsilon^2) \simeq 1$ for ε small. The point is that strategic-form trembles allow correlation between a player's tremble and his play at subsequent information sets. In the above example, if a player "trembles" onto L_1 he is very likely to play R_1' and not L_1'.

One possible response to this is that since a player's trembles may indeed be correlated, subgame perfection is too strong. Recall that subgame perfection's premise is that reasonable play in a subgame depends only on that subgame, regardless of whether that subgame is in fact the whole tree or instead can be reached only if some player i deviates from the (perfect) equilibrium strategies in a longer game. If we take the trembles story literally this premise may or may not be compelling, depending on how and why mistakes occur, and subgame perfection loses some of its persuasiveness. (At this point the reader might want to reread the examples in section 3.6.)

Selten's view in his 1975 paper was rather that the trembles are a technical device, and that they are not intended to model actual "mistakes." In that spirit, he modified his trembling-hand concept to rule out correlation and thus rule out subgame-imperfect equilibria. The modification uses the concept of the *agent-strategic form*, which treats the two choices of player 1 in figure 8.11 as made by two different players whose trembles are independent.

More precisely, in the agent-strategic form each information set is "played" by a different "agent," and the agent on move at information h has the same payoffs over terminal nodes as the player $i(h)$ on move at h in the original game. A trembling-hand perfect equilibrium in the agent-strategic form of an extensive-form game is the trembling-hand perfect equilibrium of the corresponding extensive form.

It should be clear that the equivalence between the various definitions of perfect equilibrium carries over to perfection in the agent-strategic form, as does the proof of existence. From now on, by "perfect equilibrium" we will mean "trembling-hand perfect equilibrium in the agent-strategic form" (as opposed to "strategic-form perfect equilibrium," which allows correlated trembles across information sets of the same player). Figure 8.12 displays the agent strategic form associated with the extensive form in figure 8.11a. The "first incarnation" of player 1 chooses matrices and the "second incarnation" chooses columns. Because the two incarnations have the same payoffs, we merge these payoffs in the entries of the matrices.

Definition 8.5B makes it clear why a perfect equilibrium is a sequential equilibrium. The strategies σ are, by construction, limits of totally mixed strategies σ^n. To obtain a sequential equilibrium, one must construct beliefs μ such that (σ, μ) is consistent and σ is sequentially rational given μ. Because σ^n are totally mixed, associated beliefs μ^n at information sets of any exten-

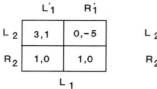

Figure 8.12

	L	R
U	1,1	0,0
D	0,0	0,0

Figure 8.13

sive form with this strategic form are uniquely defined by Bayes' rule. It then suffices to take the limit μ of a convergent subsequence μ^n. By construction, (σ, μ) is a consistent assessment. By the one-stage-deviation principle, σ_i is a best response for a "single player i" to σ_i^n, and since payoffs are continuous, (σ, μ) is sequentially rational.

A sequential equilibrium, however, need not be perfect, as is demonstrated in figure 8.13. In the simultaneous-move game with strategic form represented in this figure, the imperfect Nash equilibrium (D, R) is sequential. However, if one requires that strategies be optimal against some trembles, D and R cannot be chosen, because they are weakly dominated.

However, this game is nongeneric, because it relies on a player's (in this example, both players') being indifferent between an equilibrium strategy and a nonequilibrium strategy. Once indifference is broken by a small perturbation of payoffs, the sets of sequential and perfect equilibria coincide, as Kreps and Wilson (1982a) showed. The notion of genericity is the following: Fix an extensive form and prior beliefs and consider the family of games indexed by payoffs u at the ℓ terminal nodes. "Game u" is, by abuse of terminology, the game defined by the payoff vector u in $\mathbb{R}^{\ell \times I}$. A property is generic (satisfied for "almost all games") if the closure of the set of games that do not satisfy this property has Lebesgue measure 0 in $\mathbb{R}^{\ell \times I}$. We collect the results in theorem 8.5.

Theorem 8.5 In finite games, at least one perfect equilibrium exists (Selten 1975). A perfect equilibrium is sequential, but the converse is not true; however, for generic games the two concepts coincide (Kreps and Wilson 1982a).

The perfect-equilibrium correspondence need not be upper hemicontinuous in the payoffs. Figure 8.14 depicts a small perturbation of the

	L	R
U	1,1	0,0
D	0,0	$\frac{1}{n},\frac{1}{n}$

Figure 8.14

game defined by figure 8.13. In figure 8.14, (D, R) is a perfect equilibrium: D is a best response to a σ_2^n that assigns probability $1 - 1/n^2$ to R, and R is a best response to a σ_1^n that assigns probability $1 - 1/n^2$ to D. However, the unique perfect equilibrium of the limit game is (U, L).

We have two final notes on the idea of trembles. First, observe that trembles can be interpreted as perturbations of the players' payoff functions. In the constrained game of definition 8.5A, player i must place probability at least $\varepsilon(s_i)$ on each $s_i \in S_i$; thus, strategy s_i is effectively replaced by the mixed strategy, which assigns probability $1 - \sum_{s_i' \neq s_i} \varepsilon(s_i')$ to s_i and $\varepsilon(s_i')$ to each s_i'. Equivalently, we could leave the strategies exactly as they were originally, and define new payoff functions

$$\hat{u}_i(s_i, \sigma_{-i}) = \left(1 - \sum_{s_i' \neq s_i} \varepsilon(s_i')\right) u_i(s_i, \sigma_{-i}) + \sum_{s_i' \neq s_i} \varepsilon(s_i') u_i(s_i', \sigma_{-i}).[18]$$

Second, we should mention the work of Blume, Brandenburger, and Dekel (1990), which gives a characterization of perfect equilibrium in the strategic form in terms of "lexicographic beliefs" instead of trembles. This work stands in roughly the same relation to strategic-form perfect equilibrium as PBE does to sequential equilibrium.

8.4.2 Proper Equilibrium

Myerson (1978) considers perturbed games in which a player's second-best actions are assigned at most ε times the probability of the first-best actions, the third-best actions are assigned at most ε times the probability of the second-best actions, etc. The idea is that a player is "more likely to tremble" on an action that is not too detrimental to him, so that the probability of deviations from equilibrium behavior is inversely related to their costs.

Because a smaller set of trembles is considered, a proper equilibrium is clearly perfect in the strategic form. As we will see, proper equilibria are perfect in the agent-strategic form as well.[19]

18. We will see in chapter 12 that for generic *strategic-form* payoffs, any Nash equilibrium has a nearby Nash equilibrium in any game with nearby payoffs, so in generic strategic forms any Nash equilibrium is truly perfect. However, generic extensive-form payoffs do not generate generic strategic-form ones.

19. Although properness in the strategic form ensures backward induction, properness in the strategic form and properness in the agent-strategic form differ because two incarnations of the same player in the agent strategic form (associated with two different information sets) need not compare the payoffs of their probability-0 actions.

	L	M	R
U	1,1	0,0	-9,-9
M	0,0	0,0	-7,-7
D	-9,-9	-7,-7	-7,-7

Figure 8.15

To illustrate the notion of proper equilibrium, consider the game illustrated in figure 8.15 (due to Myerson), which adds weakly dominated strategies to the game defined in figure 8.12. This game has three pure-strategy Nash equilibria: (U, L), (M, M), and (D, R). D and R are weakly dominated strategies and therefore cannot be optimal when the other player trembles, so (D, R) is not perfect. (M, M) *is* perfect. To see this, consider the totally mixed strategy profile where each player plays M with probability $1 - 2\varepsilon$ and plays each of the other two strategies with probability ε. Deviating to U for player 1 (or to L for player 2) increases this player's payoff by $(\varepsilon - 9\varepsilon) - (-7\varepsilon) = -\varepsilon < 0$. However, (M, M) is not a proper equilibrium. Each player should put much more weight (tremble more) on his second-best strategy than on his third-best, which yields a lower payoff. But if player 1, say, puts weight ε on U and ε^2 on D, player 2 does better by playing L than by playing M, as $(\varepsilon - 9\varepsilon^2) - (-7\varepsilon^2) > 0$ for ε small. The only proper equilibrium in this game is (U, L).

Definition 8.6 An ε-*proper equilibrium* is a totally mixed strategy profile σ^ε such that, if $u_i(s_i, \sigma^\varepsilon_{-i}) < u_i(s'_i, \sigma^\varepsilon_{-i})$, then $\sigma^\varepsilon_i(s_i) \le \varepsilon \sigma^\varepsilon_i(s'_i)$. A *proper equilibrium* σ is any limit of ε-proper equilibria σ^ε as ε tends to 0.

Theorem 8.6 (Myerson 1978)   All finite strategic-form games have proper equilibria.

Proof   We first prove the existence of ε-proper equilibria. Let

$$\tilde{\Sigma}_i = \left\{ \sigma_i \in \Sigma_i^0 \mid \sigma_i(s_i) \ge \frac{\varepsilon^m}{m} \text{ for all } s_i \text{ in } S_i \right\},$$

where $m \equiv \max_i(\#S_i)$ and $0 < \varepsilon < 1$. Consider the constrained best-response correspondence of player i to strategies σ_{-i}:

$$\tilde{r}_i(\sigma_{-i}) = \{\sigma_i \in \tilde{\Sigma}_i \mid \text{if } u_i(s_i, \sigma_{-i}) < u_i(s'_i, \sigma_{-i})$$

$$\text{then } \sigma_i(s_i) \le \varepsilon \sigma_i(s'_i) \; \forall (s_i, s'_i) \in (S_i)^2\}.$$

Because $\tilde{r}_i$ is defined by a finite collection of linear weak inequalities, it is convex- and compact-valued; upper hemi-continuity of $\tilde{r}_i$ is straightforward. To prove that $\tilde{r}_i(\sigma_{-i})$ is nonempty, let $\rho(s_i)$ be the number of strategies s_i' such that $u_i(s_i, \sigma_{-i}) < u_i(s_i', \sigma_{-i})$. Then, if $\rho(s_i) > 0$, $\sigma_i \equiv \{\sigma_i(s_i)\}$, where

$$\sigma_i(s_i) = \varepsilon^{\rho(s_i)} \bigg/ \left(\sum_{s_i' \in S_i} \varepsilon^{\rho(s_i')} \right) \geq \frac{\varepsilon^m}{m},$$

belongs to $\tilde{r}_i(\sigma_{-i})$. One then applies Kakutani's fixed-point theorem in the usual way to prove the existence of an ε-proper equilibrium in $\times_i \tilde{\Sigma}_i$. Letting $\tilde{\Sigma}_i$ tend to Σ_i and taking a convergent subsequence of the associated ε-proper equilibria completes the proof.   ∎

Let us conclude with two properties of proper equilibrium.[20]

First, proper equilibrium yields backward induction without the use of the agent strategic form, because the requirement on relative trembles ensures that the players play optimally off the equilibrium path. This is illustrated in figure 8.11. The strategy (L_1, R_1') is dominated by the strategy (L_1, L_1') as long as player 2 trembles. Hence, player 1 must put almost all the weight on L_1' if his second information set is reached.

Second, Kohlberg and Mertens (1986) have shown that every proper equilibrium of a strategic-form game is sequential in every extensive form with the given strategic form. Refer back to figure 8.6, which gave two allegedly equivalent descriptions of the "same game." Player 1 playing A is a sequential equilibrium outcome in figure 8.6a, but not in figure 8.6b. However, in either game, the only proper equilibrium is (R_1, R_2). In particular, (A, L_2) is not a proper equilibrium in figure 8.6a, since player 1 must give R_1 more weight than L_1 in any ε-proper equilibrium. Kohlberg and Mertens also observe that a proper equilibrium of a strategic form need not be a trembling-hand perfect equilibrium in (the agent strategic form of) every extensive form associated with this strategic form. Figure 8.16 considers a single-decision-maker problem with three pure strategies. (L, r) is proper in

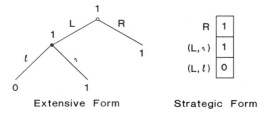

Extensive Form   Strategic Form

Figure 8.16

20. See van Damme 1987 for a more extensive and very clear discussion of proper equilibrium and several of its variants. In particular, proper equilibria do not correspond to the limits of equilibria of games in which players optimize their error probabilities in the presence of "control costs." Intuitively, if strategy s_1 is almost as good a response as strategy s_1' but neither is a best response, then when error probabilities are optimized we would expect trembles on s_1 to be almost as likely as trembles on s_1'.

the strategic form, but not (agent-strategic-form) perfect in the tree: If the player's second incarnation trembles, his first incarnation prefers playing R.

Appendix: The Structure of Sequential Equilibria[††]

We remarked in subsection 8.3.3 that, although the set of sequential equilibrium *outcomes* is finite for generic extensive-form payoffs, the set of sequential equilibrium *assessments* is, in general, infinite. This appendix develops that remark in more detail.

Consider the game illustrated in figure 8.17, taken from Kreps and Wilson 1982a. This game has two sequential equilibrium outcomes: (L, ℓ) and A. There is a unique equilibrium assessment with outcome (L, ℓ), namely $\sigma_1(L) = 1 = \sigma_2(\ell)$, and $\mu(x) = 1$. In contrast, there are two one-parameter families of equilibrium assessments with outcome A. In the first family, $\sigma_1(A) = 1, \sigma_2(\ell) = 0$, and $\mu(x) < \frac{1}{2}$; in the second, $\sigma_1(A) = 1, \sigma_2(\ell) \in [0, \frac{3}{5}]$, and $\mu(x) = \frac{1}{2}$. Projecting the equilibrium assessments onto the pairs $(\mu(x), \sigma_2(\ell))$ gives the picture in figure 8.18.

As this example illustrates, for generic payoffs the set of sequential equilibrium assessments is the union of manifolds of varying dimensions;

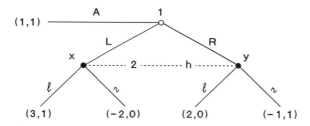

Figure 8.17

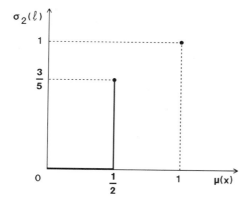

Figure 8.18

the dimensions of these manifolds is related to the number of "degrees of freedom" available in specifying the off-path strategies and beliefs. Since there are no off-path information sets in the equilibrium with outcome (L, ℓ), there are zero degrees of freedom, and the manifold associated with this outcome has dimension zero. (For the same reason, one-shot simultaneous-move games have a finite number of equilibria for generic payoffs, as we discuss in section 12.1.) In the equilibria with outcome A, player 2's information set is not reached. The horizontal segment of figure 8.18 reflects the one degree of freedom in specifying beliefs $\mu(x)$ that make $\imath$ a better choice for player 2 than ℓ; the vertical segment corresponds to the degree of freedom in specifying mixed strategies for player 2 that make player 1 prefer A to L. Since player 2 must be indifferent between ℓ and $\imath$ to randomize, the one degree of freedom in specifying beliefs must be lost to obtain the one degree of freedom in specifying mixed strategies.

Kreps and Wilson generalize these observations as follows. Let the *basis* b of an equilibrium assessment (σ, μ) be the collection of nodes and actions the assessment gives positive probability (i.e., a in $A(h)$ belongs to b iff $\sigma_{i(h)}(a|h) > 0$ and $x \in b$ iff $\mu(x) > 0$). Kreps and Wilson show that, for generic payoffs, the set of equilibria with a given basis is either empty, or is a manifold whose dimension is independent of the particular extensive-form payoffs specified.

Exercises

Exercise 8.1* Consider the public-good game of example 8.3.

(a) Show that in any PBE there exists $\hat{c}_i$ such that player i contributes if and only if $c_i \leq \hat{c}_i$. (Hint: Let z_j and z_j^{xy} denote player j's first-period probability of contributing and second-period probability of contributing conditional on whether player i contributed ($x = 1$ or 0) and player j contributed ($y = 1$ or 0) in the first period. Write the intertemporal expected payoffs for player i with type c_i from contributing and not contributing in the first period. Note that these payoffs involve second-period maximizations. What are their derivatives with respect to c_i?)

(b) Use question a to show that $\hat{c}_i < 1$ for each i. (Hint: Suppose, without loss of generality, that $\hat{c}_1 = \max\{\hat{c}_1, \hat{c}_2\} \geq 1$. Argue that player 1's not contributing in the first period induces "maximal contribution" by player 2 in the second period.)

(c) Use question b to show that $\hat{c}_i > 0$ for all i.

Exercise 8.2** In final-offer arbitration, the arbitrator is forced to choose one of the parties' offers as a settlement. Consider the following model of learning in final-offer arbitration, which is due to Gibbons (1988). In the first stage, an employer and a union simultaneously make wage offers a_e

and a_u in $\mathbb{R}$. The arbitrator then chooses $a_2 \in \{a_e, a_u\}$. The objective functions are $-Ea_2$ for the employer, $+Ea_2$ for the union, and $E[-(a_2 - \omega)^2]$ for the arbitrator, where ω, the state of nature, is the arbitrator's bliss point. The information about ω is as follows: The employer and the union receive the same signal, $z_1 = \omega + \varepsilon_1$. The arbitrator receives signal $z_2 = \omega + \varepsilon_2$. The variables ω, ε_1, and ε_2 are independent and normally distributed with means m, 0, and 0 and precisions h, h_1, and h_2. (Recall that the precision is the inverse of the variance, and that the expectation of a variable given independent normally distributed signals—including the prior—is the weighted average of the signals, where the weights are the precisions.) Show that there exists an equilibrium in which $a_e + a_u$ perfectly reveals z_1 to the arbitrator, and in which each party $i = e, u$ offers $a_i = (hm + h_1 z_1)/(h + h_1) + k_i$, where k_i is a constant.

Exercise 8.3* Gilligan and Krehbiel (1988) depict the open rule in Congress as a cheap-talk game, that is, as a signaling game in which signals are costless. As a rough approximation, the committee proposes a policy, but the floor can introduce amendments and choose the policy it likes. The open rule is depicted as a two-player game, with a single member in the committee and a single representative on the floor (who stands for the median voter). The object of the decision is a policy a_2 in $\mathbb{R}$. The outcome given policy a_2 is $x = a_2 + \omega$, where ω is a random variable uniformly distributed between 0 and 1. The committee knows ω; the floor does not. The committee moves first and suggests a policy a_1 to the floor. The preferences of both are quadratic with bliss points $x = 0$ for the floor and $x = x_c \in (0, 1)$ for the committee: $u_1(x) = -(x - x_c)^2$ and $u_2(x) = -x^2$.

(a) Show that there always exists a "babbling" equilibrium in which a_1 is uninformative and $a_2 = -\frac{1}{2}$.

(b) Look for informative perfect Bayesian equilibria. In particular, find an equilibrium in which the committee "reports low" when $\omega \in [0, \omega^*]$ and "reports high" when $\omega \in (\omega^*, 1]$.

(c)** Analyze the closed rule, in which the committee proposes a policy a_1 and the floor chooses between a_1 and a reversion or *status quo* policy a_0. (Note that this is no longer a cheap-talk game.)

See Crawford and Sobel 1982 for the first example of a cheap-talk game.

Exercise 8.4** Consider the Chatterjee-Samuelson simultaneous-offer bargaining game developed in chapter 6. Assume that the buyer's valuation v and the seller's cost c are independently and uniformly distributed on $[0, 1]$. They make simultaneous offers. They trade if the seller's bid b_1 is less than the buyer's bid b_2 at a price $p = (b_1 + b_2)/2$. Add a preplay communication stage to this Chatterjee-Samuelson model. That is, before choosing their bids, the traders simultaneously send a message to each other. These messages are costless (are cheap talk). Show that the equilibria discussed in chapter 6 are still equilibria (the messages are simply ignored). But there exist other equilibria as well. For instance, show that the following is a perfect Bayesian equilibrium: Each trader announces either "keen" or "not keen" in the preplay communication stage. The buyer announces "keen" if and only if $v > v^* = (22 + 12\sqrt{2})/49$. The seller says "keen" if and only if $c < c^* = 1 - v^*$. If they both say "not keen," they "stop bargaining" (i.e., they play the continuation equilibrium in which they make nonserious offers, such as 0 for the buyer and 1 for the seller). If one of them says "keen" and the other "not keen," they play a Chatterjee-Samuelson linear equilibrium given posterior beliefs. If they both say "keen," both bids are equal to $\frac{1}{2}$. (See Farrell and Gibbons 1989 for the answer.)

Exercise 8.5*** Exercises 8.3 and 8.4 involve a player's transmission of information that is not verifiable by the other players. This exercise involves a two-stage game of transmission of verifiable information. There are I players, $i = 1, \ldots, I$. Player i's types, θ_i, belong to some finite ordered set Θ (for instance, of elements of the real line). Types are drawn independently from the prior distribution $p(\theta) = \prod_{i=1}^{I} p_i(\theta_i)$. In period 2, the players play some simultaneous-move game that results in (reduced-form) payoffs $v_i((\mu_i, \mu_{-i}), \theta_i)$ for player i, where μ_i is the posterior beliefs about θ_i and where $\mu_{-i} \equiv \prod_{j \neq i} \mu_j$ is the posterior beliefs about θ_{-i}. (Are we allowed to write posterior beliefs as a product if we look for sequential equilibrium?)

Beliefs μ_i (first-order) stochastically dominate beliefs μ_i' if for all θ_i

$$\sum_{\tilde{\theta}_i \leq \theta_i} \mu_i(\tilde{\theta}_i) \leq \sum_{\tilde{\theta}_i \leq \theta_i} \mu_i'(\tilde{\theta}_i),$$

with a strict inequality for at least some θ_i. Assume that each player prefers his opponents to believe he has "high types": For any μ_{-i} and θ_i, if μ_i stochastically dominates μ_i',

$$v_i((\mu_i, \mu_{-i}), \theta_i) > v_i((\mu_i', \mu_{-i}), \theta_i).$$

In the first period, players simultaneously announce messages $a_i^1 \in A_i^1(\theta_i)$. Messages do not enter the players' payoff function v_i. However, they affect beliefs for the second-stage game. Suppose that, for all i and θ_i, $A_i^1(\theta_i)$ contains a message that certifies that player i's type is at least equal to θ_i. Show that in a sequential equilibrium posterior beliefs are degenerate; that is, the first-period messages are fully revealing. (See Grossman 1980, Grossman and Hart 1980, and Milgrom 1981 for this result with one informed player; see Okuno-Fujiwara et al. 1990 for the many-informed-players version.)

Exercise 8.6* Introduce asymmetric information in the stag-hunt game of chapter 1. There are two players who must decide whether to hunt the stag or the rabbit. With probability p, each player has preferences that always make him hunt the stag (he does not like rabbit, or he is able to catch the stag by himself although he would prefer to hunt with the other player); with probability q, each player always hunts the rabbit (he does not like stag); with probability $1 - p - q$, the player has the preferences described in chapter 1: He gets 1 if he hunts the rabbit, 2 if both hunt the stag, and 0 if he hunts the stag alone. Suppose that $2p > 1 - q$ and $2q > 1 - p$.

(a) Show that in the one-period version of the game there is a multiplicity of equilibria similar to the one in chapter 1 if $\max(p, q) < \frac{1}{2}$. Show that the equilibrium is unique if $p > \frac{1}{2}$ or $q > \frac{1}{2}$.

(b) Consider the two-period version of the above stag-hunt game with incomplete information. Show that for any first-period behavior the second-period behavior is uniquely determined.

(c) Assume that the discount factor between the periods is equal to 1, and that $\alpha \equiv (1 + 2p)/4 \in (p, 1 - q)$ (which implies that $p < \frac{1}{2}$). Show the existence of a symmetric equilibrium, in which each player hunts the stag with probability α in the first period. Note that the type who does not have a dominant strategy sacrifices short-run utility to build a "reputation." (Reputational phenomena are studied in much detail in chapter 9.) Show that there are exactly two other equilibria.

Exercise 8.7** Consider the following two-player three-stage game with incomplete information. In each period, player 1 has three possible actions: S (hunt the stag), R (hunt the rabbit), and H (stay home), and player 2 has two possible actions: S (hunt the stag) and R (hunt the rabbit). Player 1 has one of three equally likely types: s, r, and h. In any given period, if player 1 has type s he gets 1 if both players play S, and 0 otherwise. Similarly, type r gets 1 if both players play R and 0 otherwise, and type h gets 1 if he plays H and 0 otherwise. Player 2 has no private information. In a given period, he gets 1 if both play S or both play R and 0 otherwise. Is there a sequential equilibrium in which the following observation has positive probability: player 1 plays H in period 0, both players play S in period 1, and player 2 plays R in period 2?

Exercise 8.8*** Consider a three-player two-stage game with incomplete information. Only players 1 and 2 have private information: Player i's type, θ_i, is equal to θ_i' or θ_i'' with equal probabilities, for $i = 1, 2$. Furthermore, the prior beliefs satisfy $\mathrm{Prob}(\theta_1 = \theta_1'|\theta_2 = \theta_2') \equiv \mathrm{Prob}(\theta_1 = \theta_1''|\theta_2 = \theta_2'') = \frac{3}{4}$. Suppose that in a sequential equilibrium player i plays a_i^* in the first period whatever his type. Determine the set of player 3's joint probability distributions over (θ_1, θ_2) at the beginning of the second period that are compatible with sequential equilibrium when player 1 plays a_1^* but player 2 deviates in the first period.

Exercise 8.9*

(a) Show that (U_1, L_1) is the unique perfect equilibrium of the game illustrated in figure 8.19a.

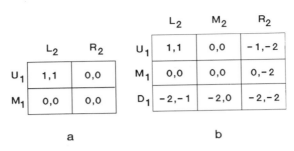

Figure 8.19

(b) Use figure 8.19b to argue that adding dominated strategies may enlarge the set of perfect equilibria. (This exercise is from van Damme 1987.)

Exercise 8.10* Consider the following signaling game. There are two players, a plaintiff and a defendant in a civil suit. The plaintiff knows whether or not he will win the case if it goes to trial, but the defendant does not have this information. The defendant knows that the plaintiff knows who would win, and the defendant has prior beliefs that there is probability $\frac{1}{3}$ that the plaintiff will win; these prior beliefs are common knowledge. If the plaintiff wins, his payoff is 3 and the defendant's payoff is -4; if the plaintiff loses, his payoff is -1 and the defendant's is 0. (This corresponds to the defendant paying cash damages of 3 if the plaintiff wins, and the loser of the case paying court costs of 1.)

The plaintiff has two possible actions: He can ask for either a low settlement of $m = 1$ or a high settlement of $m = 2$. If the defendant accepts a settlement offer of m, the plaintiff's payoff is m and the defendant's is $-m$. If the defendant rejects the settlement offer, the case goes to court. List all the pure-strategy perfect Bayesian equilibria (PBE) strategy profiles. For each such profile, specify the posterior beliefs of the defendant as a function of m, and verify that the combination of these beliefs and the profile is in fact a PBE. Explain why the other profiles are not PBE.

References

Akerlof, G. 1970. The market for "Lemons." *Quarterly Journal of Economics* 90: 629–650.

Aumann, R., and M. Machler. 1966. Game-theoretic aspects of gradual disarmament. *Mathematica ST-80*, chapter V, 1–55.

Battigalli, P. 1991. Strategic independence, generally reasonable extended assessments, and consistent assessments. Mimeo.

Blume, L., A. Brandenberger, and E. Dekel. 1990. Equilibrium refinements and lexicographic probabilities. *Econometrica*, forthcoming.

Crawford, V., and J. Sobel. 1982. Strategic information transmission. *Econometrica* 50: 1431–1452.

Farrell, J., and R. Gibbons. 1989. Cheap talk can matter in bargaining. *Journal of Economic Theory* 48: 221–237.

Fudenberg, D., and J. Tirole. 1991. Perfect Bayesian equilibrium and sequential equilibrium. *Journal of Economic Theory* 53: 236–260.

Gibbons, R. 1988. Learning in equilibrium models of arbitration. *American Economic Review* 78: 896–912.

Gilligan, T., and K. Krehbiel. 1988. Collective choice without procedural commitment. Discussion paper 88-8, Hoover Institution, Stanford University.

Grossman, S. 1980. The role of warranties and private disclosure about product quality. *Journal of Law and Economics* 24: 461–483.

Grossman, S., and O. Hart. 1980. Disclosure laws and takeover bids. *Journal of Finance* 35: 323–334.

Harsanyi, J. 1967–68. Games with incomplete information played by Bayesian players. *Management Science* 14: 159–182, 320–334, 486–502.

Kohlberg, E., and J.-F. Mertens. 1986. On the strategic stability of equilibria. *Econometrica* 54: 1003–1038.

Kreps, D., and G. Ramey. 1987. Structural consistency, consistency, and sequential rationality. *Econometrica* 55: 1331–1348.

Kreps, D., and R. Wilson. 1982a. Sequential equilibrium. *Econometrica* 50: 863–894.

Kreps, D., and R. Wilson. 1982b. Reputation and imperfect information. *Journal of Economic Theory* 27: 253–279.

Milgrom, P. 1981. Good news and bad news: Representation theorems and applications. *Bell Journal of Economics* 12: 380–391.

Milgrom, P., and J. Roberts. 1982a. Limit pricing and entry under incomplete information. *Econometrica* 50: 443–460.

Milgrom, P., and J. Roberts. 1982b. Predation, reputation, and entry deterrence. *Journal of Economic Theory* 27: 280–312.

Myerson, R. 1978. Refinements of the Nash equilibrium concept. *International Journal of Game Theory* 7: 73–80.

Myerson, R. 1985. Bayesian equilibrium and incentive compatibility: An introduction. In *Social Goals and Social Organization: Essays in Honor of Elizha Pazner*, ed. L. Hurwicz, D. Schmeidler, and H. Sonnenschein. Cambridge University Press.

Myerson, R. 1986. Multistage games with communication. *Econometrica* 54: 323–358.

Okuno-Fujiwara, M., A. Postlewaite, and K. Suzumura. 1990. Strategic information revelation. *Review of Economic Studies* 57: 25–47.

Ortega-Reichert, A. 1967. Models for competitive bidding under uncertainty. Ph.D. thesis, Stanford University.

Selten, R. 1975. Reexamination of the perfectness concept for equilibrium points in extensive games. *International Journal of Game Theory* 4: 25–55.

Spence, A. M. 1974. *Market Signalling*. Harvard University Press.

van Damme, E. 1987. *Stability and Perfection of Nash Equilibria*. Springer-Verlag.

Reputation Effects

9.1   Introduction[††]

This chapter investigates the notion that a player who plays the same game repeatedly may try to develop a reputation for certain kinds of play. The idea is that if the player always plays in the same way, his opponents will come to expect him to play that way in the future and will adjust their own play accordingly. The question then is when and whether a player will be able to develop or maintain the reputation he desires. For example, if a central bank always implements the monetary policy it announces, will traders come to believe that it will do so in the future? That is, will reputation effects allow the central bank to effectively commit itself to implementing its announcements?

To model the possibility that players are concerned about their reputations, we suppose that there is incomplete information about each player's type, with different types expected to play in different ways. Each player's reputation is then summarized by his opponents' current beliefs about his type. For example, to model a central bank's reputation for sticking to its announced monetary policy, we assign positive prior probability to a type that always implements its announcements.[1] More generally, we may suppose that each player has several different types, each of which is associated with a different kind of play, and that no player's type enters directly as an argument into any other player's utility function.[2]

Intuitively, since reputations are like assets, a player is most likely to be willing to incur short-run costs to build up his reputation when he is patient and his planning horizon is long. A player with a short horizon will be less willing to make investments, so we should expect that investments in reputation will be more likely in long relationships than in short ones, and more likely at the beginning of a game than at its end. For this reason we will follow the literature and focus on reputations in long-run relationships, although reputations can also play an important role in short-run relationships.

1. An alternative approach is to identify reputations with equilibrium strategies in a repeated game of complete information. For example, in the repeated prisoner's dilemma the equilibrium in the "grim" strategies "cooperate until an opponent defects, and then defect thereafter" can be interpreted as describing a situation where each player has a "reputation" for cooperation that vanishes the first time he defects, and in a repeated quality-choice game the strategies "expect high quality until the firm produces low quality" can be interpreted as saying that the firm begins with a reputation for high quality that it can only maintain by making high-quality output. Of course, this reinterpretation does not change the set of equilibria, and so this version of reputation does not have predictive power. Also, modeling reputations as complete-information strategies cannot capture the idea that a player's reputation corresponds to something that his opponents have learned about him.
2. This is a narrower meaning of reputation than that suggested by common usage. For example, one might speak of a worker having a "reputation" for high productivity in the Spence signaling model, and of high-productivity workers investing in this reputation by choosing high levels of education.

Our main concern, then, is when and whether a long-lived player can take advantage of a small prior probability of a certain type or reputation to effectively commit himself to playing as if he were that type. For example, what prior distributions on types imply that in equilibrium a central bank's announcements will be credible?

A related question is whether models of reputation effects provide a way to pick and choose among the many equilibria of an infinitely repeated game, and in particular whether reputation effects can provide support for the intuitions that certain of these equilibria are particularly reasonable. For example, although many papers have used the "cooperative" equilibrium of the repeated prisoner's dilemma (chapter 4) to explain trust and cooperation in long-run relationships, there is also an equilibrium where players do not cooperate. Similarly, though a rough analog of the folk theorem holds in games where a single, long-run, patient player faces a sequence of short-run opponents (see subsection 5.3.1), economic applications typically examine only the equilibrium the long-run player most prefers. For example, a long-run firm that faces a sequence of short-run consumers may choose to produce high-quality output, even though doing so is more expensive in the short run, because switching to low quality would cost it sales in the future (Dybvig and Spatt 1980; Shapiro 1982). However, there is another equilibrium where the firm's quality is always low.

This case of a single long-run player is the one in which reputation effects have the strongest and most general implications. We discuss this case in section 9.2, beginning with the work on the chain-store paradox. Since there is only one player who has an incentive to maintain a reputation, it may not be surprising that reputation effects are quite powerful: In a simultaneous-move stage game, a weak full-support distribution on the prior distribution implies that a single patient player can use reputation effects to obtain the payoff he would obtain if he could publicly commit himself to whatever strategy he most prefers.

Another case where reputation effects might be thought to allow one player to commit himself is that of a single "large" player facing a great many long-lived but "small" opponents, since the large player has much more to gain from a successful commitment than his opponents. (One reason this case of small opponents is interesting is that it may be a better description of the situation facing a government entity such as the Internal Revenue Service or the Federal Reserve than the model in which the government entity is a long-run player facing a sequence of short-run private individuals.) Whether reputation effects allow the large player to commit himself turns out to depend on the fine structure of the game, as we discuss in section 9.4.

When all players are long-run, as in the repeated prisoner's dilemma, there is no single player whose interests might be expected to dominate play, and so it would seem unlikely that reputation effects could lead to strong general conclusions. It is true that strong results can be obtained for specific prior distributions over types. For example, in the repeated prisoner's dilemma, if player 2's payoffs are known to be as in the usual complete-information case, while player 1 is either a type who always plays the strategy "tit for tat" or a type with the usual payoffs, then with sufficiently patient players and a long finite horizon every sequential equilibrium has both players cooperate in almost every period. However, other outcomes can be obtained by varying the prior distribution; in fact, any feasible, individually rational payoffs of the complete-information game can be obtained as sequential equilibrium payoffs of an incomplete-information version of the game where the payoffs are the complete-information ones with probability close to 1. This confirms the intuition that reputation effects on their own have little power when all players are long-run. Reputation effects do pick out the unique Pareto-optimal payoffs in games of pure coordination when the prior distribution on types is restricted in a particular way. Section 9.3 presents these results.

9.2   Games with a Single Long-Run Player[††]

9.2.1   The Chain-Store Game

We begin with a discussion of the work by Kreps and Wilson (1982) and Milgrom and Roberts (1982) on reputation effects in Selten's (1978) chain-store game. To set the stage for their work, we will first review a slight variant of Selten's model. A single long-run incumbent firm faces potential entry by a series of short-run firms, each of which plays only once but observes all previous play. Each period, a potential entrant decides whether to enter or stay out of a particular market. (Each entrant can enter only a single market, and the entrants' markets are distinct.) If the entrant stays out, the incumbent enjoys a monopoly in that market; if the entrant enters, the incumbent must choose whether to fight or to accommodate. The incumbent's payoffs are $a > 0$ if the entrant stays out, 0 if the entrant enters and the incumbent accommodates, and -1 if the entrant enters and the incumbent fights. The incumbent's objective is to maximize the discounted sum of its per-period payoffs; δ denotes the incumbent's discount factor. Each entrant has two possible types: tough and weak. Tough entrants always enter. A weak entrant has payoff 0 if it stays out, -1 if it enters and is fought, and $b > 0$ if it enters and the incumbent accommodates. Each entrant's type is private information, and each is tough with probability q^0 independent of the others. Thus, the incumbent has a short-run incentive to accommodate, whereas a weak entrant will enter only if it expects the probability of fighting to be less than $b/(b + 1)$.

If the game has a finite horizon, there is a unique sequential equilibrium, as Selten (1978) observed: The incumbent accommodates in the last period, so the last entrant enters whatever his type and the history of the game; thus, the incumbent accommodates in the next-to-last period, and by backward induction the incumbent always accommodates and every entrant enters. Selten called this a "paradox" because when there are a large number of entrants the equilibrium seems counterintuitive: One suspects that the incumbent would be tempted to fight to try to deter entry. Of course, no matter how often the incumbent fights, he cannot deter the "tough" entrants, so a commitment to always fight is valuable only if the resulting per-period expected payoff of $a(1 - q^0) - q^0$ exceeds the zero payoff from always accommodating. When this is the case, and when the incumbent's discount factor is close enough to 1, the infinite-horizon version of the model has an equilibrium where entry is deterred.[3]

Since there is also an infinite-horizon equilibrium where every entrant enters, this is only partial support for the intuition that entry deterrence is the reasonable outcome, and we are left with the puzzle of explaining why the entry-deterrence equilibrium is most plausible. In addition, we might believe that the outcome would be entry deterrence even with a fixed finite horizon. As we will see, allowing for reputation effects by introducing incomplete information responds to both of these points, and it does so in an intuitively appealing way: The incumbent fights to maintain its reputation for being a "tough" type that is likely to fight. After all, if the incumbent were to have fought in each of the preceding 100 periods, then it seems (to us) quite plausible that the next entrant should expect that it is likely to be fought!

To introduce reputation effects into the model, suppose that all players' payoffs are private information. With probability p^0, the incumbent is "tough," meaning that its payoffs are such that it will fight in every market along any equilibrium path.[4] The incumbent is "weak" (i.e., has the payoffs described above) with probability $1 - p^0$. And each entrant is "tough" with probability q^0, independent of the others; tough entrants enter regardless of how they expect the incumbent to respond.[5]

3. This was observed by Milgrom and Roberts (1982). One such equilibrium is for the incumbent to fight all entrants so long as it has never accommodated and to accommodate entrants if it has accommodated at least once in the past, and for the entrants to stay out if the incumbent has never accommodated and enter if it ever does accommodate. This profile is an equilibrium if $a(1 - q^0) - q^0 > (1 - \delta)/\delta$.
4. To construct such payoffs, it suffices that the tough type's payoff be equal to -1 times the number of times it fails to fight (or, more generally, the number of times it fails to follow the prescribed behavior). Alternatively, one can suppose that tough incumbents are simply unable to accommodate. Note also that the incumbent's type is chosen once and for all at the start of the game: The incumbent is either tough in all markets or tough in none of them.
5. Our presentation of the chain-store game is based on the summary by Fudenberg and Kreps (1987). Kreps and Wilson consider only the case $q^0 = 0$; Milgrom and Roberts consider a richer specification of payoffs.

To solve for the sequential equilibrium of the finite-horizon version of this game, we will first solve for the sequential equilibrium of the one-period game, then that of the two-period game, and proceed by induction to solve for the game with N periods. It is easy to determine the sequential equilibrium of a single play of this game: If there is entry, the incumbent accommodates if and only if it is weak, so that a weak entrant nets $(1 - p^0)b - p^0$ from entry. Thus, a weak entrant enters if $p^0 < b/(b + 1) \equiv \bar{p}$ and stays out if the inequality is reversed. (We ignore the knife-edge case of equality.)

Now imagine that there are two periods remaining in the game—the incumbent will play two different entrants in succession, in two different markets. Entrant 2 is faced first, and entrant 1 observes the outcome in market 2 before making its own entry decision.[6] The nature of the equilibrium depends on the prior probabilities and the parameters of the payoff functions:

(i) If $1 > a\delta(1 - q^0)$ or $q^0 > \bar{q} \equiv (a\delta - 1)/a\delta$, the maximum long-run benefit of fighting ($\delta a(1 - q^0)$) is less than its cost (which is 1), so a weak incumbent will not fight in market 2. Since the tough incumbent will fight, a weak entrant 2 enters if $p^0 < \bar{p}$ and stays out if $p^0 > \bar{p}$. A weak entrant 1 enters if the incumbent accommodates in market 2 and stays out if the incumbent fights.

(ii) If $q^0 < \bar{q}$, the weak incumbent is willing to fight in market 2 if doing so deters entry, since accommodating reveals that the incumbent is weak and causes entry to occur. In this case, if entrant 2 enters, the weak incumbent must fight with positive probability: It cannot be a sequential equilibrium for the weak incumbent to accommodate with probability 1 in market 2, as then if the incumbent fights the entrants believe he is tough, and so fighting deters entry next period.

The exact nature of the equilibrium again depends on the prior p^0 that the incumbent is tough.

(iia) If $p^0 > \bar{p}$, then, since the tough incumbent always fights, the posterior probability that the incumbent is tough given that he fights in market 2 is at least p^0, and so fighting in market 2 deters a weak entrant in market 1. Thus, the weak incumbent fights with probability 1 in market 2, the weak entrant stays out of market 2, and the weak incumbent's expected payoff is $[(1 - q^0)a - q^0] + \delta(1 - q^0)a$.

6. Example 8.1 considered a simplified version of this game in which entrant 2 has already entered, entrant 1 is assumed to always be "weak," the incumbent's decision in the final market as a function of its type has been solved out, and the discount factor δ equals 1. There we saw that if the cost of fighting today exceeds the gain from monopoly tomorrow—that is, if $a < 1$—then in the unique equilibrium the weak incumbent accommodates, whereas if $a > 1$ then in the unique equilibrium the weak incumbent fights with positive probability.

(iib) If $p^0 < \bar{p}$, it is not an equilibrium for the weak incumbent to fight with probability 1, as then the posterior probability of toughness after fighting would not deter entry, and the weak incumbent would prefer not to fight. Nor can it be an equilibrium for the weak incumbent to accommodate with probability 1, for then fighting would deter entry and the weak incumbent would prefer to fight. Thus, in equilibrium the weak incumbent must randomize, which requires that when the incumbent fights in market 2 the weak entrant 1 randomizes in a way that makes the weak incumbent indifferent in market 2. This, in turn, requires that the posterior probability that the incumbent is tough, conditional on fighting, be exactly the critical level $\bar{p} = b/(b + 1)$. If we let β be the conditional probability that a weak incumbent fights entry in market 2, and recall that the tough incumbent fights with probability 1, Bayes' rule gives

$$\text{Prob(tough|fight)} = p^0/[p^0 + \beta(1 - p^0)],$$

and for this to equal $\bar{p}$, β must equal $p^0/(1 - p^0)b$. The total probability that entry in market 2 is fought is

$$p^0 \cdot 1 + (1 - p^0) \cdot [p^0/(1 - p^0)b] = p^0(b + 1)/b,$$

so the weak entrant will stay out of market 2 if $p^0 > [b/(b + 1)]^2 = \bar{p}^2$. In this case the weak incumbent's expected average payoff is positive, whereas its payoff was 0 for the same parameters in the one-entrant game. If $p^0 < [b/(b + 1)]^2$, the weak entrant enters in market 2, and the weak incumbent's payoff is 0.

Now we can see what happens with three periods remaining: If $p^0 > [b/(b + 1)]^2$, the weak incumbent is certain to fight in market 3, and the weak entrant stays out. If p^0 is between $[b/(b + 1)]^3$ and $[b/(b + 1)]^2$, the weak incumbent randomizes and the weak entrant stays out; if $p^0 < [b/(b + 1)]^3$, the weak incumbent randomizes and the weak entrant enters. More generally, for a fixed p^0 and N entrants, the weak entrant stays out until the first period k where $p^0 < [b/(b + 1)]^k$, so that for the first $N - k$ periods the weak incumbent has expected payoff $a(1 - q^0) - q^0$ per period.

The main point of the Kreps-Wilson and Milgrom-Roberts papers is that the size of the prior p^0 required to deter entry (when q^0 is sufficiently small) shrinks as the number of periods grows; indeed, it shrinks geometrically at the rate $b/(b + 1)$. Thus, even a small amount of incomplete information can have a very large effect in long games. When $\delta = 1$, the unique equilibrium has the following form:

(a) If $q^0 > a/(a + 1)$, then the weak incumbent accommodates at the first entry, which occurs (at the latest) the first time the entrant is tough. Hence, as the number of markets N tends to infinity, the incumbent's average payoff per period goes to 0.

(b) If $q^0 < a/(a + 1)$, then for every p^0 there is a number $n(p^0)$ so that if there are more than $n(p^0)$ markets remaining, the weak incumbent's strategy is to fight with probability 1. Thus, weak entrants stay out when there are more than $n(p^0)$ markets remaining, and the incumbent's average payoff approaches $(1 - q^0)a - q^0$ as $N \to \infty$.[7]

It is easy to explain the role played by the expression $a(1 - q^0) - q^0$ in the above. Imagine that the incumbent is given a choice at time 0 of making an observed and enforceable commitment either to always fight or to always accommodate. If the incumbent always fights, its expected payoff is $a(1 - q^0) - q^0$, as it must fight the tough entrants to deter the weak ones. The asymptotic nature of the equilibrium turns exactly on whether a commitment to always fight is better than a commitment to always accommodate, which yields payoff 0. Thus, one interpretation of the results is that reputation effects allow the incumbent to credibly make whichever of the two commitments it prefers.

Note, however, that neither of these commitments need be the one the incumbent would like most. If $a(1 - q^0) > q^0$, the incumbent is willing to fight the tough entrants to deter the weak ones, but it would do even better if it could commit itself to fight with the smallest probability that deters weak entrants, which is $b/(b + 1)$. This yields it an average payoff of $a(1 - q^0) - q^0 b/(b + 1)$, which is greater than the payoff $a(1 - q^0) - q^0$ from fighting with probability 1. Of course, when the prior distribution over the incumbent's types assigns positive probability only to the weak type and to a type that fights with probability 1, the incumbent cannot develop a reputation for fighting with a positive probability less than 1, as the first time that the incumbent accommodates it is revealed to be weak and its reputation is ruined. The next subsection discusses whether it is reasonable for the incumbent to be able to maintain a reputation for playing a mixed strategy, and shows how to change the model to make mixed-strategy reputations possible.

Although the commitment interpretation of reputation suggests that reputation effects are a "good thing" for the incumbent, this depends on the exact comparison that one has in mind. It is clear that the weak incumbent cannot lose from the fact that the entrants fear that it might be tough. An alternative comparison is to hold fixed the prior probabilities at p^0 and q^0 and to compare the game described above, where each entrant observes play in all previous markets, with the situation where each stage game is played in "informational isolation," meaning that the timing of play and the payoffs are as above but entrants do not observe play in other markets.

---

7. Note that we fix p^0 and take the limit as $N \to +\infty$. For fixed N and sufficiently small p^0, the weak incumbent must accommodate in each market in any sequential equilibrium. Exercise 9.1 asks you to extend this characterization to discount factors less than but close to 1.

Under informational isolation, the weak incumbent has no chance to build a reputation, and will accommodate in each market. Yet the weak incumbent's equilibrium payoff can be higher under informational isolation than in the "informational linkage" case where each entrant observes all past play. The reason is that informational linkage imposes a cost that Fudenberg and Kreps (1987) call a loss of "strategic flexibility": Under informational linkage the weak incumbent loses the ability to deter weak entrants while accommodating tough ones. Put differently, under linkage the incumbent must fight the tough entrants to deter the weak ones. When the cost of doing so is too high, the weak incumbent may choose not to develop a reputation for toughness (and hence get payoff 0).

Even when the weak incumbent does develop a reputation for toughness his payoff can be lower under informational linkage than under informational isolation. In the simple chain-store model, this is the case when $p^0 > \bar{p}$, so that under informational isolation weak entrants do not enter and the weak incumbent has payoff $a(1 - q^0)$ per market. Under informational linkage, the weak incumbent does worse: His average payoff per market is $\max\{0, a(1 - q^0) - q^0\}$. Thus, although the incumbent may choose to develop a reputation given that markets are informationally linked, he might have been better off in a regime of informational isolation, where reputation building is not possible. More generally, informational linkage has both costs and benefits, and it is not obvious *a priori* when the benefits outweigh the costs.

9.2.2 Reputation Effects with a Single Long-Run Player: The General Case

If we view reputation effects as a way of supporting the intuition that the long-run player should be able to commit himself to any strategy he desires, the chain-store example raises several questions: Does the strong conclusion derived above depend on the fixed finite horizon, or do reputation effects have a similar impact in the infinitely repeated version of the game? Can the long-run player maintain a reputation for playing a mixed strategy when such a reputation would be desirable? How robust are the strong conclusions in the chain-store game to changes in the prior distribution to allow more possible types? And how does the commitment result extend to games with different payoffs and/or different extensive forms? What if the incumbent's action is not directly observed, as in a model of moral hazard?

To answer the first question—the role of the finite horizon—consider the infinite-horizon version of the game of the preceding subsection with $\delta > 1/(1 - q^0)(1 + a)$, so that even if the incumbent is known to be weak there is still an equilibrium where entry is deterred. If there is a prior probability $p^0 > 0$ that the incumbent is tough, entry deterrence is still an equilibrium. In this equilibrium, the weak incumbent fights all entrants, because the first time it fails to do so it is revealed to be weak and then all subsequent entrants enter and the weak incumbent accommodates from then on. However, this is not the only perfect Bayesian equilibrium of the infinite-horizon model. Here is another one: "The tough incumbent always fights. The weak incumbent accommodates the first entry, and then fights all subsequent entry if it has not accommodated two or more times in the past. Once the incumbent has accommodated twice, it accommodates all subsequent entry. Tough entrants always enter; weak entrants enter if there has been no previous entry or if the incumbent has already accommo- dated at least twice; weak entrants stay out otherwise." In this equilibrium, the weak incumbent reveals its type by accommodating in the first period; the incumbent is willing to do so because subsequent entrants stay out even after the incumbent's type is revealed.

These two equilibria (there are many more) show that reputation effects need not determine a unique equilibrium in an infinite-horizon model. At the same time, note that if the incumbent is patient it does almost as well here as in the equilibrium where all entry is deterred, so the second equilib- rium does not show that reputation effects have no force. Finally, the multi- plicity of equilibria suggests that it might be more convenient to try to characterize the set of equilibria without determining all of them explicitly.

This is the approach used by Fudenberg and Levine (1989, 1991). They extend the intuition developed in the chain-store example to general games where a single long-run player faces a sequence of short-run opponents. To generalize the introduction of a "tough type" in the chain-store game, they suppose that the short-run players assign positive prior probability to the long-run player's being one of several different "commitment types," each of which plays a particular fixed stage-game strategy in every period. The set of commitment types thus corresponds to the set of possible "rep- utations" that the long-run player might maintain. Instead of explicitly determining the set of equilibrium strategies, they obtain upper and lower bounds on the long-run player's payoff that hold in any *Nash* equilibrium of the game. (The 1991 paper allows the long-run player's actions to be imperfectly observed, as in the Cukierman-Meltzer (1986) model of the reputation of a central bank when the other players observe the realized inflation rate but not the bank's action.[8])

The upper bound on the long-run player's Nash-equilibrium payoff converges, as the number of periods grows and the discount factor goes to 1, to the long-run player's Stackelberg payoff, which is the most he could obtain by publicly committing himself to any of his stage-game strategies. If the short-run players' actions do not influence the information that is revealed about the long-run player's choice of stage-game strategy (as in a simultaneous-move game with observed actions), the lower bound on

---

8. Other models of reputation with imperfectly observed actions include those of Bénabou and Laroque (1989) and Diamond (1989).

payoffs converges to the most the long-run player could get by committing himself to any of the strategies for which the corresponding commitment type has positive prior probability. If moves in the stage game are not simultaneous, the lower bound must be modified, as we explain in subsection 9.2.3.

Consider a single long-run player 1 facing an infinite sequence of short-run player 2s in a "stage game" where players choose stage-game strategies a_i from finite sets A_i. Subsection 9.2.3 will allow the stage game to be a general, finite extensive form. This subsection treats the case where the stage game has simultaneous moves and the players' actions are revealed at the end of each period. Also, for the rest of this section, we consider infinite-horizon models; however, theorem 9.1 extends directly to the finite-horizon case. The history h^t at time t consists of past choices $(a_1^\tau, a_2^\tau)_{\tau=0,\dots,t-1}$. (Note that we now revert to counting time forward, instead of the backward counting we used in discussing the finite-horizon chain-store game. Note also that if the stage game has sequential moves it is not natural to suppose that the observed outcome at the end of period τ reveals the stage-game strategies a^τ the players used, as in a sequential-move game the a^τ prescribe play at information sets that may not be reached.) The long-run player's type, $\theta \in \Theta$, is private information; θ influences player 1's payoff but has no direct influence on player 2's payoff; θ has prior distribution p, which is common knowledge. Player 1's strategy is a sequence of maps σ_1^t from the set of possible histories H^t and the set of types Θ to the space of mixed stage-game actions $\mathcal{A}_1$; a strategy for the period-t player 2 is $\sigma_2^t: H^t \to \mathcal{A}_2$.

Since the short-run players are unconcerned about future payoffs, in any equilibrium each period's choice of mixed strategy α_2 will be a best response to the anticipated marginal distribution over player 1's actions. Let $r: \mathcal{A}_1 \rightrightarrows \mathcal{A}_2$ be the short-run player's best-response correspondence.

Two subsets of the set Θ of player 1's types are of particular interest. Types $\theta_0 \in \Theta_0$ are "sane types" whose preferences correspond to the expected discounted value of per-period payoffs $g_1(a_1, a_2, \theta_0)$. All sane types are assumed to use the same discount factor, δ, and to maximize their expected present discounted payoffs. (The chain-store papers had a single "sane type" whose probability was close to 1.) The "commitment types" are those who play the same stage-game strategy in every period; $\theta(\alpha_1)$ is the commitment type corresponding to α_1. The set of commitment strategies $C_1(p)$ are those for which the corresponding commitment strategies have positive prior probability under distribution p. We will present the case where Θ and thus C_1 are finite.

Define the *Stackelberg payoff* for $\theta_0 \in \Theta_0$ to be

$$g_1^s(\theta_0) = \max_{\alpha_1} \left[\max_{\alpha_2 \in r(\alpha_1)} g_1(\alpha_1, \alpha_2, \theta_0) \right],$$

and let the Stackelberg strategy be one that attains this maximum. This is the highest payoff type θ_0 could obtain if he could commit himself to always play any of his stage-game actions (including mixed actions). Note that the Stackelberg strategy need not be pure, as we saw in the chain-store game.

Note also that, since the long-run player's opponents are myopic, the long-run player could not do better than the Stackelberg payoff by committing himself to a strategy that varies over time with his opponents' past actions. If the opponents were themselves long-run players, player 1 might be able to do better than the Stackelberg payoff by using a strategy that induces the opponents not to play static best responses to avoid future punishment, as in the prisoner's-dilemma example we consider below. The support of p is allowed to include types who play such history-dependent strategies.

Given the set of possible (static) "reputations" $C_1(p)$, we ask which reputation from this set type θ_0 would most prefer, given that the short-run players may choose the best response that the long-run player likes least. This results in payoff

$$g_1^*(p, \theta_0) = \max_{\alpha_1 \in C_1(p)} \left[\min_{\alpha_2 \in r(\alpha_1)} g_1(\alpha_1, \alpha_2, \theta_0) \right].$$

The formal model allows for commitment types who play mixed strategies. Is this reasonable? Suppose that the incumbent has fought in 50 of the 100 periods to date where entry has occurred, and moreover that the distribution of "fight" versus "accommodate" looks consistent with the hypothesis of independent 50-50 randomization (i.e., tests based on run length do not reject independence). How then should the entrants predict that the incumbent will play? One can argue that at this point the entrants should assign a probability of about $\frac{1}{2}$ to the incumbent's fighting the next entrant, as opposed to their being certain that the incumbent will accommodate.[9]

Let $\underline{N}(\delta, p, \theta_0)$ and $\bar{N}(\delta, p, \theta_0)$ be the lowest and highest payoffs of type θ_0 in any Nash equilibrium of the game with discount factor δ and prior p.

Theorem 9.1 (Fudenberg and Levine 1991) Suppose that the long-run player's choice of a_1 is revealed at the end of each period. Then for all θ_0 with $p(\theta_0) > 0$, and all $\lambda > 0$, there is a $\underline{\delta} < 1$ such that, for all $\delta \in (\underline{\delta}, 1)$,

---

9. For those who are uncomfortable with the idea of types who "like" to play mixed strategies, an equivalent model identifies a countable set of types with each mixed strategy of the incumbent. Thus, one type always plays fight, the next accommodates in the first period and fights in all others, another fights at every other opportunity, and so on—one type for every sequence of fight and accommodate. Thus, every type plays a deterministic strategy, and by suitably choosing the relative probabilities of the types the aggregate distribution induced by all of the types will be the same as that of the given mixed strategy.

$$(1 - \lambda)g_1^*(p, \theta_0) + \lambda \min_{\alpha} g_1(\alpha_1, \alpha_2, \theta_0) \leq \underline{N}(\delta, p, \theta_0) \qquad (9.1a)$$

and

$$\bar{N}(\delta, p, \theta_0) \leq (1 - \lambda)g_1^s(\theta_0) + \lambda \max_{\alpha} g_1(\alpha_1, \alpha_2, \theta_0). \qquad (9.1b)$$

Remarks

• The theorem says that if type θ_0 is patient he can obtain about his commitment payoff relative to the prior distribution, and that regardless of the prior probability distribution a patient type cannot obtain much more than his Stackelberg payoff. Note that the lower bound depends only on which feasible reputation type θ_0 wants to maintain, and is independent of the other types to which p assigns positive probability and of the relative likelihood of different types.

• Of course, the lower bound depends on the set of possible commitment types: If no commitment types have positive probability, reputation effects have no force! For a less trivial illustration, consider a modified version of the chain-store game presented in subsection 9.2.1, where each period's entrant, in addition to being tough or weak, is one of three "sizes" (large, medium, and small), and the entrant's size is public information. It is easy to specify payoffs so that the incumbent's best pure-strategy commitment is to fight the small and medium entrants and accommodate the large ones. The theorem shows that the sane incumbent can achieve the payoff associated with this strategy *if* the entrants assign it positive prior probability. However, if the entrants assign positive probability to only two types, one which is "weak" and another which fights all entrants regardless of size, then the incumbent cannot maintain a reputation for fighting only the small and medium entrants, for the first time it accommodates a large entrant it reveals that it is weak.

• For a fixed prior distribution p, the upper and lower bounds can have different limits as $\delta \to 1$ even if the Stackelberg type belongs to the prior distribution. Fudenberg and Levine (1991) show that in generic[10] simultaneous-move games, $g_1^*(p, \theta_0) = g_1^s(\theta_0)$ when the prior assigns a positive density to every commitment strategy.

• The Stackelberg payoff supposes that the short-run players correctly forecast the long-run player's stage-game action. The long-run player can obtain a higher payoff if his opponents mispredict his action. For this reason, for a fixed discount factor less than 1, some types of the long-run player can have an equilibrium payoff that strictly exceeds their Stackelberg level, as the short-run players may play a best response to the equilibrium actions of other types.

---

10. Genericity is needed to ensure that, by changing α_1 a bit, player 1 can always "break ties" in the right direction in the definition of $g_1^*(p, \theta_0)$, so that $g_1^*(p, \theta_0) = g_1^s(\theta_0)$.

For example, suppose that in the finite-horizon chain-store game

$$a(1 - q^0) < q^0 b/(b + 1),$$

so that the weak incumbent's Stackelberg payoff is 0, and suppose that the prior probability of the "tough" type is greater than $b/(b + 1)$. Then the equilibrium is for the weak incumbent to always accommodate, and the weak entrants stay out until they have seen a tough entrant enter and the incumbent accommodate. Then the weak incumbent's equilibrium normalized payoff is

$$\frac{a(1 - \delta)(1 - q^0)}{1 - \delta(1 - q^0)} > 0.$$

For $\delta = 0$, the weak incumbent's payoff is $a(1 - q^0)$ (which is higher than the Stackelberg payoff for any value of q^0): If the first entrant is weak it stays out, and if the first entrant is tough it enters and the incumbent accommodates. However, as $\delta \to 1$ the weak incumbent's payoff converges to its Stackelberg payoff of 0. Intuitively, the "supernormal" payoffs of the weak type are informational rents that come from the short-run players' not knowing its type. In the long run, the short-run players cannot be repeatedly "fooled" about the long-run player's play (unless $q^0 = 0$, in which case the long-run player's weakness is never tested), and the long-run player will have to bear the cost of fighting to maintain its reputation. This is why a patient long-run player cannot do better than its Stackelberg payoff. Reputation effects can serve to make commitments credible, but in the long run this is all they do.

• Although the theorem is stated for the limit $\delta \to 1$ in an infinite-horizon game, the same result covers the limit, as the horizon grows to infinity, of finite-horizon games with time-average payoffs.

• The key property of short-run players for the proof of theorem 9.1 is that they always play short-run best responses to the anticipated play of their opponents. Consider a single long-run "big" player facing a continuum of long-run "small" opponents in a repeated game. Suppose further that the various small players are anonymous, and that each player observes only the play of the big player and the play of subsets of small players of positive measure (see section 4.7 for a discussion of these assumptions). In this case the small players will play myopically, so the situation is equivalent to the case of short-run players and theorem 9.1 should be expected to apply. (At this writing no one has worked out a careful version of the argument, attending to the niceties of a continuum-of-players model.) It would be interesting to know if this observation extends to a limit result as the number of players grows. Section 9.4 discusses a game in which the small players are not anonymous and can try to maintain their own reputations; here the conclusions are much less sharp.

Sketch of Proof We will give an overview of the general argument and a detailed sketch for the case of commitment to a *pure strategy*. Fix a Nash equilibrium $(\hat{\sigma}_1, \hat{\sigma}_2)$ (Recall that σ denotes overall strategies.) This generates a joint probability distribution π over Θ and histories h^t for each t. The short-run players will use π to compute their posterior beliefs about θ at every history that π assigns positive probability. Now consider a type $\bar{\theta}$ with $p(\bar{\theta}) > 0$, and imagine that player 1 chooses to play type $\bar{\theta}$'s equilibrium strategy, which we denote by $\bar{\sigma}_1$. This generates a sequence of actions with positive probability under π.

Since the short-run players are myopic, and best-response correspondences are upper hemi-continuous, Nash equilibrium requires that the short-run players' action be close to a best response to $\bar{\sigma}_1$ in any period where the observed history has positive probability and they expect the distribution over outcomes to be close to that generated by $\bar{\sigma}_1$. Because the short-run players have a finite number of actions in the stage game, this conclusion can be sharpened: If the expected distribution over outcomes is close to that generated by $\bar{\sigma}_1$, the short-run players must play a best response to $\bar{\sigma}_1$.

More precisely, for any h^t with $\pi(h^t) > 0$, let $\rho(h^t) = \pi[\alpha_1^t = \bar{\sigma}_1^t(\cdot | h^t) | h^t]$.

Claim For any $\bar{\theta}$, there is a $\bar{\rho} < 1$ such that $\hat{\sigma}_2^t \in r(\bar{\sigma}_1^t(\cdot | h^t))$ whenever $\rho(h^t) > \bar{\rho}$. (Proving this is exercise 9.2.)

Conversely, in any period in which the short-run players do not play a best response to $\bar{\sigma}_1$, when player 1's action is observed there is a non-negligible probability that they will be "surprised" and will increase the posterior probability that player 1 is type $\bar{\theta}$ by a nonnegligible amount. After sufficiently many of these surprises, the short-run players will attach a very high probability to player 1's playing $\bar{\sigma}_1$ for the remainder of the game. In fact, one can show that for any ε there is a $K(\varepsilon)$ such that with probability $1 - \varepsilon$ the short-run players play best responses to $\bar{\sigma}_1$ in all but $K(\varepsilon)$ periods, and that this $K(\varepsilon)$ holds uniformly over all equilibria, all discount factors, and all priors p with the same prior probability of $\bar{\theta}$.

Once one obtains a $K(\varepsilon)$ that holds uniformly, one derives the lower bound on payoffs by considering $\bar{\theta}$ to be a commitment type that has positive prior probability and observing that type θ_0 gets at least the corresponding commitment payoff whenever the short-run players play a best response to $\bar{\sigma}_1$. To obtain the upper bound, let $\bar{\theta} = \theta_0$, so that type θ_0 plays his own equilibrium strategy. Whenever the short-run players are approximately correct in their expectations about the marginal distribution over actions, type θ_0 cannot obtain much more than his Stackelberg payoff.

In general, the stage-game strategies prescribed by $\bar{\sigma}_1$ may be mixed. Obtaining the bound $K(\varepsilon)$ on the number of "surprises" is particularly simple when $\bar{\sigma}_1$ prescribes the same pure strategy $\bar{a}_1$ in every period for every history. Fix an $\bar{a}_1$ such that the corresponding commitment type $\bar{\theta}$

has positive prior probability, and consider the strategy for player 1 of always playing $\bar{a}_1$. By upper hemi-continuity, there is a $\bar{\rho}$ such that, in any period where the player 2s do not play a best response to $\bar{a}_1$, $\rho(h^t) < \bar{\rho}$. We show that when player 1 plays $\bar{a}_1$ in every period there can be at most $\ln(p(\bar{\theta}))/\ln(\bar{\rho})$ periods where this inequality obtains. To see this, note that $\rho(h^t) \geq \mu(\bar{\theta}|h^t)$, because $\bar{\theta}$ always plays $\bar{a}_1$. Along any history with positive probability, Bayes' rule implies that

$$\mu(\bar{\theta}|h^{t+1}) = \mu(\bar{\theta}|(h^t, a^t)) = \frac{\pi(a^t|h^t, \bar{\theta})\mu(\bar{\theta}|h^t)}{\pi(a^t|h^t)}. \tag{9.2}$$

Then, since player 2's play is independent of θ, and the choices of the two players at time t are independent conditional on h^t,

$$\pi(a^t|h^t) = \pi(a_1^t|h^t) \cdot \pi(a_2^t|h^t)$$

and

$$\pi(a^t|\bar{\theta}, h^t) = \pi(a_1^t|\bar{\theta}, h^t) \cdot \pi(a_2^t|h^t).$$

If we now consider histories where $a_1^t = \bar{a}_1$ for all t,

$$\pi(a_1^t|\bar{\theta}, h^t) = 1,$$

and equation 9.2 simplifies to

$$\mu(\bar{\theta}|h^{t+1}) = \frac{\mu(\bar{\theta}|h^t)}{\pi(a_1^t|h^t)}. \tag{9.3}$$

Consequently $\mu(\bar{\theta}|h^{t+1})$ is nondecreasing, and increases by at least $1/\bar{\rho}$ whenever a best response to $\bar{a}_1$ is not played, as then $\pi(a_1^t|h^t) \leq \bar{\rho}$. Thus, there can be at most $\ln(p(\bar{\theta}))/\ln(\bar{\rho})$ periods where $\pi(a_1^t|h^t) \leq \bar{\rho}$, and the lower bound on payoffs follows. (The additional complication posed by types $\bar{\theta}$ that play mixed strategies is that $\mu(\bar{\theta}|h^t)$ need not evolve deterministically when player 1 uses type $\bar{\theta}$'s strategy.) ∎

Note that the proof does not assert that $\mu(\bar{\theta}|h^t)$ converges to 1 when player 1 uses type $\bar{\theta}$'s strategy. This stronger assertion is not true. For example, in a pooling equilibrium where all types play the same strategy, $\mu(\bar{\theta}|h^t)$ is equal to the prior probability in every period. Rather, the proof shows that if player 1 always plays like type $\bar{\theta}$, eventually the short-run players become convinced that he will play like $\bar{\theta}$ in the future.

9.2.3   Extensive-Form Stage Games[†††]

Theorem 9.1 assumes that the long-run player's choice of stage-game strategy is revealed at the end of each period, as in a simultaneous-move game. The following example shows that the long-run player may do much less well than is predicted by theorem 9.1 if moves in the stage game are sequential. This may seem surprising, because the chain-store game con-

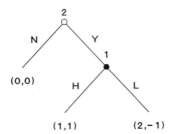

Figure 9.1

sidered by Kreps and Wilson (1982) and Milgrom and Roberts (1982) has sequential moves. Indeed, it has the same game tree as our example, but with different payoffs.

In figure 9.1, player 2 begins by choosing whether or not to purchase a good from player 1. If he does not buy, both players receive 0. If he buys, player 1 must decide whether to produce low or high quality. High quality gives each player a payoff of 1; low quality gives player 1 a payoff of 2 and gives -1 to player 2. If player 2 does not buy, player 1's (contingent) choice of quality is not revealed.

If player 1 could commit to high quality, all player 2s would purchase. Thus, if theorem 9.1 extended to this game it would say that if there is positive probability p^* that player 1 is a type who always produces high quality, then the Nash-equilibrium payoffs of a sane type θ_0 of player 1 (whose payoffs are as in figure 9.1) are bounded below by an amount that converges to 1 as the discount factor δ goes to 1.[11]

This extension is false, as the following infinite-horizon example shows. Take $p(\theta_0) = 0.99$ and $p^* = 0.01$, and consider the following strategies: The high-quality type always produces high quality. The "sane" type, θ_0, produces low quality if no more than one short-run player has ever made a purchase. Beginning with the second time a short-run player buys, type θ_0 produces high quality, and it continues to do so as long as its own past actions conform to this rule. If type θ_0 deviates and produces low quality more than once, it produces low quality forever afterward. The short-run players do not buy unless a previous short-run player has already bought, in which case they buy so long as all short-run purchasers but the first have received high quality. These strategies give type θ_0 a payoff of 0; exercise 9.3 asks you to verify not only that these strategies are a Nash equilibrium but also that they can be combined with consistent beliefs to form a sequential equilibrium.[12]

---

11. The Stackelberg strategy here is not "always H" but "H with probability $\frac{1}{2}$."
12. These strategies are not a sequential equilibrium if the horizon is finite. They thus do not form a counterexample to the sequential-equilibrium version of theorem 9.1 for finite-horizon games. (Theorem 9.1 is stated for an infinite horizon, but it holds equally well for large, finite horizons as long as δ is close to 1.) Kim (1990) has shown that, when this game is played with

The reason that reputation effects fail in this example is that when the short-run players do not buy, player 1 does not have an opportunity to signal his type. This problem did not arise in the chain-store game, for there the one action the entrant could take that "hid" the incumbent's action—to stay out—was precisely the action the incumbent wished to be played. One response to the problem posed by the example is to assume that some consumers always purchase, so that there are no probability-0 information sets.

A second response is to weaken the theorem. Let the stage game be a finite extensive form of perfect recall without moves by nature. As in the example, the play of the stage game need not reveal player 1's choice of stage-game strategy a_1 (since the stage game need not be a simultaneous-move game, a_1 may be a contingent plan rather than an action). However, when both players use pure strategies the information revealed about player 1's play is deterministic. Let $0(a_1, a_2)$ be the subset of A_1 corresponding to strategies a_1' of player 1 such that (a_1', a_2) leads to the same terminal node as (a_1, a_2). We will say that these strategies are *observationally equivalent*. For each a_1, let $w(a_1)$ satisfy

$$w(a_1) = \{a_2 | \text{ for some } \alpha_1' \text{ with support in } 0(a_1, a_2), a_2 \in r(\alpha_1')\}. \qquad (9.4)$$

In words, $w(a_1)$ is the set of pure-strategy best responses for player 2 to beliefs about player 1's strategy that are consistent with the true strategy being a_1 and with the information revealed when player 2's response is played. Then, if δ is near 1, player 1's equilibrium payoff should not be much less than

$$g_1^*(\theta_0) = \max_{a_1} \min_{a_2 \in w(a_1)} g_1(a_1, a_2, \theta_0). \qquad (9.5)$$

This is verified in Fudenberg and Levine 1989.

This result, though not as strong as the assertion in theorem 9.1 that player 1 can pick out his preferred payoff in the graph of r, does suffice to prove that player 1 can develop a reputation for "toughness" in the sequential-move version of the chain-store game described in subsection 9.2.1, even if $q^0 = 0$ so that there are no "tough" entrants. In this game, $r(\text{fight}) = \{\text{stay out}\}$ and $r(\text{accommodate}) = \{\text{enter}\}$. Also, $0(\text{fight, stay out}) = 0(\text{accommodate, stay out}) = \{\text{accommodate, fight}\}$, whereas $0(\text{fight, enter}) = \{\text{fight}\}$ and $0(\text{accommodate, enter}) = \{\text{accommodate}\}$. First, we argue that $w(\text{fight}) = r(\text{fight})$. To see this, observe that $w(\text{fight})$ is at least as large as $r(\text{fight}) = \{\text{stay out}\}$. Moreover, "enter" is not a best response to "fight," and "accommodate" is not observationally equivalent to "fight"

---

a long but finite horizon, there is a unique sequential equilibrium, in which the firm does maintain a reputation for high quality. Kim is currently working on the question of the best lower bound for sequential-equilibrium payoffs in finite repetitions of general stage games with reputation effects.

when player 2 plays "enter." Consequently, no strategy placing positive weight on "enter" is in w(fight). Since player 1's Stackelberg action with observable strategies is fight, and w(fight) $= r$(fight), the generalized Stackelberg payoff and the usual one coincide in this game.

9.3 Games with Many Long-Run Players[††]

9.3.1 General Stage Games and General Reputations

Section 9.2 showed how reputation effects can allow a single, "long-run," or patient player to commit himself to his preferred strategy. Of course, there are also incentives to maintain reputations when all players are equally patient, but here it is difficult to draw general conclusions about how reputation effects influence play.

Kreps et al. (1982) analyze reputation effects in the finitely repeated prisoner's dilemma. They consider a game in which each player, if "sane," has payoffs corresponding to the expected average value of the per-period payoffs shown in figure 9.2. If both types are sane with probability 1, then the unique Nash equilibrium of the game is for both players to defect in every period, but intuition and experimental evidence suggest that even with a fixed finite horizon players may tend to cooperate. To explain this intuition, Kreps et al. introduced incomplete information about player 1's type, with player 1 either "sane" or a type who plays the strategy "tit for tat," which is "I play today whichever action you played yesterday." They showed that, for any fixed prior probability ε that player 1 is "tit for tat," there is a number K independent of the horizon length T such that, in any sequential equilibrium, both players must cooperate in almost all periods before date $T - K$, so that if T is sufficiently large the equilibrium payoffs will be close to those if the players always cooperated. The point is that a sane player 1 has an incentive to maintain a reputation for being "tit for tat," because if player 2 were convinced that player 1 plays "tit for tat" player 2 would cooperate until the last period of the game.

Just as in the chain-store game, adding a small amount of the right sort of incomplete information yields the "intuitive" outcome as the essentially unique prediction of the model with a long finite horizon. However, in contrast with games having a single long-run player, the resulting equi-

	Cooperate	Defect
Cooperate	2,2	−1,3
Defect	3,−1	0,0

Figure 9.2

librium is very sensitive to the exact nature of the incomplete information specified (Fudenberg and Maskin 1986).

Fix a two-player stage game g, and let V^* be the set of feasible, individually rational payoffs. Now consider repeated play of g with a fixed finite horizon T. Call player i "sane" if his payoff is the expected value of the sum of g_i. (Without loss of generality, we take $\delta = 1$ instead of "δ close to 1" because we consider a large but finite horizon.)

Theorem 9.2 (Fudenberg and Maskin 1986) For any $v = (v_1, v_2) \in V^*$ and any $\varepsilon > 0$, there exists a $\underline{T}$ such that, for all $T > \underline{T}$, there exists a T-period game such that each player i has probability $1 - \varepsilon$ of being sane, independent of the other, and there exists a sequential equilibrium of the game where player i's expected average payoff if sane is within ε of v_i.

Remark This theorem asserts the existence of a game and of an equilibrium; it does not say that all equilibria of the game have payoffs close to v. Note also that no restrictions are placed on the form of the payoffs that players have when they are not sane, i.e., on the support of the distribution of types: No possible types are excluded, and there is no requirement that certain types have positive prior probability. However, the theorem can be strengthened to assert the existence of a game with a strict equilibrium (subsection 1.2.1) where the sane types' payoffs are close to v; and a strict equilibrium of a game remains strict when additional types are added whose prior probability is sufficiently small. (Chapter 11 discusses this kind of robustness question.)

Partial Proof We will prove only the weaker theorem that any payoffs that Pareto dominate those of a static equilibrium can be approximated. Let e be a static-equilibrium profile with payoffs $y = (y_1, y_2)$, and let v be a payoff vector that Pareto dominates y. To avoid a discussion of public randomizations, assume that payoffs v can be attained with a pure-action profile a, i.e., $g(a) = v$.

Now consider a T-period game in which each player i has two possible types, "sane" and "crazy," and crazy types have payoffs that make the following strategy weakly dominant: "Play a_i as long as no deviations from a have occurred in the past; otherwise play e_i."

Let $\bar{g}_i = \max_a g_i(a)$ be player i's highest feasible stage-game payoff, and let $\underline{g}_i = \min_a g_i(a)$ be player i's lowest feasible stage-game payoff. Set

$$\underline{T} > \max_i \left(\frac{\bar{g}_i - (1 - \varepsilon)\underline{g}_i - \varepsilon y_i}{\varepsilon(v_i - y_i)} \right). \tag{9.6}$$

Consider the extensive-form game corresponding to $T = \underline{T}$. This game has at least one sequential equilibrium for any specification of beliefs; pick one and call it the "endgame equilibrium."

Now consider $T > \underline{T}$. It will be convenient to number periods backwards, with period T the first one played and period 1 the last. Consider strategies that specify that profile a is played for all $t > \underline{T}$, and that if a deviation does occur at some $t > \underline{T}$ (i.e., "before $\underline{T}$") then e is played for the rest of the game, whereas if a is played in every period until $\underline{T}$ play follows the endgame equilibrium corresponding to prior beliefs. Let the beliefs prescribe that if any player deviates before $\underline{T}$ that player is believed to be sane with probability 1, and that if there are no such deviations before $\underline{T}$ then the beliefs are the same as the prior until $\underline{T}$ is reached.

We claim that these strategies form a sequential equilibrium. First, the beliefs are clearly consistent in the Kreps-Wilson sense.[13] They are sequentially rational by construction in the endgame equilibrium, and are also sequentially rational in all periods following a deviation before $\underline{T}$, where both types of both players play the static-equilibrium strategies.

It remains only to check that the strategies are sequentially rational along the path of play before $\underline{T}$. Pick a period $t > \underline{T}$ in which there have been no deviations to date. If player i plays anything but a_i, he receives at most $\bar{g}_i$ today and at most y_i thereafter, for a continuation payoff of

$$\bar{g}_i + (t - 1)y_i. \tag{9.7}$$

If instead he follows the (not necessarily optimal) strategy of playing a_i each period until his opponent deviates and playing e_i thereafter, his expected payoff will be at least

$$\varepsilon t v_i + (1 - \varepsilon)[\underline{g}_i + (t - 1)y_i], \tag{9.8}$$

as this strategy yields tv_i if his opponent is crazy and at least $\underline{g}_i + (t - 1)y_i$ if his opponent is sane. The definition of $\underline{T}$ has been chosen so that quantity 9.8 exceeds quantity 9.7 for $t > \underline{T}$, which shows that player i's best response to player j's strategy must involve playing a_i until $\underline{T}$. (A best response exists by standard arguments.) The key in the construction is that when players respond to deviations as we have specified, any deviation before $\underline{T}$ gives only a one-period gain (relative to y_i), whereas playing a_i until $\underline{T}$ gives probability ε of a gain ($v_i - y_i$) that grows linearly in the time remaining, and risks only a one-period loss. This is why even a very small ε makes a difference when the horizon is sufficiently long.[14]   ∎

9.3.2   Common-Interest Games and Bounded-Recall Reputations[†††]

Aumann and Sorin (1989) consider reputation effects in the repeated play of two-player stage games of "common interests," which they define as stage

13. This is a two-types-per-player game of incomplete information. From chapter 8, we know that beliefs that are updated from one period to the next using Bayes' rule such that the updating about a player's type does not depend on the other player's action are consistent.
14. Note once again that as ε tends to 0 the $\underline{T}$ of the theorem tends to ∞, and that for a fixed horizon T a sufficiently small ε has no effect.

	L	R
U	9,9	0,8
D	8,0	7,7

Figure 9.3

games in which there is a payoff vector that strongly Pareto dominates all other feasible payoffs. In these games the Pareto-dominant payoff vector corresponds to a static Nash equilibrium; however, there can be other equilibria, as in the game illustrated in figure 9.3. This is the game we used in chapter 1 to show that even a unique Pareto-optimal payoff need not be the inevitable result of preplay negotiation: Player 1 should play D if he believes the probability that player 2 will play R is more than $\frac{1}{8}$. Also, player 1 would like player 2 to play L regardless of how player 1 intends to play. Thus, when the players meet, each will try to convince the other that he will play his first strategy, but these statements need not be compelling.

Aumann and Sorin show that when the possible reputations (i.e., crazy types) are all "pure strategies with bounded recall" (to be defined shortly) then reputation effects pick out the Pareto-dominant outcome so long as only pure-strategy equilibria are considered. A pure strategy for player i has recall k if it depends only on the last k choices of his opponent, that is, if all histories where i's opponent has played the same actions in the last k periods induce the same action by player i. (Note that when player i plays a pure strategy and does not contemplate deviations, conditioning on his own past moves is redundant.) When k is large this condition may seem innocuous, but it does rule out "grim" or "unrelenting" strategies that prescribe, e.g., reversion to the worst static Nash equilibrium for player i if player i ever deviates.

Aumann and Sorin consider perturbed games with independent types, where each player's type is private information, each player's payoff function depends only on his own type, and types are independently distributed. The prior p_i about player i's type is that player i is either the "sane" type θ_0, with the same payoffs as in the original game, or a type that plays a pure strategy with recall less than some bound ℓ. Moreover, p_i is required to assign positive probability to the types corresponding to each pure strategy of recall 0. These types play the same action in every period regardless of the history, just like the commitment types of Fudenberg and Levine. Such priors correspond to "admissible perturbations of recall ℓ," or "ℓ-perturbations" for short. Say that a sequence p^m of ℓ-perturbations supports a game G if $p^m(\theta_0^i) \to 1$ for all players i as $m \to \infty$ and if the conditional distribution $p^m(\theta^i | \theta^i \neq \theta_0^i)$ is constant.

Theorem 9.3 (Aumann and Sorin 1989)   Let the stage game g be a game of common interests, and let z be its unique Pareto-optimal payoff vector. Fix a recall length ℓ, and let p^m be a sequence of ℓ-perturbations that support the associated discounted repeated game $G(\delta)$. Then the set of pure-strategy Nash equilibria of the games $G(\delta, p^m)$ is not empty, and the pure-strategy equilibrium payoffs converge to z for any sequence (δ, m) converging to $(1, \infty)$.

Idea of Proof   We give a partial intuition for the convergence of equilibrium payoffs for the case in which δ goes to 1 much faster than m goes to ∞ (the theorem holds uniformly over sequences (δ, m)). Suppose more strongly that the game is symmetric and that a symmetric pure-strategy equilibrium exists. Fix $\varepsilon > 0$ and suppose further that, even when the probability of a sane type is very close to 1, a sane type's payoff is less than $(z - \varepsilon)$, where z is now the symmetric Pareto-optimal payoff. Since the equilibrium is pure, then, conditional on both types being sane, there must be some period in which the players fail to play the symmetric action $a(z)$ with payoff z. Then if player 1, say, adopts the strategy of always playing the action $a(z)$ corresponding to z, he will reveal that he is not sane. Suppose a pure-strategy equilibrium exists, and suppose its payoff is less than z. Consider the strategy for player 1 of always playing the action $a_1(z)$ corresponding to z. Since the equilibrium is pure, this strategy is certain to eventually reveal that player 1 is not of type θ_0. The commitment type $\theta_1(z)$ corresponding to $a_1(z)$ has positive probability by assumption, so if $\ell = 0$ player 2 will infer that player 1 is $\theta_1(z)$ and will play $a_2(z)$ from then on (because crazy types play constant strategies when $\ell = 0$). However, player 1 could be some other type with memory longer than 0, and to learn player 1's type will require player 2 to "experiment" to see how player 1 responds to different actions. Such experiments could be very costly if they provoked an unrelenting punishment by player 1; however, since player 1's crazy types all have recall at most ℓ, player 2's potential loss (in normalized payoff) from experimentation goes to 0 as δ goes to 1. Thus, if δ is sufficiently large we expect player 2 to eventually learn that player 1 has adopted the strategy "always play $a_1(z)$," and so when δ is close to 1 player 1 can obtain approximately z by always playing $a_1(z)$.

Remark   Aumann and Sorin give counterexamples to show that the assumptions of bounded recall and full support on recall 0 are necessary, and to also show that there can be mixed-strategy equilibria whose payoffs are bounded away from z. They interpret the necessity of the bounded recall assumption with the remark that "in a culture in which irrational people have long memories, rational people are less likely to cooperate." Note that the theorem concerns the case where δ is large in comparison with the recall length ℓ, though one might expect that a more patient player would tend to have a longer memory. This is important for the proof: It is not clear that if ℓ grew with δ player 2 would try to learn player 1's strategy.

9.4 A Single "Big" Player against Many Simultaneous Long-Lived Opponents[†††]

Section 9.2 showed how reputation effects allow a single long-lived player to commit himself when facing a sequence of short-run opponents. An obvious question is whether a similar result obtains for a single "big" player who faces a large number of small but long-lived opponents. For example, one might ask if a large "government" or "employer" could maintain its desired reputation against small agents whose lifetimes are of the same order as the large player's. We will give an informal sketch of some of the issues involved based on the formal treatment of Fudenberg and Kreps (1987), who consider a special case where the large player plays each of the small ones in separate versions of the two-sided concession game studied by Kreps and Wilson (1982), which is essentially a continuous-time version of the chain-store game presented above.[15]

In the concession game, time is counted backward as in section 9.2. Thus, if $t \in [0, 1]$, time 0 is the final date. At each instant t, both players decide whether to "fight" or to "concede." The "tough" types always fight; the "weak" ones find fighting costly but are willing to fight to induce their opponent to concede in the future. More specifically, both weak types have a cost of 1 per unit time of fighting. If the entrant concedes first at t, the weak incumbent receives a flow of a per unit time until the end of the game, so the weak incumbent's payoff is $at - (1 - t)$ and the weak entrant's payoff is $-(1 - t)$. If the weak incumbent concedes first at t, the weak incumbent's payoff is $-(1 - t)$ and the weak entrant's payoff is $bt - (1 - t)$, where b is the entrant's flow payoff once the incumbent concedes. Thus, each weak player would like its opponent to concede, and each weak player will concede if it thinks its opponent is likely to fight until the end. The unique equilibrium involves the weak type of one player conceding with positive probability at date 0 (so the corresponding distribution of stopping times has an "atom" at 0); if there is no concession at date 0, both players concede according to smooth density functions thereafter.

Now suppose that a "large" incumbent is simultaneously involved in N such concession games against N different opponents, each of which plays only against the incumbent. The incumbent's type is perfectly correlated across games, in that the incumbent is tough in all the games with prior probability p^0 and weak in all of them with complementary probability $1 - p^0$. Each entrant is tough with probability q^0, independent of the

15. The concession game is also a variant of the incomplete-information war of attrition, studied in chapter 6.

others. Since the entrants are long-run, each has its own reputation to worry about.

The nature of the equilibrium depends on whether an entrant is allowed to reenter its market and resume fighting after it has dropped out. In the "captured contests" version of the game, if an entrant has ever conceded, it must concede from then on; the "reentry" version allows the entrant to revert to fighting after it has conceded. Note that when there is only one entrant, the "captured contests" and "reentry" versions have the same sequential equilibrium, as once the entrant chooses to concede it receives no subsequent information about the incumbent's type and thus will choose to concede from then on.[16]

One might guess that if there are enough entrants, the large incumbent can deter entry in either version of the game. This turns out not to be the case. Specifically, under captured contests, when each entrant has the same prior probability of being tough, no matter how many entrants the incumbent faces, equilibrium play in each market is exactly as if the incumbent played against only that entrant. To see why, suppose that there are N entrants, and that $N - k$ of them have conceded at time t, so that there are k entrants still fighting. If the equilibrium is symmetric (one can show that it must be), the incumbent then has the same posterior beliefs q^t about the type of each active entrant. Further, if the incumbent is randomizing at date t, it must be indifferent between conceding now (in which case it receives a continuation payoff of 0 in the remaining markets) and fighting on for a small interval dt and then conceding. The key is that, whatever happens in the active markets, the captured markets remain captured, so the incumbent does not consider them in making its current plans. If we denote the probability that each entrant concedes between t and $t - dt$ by σ^t, we have

$$0 = -k + k(1 - q^t)\sigma^t at. \tag{9.9}$$

Note that the number of active entrants, k, factors out of this equation, so that it is the same equation we have for the one-entrant case. This is why adding more entrants has no effect on equilibrium play.

In contrast, when reentry is allowed and there are many entrants, reputation effects can be shown to enable the incumbent to obtain approximately its commitment payoff. We say "can," rather than "will," because here the equilibrium is not unique; in one of the equilibria the incumbent can commit itself, but in another it cannot. The multiplicity comes from the fact

16. If there are several entrants and the incumbent plays them in succession, so that $t \in [0, 1]$ is against the first entrant, $t \in [1, 2]$ against the second, and so on, the first entrant might regret having conceded if it sees the incumbent concede to a subsequent entrant, but at that point the first entrant's contest is over, and once again the captured contests and reentry versions have the same equilibrium.

that in subgames where the weak incumbent has conceded and is thus revealed to be weak, the symmetric-information wars of attrition between the weak incumbent and the weak entrants who have previously conceded have multiple equilibria.

Fudenberg and Kreps focus on the equilibrium where, once the incumbent has conceded in any market, it concedes in all markets and all entrants who had previously conceded reenter. (This is the unique sequential equilibrium in the finite-horizon, discrete-time version of the game.) In this case, when the incumbent has captured a number of markets it has a great deal to lose by conceding. Here the incumbent's myopic incentive is to concede to entrants who have fought a long time and thus are likely to be tough, but the incumbent lacks the flexibility to concede to these active entrants without also conceding to the entrants who have already been revealed to be weak, and this lack of flexibility enables the incumbent to commit itself to tough play.

In contrast, if we specify that even if it is revealed to be weak the incumbent keeps control of any market in which an entrant has conceded, then, as in the case of captured contests, play in each market is exactly as if there were only one entrant, so that facing more entrants does not make the incumbent tougher. The reason is that since the incumbent keeps the flexibility to concede to the active entrants while threatening to fight the inactive ones, the presence of more entrants does not "stiffen the incumbent's backbone."

The moral of these observations is that the workings of reputation effects with one big player facing many small long-run opponents can depend on aspects of the game's structure that would be irrelevant if the small opponents were played sequentially. Thus, in applications of game theory one should be wary of blanket assertions that reputation effects will allow a large player to make its commitments credible.

An open question in this field is what happens when the incumbent's type need not be the same in each contest, so that the incumbent can be tough in some contests and weak in others.

Exercises

Exercise 9.1* Characterize the equilibria of the chain-store game of subsection 9.2.1 in the limit $N \to \infty$ for discount factors δ close to 1.

Exercise 9.2** Prove the claim in the proof of theorem 9.1.

Exercise 9.3** In subsection 9.2.3, check that the strategies yielding no purchase in the repeated quality game of figure 9.1 form a sequential equilibrium.

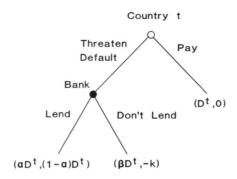

Figure 9.4

Exercise 9.4* Consider the chain-store game as described in subsection 9.2.1. Suppose that there is a single potential entrant, two markets (A and B), and two periods. The entrant can enter each market at most once and can enter at most one market per period, but he can choose which market to enter first. The incumbent is either tough in both markets or weak in both; the entrant is weak with probability 1. The tough incumbent always fights. Payoffs for the weak players in market A are as in subsection 9.2.1: The incumbent gets a if no entry, 0 if accommodate, -1 if fight; the entrant gets b if accommodate, 0 if no entry, -1 if fight. In market B, which is "big," all these payoffs are multiplied by 2. Which market should the entrant enter first? (Hint: Why might entering both markets at once, if feasible, be better than sequential entry?)

Exercise 9.5* Consider the following model of international debt repayment: A bank (representing the coalition of creditors) faces two countries sequentially. At date $t \in \{1, 2\}$, country t decides whether to pay its debt, D^t, or to threaten default. If it threatens default, the bank can either lend (or reschedule the debt) or not lend; the latter results in default. The stage game is illustrated in figure 9.4, where the first payoff is the bank's and the second is the country's. (See Armendariz de Aghion 1990 for more motivation.) Assume $1 > \alpha > 0$, $\alpha > \beta$, and $k > 0$. The bank can be "soft" (have payoffs as in figure 9.4) or "tough" (never lend, because of pessimism about future repayment, or because of costly acquisition of cash reserves). Only the bank knows whether it is soft or tough. Assume that the bank's discount factor is equal to 1, and that $(1 - p)(1 - \alpha)D^t - pk > 0$ for $t = 1, 2$, where p is the prior probability that the bank is tough.

Solve for the equilibrium of this two-period game. If the bank had the choice between facing the low-debt country or the high-debt country first, which one would it choose? (Compare your answer to that of exercise 9.4.)

Exercise 9.6** Consider the following two-period repeated game between a supervisor and an agent. In period $t = 1, 2$, the agent has type $\theta^t = 1$ with

Supervisor

	Denounce	Don't
Jam	$\theta^t - j, 0$	$\theta^t - j, 0$
Don't	$0, w^t$	$\theta^t, 0$

Agent (label at left, rows: Jam, Don't)

Figure 9.5

probability α, and $\theta^t = 0$ with probability $1 - \alpha$. θ^t is learned by the agent at the beginning of the period, and θ^1 and θ^2 are independent. At cost $j > 0$ (for "jamming"), the agent can prevent the supervisor from observing θ^t. If the agent does not jam, the supervisor observes θ^t and chooses whether or not to report the agent's type to the manager. If the supervisor reports the agent's type, the manager adjusts the agent's contract to extract his rent, and the agent then gets payoff 0. If the supervisor does not report, the agent receives rent θ^t. (Think of θ^t as the agent's productivity.) The supervisor is sane (has payoffs as indicated in figure 9.5) with probability r, and "pro-agent" (never reports) with probability $1 - r$. Assume that $r > j$, so that the agent jams when he is of type 1 in the one-period version of the game. The discount factor is equal to 1.

(a) Assume that $w^1 > \alpha w^2$. Show that the agent "experiments" (i.e., does not jam) in period 1 when $\theta^1 = 1$ if and only if $r < j + \alpha(1 - r)j$. Interpret this condition.

(b) Assume that $0 < w^1 < \alpha w^2$. Show that the agent experiments in period 1 if and only if $1 - j < (1 - r)/(1 - j)$, and that the sane supervisor builds a reputation for trust with positive probability.

(Aghion and Caillaud (1988) develop a richer model of reputation and draw some inferences for organizational design.)

Exercise 9.7** This exercise (which concerns commitment in monetary policy) considers a central bank which chooses the level of the money supply as in the discussion of "time consistency" in chapter 3. The new wrinkles here are that the bank's preferences are private information and that the link between the money supply and inflation is stochastic. Specifically, suppose that the central bank's payoff in each period is $\theta N - \pi^2/2$, where N is the level of employment, π is the rate of inflation, and θ is a taste parameter.

The payoff functions of the "public" generate a link between employment and inflation given by a "Phillips curve,"

$$N = \alpha(\pi - \pi^e),$$

where π^e is the rate of inflation the public expects to occur. Thus, the

Phillips curve corresponds to the short-run reaction correspondence of the unmodelled economic agents.

The realized level of inflation depends on the central bank's action a and a random disturbance ε:

$$\pi = a + \varepsilon,$$

where ε has a normal distribution with mean 0 and variance v_ε.

(a) Show that if the game is played only once and θ is public information, the unique equilibrium is $a = \alpha\theta$. What is the central bank's equilibrium payoff?

(b) What is the Stackelberg action for the bank? What is the Stackelberg payoff?

(c) Still in the one-shot game, suppose that θ is private information for the bank, and that the prior distribution on θ is normal with mean $\bar{\theta}$ and variance v_θ. Show that the bank's equilibrium payoff as a function of θ is

$$[\theta^2\alpha^2 - v_\varepsilon]/2 - \alpha^2\theta\bar{\theta}.$$

Why do types with $\theta > 0$ prefer $\bar{\theta}$ to be very negative?

(d) Now consider a two-period version of this game, with prior beliefs as in question c. At the end of the first period, the public observes the realized inflation π^1 but not the bank's action a^1. Suppose the bank maximizes the discounted sum of its per-period payoffs with discount factor δ. Look for an equilibrium where the bank's first-period action a^1 is a linear function of θ: $a^1 = K\theta$. Show that K is defined implicitly by

$$K = \alpha[1 - \delta K v_\theta/(K^2 v_\theta + v_\varepsilon)].$$

Hint: If the bank uses a linear first-period strategy, the public's second-period beliefs about θ have a normal distribution, with a variance independent of first-period inflation π^1 and a mean equal to

$$\bar{\theta} + [K^2 v_\theta/(K^2 v_\theta + v_\varepsilon)](K\bar{\theta} - \pi^1)/K.$$

How does K depend on δ, v_ε, and v_θ? Explain. Is any type's equilibrium payoff above its Stackelberg level? Why?

(e) Suppose that the game is repeated infinitely often with discount factor near 1, and that θ has a finite support including $\theta = 0$. Suppose that type 0 always sets $a = 0$. Characterize each type's equilibrium payoff. (This exercise is based on Cukierman and Meltzer 1986. See Cukierman 1990 for more on central banks' reputations.)

References

Aghion, P., and B. Caillaud. 1988. On the role of intermediaries in organizations. In B. Caillaud, Three Essays in Contract Theory: On the Role of Outside Parties in Contractual Relationships. Ph.D. thesis, Massachusetts Institute of Technology.

Armendariz de Aghion, B. 1990. International debt: An explanation of the commercial banks' lending behavior after 1982. *Journal of International Economics* 28: 173–186.

Aumann, R., and S. Sorin. 1989. Cooperation and bounded recall. *Games and Economic Behavior* 1: 5–39.

Bénabou, R., and G. Laroque. 1989. Using privileged information to manipulate markets. Working paper 137/930, INSEE.

Cukierman, A. 1990. *Central Bank Behavior, Credibility, Accommodation and Stabilization.* Forthcoming.

Cukierman, A., and A. Meltzer. 1986. A theory of ambiguity, credibility and inflation under discretion and asymmetric information. *Econometrica* 54: 1099–1021.

Diamond, D. 1989. Reputation in acquisition and debt markets. *Journal of Political Economy* 97: 828–862.

Dybvig, P., and C. Spatt. 1980. Does it pay to maintain a reputation? Mimeo.

Fudenberg, D., and D. Kreps. 1987. Reputation and simultaneous opponents. *Review of Economic Studies* 54: 541–568.

Fudenberg, D., D. Kreps, and D. Levine. 1988. On the robustness of equilibrium refinements. *Journal of Economic Theory* 44: 354–380.

Fudenberg, D., and D. Levine. 1989. Reputation and equilibrium selection in games with a patient player. *Econometrica* 57: 759–778.

Fudenberg, D., and D. Levine. 1991. Maintaining a reputation when strategies are not observed. *Review of Economic Studies*, forthcoming.

Fudenberg, D., and E. Maskin. 1986. The folk theorem in repeated games with discounting or with incomplete information. *Econometrica* 54: 533–554.

Kim, Y.-S. 1990. Characterization and properties of reputation effects in finitely-repeated extensive form games. Mimeo, University of California, Los Angeles.

Kreps, D., P. Milgrom, J. Roberts, and R. Wilson. 1982. Rational cooperation in the finitely repeated prisoners' dilemma. *Journal of Economic Theory* 27: 245–252, 486–502.

Kreps, D., and R. Wilson. 1982. Reputation and imperfect information. *Journal of Economic Theory* 27: 253–279.

Milgrom, P., and J. Roberts. 1982. Predation, reputation and entry deterrence. *Journal of Economic Theory* 27: 280–312.

Selten, R. 1978. The chain-store paradox. *Theory and Decision* 9: 127–159.

Shapiro, C. 1982. Consumer information, product quality, and seller reputation. *Bell Journal of Economics* 13: 20–35.

10 Sequential Bargaining under Incomplete Information

10.1 Introduction[††]

A bargaining situation involves players who must reach an agreement in order to realize gains from trade. The standard example is the problem of sharing a pie. No player can have any pie until they all agree about the shares each will receive. Negotiating about the shares is costly, and the pie may decay or disappear if the negotiations go on for very long.

At least since Edgeworth (1881) bargaining has been perceived as an important question in economics and political science. The first efforts to predict bargaining outcomes used the framework of "cooperative games." In this framework, axioms are developed on the result of the bargaining process, and in particular on how the result should vary with changes in the set of feasible utilities; these axioms are typically defended on normative grounds as well as positive ones. Cooperative game theory's use of axioms on outcomes distinguishes it from the noncooperative approach developed in this book, where outcomes depend explicitly on behavior and behavior is assumed to correspond to equilibrium play in exogenously determined games.

Nash (1950, 1953) used both the cooperative or axiomatic approach and the noncooperative one in his work on bargaining; he first characterized the unique outcome satisfying a set of axioms, and then proposed a non-cooperative game whose equilibrium was precisely this outcome.[1] However, Nash's noncooperative model assumed that players had only one chance to reach an agreement, and that if they failed to do so they were unable to continue negotiating (see exercise 1.6). This game seemed too simple to capture the richness of bargaining, and (perhaps as a result) the noncooperative approach to bargaining received little attention until the 1970s.

The model of Ståhl (1972) and Rubinstein (1982), described in chapter 4, was the first bargaining model to reflect the fact that bargaining is a typically dynamic process involving offers and counteroffers. Ståhl and Rubinstein considered bargaining under complete information and found that sequential bargaining yields a unique, Pareto-efficient outcome in which the bargainers reach an efficient agreement without haggling. Ståhl and Rubinstein also gave some useful intuitions about what determines bargaining power; for instance, players who are more patient do better.

It is worth explaining the importance of the uniqueness and efficiency results. First, the interest of the uniqueness result derives in part from its contradicting the conventional wisdom that bargaining outcomes are arbitrary and that an outside observer is unable to predict which point of the

---

1. Actually, he considered a sequence of games whose equilibrium outcomes converged to this point in the limit.

Pareto frontier (if any) will prevail. Second, the Coase (1960) theorem makes efficiency a central issue. This theorem asserts that the distribution of ownership in an economy has no relevance for efficiency as long as the "transaction costs" are negligible, in the sense that bargaining results in efficient outcomes. Although neither efficiency nor equilibrium uniqueness obtains in all sequential-bargaining games with complete information (see for instance exercises 4.3 and 4.9), Ståhl and Rubinstein defined a class of games for which these obtain.

Since the early 1980s, a number of authors have developed models of sequential bargaining with incomplete information. It was clear from the start that the introduction of incomplete information would tend to introduce inefficiencies. As is noted in chapter 7, the simplest example of a bargaining process is monopoly pricing, in which a seller makes a "take it or leave it" offer to a buyer (or several buyers) who then decides whether to purchase the good. If the seller does not know the buyer's valuation for the good, suboptimal trade results. Because the seller charges a price above the marginal cost, trade does not take place when the buyer's valuation exceeds the marginal cost but is lower than the monopoly price, even though such trade would be efficient. Similar inefficiencies seem likely in more complex bargaining games where the buyer may have an incentive to reject a price that is below his valuation in the hope of obtaining a better price later on. Indeed, Myerson and Satterthwaite (1983) (discussed in subsection 7.4.4) give general sufficient conditions for all equilibria of a bargaining game to be inefficient when neither player knows the valuation of the other.[2] When bargaining can be inefficient, the choice of economic institutions—that is, the rules of the game—can influence the efficiency of the outcome. For instance, labor disputes may be explained as resulting from incomplete information about the firm's profitability, and arbitration clauses and labor law can influence the likelihood of strikes and lockouts.[3] Similarly, the distribution of ownership, by determining residual rights of control and therefore the *status quo* allocation in bargaining, has an effect on the efficiency of bargaining between two units.

Though we favor the noncooperative approach over the axiomatic approach, we should point out that the noncooperative approach has so far been unsuccessful in "solving the bargaining problem." There are two unresolved difficulties. The first is that, in both complete-information and incomplete-information models, the equilibrium outcomes are very sensitive to the choice of the extensive form. Even with complete information, any split of the pie can be obtained by changing the extensive form for

2. When only the buyer's valuation is private information, the game in which the buyer makes a "take it or leave it" offer to the seller leads to efficient trade.
3. Fudenberg, Levine, and Ruud (1985), Kennan and Wilson (1989, 1990), and Cramton and Tracy (1990) offer empirical analyses of strikes based on models of bargaining with incomplete information.

bargaining. This is worrisome because we, as outside observers of a bargaining process, usually have little information about which extensive form is being played, and furthermore the extensive form is likely to vary from one situation to the next. Of course, in any application of game theory the conclusions can vary with the extensive form chosen, but the issue seems more serious here than in other contexts, in which we may be able to limit attention to a smaller set of extensive forms.

The second difficulty with the noncooperative approach to bargaining is more specific to incomplete information. It was soon realized (Fudenberg and Tirole 1983; Cramton 1984; Rubinstein 1985) that games in which a bargainer with private information can propose agreements can have a great many perfect Bayesian equilibria. (This will not surprise the reader of the section on signaling games in chapter 8.) Thus, bargaining theory seems unlikely to offer unique predictions even if one knows the extensive form. Several authors have tried to select particular equilibria either on *a priori* grounds or by using stronger equilibrium refinements. (Chapter 11 discusses some, but far from all, of the refinements that have been used.) The theory of bargaining under incomplete information is currently more a series of examples than a coherent set of results. This is unfortunate because bargaining derives much of its interest from incomplete information.

Though many incomplete-information bargaining models do have many equilibria, strong results can be obtained in the special class of "one-sided-offer" bargaining games (section 10.2). A seller, who has one unit of a good and whose cost is common knowledge, makes sequential offers to a buyer, who has private information about his willingness to pay for the unit. Bargaining stops after the buyer has accepted an offer. With the buyer's strategy space restricted to "yes" and "no" at each stage, the issue of updating of beliefs about the buyer's valuation off the equilibrium path does not arise. Because this model avoids the multiplicity of equilibria associated with updating of beliefs, and because it illustrates most of the insights that have been obtained in bargaining theory, we devote a disproportionate amount of attention to it, but this extensive form is very special.

This "single-sale" model assumes that the seller sells the good once and for all to the buyer. (This does not mean that the good is consumed instantaneously by the buyer.) Section 10.3 takes up the case of repeated bargaining for a perishable good which the buyer must purchase anew each period. One interpretation of this model is that the perishable good is the current period's flow of service from a durable asset that belongs to the seller, so that each period's bargaining is over the current rental price.

Section 10.4 returns to the single-sale model and takes up the case of more complex bargaining processes, such as alternating-offer bargaining. It illustrates the difficulty in making predictions when an informed player makes offers. It also draws the link between static mechanism design (studied in chapter 7) and sequential bargaining. In particular, it discusses which incentive-compatible and individually rational outcomes can arise as *some* equilibrium of *some* sequential-bargaining game.

10.2   Intertemporal Price Discrimination: The Single-Sale Model[††]

10.2.1   The Framework

A seller and a buyer bargain over the trade of one unit of a good. The seller has known production (or opportunity) cost c incurred when transfer takes place. The buyer has valuation v for the good. In the single-sale model, v and c are stock variables; in particular, if the good is durable, v is the present discounted value of the buyer's per-period benefit from the date of purchase.

The seller makes offers at dates $t = 0, 1, \ldots, T$, where $T \leq +\infty$. In each period, the buyer says yes or no. In the single-sale model, an offer at date t is a purchase price m^t. A strategy for the seller is thus a sequence of prices m^t that are charged at date t conditional on the rejection of all previous offers. A strategy for the buyer is a choice of "accept" or "reject" in each period, and is conditional on the sequence of past and current offers. If $\delta \in (0, 1)$ denotes the (common) discount factor, the payoffs are $u_s = \delta^t(m^t - c)$ for the seller and $u_b = \delta^t(v - m^t)$ for the buyer if agreement is reached at date t at price m^t.

Two formulations of the asymmetry of information have been considered in the literature:

In the *two-type case*, v takes value $\bar{v}$ with probability $\bar{p}$ and value $\underline{v}$ with probability $\underline{p}$ (such that $\bar{p} + \underline{p} = 1$), where $\bar{v} > \underline{v} > c$.

In the *continuum-of-types case*, v takes a value in some interval $[\underline{v}, \bar{v}]$ with cumulative distribution function $P(\cdot)$ and continuous density $p(\cdot) > 0$ for all v, and $\bar{v} > c$. This case is divided into two subcases: the *gap case*, $\underline{v} > c$ (gains from trade bounded away from 0), and the *no-gap case*, $\underline{v} \leq c$ (there may not exist gains from trade).[4]

With any specification of the distribution of types, the model can be interpreted either as having a single buyer whose type is unknown (the "bargaining model") or as having a continuum of infinitesimal consumers, with the distribution of their willingness to pay given by $P(\cdot)$ (the "durable-good monopoly"). In the latter case, we suppose that the seller cannot tell the consumers apart, and that the seller observes only the measures of the sets who accept and reject.

In keeping with our focus on bargaining, we will assume in most of the chapter that there is a single buyer. However, because of the importance of

---

4. The assumption of a positive continuous density at $v = c$ is important. For instance, if there is no v in $[c - \varepsilon, c + \varepsilon]$, the no-gap case is equivalent to the gap case, because the seller never sells to a buyer with valuation under c.

the durable-good interpretation, we show how to switch from one interpretation to the other in our discussion of the example in subsection 10.2.3.

Whether the gap case or the no-gap case is more descriptive may depend on the context under consideration. Neither case is completely satisfactory, as both ignore the possibility that one party or the other may break off negotiations to bargain with a third party (in the single-buyer interpretation) and the possibility that there may be a steady influx of new potential buyers (in the continuum-of-buyers interpretation). We say more about these extensions, and about the relative merits of the gap and no-gap assumptions, in subsection 10.2.7.

From now on, we will simplify notation by assuming $c = 0$.

Our focus is on whether the equilibria display various properties implicitly and/or explicitly discussed in Coase's (1972) analysis of pricing by a durable-good monopolist.

The first set of properties relates to the dynamics of equilibrium behavior.

Coasian Dynamics

Skimming Property In a perfect Bayesian equilibrium, higher-valuation types of buyer buy earlier because they are more impatient to consume. (As we will see later, this property is a straightforward consequence of the sorting condition defined in chapter 7.)

Monotonicity of Prices The equilibrium path exhibits a weakly decreasing sequence of prices until one price is accepted. (As we will see, this property requires a stationarity assumption on strategies in the no-gap case).[5]

The second set of properties concerns the limit of the equilibrium outcomes as the time period between offers shrinks to 0, so that the per-period discount factor δ tends to 1. These properties were conjectured by Coase (1972).

Coase Conjecture

When offers take place very quickly ($\delta \to 1$),

Zero Profit The seller's profit tends to 0 and

Efficiency All potential gains from trade are realized almost instantaneously.

To study the Coase conjecture, we let r denote the rate of interest per unit of time and Δ be the length of time between offers. Hence, $\delta = e^{-r\Delta}$.

---

5. It may be useful here to distinguish between "offers" and "serious offers" (which are prices offered and accepted with positive probability). It can be seen that in any pure-strategy PBE the equilibrium sequence of serious offers is strictly decreasing even if $\underline{v} \leq c$ and no stationarity assumption is made (a buyer who rejects offer m^t at date t and accepts $m^{t+\tau} \geq m^t$ at date $t + \tau$, where $\tau > 0$, would be better off accepting m^t). The stationarity assumption implies that sales have a positive probability in every period (see note 19).

The focus of the Coase-conjecture analysis is thus the equilibrium behavior as Δ converges to 0.

We start with a two-period example that illustrates Coasian dynamics and an infinite-horizon example that satisfies the Coase conjecture. These examples may be skipped by a reader who has some familiarity with the topic. We then tackle the Coase conjecture in the two-type case, and more generally the gap case. We do the same in the no-gap case, and we conclude the section with some extensions of the sale model.

10.2.2 A Two-Period Introduction to Coasian Dynamics

Let $T = 1$, and let $v = \bar{v}$ with probability $\bar{p}$ and $\underline{v}$ with probability $\underline{p}$. Let m^0 denote a first-period price, let $\bar{\mu}(m^0)$ denote the posterior beliefs that $v = \bar{v}$ conditional on the rejection of offer m^0 in period 0, and define $\underline{\mu}(m^0) \equiv 1 - \bar{\mu}(m^0)$.

Because period 1 is the last period, the seller with beliefs $\bar{\mu}$ that $v = \bar{v}$ makes a "take it or leave it" offer m^1 so as to maximize that period's profit. The buyer will accept if and only if his valuation is at least m^1.[6] It is clear that the optimal offer is either $\bar{v}$ or $\underline{v}$. By charging $m^1 = \underline{v}$, the seller sells for sure and obtains $\underline{v}$; by charging $m^1 = \bar{v}$, the seller sells with probability $\bar{\mu}$ and has second-period profit $\bar{\mu}\bar{v}$. Therefore, the seller's optimal strategy at date $t = 1$ is

$$
m^1 = \begin{cases} \underline{v} \text{ if } \bar{\mu} < \alpha \\ \bar{v} \text{ if } \bar{\mu} > \alpha \\ \text{any randomization between } \underline{v} \text{ and } \bar{v} \text{ if } \bar{\mu} = \alpha, \end{cases}
$$

where $\alpha \equiv \underline{v}/\bar{v}$. We can rewrite this optimal strategy by introducing the probability x that the seller charges $\underline{v}$ in the second period:

$$
x = \begin{cases} 1 \text{ if } \bar{\mu} < \alpha \\ 0 \text{ if } \bar{\mu} > \alpha \\ \in [0, 1] \text{ if } \bar{\mu} = \alpha. \end{cases}
$$

Note that type $\underline{v}$ never obtains a surplus in the second period and, therefore, will behave myopically in the first period. Type $\bar{v}$ obtains a surplus only if the seller is sufficiently convinced that the type is $\underline{v}$. Consider now the buyer's behavior at $t = 0$ when offered price $m^0 \in [\underline{v}, \bar{v}]$ (it is straightforward to check that prices outside this interval are irrelevant). Price $m^0 = \underline{v}$ is accepted by both types, as they will not face a more favorable price at

---

6. Each type is actually indifferent between accepting and rejecting a price m^1 that exactly equals the type's valuation. However, if the supremum of the seller's payoff is attained in the limit of prices $m^1 = v - |\varepsilon|$ as $\varepsilon \to 0$, then existence of an equilibrium given the seller's beliefs requires that type v accept $m^1 = v$, and whether the other type accepts a price equal to its valuation is irrelevant.

$t = 1.$[7] Now consider $m^0 > \underline{v}$. The low-valuation type rejects this offer because buying would give him a negative surplus. The interesting part is type $\bar{v}$'s behavior.

Suppose, first, that rejection of m^0 generates "optimistic beliefs," meaning $\bar{\mu}(m^0) > \alpha$. Then the seller charges $m^1 = \bar{v}$, and type $\bar{v}$ has no second-period surplus. Therefore, type $\bar{v}$ is better off accepting m^0. And since m^0 is rejected by type $\underline{v}$, Bayes' rule yields $\bar{\mu}(m^0) = 0$, a contradiction.

Suppose, second, that rejection of m^0 generates "pessimistic beliefs": $\mu(m^0) < \alpha$. The seller then charges $m^1 = \underline{v}$ at date 1. Therefore, type $\bar{v}$ should accept m^0 only if

$$\bar{v} - m^0 \geq \delta(\bar{v} - \underline{v})$$

or

$$m^0 \leq \tilde{v} \equiv (1 - \delta)\bar{v} + \delta\underline{v}.$$

If $m^0 > \tilde{v}$, rejecting m^0 is optimal for type $\bar{v}$ (as it is for type $\underline{v}$), and therefore Bayes' rule yields $\bar{\mu}(m^0) = \bar{p}$ (the posterior beliefs coincide with the prior beliefs).

We are thus led to consider two cases:

$\bar{p} < \alpha$ In this case, for any $m^0 > \underline{v}, \bar{\mu}(m^0) \leq \bar{p} < \alpha$, and therefore the seller always charges $m^1 = \underline{v}$ at date 1. Type $\bar{v}$ accepts m^0 if and only if $m^0 \leq \tilde{v}$. The seller's optimal first-period strategy is either to charge $m^0 = \underline{v}$ and have payoff $U_s = \underline{v}$ or to charge $m^0 = \tilde{v}$ and have payoff $U_s = \bar{p}\tilde{v} + \delta\underline{p}\underline{v}$. If $\underline{v} > \bar{p}\tilde{v} + \delta\underline{p}\underline{v}$, no "price discrimination" occurs and agreement is reached instantaneously. If $\underline{v} < \bar{p}\tilde{v} + \delta\underline{p}\underline{v}$, price discrimination occurs, in that the seller first sells at price $\tilde{v}$ to the high-valuation type and later sells at the lower price $\underline{v}$ to the low-valuation type. This is our first example of Coasian dynamics.

$\bar{p} > \alpha$ Then, when $m^0 \in (\tilde{v}, \bar{v}]$, in equilibrium type $\bar{v}$ cannot reject m^0 with probability 1, because in that case we would have $\bar{\mu}(m^0) = \bar{p} > \alpha$ and the seller charging $m^1 = \bar{v}$, so type $\bar{v}$ would be better off accepting m^0. But we already saw that type $\bar{v}$ cannot accept such an m^0 with probability 1 either. Hence, in equilibrium type $\bar{v}$ must randomize and the posterior probability must satisfy $\bar{\mu}(m^0) = \alpha$. Let $y(m^0)$ denote the probability that type $\bar{v}$ accepts m^0; $\bar{\mu}(m^0) = \alpha$ is equivalent to

$$\frac{\bar{p}(1 - y(m^0))}{\bar{p}(1 - y(m^0)) + \underline{p}} = \alpha,$$

which defines a unique $y(m^0) = y$ in $[0, 1]$. Note that $y(m^0)$ is independent of m^0, a fact we will comment on later.

7. As in note 6, type $\underline{v}$ is actually indifferent between accepting and rejecting $m^0 = \underline{v}$, but our assumption involves no loss of generality.

Furthermore, in order for type $\bar{v}$ to be indifferent between accepting and rejecting m^0, it must be the case that $\bar{v} - m^0 = \delta x(m^0)(\bar{v} - \underline{v})$, which defines a unique probability $x(m^0)$ for $m^0 \in (\tilde{v}, \bar{v}]$.

Thus, when $\bar{p} > \alpha$, the seller's optimal price in the first period is one of the following:

$m^0 = \underline{v}$, which generates payoff $U_s = \underline{v}$,

$m^0 = \tilde{v}$, which generates payoff $U_s = \bar{p}\tilde{v} + \delta\underline{p}\underline{v}$,

$m^0 = \bar{v}$, which generates payoff $U_s = \bar{p}y\bar{v} + \delta(\bar{p}(1 - y) + \underline{p})\underline{v}$,

where the third payoff is computed using the fact that, for posterior beliefs α, $m^1 = \underline{v}$ is an optimal price in period 1 for the seller. Any of these payoffs can be highest, depending on the parameters. Note that if the third payoff is highest, the seller never sells to the low-valuation type (as $x(\bar{v}) = 0$).

We thus conclude that for generic values of the parameters there exists a unique perfect Bayesian equilibrium, and that this equilibrium exhibits Coasian dynamics—that is, $\bar{\mu}(m^0) \le \bar{p}$ for all m^0, so the seller becomes more pessimistic over time, and $m^1 \le m^0$, so the seller's price decreases over time.

Fudenberg and Tirole (1983) characterize the set of equilibria of two-period bargaining games when the seller and the buyer each have two potential types (two-sided incomplete information), when the seller makes the two offers, and when the players alternate making offers. As is mentioned above, the fact that a player's offer can then signal his private information leads to a continuum of perfect Bayesian equilibria, as in the similar examples of chapter 8.

In contrast, when an uninformed seller makes offers to an informed buyer, the buyer has comparatively little scope to signal his type (he can do so only through his acceptance decision), and thus the leeway in specifying the beliefs after a probability-0 action has much less impact on the set of equilibria.

The two-period model raises the question of why the parties stop bargaining at the end of the second period if both offers have been rejected. The existence of unrealized gains from trade suggests that the parties would be better off if they continued to bargain. Thus, it would seem that a natural model of bargaining should have an infinite horizon unless one of the parties must quit for some reason. In practice, however, the players are likely to stop bargaining after some time even if they have not exhausted the gains from trade. It may be that they face a deadline for agreement (imposed by production constraints, for instance). Alternatively, the good may become obsolete at date 2 because of the introduction of a superior product at that date. The two-period model developed above applies directly to these two exogenous-horizon situations, with the minor modification that if the good becomes obsolete the buyer's willingness to pay is lower in period 1 than in period 0, as the good is enjoyed only for one period instead of two. (This modification introduces a quantitative difference, but not a qualitative one.)

A more complex explanation for a finite horizon is that the players have a fixed bargaining cost per period or have outside opportunities, so that they may decide to stop bargaining or to bargain with someone else if they become sufficiently pessimistic about the gains from trade with their current partner. The bargaining horizon is then endogenously finite. The endogenous-horizon model is more complex than the model with an exogenously finite horizon. In particular, the endogenous-horizon model tends to have multiple equilibria, whereas equilibrium is unique in the exogenous-horizon model. To see this, note that if the seller expects the buyer to concede (buy) quickly, he becomes very pessimistic, so that if several offers are rejected he will break negotiation with the buyer and exercise his outside opportunity (which might be to sell to an alternative buyer). Also, if the seller is "switch-happy," the buyer is in a weak position and concedes quickly. Thus, fast concessions and fast switching are self-fulfilling; so are slow concessions and slow switching, which is why there are multiple equilibria (Fudenberg, Levine, and Tirole 1987).

10.2.3 An Infinite-Horizon Example of the Coase Conjecture

In this subsection we adopt the interpretation that a single seller faces a continuum of infinitesimal buyers. As is mentioned above, under this interpretation we suppose that the seller cannot distinguish between the buyers, and only observes the measures of the sets who accept and reject. We also explain how to interpret the model as having a single buyer.

Sobel and Takahashi (1983) study the following "linear demand curve" model.[8] The seller and the buyers are infinitely-lived, and the bargaining process has $T = +\infty$. The buyers' valuations are uniformly distributed on $[0, 1]$. (Sobel and Takahashi consider the more general distribution $P(v) = (v/\bar{v})^\beta$, where $\beta > 0$; the uniform distribution is $\beta = 1$.)

We look for an equilibrium with the following properties:

(i) If price m^t is offered at date t, types $v \geq w(m^t) = \lambda m^t$, where $\lambda > 1$, buy (if they have not purchased before) and types $v < w(m^t)$ do not.

(ii) If at some date t types greater than κ have purchased before and types less than κ have not (so that the seller's posterior beliefs are represented by the truncated uniform distribution on $[0, \kappa]$), the seller charges $m^t(\kappa) = \gamma\kappa$, where $0 < \gamma < 1$.

As mentioned above, the skimming property, which is proved in lemma 10.1 below, ensures that the buyers always use a cutoff rule of the form "accept the current offer if and only if the valuation exceeds some (possibly history-dependent) number." Hence, the real force of condition i is to

---

8. For early work on the Coase conjecture, see Bulow 1982 and Stokey 1981.

require that the cutoff valuation be stationary (it depends only on the current price and not on previous price offers) and linear (λ is independent of m'). Condition ii requires that the seller's strategy also be stationary and linear. Note that because all players use a stationary strategy, each player loses nothing by using a stationary strategy himself.[9]

Let $U_s(\kappa)$ denote the seller's present discounted value of profits when the posterior beliefs are uniform on $[0, \kappa]$. From dynamic programming, $U_s(\cdot)$ must satisfy

$$U_s(\kappa) = \max_m \{(\kappa - \lambda m)m + \delta U_s(\lambda m)\}. \tag{10.1}$$

With a continuum of buyers, the term $(\kappa - \lambda m)$ in equation 10.1 is interpreted as the fraction of the population that will accept offer $m \leq \kappa/\lambda$, and $U_s(\lambda m)$ is the continuation present discounted value of profits. With a single buyer, equation 10.1 still holds as long as $U_s(\kappa)$ is interpreted as the product of the probability κ that the buyer has type below κ and the continuation expected present discounted value of profits. The term $(\kappa - \lambda m)$ in equation 10.1 is the probability that offer m is accepted by the buyer. Equations 10.5 and 10.6 are also valid in the single-buyer case.

If U_s is assumed to be differentiable, the maximization with respect to m yields

$$\kappa - 2\lambda m + \delta\lambda U_s'(\lambda m) = 0. \tag{10.2}$$

On the other hand, the envelope theorem can be applied to equation 10.1:

$$U_s'(\kappa) = m(\kappa) = \gamma\kappa. \tag{10.3}$$

Substituting equation 10.3 into equation 10.2 and eliminating κ yields

$$1 - 2\lambda\gamma + \delta\lambda^2\gamma^2 = 0. \tag{10.4}$$

We now look at the buyer's optimization. For type λm to be indifferent between accepting m and waiting one period and buying at price $\gamma\lambda m$, it must be the case that

$$\lambda m - m = \delta(\lambda m - \gamma\lambda m), \tag{10.5}$$

or

$$\lambda - 1 = \delta\lambda(1 - \gamma).^{10} \tag{10.6}$$

9. Condition ii is used only for convenience, as it is implied by condition i. To see this, note that, given condition i, *an* optimal policy for the seller is to use a stationary and linear strategy, as we show below. Furthermore, the valuation function $U_s(\cdot)$ is the quadratic function derived below (applying Blackwell's theorem—see, e.g., Stokey and Lucas 1989—to equation 10.1 shows that this valuation function is unique). The maximization in equation 10.1 then yields a unique optimal price, which therefore is a stationary and linear function of the cutoff type.

10. We can now check the second-order condition in the maximization in equation 10.1: $-2\lambda + \delta\lambda^2\gamma \leq 0$, which is implied by equation 10.6.

Equations 10.4 and 10.6 yield

$$\lambda = \frac{1}{\sqrt{1-\delta}}$$

and

$$\gamma = \frac{\sqrt{1-\delta}-(1-\delta)}{\delta}.$$

This perfect Bayesian equilibrium exhibits Coasian dynamics. Furthermore, it satisfies the Coase conjecture. When offers take place very quickly, γ tends to 0. Hence, even the first offer m^0, which is the highest offer, converges to 0, and so does the seller's expected profit $U_s(1)$. To see that all potential gains from trade are realized almost instantaneously, consider a valuation v. By purchasing at or after *real* time $\tau > 0$, this type has a utility of at most $e^{-r\tau}v$, where r is the rate of interest. By buying in the first period, he gets $v - m^0(\delta)$, where $m^0(\delta) = \gamma(\delta) \to 0$. Hence, for any given τ, any type v buys before real time τ if δ is sufficiently close to 1.

10.2.4   The Skimming Property[†††]

We now return to the single-buyer interpretation, and give a characterization of equilibrium.

The following lemma[11] considerably simplifies the study of buyer behavior:

Lemma 10.1 (skimming or cutoff-rule property)   Suppose that the buyer accepts price m^t at date t when he has valuation v. Then he accepts price m^t with probability 1 when he has valuation $v' > v$.

Proof   Let $h^t = (m^0,\ldots,m^{t-1})$ denote the history at date t (where the fact that the buyer has rejected all offers is implicit). Type v accepts m^t only if

$$v - m^t \geq \delta U_b(v,(h^t,m^t))$$

or

$$v - m^t \geq \mathrm{E}\sum_{\tau=1}^{T-t}[\delta^\tau(v - m^{t+\tau}(h^{t+\tau}))I^{t+\tau}(h^{t+\tau},m^{t+\tau},v)|(h^t,m^t)],$$

where $U_b(v,(h^t,m^t))$ is the continuation valuation of type v and where $I^{t+\tau}(h^{t+\tau},m^{t+\tau},v)$ is an indicator function indicating whether type v buys ($I = 1$) or not ($I = 0$) at price $m^{t+\tau}(h^{t+\tau})$ at date $t + \tau$. The random variables $m^{t+\tau}(h^{t+\tau})$ and $I^{t+\tau}(h^{t+\tau},m^{t+\tau},v)$ are determined by the history (h^t,m^t) at the end of the period if m^t is rejected and by the subsequent equilibrium strategies. Because the expected discounted volume of trade is always less than 1 and because type v' can mimic type v's bargaining strategy (and conversely),

---

11. Fudenberg et al. 1985.

$$|U_b(v',(h^t,m^t)) - U_b(v,(h^t,m^t))| \leq |v' - v|.$$

Therefore, for $v' > v$,

$$v' - m^t - \delta U_b(v',(h^t,m^t))$$
$$\geq (v' - v) - \delta(U_b(v',(h^t,m^t)) - U_b(v,(h^t,m^t))) > 0. \quad \blacksquare$$

Lemma 10.1 shows that because higher-valuation types are more eager to buy, they buy earlier. In particular, if v is drawn from a continuous distribution, the buyer's behavior is fully described by the cutoff rule $\kappa(\cdot)$: At date t, the buyer buys if $v > \kappa(h^t,m^t)$ and does not buy if $v < \kappa(h^t,m^t)$. (The type $v = \kappa(h^t,m^t)$ has probability 0, and the resolution of his in-difference is irrelevant. Of course, with atoms in the distribution of types, as in the example of subsection 10.2.2, the cutoff rule still holds but the mixing behavior of the cutoff type becomes important.)

10.2.5 The Gap Case[†††]

We now make the following assumptions on the distribution of types:

(G) $\underline{v} > 0$

(R) Either $P(\underline{v}) > 0$ or P admits a strictly positive and continuous density at $\underline{v}$.

Condition G asserts that there is a gap between the lowest valuation and the seller's cost. The regularity condition R allows either an atom or a strictly positive density at the lowest valuation.

Under assumption G, the Coase conjecture takes the following form: When $\delta \to 1$,

(c') the seller's profit tends to $\underline{v}$ and

(d) all gains from trade are realized almost instantaneously.

Next, we introduce a condition on the buyer's equilibrium strategy. This condition is satisfied by any equilibrium in the gap case; it will be imposed as an assumption in the no-gap case.

(S) The buyer's strategy satisfies property S if $\kappa(h^t,m^t) = \kappa(\tilde{h}^t,m^t)$ when m^t is lower than any price offered in histories h^t and $\tilde{h}^t$. That is, if the price is lower than in the past, the buyer's behavior is independent of previous prices.

Property S, which can be called "stationarity" or "the strong cutoff rule property," has a Markov flavor.[12]

Theorem 10.1 (Fudenberg, Levine, and Tirole 1985; Gul, Sonnenschein, and Wilson 1986) Suppose that the distribution of the buyer's type satisfies

---

12. In contrast with the Markov concept (see chapter 13 for the definition of Markov perfect equilibrium in games of complete information; see Maskin and Tirole 1989 for a definition of Markov perfect Bayesian equilibrium), property S cannot be required in states that are not reached in equilibrium.

conditions G and R. Then

(i) a perfect Bayesian equilibrium exists and is generically unique,

(ii) the equilibrium satisfies the Coase conjecture as $\delta \to 1$,

(iii) the equilibrium satisfies condition S, and

(iv) when $\underline{v} \to 0$, the equilibrium converges to a perfect Bayesian equilibrium of the no-gap case (that satisfies the Coase conjecture).[13]

Instead of proving this theorem, we give the flavor of the argument by analyzing the *two-type* case.[14] Suppose that $v = \bar{v}$ with probability $\bar{p}$, and $\underline{v}$ with probability $\underline{p}$. Let $\bar{\mu}^t$ denote the date-t posterior probability of $\bar{v}$, conditional on the history h^t of rejected prices. The first step in the proof is to show that the seller never makes an offer less than $\underline{v}$. Let $\underline{m}$ denote the infimum of equilibrium offers made by the seller for any period and history, and suppose that $\underline{m} < \underline{v}$.[15] Then we claim that $\underline{m}$ or any offer close to it is accepted with probability 1 by both types, as the most favorable offer made by the seller in the future is at best $\underline{m}$ (i.e., $v - \underline{m} > \delta(v - \underline{m})$ for all v). Hence, the seller can raise his price by a discrete amount above $\underline{m}$ and still have his offer accepted with probability 1, which means that prices arbitrarily close to $\underline{m}$ cannot be optimal after all. So $\underline{m} \geq \underline{v}$. This implies that type $\underline{v}$ accepts all prices below $\underline{v}$, so that for any history h^t the seller can guarantee himself present payoff $\underline{v}$ by offering $\underline{v}$.

With these preliminary observations in hand, we turn to the heart of the proof, which uses an "upward induction on beliefs":

• If $\bar{\mu}^t \leq \alpha \equiv \underline{v}/\bar{v}$, the seller's maximum profit from date t on is his "monopoly profit" $\underline{v}$. To see this, note that if the seller could commit himself to a single price, he would either choose $\underline{v}$ and get $\underline{v}$ or choose $\bar{v}$ and get $\bar{\mu}^t \bar{v}$, and because $\bar{\mu}^t \bar{v} \leq \underline{v}$, the optimal "commitment price" is $\underline{v}$. We saw in section 7.3 that commitment to a single price is an optimal mechanism for the seller; in particular, it weakly dominates the direct-revelation mechanism associated with the perfect Bayesian equilibrium of the bargaining game. Hence, $\underline{v}$ is an upper bound on the seller's profit from t on, and, furthermore, the seller can guarantee himself this upper bound by charging $\underline{v}$.

• Suppose now that $\bar{\mu}^t > \alpha$. Will the seller make an offer $m^t > \underline{v}$? If this offer results in posterior belief $\bar{\mu}^{t+1}(h^t, m^t) < \alpha$ (which implies that it is

13. Fudenberg et al. (1985) prove theorem 10.1 under assumption G and the following stronger assumption (R′): The distribution of types is smooth and has a density that is bounded and bounded away from zero ($0 < p_{\min} \leq p(v) \leq p_{\max}$ for all $v \in [\underline{v}, \bar{v}]$). Gul et al. (1986, theorem 1) generalize parts i–iii of their result by showing that a slightly weaker version of assumption R suffices; they also show that the seller does not randomize on the equilibrium path (although he does randomize off the path). Ausubel and Deneckere (1989a, theorem 4.2) obtain iv without assumptions on the distribution of types.

14. Hart 1989 provides more details of the two-type case.

15. That $\underline{m}$ is not $-\infty$ results from the fact that the buyer could guarantee himself a surplus close to $+\infty$ when such offers are made. Because aggregate gains from trade are finite, the seller would make a negative profit, which is impossible.

accepted by type $\bar{v}$ with positive probability), then, from our previous characterization, $m^{t+1}(h^t, m^t) = \underline{v}$. Therefore, $\bar{v} - m^t \geq \delta(\bar{v} - \underline{v})$ or $m^t \leq \tilde{v}_1 = \tilde{v} = \bar{v} - \delta(\bar{v} - \underline{v})$. Conversely, any $m^t \leq \tilde{v}$ is accepted by type $\bar{v}$, as the best offer he will get in the future is $\underline{v}$. Furthermore, when $\bar{\mu}^t \geq \alpha$, the seller always prefers to charge $\tilde{v}$ rather than $\underline{v}$, as the payoff from doing so,

$$\bar{\mu}^t(\bar{v} - \delta(\bar{v} - \underline{v})) + \delta(1 - \bar{\mu}^t)\underline{v},$$

exceeds $\underline{v}$ whenever $\bar{\mu}^t > \alpha$.

Now that we have established that for $\bar{\mu}^t > \alpha$ bargaining has at least two effective rounds, let us show that for $\bar{\mu}^t \in [\alpha, \alpha + \varepsilon]$, where ε is positive and small, the seller charges $\tilde{v}$ and then $\underline{v}$. Because $\bar{\mu}^{t+1} \geq \alpha$ if $m^t > \tilde{v}$, the seller's maximum profit from date t on, U_s^{sup}, over all beliefs in $[\alpha, \alpha + \varepsilon]$ and all possible equilibria, satisfies

$$U_s^{\text{sup}} \leq \max\left[\bar{\mu}^t \tilde{v} + \delta(1 - \bar{\mu}^t)\underline{v}, \frac{\bar{\mu}^t - \alpha}{1 - \alpha}\bar{v} + \delta\left(1 - \frac{\bar{\mu}^t - \alpha}{1 - \alpha}\right) U_s^{\text{sup}} \right].$$

$$(10.7)$$

Clearly, for ε small, the max is obtained for the first term, and therefore it is optimal to charge $\tilde{v}$ at date t. Having pinned down equilibrium when $\bar{\mu}^t \in [\alpha, \alpha + \varepsilon]$, one determines equilibrium for $\bar{\mu}^t \in [\alpha + \varepsilon, \alpha + 2\varepsilon]$, etc., until the first beliefs μ_2 such that the seller prefers to charge a price above $\tilde{v}$.[16]

Proceeding by upward induction on beliefs, one finds cutoff beliefs $\mu_0 = 0 < \mu_1 = \alpha < \mu_2 < \cdots$ such that, if $\bar{\mu}^t \in [\mu_n, \mu_{n+1})$, there are $n + 1$ effective bargaining rounds (i.e., the seller charges strictly above $\underline{v}$ for n periods and then offers $\underline{v}$). Posterior beliefs on the equilibrium path are decreasing (as the skimming property requires): $\bar{\mu}^{t+1} = \mu_{n-1}$, $\bar{\mu}^{t+2} = \mu_{n-2}, \ldots, \bar{\mu}^{t+n} = 0$. Prices are decreasing on the equilibrium path. They are given by type $\bar{v}$'s indifference between accepting in a given period and accepting in the following period: $m^{t+n} = \underline{v}$, $m^{t+n-1} = \tilde{v}$, and, more generally,

$$\bar{v} - m^{t+k} = \delta(\bar{v} - m^{t+k+1}) \text{ or } m^{t+k} = \bar{v} - \delta^{n-k}(\bar{v} - \underline{v}). \qquad (10.8)$$

To check the Coase conjecture, it suffices to show that for any $\bar{p}$ there exists n such that $\mu_n > \bar{p}$ for any δ.[17] Equation 10.8 together with the last offer's being $\underline{v}$ then implies that, for δ close to 1, all offers in the effective length of bargaining are close to $\underline{v}$. Furthermore, because there are at most $n + 1$ offers, agreement is reached almost instantaneously.

---

16. It is easily checked that μ_2 satisfies

$$\frac{\mu_2 - \alpha}{1 - \alpha}[\bar{v} - \delta^2(\bar{v} - \underline{v})] + \left(1 - \frac{\mu_2 - \alpha}{1 - \alpha}\right)\delta\underline{v} = \mu_2(\bar{v} - \delta(\bar{v} - \underline{v})) + (1 - \mu_2)\delta\underline{v}.$$

17. For instance, check that μ_2 does not converge to α when δ converges to 1.

10.2.6　The No-Gap Case[†††]

Assume now that

(NG)　$\underline{v} = 0$.

Statement iv of theorem 10.1 shows that there exists an equilibrium satisfying condition S and the Coase conjecture. However, Gul, Sonnenschein, and Wilson (1986) show that there exist other equilibria.[18] They characterize the set of perfect Bayesian equilibria satisfying condition S as follows.

Theorem 10.2 (Gul, Sonnenschein, and Wilson 1986)　Assume conditions NG and R. Then any perfect Bayesian equilibrium satisfying condition S satisfies the Coase conjecture.

We do not give the proof of theorem 10.2, which is complex. But it is worth sketching it, because the proof highlights the logic of the Coase conjecture better than the proof of theorem 10.1, which constructs the equilibrium and checks that it satisfies the conjecture. Under assumption S, the buyer's cutoff valuation $\kappa(\cdot)$ is independent of history. Therefore, the seller's expected profit from date t onward, given current beliefs $P(v)/P(\kappa^t)$ on $[0, \kappa^t]$ at the beginning of period t, depends only on the current cutoff valuation, κ^t, and not on history. Denote this payoff by $U_s(\kappa^t)$. Suppose for simplicity that the sequence of cutoffs κ^t is deterministic.

Fix a real time $\varepsilon > 0$, and let the time period Δ tend to 0 (so that there is a large number of offers between 0 and ε). For any $\eta > 0$, there exists a Δ sufficiently small and a $t \leq \varepsilon/\Delta - 2$ such that

$$P(\kappa^t) - P(\kappa^{t+2}) < \eta,$$

that is, the seller sells with low probability between t and $t + 2$. (Because of the large number of offers between 0 and ε, there must be some periods in which he sells with low probability.) The intuition for the Coase conjecture is that the seller would be eager to speed up the sale process at t if his continuation value at time ε were not negligible. To see this, note that the seller could offer m^{t+1} at t and thus create posterior beliefs κ^{t+2} at $t + 1$, so

$$[P(\kappa^t) - P(\kappa^{t+1})]m^t + \delta[P(\kappa^{t+1}) - P(\kappa^{t+2})]m^{t+1} + \delta^2 P(\kappa^{t+2})U_s(\kappa^{t+2})$$

$$\geq [P(\kappa^t) - P(\kappa^{t+2})]m^{t+1} + \delta P(\kappa^{t+2})U_s(\kappa^{t+2}). \tag{10.9}$$

(This is where the stationarity assumption is crucial: The seller gets $U_s(\kappa^{t+2})$

---

18. Note that this multiplicity as well as that in theorem 10.3 depends on the set of possible valuations' being an interval. If the set of possible valuations is discrete and does not include the seller's cost, the no-gap model is equivalent to the gap model even if some types of buyers have valuation lower than cost. We know from theorem 10.1 that the equilibrium is then (generically) unique. (When the distribution is discrete and there is a type whose valuation is equal to cost, the uniqueness of equilibrium depends on what the seller is assumed to do when he has sold to all types with valuation above his cost.)

when beliefs are κ^{t+2} independent of the history leading to those beliefs.) But, by definition of the cutoff valuations,

$$\kappa^{t+1} - m^t = \delta(\kappa^{t+1} - m^{t+1}). \tag{10.10}$$

Combining equations 10.9 and 10.10 and using the fact that $t + 2 \le \varepsilon/\Delta$ and that posteriors are decreasing, we get

$$[P(\kappa^t) - P(\kappa^{t+1})]\kappa^{t+1} - [P(\kappa^t) - P(\kappa^{t+2})]m^{t+1}$$
$$\ge \delta P(\kappa^{t+2})U_s(\kappa^{t+2}) \ge \delta P(\kappa^{\varepsilon/\Delta})U_s(\kappa^{\varepsilon/\Delta}). \tag{10.11}$$

The left-hand side of equation 10.11 is very small if η is small, and therefore $P(\kappa^{\varepsilon/\Delta})U_s(\kappa^{\varepsilon/\Delta})$ is very small. Thus, for any real time ε, the seller's profit from ε on tends to 0. This does not yet imply that the price at any time $\varepsilon > 0$ tends to 0, as small profits can arise either from low prices or from slow (i.e., delayed) sales. But it can be shown that, for any real time $\varepsilon > 0$, the price tends to 0 as Δ tends to 0.[19] This, in turn, implies trivially that profit at date 0, and not only profit as positive dates, tends to 0 (the buyers do not buy at date 0 at a noninfinitesimal price if they expect infinitesimal prices in the near future).

Ausubel and Deneckere (1989a) show that, without the stationarity assumption, the Coase conjecture does not hold, and, worse still, "anything" is an equilibrium of the no-gap case when $\delta \to 1$:

Theorem 10.3 (Ausubel and Deneckere 1989a)[20]   Assume NG and that there exists $L > M > 0$ such that, for all $v \in [0, \bar{v}]$, $Lv \le P(v) \le Mv$. Let $U_s^* \equiv \sup_m[m(1 - P(m))]$ denote the monopoly profit. Then, for any $\varepsilon > 0$, there exists $\Delta(\varepsilon) > 0$ such that for any $\Delta \le \Delta(\varepsilon)$, for any $U_s \in [\varepsilon, U_s^* - \varepsilon]$,

19. A sketch of the argument follows: Fix a real time $\varepsilon > 0$, and suppose that there exists a sequence $\Delta \to 0$ such that the price at date ε does not converge to 0: $m^{\varepsilon/\Delta} \ge \bar{m} \ge 0$. The proof that this is impossible has four steps: (1) Because prices decrease over time and because the profit from real time Δ (i.e., period 1) on tends to 0, the probability of a sale between Δ and ε must converge to 0 with Δ. (2) Unless he has charged 0 in the past, the seller sells with strictly positive probability in each period. If not, his continuation valuation would be 0, as under assumption S the seller's optimal strategy can be taken to be stationary and therefore to generate no sale forever. However, he would be better off charging a price slightly above 0, as such a price would be accepted with positive probability because of discounting and because of the fact that the seller never charges prices below 0. (3) Because the probability of sale between Δ and ε tends to 0, $m^1 - m^{\varepsilon/\Delta}$ tends to 0. (The price schedule is almost flat, as one can see by using equation 10.10 at periods 1 and ε/Δ.) Let m^* denote the common limit of m^0 and $m^{\varepsilon/\Delta}$. ($m^0 - m^{\varepsilon/\Delta}$ tends to 0 by equation 10.10 applied to $t = 0$.) (4) Because $\bar{v}$ purchases at date 0, $\bar{v} \ge m^*$. Also, $e^{-r\varepsilon}(\kappa^{\varepsilon/\Delta+1} - m^{\varepsilon/\Delta}) \ge \kappa^{\varepsilon/\Delta+1} - m^0$, where $\kappa^{\varepsilon/\Delta+1}$ is the cutoff type at period $\varepsilon/\Delta + 1$, and hence $m^* = \bar{v}$, as $\kappa^{\varepsilon/\Delta+1} \to \bar{v}$ in the limit. This means that type $\bar{v}$ has utility 0 in the limit, and so have the other types (who have lower utility than type $\bar{v}$). Therefore, the (stationary) strategy of each type is to accept any offer less than their type in the limit, and the profit of the seller from any date on cannot go to 0.
20. Ausubel and Deneckere (1989b, theorem 2) show not only that any profit is an equilibrium profit when $\Delta \to 0$, but also that any vector $\{U_s, U_b(\cdot)\}$ of expected utilities for the seller and each type of buyer that is feasible (i.e., is individually rational for all, is incentive compatible for the buyer, and corresponds to a feasible trade policy) can be approximately obtained as a perfect-equilibrium payoff vector of the sale model when $\Delta \to 0$.
    Ausubel and Deneckere (1987) and Gul (1987) derive theorems similar to theorem 10.3 for durable-good oligopolists.

there exists a perfect Bayesian equilibrium of the sale model such that the seller's profit is equal to U_s.

The intuition for this theorem is that one can use a Coase path associated with some equilibrium satisfying condition S as a "threat" that prevents the seller from deviating from a given price path. To illustrate the idea, take the linear-demand case of the example in subsection 10.2.3. There, we derived an equilibrium satisfying condition S. The seller's valuation for current cutoff κ was $U_s^C(\kappa) = \gamma(\delta)\kappa^2/2$, with $\lim_{\delta \to 1} \gamma(\delta) = 0$ (where C stands for "Coase"). (Because subsection 10.2.3 was interpreted with a continuum of buyers, the valuations $U_s^*(\cdot)$ and $U_s^C(\cdot)$ in this proof correspond to the continuum-of-buyers case. See subsection 10.2.3 for the link with the single-buyer case.) Let us show that there is a perfect Bayesian equilibrium in which the seller makes a profit close to the monopoly profit $\frac{1}{4}$ (which is obtained by charging the monopoly price $\frac{1}{2}$). Consider the following exponential price path in *real* time (where τ denotes real time and t, as before, denotes periods):

$$m^\tau = \tfrac{1}{2}e^{-\eta\tau}.$$

That is, the price starts at the monopoly price $\frac{1}{2}$ and decreases exponentially. In period t in discrete time, one has

$$m^t = \tfrac{1}{2}e^{-\eta t\Delta}.$$

Note that if η is close to 0, and if $m^{\cdot}$ is the equilibrium path, all types $v \geq \frac{1}{2} + \varepsilon$ buy at date 0 for ε small (as $v - \frac{1}{2} = \sup_\tau e^{-r\tau}(v - \frac{1}{2}e^{-\eta\tau})$). Hence, the seller makes almost his monopoly profit (as we noted in chapter 7, he cannot make more than the monopoly profit, because the equilibrium must respect the buyers' incentive-compatibility and individual-rationality constraints).

Consider the following strategies: "The seller charges $m^t = \frac{1}{2}e^{-\eta t\Delta}$; type v chooses t so as to maximize $e^{-rt\Delta}(v - \frac{1}{2}e^{-\eta t\Delta})$. If the seller deviates from the above path, the equilibrium switches immediately to the Coase equilibrium obtained in subsection 10.2.3." Clearly, the buyers' behavior is optimal in view of the seller's. Would the seller ever deviate? Clearly not at date 0; but if we assume that η is very small (as we want to ensure that types above $\frac{1}{2}$ buy at date 0), it might be the case that sales are so slow later that the seller may want to switch to the Coase path. In this case, the high-price path would not be credible and the buyers would refrain from buying because they would expect a switch to the low-price Coase path. To make sure that this does not occur, let us fix η small, take Δ to 0, and show that the seller never wants to deviate from the exponential path.

Note that along the equilibrium path the cutoff valuation κ^{t+1} is given by indifference between buying today and buying tomorrow:

$$\kappa^{t+1} - \tfrac{1}{2}e^{-\eta t\Delta} = e^{-r\Delta}(\kappa^{t+1} - \tfrac{1}{2}e^{-\eta(t+1)\Delta}), \tag{10.12}$$

and therefore

$$\kappa^t - \kappa^{t+1} \simeq \frac{\eta(r + \eta)}{2r} e^{-\eta t \Delta} \Delta \tag{10.13}$$

for Δ small. That is, sales per unit of time decline exponentially at rate η with real time ($t\Delta$). Furthermore, κ^t is approximately proportional to $e^{-\eta(t\Delta)}$. Integrating price times sales over the remaining horizon, one can see that the expected profit from date t on, $U_s^*(\kappa^t)$, is approximately $v(\Delta)(\kappa^t)^2$, where $\lim_{\Delta \to 0} v(\Delta) = v > 0$. Therefore, for Δ sufficiently small, $U_s^*(\kappa^t) > U_s^C(\kappa^t)$, so deviations to the Coase path are not profitable to the seller.

Clearly, the derivations for profits smaller than the monopoly profit follow the same lines.

Remark   This equilibrium has the same bootstrap feature as the equilibria derived in the folk theorem for infinitely repeated games (see chapter 5): A player is threatened by a move to an equilibrium that is bad for him if he deviates. It should, therefore, not come as a surprise that with a finite "real-time" horizon, the Coase conjecture holds in the limit of shorter time periods even without assumption S: See exercise 10.1.

10.2.7   Gap vs. No Gap and Extensions of the Single-Sale Model[†††]

After this detailed examination of the equilibrium set, it is time to step back and discuss some modeling issues.

Subsections 10.2.4 and 10.2.5 showed how the gap and no-gap cases have different implications. In the gap case negotiations end after a fixed finite number of offers; in the no-gap case negotiations can continue forever. As we suggest in subsection 10.2.2, finite negotiations seem the more reasonable prediction in many cases. If the seller has an "outside option"—an opportunity to sell to another buyer—he will take it when he becomes sufficiently pessimistic about the current buyer's willingness to trade. Although this may be an argument for the gap case, it seems unnatural to suppose that all potential buyer types have gains from trade. A more descriptive model might have a positive probability that $v < c$, i.e., that gains from trade are negative, coupled with a decision by the buyers about whether to enter negotiations.[21]

---

21. The simplest version of such a model, which supposes that all buyers have the same positive cost of entering negotiations, runs into a kind of market shutdown; the seller will never charge a price below the lowest valuation that chooses to enter negotiations, so this lowest-valuation type has a negative payoff from negotiations, and there is an equilibrium where *no* types pay the entry fee. (See Fudenberg and Tirole 1983, Perry 1986, Cramton 1990, and exercise 10.6.) To avoid this market closure, a good model of endogenous entry into negotiations must be even more complicated; one possibility is to make the buyer's entry cost private information as well, with a positive probability that his cost is negative. We would be interested in knowing what such a model implies.

When one interprets the single-sale model as representing sales by a durable-good monopolist, one supposes that all of the monopolist's potential customers are present and actively "shopping" at the beginning of the first period. Sobel (1990) extends the model to allow for a regular flow of new consumers. Because of this inflow, the distribution of consumer types does not deteriorate monotonically as in the equilibria with a fixed stock of consumers. Sobel's model is a generalization of the two-type model discussed in subsection 10.2.5. At each period $t = 0, 1, \ldots$, a new group of buyers enters the market; a proportion $\bar{p}$ of them have valuation $\bar{v}$ and a proportion p have valuation $\underline{v}$. (There is a continuum of "small" consumers of each type, and the aggregate system is modeled as deterministic—if the high-value buyers play a mixed strategy of accepting with probability $\frac{1}{3}$, then a fraction of exactly $\frac{1}{3}$ will accept and the others will reject.) The size of the inflow is the same in each period. The buyers who entered earlier and have not purchased yet are still in the market.

Because the inflow of new consumers prevents the distribution from becoming concentrated at $\underline{v}$, the backward-induction argument that yields uniqueness in the gap case of theorem 10.1 cannot be applied. And indeed there are multiple equilibria in this model; in fact, any payoff between $\underline{v}$ and the monopoly profit can be sustained by a perfect Bayesian equilibrium.

This leads Sobel to consider a restriction to stationary strategies, similar to that proposed by Gul et al. (1986). Here the state of the system includes the numbers of low- and high-valuation buyers currently in the market, so a stationary strategy for the buyer can depend on this aggregate state as well as on the current price and his own valuation.[22]

Sobel shows that equilibria in stationary strategies exist, and provides a partial characterization. With stationary strategies, the seller cannot sell at prices significantly greater than $\underline{v}$ when the period length shrinks to 0. Moreover, in a stationary equilibrium prices must cycle: As long as the seller charges prices above $\underline{v}$, the number of low-valuation buyers in the market increases both in absolute and in relative terms. At some point, it pays the seller to have a "sale," that is, to charge $\underline{v}$ (in that case, there are only new buyers in the following period). The price path is thus an infinite replica of the one obtained in subsection 10.2.5. The price decreases within each cycle according to the formula $m = \bar{v} - \delta^n(\bar{v} - \underline{v})$, n periods before the next sale. Sobel shows that there is an upper bound n^* independent of the discount factor such that there are never more than n^* periods until the next sale, so that as the discount factor tends to 1 (i.e., the time intervals

22. Contrast this with the continuum case in the traditional model, where the current price determines the next cutoff valuation and thus makes the information about the current cutoff "irrelevant."

shrink to 0) the next sale is always "soon," and prices converge to $\underline{v}$, as in the Coase conjecture.[23]

The variants of the single-sale model we have discussed so far all suppose that the seller has no private information. This may account for the striking force of the Coase conjecture, which shows that the buyer (who does have private information) has all of the bargaining power when the discount factor is close to 1, at least when attention is restricted to stationary strategies. Since this conclusion seems unrealistic, researchers have studied models in which the seller has private information as well.

One way to introduce private information is to suppose that the buyer does not know the seller's cost, or, more generally, the seller's willingness to sell at a given price. As in the models of chapters 8 and 9, the seller may have an incentive to build a reputation for high cost or "toughness" by charging a high price, just as the buyer has an incentive to build a reputation for a low valuation by rejecting offers. Since the Coase conjecture implies that bargaining is at least approximately efficient and bargaining with two-sided incomplete information tends to be inefficient (subsection 7.4.4), the Coase conjecture should not be expected to obtain here. We say more about this case in section 10.4.

An alternative way to introduce private information for the seller is to allow him to have private information about the quality of the good, as in Evans 1989 and Vincent 1989. If the sellers' cost of production is increasing in quality, or if the good has already been produced and the seller receives a flow of utility from owning the good, the seller's willingness to sell at a given price decreases with quality, which leads to inefficiency in the same way as in Akerlof's (1970) lemons model.

10.3 Intertemporal Price Discrimination: The Rental or Repeated-Sale Model[†††]

Now consider a seller who wishes to rent a service or a good to a buyer. The buyer is free to rent the good today but not rent it tomorrow, so the game is not over the first time the buyer accepts an offer. In this model, it is more convenient to define v and c as flow variables, i.e., the per-period valuation and cost. We consider two variants of the rental model. In the short-term-contracts variant, the seller makes offers for rental in the current

---

23. There are other interesting variants. Bond and Samuelson (1984, 1987) allow the good to depreciate. Buyers therefore come back to the seller after a while. Under a stationarity assumption, a form of the Coase conjecture holds, but monopoly profits can be sustained with nonstationary perfect Bayesian equilibria. The results thus have the same flavor as those of Sobel (1990) and those of subsections 10.2.5 and 10.2.6.

Durable-good monopolists with either decreasing returns in each period (Kahn 1986) or learning by doing (Olsen 1988; see exercise 10.5 below) have also been studied. With decreasing returns, the Coase conjecture does not hold (an extreme case of decreasing returns is a capacity constraint in each period, which acts as a commitment not to "flood the market"). With learning by doing, only part of the Coase conjecture holds.

period only, and is not allowed to propose prices for rentals in future periods. The other variant supposes that the seller is allowed to propose long-term contracts, but that the seller cannot commit not to try to re-negotiate these contracts in the future.

Before treating these two variants, it is interesting to note that the formulations of a single buyer and a continuum of buyers, which coincide in the sale model, are dramatically different in the rental model. A continuum of buyers who are treated anonymously (i.e., cannot be told apart by the seller) do not make strategic use of their information in the rental model, and so the outcome is the usual monopoly outcome. In contrast, we will see that a single buyer with a continuum of possible valuations zealously guards knowledge of his valuation in the rental model. We will pursue the single-buyer interpretation in this section.

10.3.1 Short-Term Contracts

Suppose first that the seller can make offers only for the current period. Thus, at date $t = 0, 1, \ldots, T$, the seller offers a rental price r^t. The buyer with valuation v has utility $v - r^t$ during period t if he accepts the offer, and 0 otherwise, where v is now a flow utility. The history of the game at date t is the sequence of the previous price offers and whether those were accepted. The price and the acceptance decision at date t depend on the history at that date (and the acceptance decision depends also on the current price). We again normalize the production cost at 0.[24]

When T is large, the seller gets almost no profit in this game:

Theorem 10.4 (no price discrimination) Suppose that the buyer has n possible valuations $0 < v_1 = \underline{v} < v_2 < \cdots < v_n = \bar{v}$. And assume $\delta > \frac{1}{2}$ and $T < +\infty$.

• Let $n = 2$. There exist T_0 and T_1 such that, for any $T \geq T_0$ and any perfect Bayesian equilibrium of the rental game, the seller charges $r^t = \underline{v}$ for all $t = 0, 1, \ldots, T - T_1$. (Hart and Tirole 1988)
• Let $n \geq 2$. There exist T_0 and T_1 such that, for any $T \geq T_0$ and any Markov perfect Bayesian equilibrium[25] of the rental game, the seller charges $r^t = \underline{v}$ for all $t = 0, 1, \ldots, T - T_1$. (Schmidt 1990)

Theorem 10.4 shows that when the horizon is long, the seller charges the low price $\underline{v}$ for all periods but the last. Consequently, his expected

24. With positive costs, one must distinguish variable or flow costs from fixed or one-time costs. The analysis below extends immediately to the case of a flow cost. With a one-time cost, the continuation equilibrium after the first period of actual rental is the one described below but the analysis must be extended to include the game before the first rental.
25. Schmidt (1990) uses the Markov perfect Bayesian equilibrium (MPBE) concept defined by Maskin and Tirole (1989): An MPBE is a limit of a sequence of approximate strong MPBE, in which the strategies of the players depend on their private information and the common beliefs. Strong MPBE do not always exist, but MPBE always exist. For example, the unique equilibrium of subsection 10.2.2 is not strong Markov.

discounted profit converges to $\underline{v}/(1 - \delta)$ as T goes to infinity. Theorem 10.4 is similar to the Coase conjecture in that the seller cannot price discriminate and makes a profit approximately equal to the lowest (per-period) valuation. It differs from the Coase conjecture in that the result holds for any $\delta > \frac{1}{2}$, whereas the Coase conjecture requires δ near 1. The additional strength comes from a fixed finite horizon. Indeed, the rental model can give weaker conclusions than the sale model with the infinite horizon considered in the last section. For example, with only one valuation for the buyer, this is a standard repeated game, and the folk theorem obtains.

We should also comment on the Markov assumption with more than two potential valuations. A step in the proof that the seller does not price discriminate shows that the seller never offers $r^t < \underline{v}$ for any t, because this offer is dominated by $r^t = \underline{v}$. The difficulty in showing this property comes from the fact that there might be continuation equilibria from $t + 1$ on that yield different continuation payoffs for the seller. Therefore, it might be the case that the seller charges $r^t < \underline{v}$ rather than $r^t = \underline{v}$ because the continuation equilibria (which correspond to the same posterior beliefs) differ. Now, with $n = 2$, one can show that (perfect Bayesian equilibrium) continuation payoffs for the seller are unique for given posterior beliefs (see exercise 10.4) and therefore the issue does not arise. On the other hand, for any n, the Markov assumption guarantees that the seller's continuation valuation is the same whether $r^t < \underline{v}$ or $r^t = \underline{v}$, so that he strictly prefers $r^t = \underline{v}$. It is not known whether the Markov assumption is needed in general.

To obtain some intuition about the result, consider the case of an infinite horizon and two types, and the following strategies and beliefs: "At any t, the seller offers $r^t = \underline{v}$ if the posterior probability of a high-valuation buyer, $\bar{\mu}^t$, is less than 1, and $r^t = \bar{v}$ if $\bar{\mu}^t = 1$; when $\bar{\mu}^t < 1$, the buyer accepts all offers below or at $\underline{v}$, and rejects all other offers whatever his type; if $\bar{\mu}^t = 1$, the $\bar{v}$-type accepts $r^t \leq \bar{v}$ and rejects higher offers and the $\underline{v}$-type accepts $r^t \leq \underline{v}$ and rejects higher offers. Last, if $\bar{\mu}^t = 1$, then $\bar{\mu}^{t+1} = 1$; if $\bar{\mu}^t < 1$ and $r^t \leq \underline{v}$, then $\bar{\mu}^{t+1} = \bar{\mu}^t$; if $r^t > \underline{v}$, then $\bar{\mu}^{t+1} = 1$ if r^t is accepted and $\bar{\mu}^{t+1} = \bar{\mu}^t$ if r^t is rejected." In words, the buyer always refuses any offer above $\underline{v}$ whatever his type, and if he were to accept such an offer he would be identified as a high-valuation buyer. The seller always charges $\underline{v}$. It is easy to see that these strategies form a perfect Bayesian equilibrium of the infinite-horizon game: If type $\bar{v}$ accepts an offer above $\underline{v}$, he gets current payoffs at most equal to $\bar{v} - \underline{v}$. However, the seller learns his identity and charges $\bar{v}$ forever. In contrast, if $\bar{v}$ rejects the offer, he will be able to continue buying at price $\underline{v}$ forever. Hence, if

$$\bar{v} - \underline{v} < (\delta + \delta^2 + \cdots)(\bar{v} - \underline{v}) \Leftrightarrow \delta > \tfrac{1}{2},$$

type $\bar{v}$ prefers to reject offers above $\underline{v}$. Knowing this, the seller never tries to offer prices above $\underline{v}$.

When $T < \infty$, the unique equilibrium is close to the infinite-horizon one described above.[26] The high-valuation buyer's payoff is at most $\bar{v} - \underline{v}$ if he accepts today, and is close to $[\delta/(1 - \delta)](\bar{v} - \underline{v})$ if he rejects and the horizon is sufficiently long. The seller does not attempt to price discriminate (charge prices above $\underline{v}$) until the final periods of their relationship.

The method of proof used by Hart and Tirole is quite different from that of Schmidt. Hart and Tirole use a reasoning of upward induction on beliefs very similar to that of subsection 10.2.5. Schmidt's method of proof resembles the one used by Fudenberg and Levine (1989) in their paper on the reputation of a long-term player facing a sequence of short-term players (see theorem 9.1).[27] He first shows that there is a strictly positive minimum probability of acceptance of offers $r^t > \underline{v}$ that are made on the equilibrium path. He then shows that, because types $v > \underline{v}$ can build a reputation for being of type $\underline{v}$, they will do so, as the revelation of their type would be very costly.

Remark   Because the seller is a long-run player, he may have an incentive to maintain a reputation of his own if the prior distribution makes this possible. For example, if the seller has private information about his cost, he may try to maintain a reputation for high cost by charging high prices. Thus, the results here about limits of equilibria with a long, finite horizon may be sensitive to the introduction of a small probability that the seller has high cost. An interesting question is whether a small probability of high cost would be swamped by a larger probability that the buyer is type $\underline{v}$, so that the equilibrium outcome would be close to that when the seller's cost is known. This has not been worked out, to the best of our knowledge.

10.3.2   Long-Term Contracts and Renegotiation

Now we suppose that the seller is able to offer long-term rental agreements to the buyer, and that the seller cannot commit not to renegotiate a contract later; that is, although a long-term contract is enforced if one of the parties wants it to be enforced, the parties can agree to replace an old contract with a new one if they both benefit from doing so.[28] As before, we suppose that the seller makes all the offers, including offers to renegotiate existing contracts. We also restrict our attention to the case where the buyer has

26. There are many other perfect Bayesian equilibria in the infinite-horizon game, as the "threat" of reverting to this equilibrium can be used to support price paths where the seller's profit is higher.
27. Fudenberg and Levine's "Stackelberg type," i.e., the type that the buyer would like the seller be convinced of, is type $\underline{v}$ here. Note that theorem 9.1 does not apply for two reasons. First, the rental game has two long-term players rather than a long-term player facing a sequence of short-term players. Second, theorem 9.1 covers both finite-horizon and infinite-horizon games, and has force only in the limit of discount factors $\delta \to 1$. The assumption of a long but finite horizon permits a backward-induction argument that yields a strong conclusion for any $\delta > \frac{1}{2}$.
28. Equivalently, they can keep the old contract and "cancel" its effects through an additional contract.

two possible types, $\underline{v}$ and $\bar{v}$. The space of contracts that can be signed at date t is quite large: A long-term contract signed at date t specifies probabilities of consumption $\{x^{t+\tau}\}_{\tau=0}^{T-t}$ by the buyer from t to T and transfers $\{r^{t+\tau}\}_{\tau=0}^{T-t}$ from the buyer to the seller. The numbers $x^{t+\tau}$ and $r^{t+\tau}$ depend on messages sent by the buyer at each date up to $t + \tau$.[29] Note that the short-term contracts of subsection 10.3.1 are long-term contracts with $x^{t+\tau} = r^{t+\tau} = 0$ for $\tau > 0$.

Although the space of feasible long-term contracts is very large, only one kind of long-term contract is actually used in equilibrium: long-term leasing contracts signed at t in which the seller commits to supply from t to T. (Such contracts can be called "sale contracts" even though the good is a rental good.) As a consequence, the pattern of consumption is the same as if the good were durable. Another way to describe the result is as follows: If the only feasible long-term contract were a sale contract, the equilibrium would be the one of the sale model described in section 10.2, and introducing other contracts does not affect the equilibrium allocation (the time pattern of consumption and utilities).

Theorem 10.5 (Hart and Tirole 1988)   Suppose that the buyer has valuation $\underline{v}$ or $\bar{v}$. Then the outcome of the rental model with (possibly renegotiated) long-term contracts coincides with that of the durable-good model.

To obtain some intuition about this result,[30] it is useful to recall the basic conflict between efficiency and rent extraction in mechanism design (see section 7.3 and the price-discrimination example in subsection 7.1.1). Let $\underline{x}^t$ and $\bar{x}^t$ denote the probabilities of consumption at date t of types $\underline{v}$ and $\bar{v}$, and let $\underline{X} \equiv \mathrm{E}(\sum_{t=0}^{T} \delta^t \underline{x}^t)$ and $\bar{X} \equiv (\sum_{t=0}^{T} \delta^t \bar{x}^t)$ denote the expected discounted consumptions of the two types, where $0 \leq \underline{X}$ and $\bar{X} \leq 1 + \delta + \cdots + \delta^T$. The total social surplus is $\underline{p}\underline{X}\underline{v} + \bar{p}\bar{X}\bar{v}$. If $\underline{U}_b$ and $\bar{U}_b$ denote the expected utilities of the two types, and U_s the seller's expected profit, that the social surplus equals the buyer surplus plus the seller surplus implies that

$$U_s = \underline{p}(\underline{X}\underline{v} - \underline{U}_b) + \bar{p}(\bar{X}\bar{v} - \bar{U}_b).$$

But, in any mechanism or game, type $\bar{v}$ can always pretend to be type $\underline{v}$:

$$\bar{U}_b \geq \underline{X}(\bar{v} - \underline{v}) + \underline{U}_b.$$

Hence,

---

29. This is the appropriate version of the revelation principle for this model. The standard revelation principle, that the buyer announces his type truthfully to the seller when signing a contract, does not hold because of the possibility of renegotiation.
30. Laffont and Tirole (1990) extend this result to continuous consumption per period in the two-period case.

$$U_s \leq \underline{U}_b + \bar{p}\bar{X}\bar{v} + \underline{X}(\underline{v} - \bar{p}\bar{v}).$$

If the seller could commit to an allocation at date 0, he would clearly choose $\underline{U}_b = 0$, $\bar{X} = 1 + \delta + \cdots + \delta^T$ (efficient consumption for type $\bar{v}$), and $\underline{X} = 0$ if $\underline{v} < \bar{p}\bar{v} \Leftrightarrow \underline{v}/\bar{v} \equiv \alpha < \bar{p}$, or $\underline{X} = 1 + \delta + \cdots + \delta^T$ if $\alpha > \bar{p}$. That is, the seller would either "sell" at date 0 or "not sell" at all.

Assume that the seller would like to price discriminate, i.e., that $\bar{p} > \alpha$. (If $\alpha \geq \bar{p}$, the seller can guarantee himself the monopoly profit by offering the sale contract at date 0 at price $r^t = \underline{v}$ for all t. The interesting case, and the general one with a general distribution, is that in which profit maximization conflicts with efficiency.) Suppose that the seller tries to get his monopoly profit by offering the two long-term contracts ($x^t = 1$ and $r^t = \bar{v}$ for all t) and ($x^t = 0$ and $r^t = 0$ for all t) in period 0, and by claiming that no other contract will be offered in the future. If in equilibrium type $\bar{v}$ chooses the first contract and type $\underline{v}$ chooses the second contract (as the seller would want them to do), then if the buyer chooses the second contract at period 0, revealing that he is type $\underline{v}$, in period 1, seller would want to offer ($x^t = 1$ and $r^t = \underline{v}$ for all $t \geq 1$). Anticipating this, type $\bar{v}$ would not want to take the first contract.

As in static mechanism design, the binding incentive constraint here is to induce the high-valuation buyer to reveal his type. When he reveals his type, he must consume with probability 1 in all subsequent periods, because only the efficient contracts are renegotiation-proof under symmetric information. This suggests that a sale contract will be offered in each period (the game being over if it is accepted, as efficient contracts are always renegotiation-proof). Consider now the contract that is chosen at date t by type $\underline{v}$ (one can show that it is unique). Either it is efficient and the game is over; or it is inefficient because the seller prefers to extract rent, and the linearity of the tradeoff between efficiency and rent extraction suggests that $x^t = 0$. That is, the seller can wait at least one more period to offer a contract that is accepted by type $\underline{v}$. In equilibrium, the seller offers only one contract, a sale contract, in each period, and type $\bar{v}$ randomizes between accepting and refusing it, until the seller offers the sale contract at per-period price $\underline{v}$. The nature of renegotiation-proof contracts with several types is still unknown.

10.4 Price Offers by an Informed Player[†††]

In sections 10.2 and 10.3 we obtained a coherent set of results by making the strong assumption that only one party has private information and that the other party has all the bargaining power. Because we have little information about the extensive form that is played in practice, we must consider alternative bargaining processes. Furthermore, both parties may have private information. Changing the model in either of these directions introduces a multiplicity of equilibria. For instance, even in a two-period model, if the informed party makes the first-period offer (and the uninformed party makes the second-period offer) or if both parties have private information, there may exist continua of pooling, separating, and hybrid equilibria.[31] Needless to say, this feature carries over to any horizon. The reason for the large multiplicity of equilibria is the same as in the case of the Spence signaling game. Here an off-the-equilibrium-path offer may be interpreted by the other party as stemming from a "weak type" (high-valuation buyer, low-cost seller) who is eager to reach an agreement and will concede fast in the future; such beliefs, in turn, make the offers unattractive to the party that proposes because they are unlikely to be accepted and will induce the other party to take a tough stance in the future.

The literature on bargaining with price offers by an informed player is large, and because of the limited scope of this chapter we do not attempt to review it thoroughly. It is divided along several lines: whether one player makes all the offers (as in the sale and rental models) or another bargaining process is used (typically, the alternating-move model), whether there is asymmetric information on one side or both, and whether the paper tries to characterize the equilibrium set or uses a refinement to reduce its size. We start with the one-sided-offers, bilateral-asymmetric-information model to highlight some new features relative to the previous sections, and then move on to the alternative-offer, one-sided-asymmetric-information model. Last, we give a few results using mechanism design to characterize equilibria of bargaining processes.

10.4.1 One-Sided Offers and Bilateral Asymmetric Information

Consider the durable-good model of section 10.2, but let the seller have private information about his cost $c \in [\underline{c}, \overline{c}]$. As before, the buyer has private information about his valuation $v \in [\underline{v}, \overline{v}]$. Assume that the horizon is infinite and the seller makes all offers, denoted $\{m^t\}_{t=0}^{\infty}$.

There are many perfect Bayesian equilibria for this model. Assume that the distributions over types are continuous and that $\underline{v} < \overline{c}$. Cramton (1984) constructs an equilibrium for the double-uniform case in which the seller sells only after having revealed his cost, and, as the time Δ between successive offers converges to 0, the initial revealing offer converges to his cost. Thus, the first part of the Coase conjecture (that the seller's profit converges to 0) is satisfied. Cho (1990) obtains an equilibrium in which this property applies only to the seller with type $\underline{c}$, who reveals his type in the first period and makes an offer that converges to $\underline{c}$ as $\Delta \to 0$. Buyers reject offers much above $\underline{c}$, so sellers with costs much above $\underline{c}$ are unlikely to sell, and, when $\Delta \to 0$, the probability of sale goes to 0 as well! This conclusion may seem perverse, but Cho shows that it applies to any perfect Bayesian equilibrium

31. See Fudenberg and Tirole 1983.

in which the strategies satisfy a form of stationarity and in which the seller's offer in any period, on or off the equilibrium path, is taken to be a perfectly revealing signal of his type.[32]

But one can also construct equilibria in which the seller makes approximately the monopoly profit corresponding to his cost if Δ is close to 0 (Ausubel and Deneckere 1990a). The idea is clear from theorem 10.3: If the seller detectably deviates from the path that gives him almost his monopoly profit, he is thought of as being of type $\underline{c}$, and the continuing path is the Coase-conjecture path corresponding to the buyer's being informed that $c = \underline{c}$.

Ausubel and Deneckere (1990a) give two properties of equilibria of the one-sided-offer, bilateral-asymmetric-information model. First, they show that trade never occurs between a seller of type c and a buyer of type $v < c$. This intuitive property can be obtained as follows: If type c made an offer m^t that was accepted with positive probability by type $v < c$ (and was therefore no higher than c), the seller would lose money on this offer. Furthermore, only types less than v would remain from $t + 1$ on (from the successive-skimming lemma 10.1). Because the buyer never accepts offers above his valuation, type c does not make a profit from $t + 1$ on; therefore, he strictly loses money from date t on, and he would do better to offer prices above $\bar{v}$ (even though these would be rejected). Second, and more important, Ausubel and Deneckere show that if the supports of the distributions are common ($\underline{c} = \underline{v}, \bar{c} = \bar{v}$) and if the Coase conjecture holds for type $\underline{c}$ (that is, his initial offer converges to $\underline{c}$ as $\Delta \to 0$), then the expected amount of "discounted trade" converges to 0 as $\Delta \to 0$.[33] More precisely, fix an equilibrium, and let $x(c, v)$ be the expected discounted value of the indicator function, which has value 1 when trade occurs. Then the expectation of $x(c, v)$ over c and v—the ex $ante$ expected discounted trade—converges to 0. The intuition for this result is clear. Let $X(c)$ denote the expected discounted trade for type c ($X(c) = E_v(x(c, v))$), and let $U_s(c)$ denote the expected utility of type c in the equilibrium. Because type $c - dc$ can always mimic type c, $dU_s/dc = -X(c)$ (see chapter 7). Hence,

$$U_s(\underline{c}) = U_s(c) + \int_{\underline{c}}^{c} X(\gamma) \, d\gamma \text{ for all } c.$$

Now, if the Coase conjecture holds for type $\underline{c}$, $U_s(\underline{c}) \to 0$ and $U_s(c) \geq 0$ for all c implies that $X(c) \to 0$ for all c. Thus, there is an inherent conflict between the Coase conjecture and the existence of trade in equilibrium.[34]

---

32. More precisely, the period-t offer m^t reveals that the seller's type is $c^{-1}(m^t)$ $even$ if the seller has previously been revealed to be a different type.
33. More generally, if $\underline{c} < \underline{v}$, the ex $ante$ expected probability of trade is bounded above by the probability that $c \leq \underline{v}$. (The case $\underline{c} > \underline{v}$ is, of course, equivalent to the case $\underline{c} = \underline{v}$.)
34. Thus, in the Cramton and Cho equilibria, all trade is deferred far into the future.

10.4.2 Alternating Offers and One-Sided Asymmetric Information

Ausubel and Deneckere (1989b), Chatterjee and Samuelson (1987), Grossman and Perry (1986a), Gul and Sonnenschein (1988), and Rubinstein (1985) have studied the alternating-offer model with one-sided asymmetric information, which has many equilibria for the now-familiar reason. The model is that of Rubinstein and Ståhl (see chapter 4 above), except that one party (the buyer, say) has private information.

What is the equilibrium set for this game? This question is easiest to answer in the case of time period $\Delta \to 0$. (If $\Delta = +\infty$, i.e., $\delta = 0$, the first player to make an offer obtains his monopoly profit.) As is noted above, a perfect Bayesian equilibrium of the bargaining game gives rise to two functions of the buyer's valuation, v: $M(\cdot)$ and $X(\cdot)$. First,

$$X(v) = E\left(\sum_{t=0}^{\infty} \delta^t x^t(h^t, m^t, v) \right)$$

is the expected discounted trade of type v in this equilibrium (where $x^t(h^t, m^t, v)$ is the probability of trade at date t, and $0 \leq x^t(h^t, m^t, v) \leq 1$ for all t implies $0 \leq X(v) \leq 1$). And

$$M(v) = E\left(\sum_{t=0}^{\infty} \delta^t m^t x^t(h^t, m^t, v) \right)$$

is the expected discounted transfer from the buyer to the seller. One can view $\{M(\cdot), X(\cdot)\}$ as a *mechanism* in the sense of chapter 7.

The mechanism $\{M(\cdot), X(\cdot)\}$ is *feasible* if it satisfies

$$0 \leq X(v) \leq 1 \text{ for all } v$$

(IR_B) $X(v)v - M(v) \geq 0$ for all v

(IR_S) $E_v M(v) \geq 0$

(IC) $X(v)v - M(v) \geq X(\hat{v})v - M(\hat{v})$ for all $(v, \hat{v})$.

That is, a feasible mechanism is individually rational and incentive compatible. Note that we keep normalizing the seller's cost to be 0.

Any perfect Bayesian equilibrium of the alternating-offer bargaining game (or, more generally, of any sequential bargaining game) with equal discount factors must give rise to a feasible mechanism. First, it must satisfy individual rationality, because each party would not bargain (or would reject offers and make outrageous offers himself) if he expected a negative payoff from bargaining. Second, any type v of buyer can always adopt the strategy of another type $\hat{v}$, and therefore the equilibrium must satisfy IC.

Conversely, we may ask: When is a feasible mechanism the outcome (or approximate outcome) of a perfect Bayesian equilibrium of the alternating-offer game? An answer is found in the following theorem.

Theorem 10.6 (Ausubel and Deneckere 1989b)   Assume $\underline{v} = 0$. A feasible mechanism $\{M(\cdot), X(\cdot)\}$ is implementable by a perfect Bayesian equilibrium of the alternating-offer bargaining game with one-sided asymmetric information (in the sense that for any $\varepsilon > 0$ there exists $\Delta_0 > 0$ such that for any $\Delta \leq \Delta_0$ there exists a perfect Bayesian equilibrium with payoffs U_s for the seller and $U_b(\cdot)$ for the buyer such that $|[X(v)v - M(v)] - U_b(v)| < \varepsilon$ and $|E_v M(v) - U_s| < \varepsilon$) if and only if

$$\bar{v} X(\bar{v}) - M(\bar{v}) \geq \bar{v}/2.$$

That is, any feasible mechanism is an equilibrium outcome as long as the highest-valuation buyer obtains at least his full information payoff for Δ close to 0 (which is approximately the Nash bargaining solution $\bar{v}/2$; see chapter 4). The intuition for this necessary condition is that buyer $\bar{v}$ is at worst thought of as being weak, i.e., as having type $\bar{v}$. Even in this case, he can guarantee himself the symmetric-information payoff.

Sketch of Proof   Let us first show that in equilibrium $X(\bar{v})\bar{v} - M(\bar{v}) \geq \bar{v}/2$. (The following reasoning is due to Grossman and Perry (1986a).) Let $\bar{m}$ denote the highest price that the seller gets (which is either accepted or offered by some type of buyer with some positive probability) in *any* equilibrium after *any* history. Suppose for instance that, for some equilibrium and history, the seller offers $m^t = \bar{m}$ (or close to $\bar{m}$ if $\bar{m}$ is a supremum rather than a maximum) and that at least some type v of buyer accepts $\bar{m}$. This type obtains utility $v - \bar{m}$ from date t on. But he could reject the offer and offer price $m^{t+1} = \delta\bar{m} + \varepsilon$ next period to the seller. The seller would accept m^{t+1} with probability 1, because he would never get more than $\bar{m}$ in the future. Therefore, this deviation yields type v utility $\delta(v - \delta\bar{m} - \varepsilon)$. Now, if $\bar{m} > \bar{v}/(1 + \delta)$,

$$v - \bar{m} - \delta(v - \delta\bar{m} - \varepsilon) = v(1 - \delta) - \bar{m}(1 - \delta^2) + \delta\varepsilon$$

$$< (1 - \delta)(\bar{v} - (1 + \delta)\bar{m}) + \delta\varepsilon$$

$$< 0$$

if ε is close to 0. Hence, $\bar{m} \leq \bar{v}/(1 + \delta)$. Therefore, the type-$v$ buyer can always guarantee himself $v - (\bar{v}/(1 + \delta))$ when it is his turn to make an offer. In particular, type $\bar{v}$ gets at least $\delta\bar{v}/(1 + \delta) \simeq \bar{v}/2$ for δ close to 1. The same reasoning applies if $\bar{m}$ is attained for an offer by the buyer.

Second, to show that any feasible outcome with utility at least $\bar{v}/2$ for type $\bar{v}$ can occur, one can "embed" the seller-offer equilibria of theorem 10.3 with time interval 2Δ between the seller's offers into the alternating-move game with time interval Δ. The idea is to look for equilibria in which the buyer offers nonserious (negative) prices when it is his turn to make an offer. In such periods, beliefs remain the same. However, if the buyer were to deviate and offer a positive price, he would be thought of as the

$\bar{v}$ type, and the only feasible agreement in the future would be the "full-information" one at price approximately equal to $\bar{v}/2$. Such optimistic beliefs of the seller prevent the buyer from making serious offers. When it is the seller's turn to make offers, the strategies are those of the proof of theorem 10.3. The seller offers a price on an exponential path, say, and if he deviates the equilibrium switches to one in which the Coase conjecture holds.   ∎

Several authors have imposed additional restrictions to narrow down the set of equilibria with alternating offers. In the two-type case, Rubinstein (1985) looks for equilibria with the following properties:

(i) If the buyer rejects the seller's offer, and if his next counteroffer, when accepted, yields less utility than the acceptance of the seller's offer to the high-valuation type and more utility to the low-valuation type, then the seller assigns probability 1 to the buyer's having a low valuation.[35]

(ii) If the counteroffer, when accepted, yields more utility to both types, then the seller's belief that the buyer has low valuation does not increase.

Grossman and Perry (1986a) restrict the set of equilibria by imposing the notion of perfect sequential equilibrium developed in their 1986b paper to the alternating-offer model. They require that if the seller receives an out-of-equilibrium offer m^t he attempts to find a set of valuations with the property that, if the seller believes that the buyer's type is in this set, then this set is indeed the set of buyer's types that are better off than had the equilibrium path been followed. Gul and Sonnenschein (1988), under assumption G, show that the Coase conjecture holds in the class of pure-strategy equilibria that have the property of stationarity of the buyer's strategy, have the monotonicity property that the possibility of additional high-valuation buyers does not lead a low-valuation buyer to lower his acceptance price, and are such that nonserious offers convey no information beyond the fact that the seller is unwilling to trade in the current period.[36,37]

Admati and Perry (1987) consider a different extensive form for alternating offers, in which a player who receives an offer chooses how much time elapses before he makes a counteroffer, subject to the constraint that the time between the two offers exceeds some fixed number (the other party

35. This property has some of the flavor of the forward-induction idea developed in chapter 11, but is different.
36. That is, the offer of a buyer can influence the seller's beliefs only if the strategies specify that the offer be accepted (with probability 1 because of pure strategies).
37. Ausubel and Deneckere (1990c) show that, under similar assumptions (stationarity, monotonicity, pure strategies, and nonserious offers do not affect relative beliefs over the set of types making them), the buyer never makes a serious offer. That is, all information revelation occurs through passive responses by the informed party to offers of the uninformed party. Ausubel and Deneckere argue that this result provides a justification for the one-sided incomplete-information model of section 10.2, in which only the uninformed party is permitted to make offers.

cannot make another offer in the meantime).[38] Thus, the low-valuation buyer can use delay to signal his valuation to the seller. Admati and Perry use a variation on the Cho-Kreps and Banks-Sobel refinements studied in chapter 11 below, and find that the equilibrium delay does not disappear as the minimum time between offers tends to 0.[39]

There is a small literature on bargaining with two-sided asymmetric information and alternating moves: Chatterjee and Samuelson (1987, 1988), Cho (1990), and Cramton (1987) discuss bargaining with an infinite horizon, and Fudenberg and Tirole (1983) discuss the case of two periods.

10.4.3 Mechanism Design and Bargaining

Cramton (1985), Wilson (1987a,b) and Ausubel and Deneckere (1989b; 1990a,b) have emphasized the link between mechanism design and bargaining. Recall from chapter 7 that Myerson and Satterthwaite showed that if the seller's cost c and the buyer's valuation v are continuously distributed on $[\underline{c}, \overline{c}]$ and $[\underline{v}, \overline{v}]$, respectively, and if $\underline{v} < \overline{c}$, there exists no incentive-compatible and individually rational mechanism that realizes all gains from trade. This result implies that perfect Bayesian equilibria of bargaining games are, in general, inefficient. However, one may ask whether they can be "constrained efficient"—that is, can their outcomes coincide with those of static mechanisms that maximize a convex combination of the seller's and the buyer's *ex ante* payoffs subject to the incentive-compatibility and individual-rationality constraints? Ausubel and Deneckere (1990b) pose this question and provide a partial answer. They show that, under certain distributional assumptions when $\underline{c} = \underline{v}$ and $\overline{c} = \overline{v}$ (common support), all the *ex ante* (constrained) efficient allocations can be attained by perfect Bayesian equilibria of the two infinite-horizon, frequent-one-sided-offers bargaining games (i.e., the seller makes all offers or the buyer makes all offers and Δ goes to zero).[40] Of course, there is no guarantee that the players will coordinate on an *ex ante* efficient equilibrium if such an equilibrium

---

38. A party can also delay the bargaining process in the models of Perry and Reny (1989) and Stahl (1990). (Unlike the model of Admati and Perry, these are complete-information models.)
39. Cramton (1987) extends the Admati-Perry model to two-sided uncertainty. In his equilibrium, both players delay making an initial offer. Eventually, either the players realize that there are no gains from trade, and so terminate negotiations, or the more patient player makes a revealing offer, at which point the equilibrium path is the same as with one-sided uncertainty.
40. A mechanism specifies discounted probabilities of trade $x(\cdot, \cdot)$ and expected discounted transfers $m(\cdot, \cdot)$. Let $X_s(c) = E_v x(c, v)$, $X_b(v) = E_c x(c, v)$, $M_s(c) = E_v m(c, v)$, and $M_b(v) = E_c m(c, v)$. A mechanism is feasible if it satisfies individual rationality and incentive compatibility:

$$M_s(c) - c X_s(c) \geq 0$$

$$\geq M_s(\hat{c}) - c X_s(\hat{c}) \text{ for all } (c, \hat{c}),$$

$$v X_b(v) - M_b(v) \geq 0$$

$$\geq v X_b(\hat{v}) - M_b(\hat{v}) \text{ for all } (v, \hat{v}).$$

exists, but it is interesting to know that simultaneous offers, as is implied by the revelation principle, may not be necessary in order to obtain *ex ante* efficient outcomes.

Exercises

Exercise 10.1** Consider the sale model with $c = 0$ and v uniformly distributed on $[0, 1]$. Prove the Coase conjecture without a stationarity assumption when the real time horizon τ is finite (so there are $T + 1$ periods of price offers where $T = \tau/\Delta$). Hint: Solve for the last-period price as a function of the beliefs in that period. Prove by backward induction that the seller's expected profit at date t, $U_s^t(\kappa^t)$, is quadratic in the current cutoff valuation κ^t; that the seller charges $\gamma^t \kappa^t$; that the buyer accepts m^t if his valuation exceeds $\lambda^t \kappa^t$; and that the coefficients $\gamma^{w/\Delta}$ (for real time w less than τ) tend to 0.

Exercise 10.2***
(a) Solve the rental model of subsection 10.3.1 with two periods when the buyer has two types, and when he has a continuum of types uniformly distributed on $[0, 1]$, say. In particular show that, in the continuum case, there is a "truncation equilibrium" (in which the buyer accepts in period 0 if and only if his valuation exceeds some number).
(b) Show that with a continuum of types and three periods, no truncation equilibrium exists if δ is sufficiently large.
For more on this, see Fernandez-Arias and Kofman 1989.

Exercise 10.3** Use theorem 10.4 to show that in the rental model with $\delta > \frac{1}{2}$, T large, two types, and $\bar{p} > \alpha = \underline{v}/\bar{v}$ there is positive probability in equilibrium that the buyer does not consume in a period even though he consumed in all earlier periods.

Exercise 10.4*** Show that in the two-type rental model the seller never charges less than $\underline{v}$ and his expected payoff is unique. Hint: Prove by induction that (1) the seller never charges $m^t < \underline{v}$ and type $\underline{v}$ always has payoff 0, (2) the high-valuation buyer's continuation utility at t, $\bar{U}^t(\bar{\mu}^t)$, is a step function ($\exists\, 0 < \bar{\mu}_1^t < \cdots < \bar{\mu}_K^t < 1$ and $\bar{U}_1^t > \bar{U}_2^t > \cdots > \bar{U}_{K+1}^t$ such that $\bar{\mu}_k^t < \bar{\mu}^t < \bar{\mu}_{k+1}^t \Rightarrow \bar{U}^t(\bar{\mu}^t) = \bar{U}_{k+1}^t$ and $\bar{\mu}^t = \bar{\mu}_k^t \Rightarrow \bar{U}^t(\bar{\mu}^t) \in [\bar{U}_{k+1}^t, \bar{U}_k^t]$, (3) the seller's continuation valuation at t is unique and is weakly increasing in $\bar{\mu}^t$, and (4) the expected discounted trades $\bar{X}^t(\bar{\mu}_k^t)$, $\underline{X}^t(\bar{\mu}_k^t)$ and type $\bar{v}$'s

---

A mechanism is *ex ante* efficient if it maximizes

$$\lambda E_c[M_s(c) - cX_s(c)] + (1 - \lambda)E_v[vX_b(v) - M_b(v)]$$

over the set of feasible mechanisms, for some $\lambda \in [0, 1]$.

continuation payoff from t on are weakly decreasing in $\bar{\mu}_k^t$. Use a revealed preference argument.

Exercise 10.5** Generalize the Sobel-Takahashi infinite-horizon linear equilibrium of the sale model (subsection 10.2.3) to learning by doing (this makes sense only for the durable-good interpretation of the model). The buyers' valuations are uniformly distributed on $[0, 1]$. The horizon is infinite: $t = 0, 1, \ldots$. The seller's unit cost in a given period is constant with respect to the period's output and decreases proportionally with the volume of past sales: $c = c_0 v$, where v is the cutoff valuation at the beginning of the period.

(a) Look for a linear equilibrium in which the seller charges γv when the cutoff valuation is v (where $\gamma < 1$) and the buyers accept m when their valuation exceeds λm (where $\lambda > 1$). Show that

$$\lambda - 1 = \delta\lambda(1 - \gamma)$$

and

$$1 - 2\lambda\gamma + \delta\lambda^2\gamma^2 = c_0[\delta\lambda^2\gamma(2 - \delta\lambda\gamma) - \lambda].$$

(b) Show that only part of the Coase conjecture holds. When $\delta \to 1$, $\gamma \to c_0$ (price converges to marginal cost), and $\lambda \to 1/c_0$. Show that the market is not dissipated immediately (neither would it be if the seller could commit to an intertemporal price path). Compute the rate of sale per unit of time in the limit as the time between offers tends to 0.

(This exercise is from Olsen 1988.)

Exercise 10.6* Consider the sale model (in which the seller makes all the offers). Suppose that the buyer has cost $\varepsilon > 0$ of starting to bargain, which must be paid before the seller makes the first offer. Show that in a perfect Bayesian equilibrium bargaining never takes place, i.e., no buyer pays the entry fee. (Hint: Consider the lowest type who enters the bargaining process.) Is this conclusion robust to the extensive form? (Hint: Suppose that the buyer makes an offer every other period, and take ε to 0.)

Exercise 10.7** The bargaining games we have considered in this chapter have private values: The players do not care about their opponents' private information *per se* (but they try to learn about it to predict their opponents' future behavior). Some interesting bargaining games exhibit common values (see also exercise 10.8). Consider the following model of out-of-court settlement, due to Spier (1989).[41] Player 1, the plaintiff, is uninformed and makes settlement offers m^t at dates $t = 0, \ldots, T - 1$. Player 2, the defendant, knows the expected damage x that he will have to pay player 1 if the two

---

41. Other models of bargaining before litigation are offered by Bebchuk (1984), Nalebuff (1987), Ordover and Rubinstein (1986), Reinganum and Wilde (1986), and Spulber (1989).

parties do not settle out of court, which results in litigation at date T. The discount factor is δ, and the parties have fixed costs $c_1 > 0$ and $c_2 > 0$ of going to court. Thus, if the parties agree on damage m^t at date t, the payoffs are $u_1 = \delta^t m^t$ for the plaintiff and $u_2 = -\delta^t m^t$ for the defendant; if the defendant rejects all the plaintiff's offers $m^0, \ldots, m^{T-1}$, the payoffs are $u_1 = \delta^T(x - c_1)$ and $u_2 = \delta^T(-x - c_2)$. Assume $c_1 + c_2 < 1$.

(a) Suppose that $T = 1$ (single offer), and that x is uniformly distributed on $[0, 1]$. Show that the plaintiff offers $m^0 = \delta(1 - c_1)$ and that there is positive probability that the case is litigated.

(b) Assume that $x = \underline{x}$ with probability $\underline{p}$ and $x = \bar{x}$ with probability $\bar{p}$. Show that for any T there exists a unique perfect Bayesian equilibrium of the pretrial-bargaining game.

Exercise 10.8** Consider bargaining between a seller and a buyer over one unit of good, and suppose that the buyer's valuation is perfectly correlated with the seller's cost[42]: The cost c can take one of two values, $\underline{c}$ and $\bar{c}$, with equal probabilities, where $\underline{c} = 0$. The buyer's valuation is $\underline{v} = 0$ if $c = \underline{c}$ and $\bar{v}$ if $c = \bar{c}$. Assume $\bar{v} > \bar{c}$, so that there exist potential gains from trade. The seller knows c, but the buyer does not (and therefore does not know his valuation). The interpretation is that the seller's cost is related to the quality of his good (the cost may be either a production cost or an opportunity cost). Both parties are risk neutral, and $\bar{v} < 2\bar{c}$.

Show that for *any* bargaining process in which each party can refuse to trade, and both parties have the same discount factor, there is no trade in equilibrium. Hint: Use the mechanism-design approach mentioned in section 10.4. Let $\underline{X}$ and $\bar{X}$ denote the expected discounted volumes of trade of the two types of sellers, and let $\underline{M}$ and $\bar{M}$ denote their expected discounted revenues. Write the seller's incentive-compatibility and individual-rationality constraints. Show that no trade ($\underline{X} = \bar{X} = \underline{M} = \bar{M} = 0$) is interim efficient, i.e., that no trade maximizes the buyer's expected utility subject to the above constraints. Conclude that equilibrium involves no trade.

Exercise 10.9** In subsection 10.2.2 it was mentioned that if the seller has the opportunity to switch and bargain with another buyer, he will in general quit the relationship in finite time before having realized all gains from trade with this buyer. The same phenomenon occurs when the seller can break off the negotiations and consume the good himself. Here is an example of the multiplicity of equilibria associated with an endogenous finite horizon.

Consider the sale model. The seller makes offers at $t = 0, 1, \ldots$. The seller's cost is 0. The seller decides in each period whether to consume the good (in which case he does not get to make an offer in this period or any

42. Evans (1989) and Vincent (1989) present such models.

subsequent period) or to make an offer to the buyer. Consumption gives him current utility w. The buyer's valuation, which is private information, takes one of three values, v_1, v_2, and v_3, with probabilities p_1, p_2, and p_3, where $0 \leq v_1 < w < v_2 < v_3$ and $p_1 + p_2 + p_3 = 1$. Assume that $w > (p_2 v_2 + \delta p_1 w)/(p_2 + p_1)$; that $p_3 \tilde{v} + \delta p_2 w < (p_3 + p_2)v_2 < p_3 v_3 + \delta p_2 w$, where $\tilde{v}$ satisfies $v_3 - \tilde{v} = \delta(v_3 - v_2)$; and that $(1 - p_1)v_2 \geq w(1 - \delta p_1)$. Show that the following are equilibria:

(a) Type v_3 accepts any offer $m^t \leq v_3$. The seller charges $m^t = v_3$ if $\min_{\tau < t} m^\tau > v_3$, and consumes otherwise. (Hence, the equilibrium path has an offer at v_3 at $t = 0$ followed by consumption at $t = 1$ if the offer is rejected.)

(b) Type v_3 accepts any offer $m^t \leq \tilde{v}$ (and rejects offers $m^t > \tilde{v}$ if $\min_{\tau < t} m^\tau > \tilde{v}$). The seller charges $m^t = v_2$ if $\min_{\tau < t} m^\tau > \tilde{v}$, and consumes otherwise. (Hence, the equilibrium path has one offer at v_2 at $t = 0$ followed by consumption at $t = 1$ if the offer is rejected.)

Explain the multiplicity of equilibria.

Exercise 10.10** The point of this exercise[43] is to show that part b of Coasian dynamics for the sale model (that the seller's price decreases over time) does not necessarily hold once the seller as well as the buyer has private information. Suppose there are two equally likely types of buyer, $\underline{v} < \overline{v}$, and two equally likely types of seller, $\underline{c} < \overline{c}$. Suppose that (in a one-period model) one type of seller is soft and the other tough: $\underline{v} - \underline{c} \geq (\overline{v} - \underline{c})/2$ and $\underline{v} - \overline{c} < (\overline{v} - \overline{c})/2$.

The bargaining process has two periods, and the seller makes both offers, as in subsection 10.2.2. Let $\tilde{m}$ and $\hat{m} > \tilde{m}$ be defined by $\tilde{m} = \overline{v} - \delta(\overline{v} - \underline{v})$ and $\hat{m} = \overline{v} - (\delta/2)(\overline{v} - \underline{v})$, where δ is the discount factor. Let $x \equiv (\overline{v} + \overline{c} - 2\underline{v})/(\overline{v} - \underline{v})$. Assume that, for $c = \underline{c}, \overline{c}$,

$$\frac{x}{2}(\hat{m} - c) + \delta(1 - x/2)(\underline{v} - c)$$

$$\geq \max\left(\tfrac{1}{2}(\tilde{m} - c) + \frac{\delta}{2}(\underline{v} - c), \underline{v} - c, \frac{\delta}{2}(\overline{v} - c)\right).^{[44]}$$

Show that there exists a pooling equilibrium in which both types of seller charge $\hat{m}$ at $t = 0$, which is rejected by type $\underline{v}$ and accepted with probability x by type $\overline{v}$. After rejection of $\hat{m}$, type $\underline{c}$ charges $\underline{v}$ and type $\overline{c}$ raises his price to $\overline{v}$ at $t = 1$. Comparing this equilibrium with the one-sided-asymmetric-information one of subsection 10.2.2, show that the seller gains from having private information about cost when he is of type $\underline{c}$ and loses when he is of type $\overline{c}$.

43. Drawn from Fudenberg and Tirole 1983.
44. For instance, $\{\underline{v} = 1, \overline{v} = 2, \underline{c} = 0, \overline{c} = 1, \delta = 1\}$ satisfies these assumptions.

References

Admati, A. R., and M. Perry. 1987. Strategic delay in bargaining. *Review of Economic Studies* 54: 345–364.

Akerlof, G. 1970. The market for lemons: Qualitative uncertainty and the market mechanism. *Quarterly Journal of Economics* 84: 488–500.

Ausubel, L., and R. Deneckere. 1987. One is almost enough for monopoly. *Rand Journal of Economics* 18: 255–274.

Ausubel, L., and R. Deneckere. 1989a. Reputation in bargaining and durable goods monopoly. *Econometrica* 57: 511–531.

Ausubel, L., and R. Deneckere. 1989b. A direct mechanism characterization of sequential bargaining with one-sided incomplete information. *Journal of Economic Theory* 48: 18–46.

Ausubel, L., and R. Deneckere. 1990a. Durable goods monopoly with incomplete information. Mimeo, Northwestern University.

Ausubel, L., and R. Deneckere. 1990b. Efficient sequential bargaining. Mimeo, Northwestern University.

Ausubel, L., and R. Deneckere. 1990c. Bargaining and the right to remain silent. Mimeo, Northwestern University.

Bebchuk, L. 1984. Litigation and settlement under imperfect information. *Rand Journal of Economics* 15: 404–415.

Bond, E., and L. Samuelson. 1984. Durable good monopolies with rational expectations and replacement sales. *Rand Journal of Economics* 15: 336–345.

Bond, E., and L. Samuelson. 1987. The Coase conjecture need not hold for durable good monopolies with depreciation. *Economics Letters* 24: 93–97.

Bulow, J. 1982. Durable goods monopolists. *Journal of Political Economy* 90: 314–322.

Chatterjee, K., and L. Samuelson. 1987. Bargaining with two-sided incomplete information: An infinite horizon model with alternating offers. *Review of Economic Studies* 54: 175–192.

Chatterjee, K., and L. Samuelson. 1988. Bargaining under two-sided incomplete information: The unrestricted offers case. *Operations Research* 36: 605–618.

Cho, I.-K. 1990. Uncertainty and delay in bargaining. *Review of Economic Studies* 57: 575–596.

Cho, I.-K. 1990. Characterization of stationary equilibria in bargaining models with incomplete information. Mimeo, University of Chicago.

Coase, R. 1960. The problem of social cost. *Journal of Law and Economics* 3: 1–44.

Coase, R. 1972. Durability and monopoly. *Journal of Law and Economics* 15: 143–149.

Cramton, P. 1984. Bargaining with incomplete information: An infinite-horizon model with continuous uncertainty. *Review of Economic Studies* 51: 579–593.

Cramton, P. 1985. Sequential bargaining mechanisms. In *Game-Theoretic Models of Bargaining*, ed. A. Roth. Cambridge University Press.

Cramton, P. 1987. Strategic Delay in Bargaining with Two-Sided Uncertainty. *Review of Economic Studies*, forthcoming.

Cramton, P. 1990. Dynamic bargaining with transaction costs. Mimeo, Yale University.

Cramton, P., and J. Tracy. 1990. Strikes and delays in wage bargaining: Theory and data. Mimeo, School of Management, Yale University.

Edgeworth, F. 1881. *Mathematical Psychics: An Essay on the Application of Mathematics to the Moral Sciences.* Harvard University Press.

Evans, R. 1989. Sequential bargaining with correlated values. *Review of Economic Studies* 56: 499–510.

Fernandez-Arias, E., and A. Kofman. 1989. Equilibrium characterization in finite-horizon games of reputation. Mimeo, University of California, Berkeley.

Fudenberg, D., and D. Levine. 1989. Reputation and equilibrium selection in games with a patient player. *Econometrica* 57: 759–778.

Fudenberg, D., D. Levine, and P. Ruud. 1985. Strike activity and wage settlements. Mimeo, Massachusetts Institute of Technology.

Fudenberg, D., D. Levine, and J. Tirole. 1985. Infinite-horizon models of bargaining with one-sided incomplete information. In *Game-Theoretic Models of Bargaining*, ed. A. Roth. Cambridge University Press.

Fudenberg, D., D. Levine, and J. Tirole. 1987. Incomplete information bargaining with outside opportunities. *Quarterly Journal of Economics* 102: 37–50.

Fudenberg, D., and J. Tirole. 1983. Sequential bargaining with incomplete information. *Review of Economic Studies* 50: 221–247.

Grossman, S., and M. Perry. 1986a. Sequential bargaining under asymmetric information. *Journal of Economic Theory* 39: 120–154.

Grossman, S., and M. Perry. 1986b. Perfect sequential equilibrium. *Journal of Economic Theory* 39: 97–119.

Gul, F. 1987. Noncooperative collusion in durable goods oligopoly. *Rand Journal of Economics* 18: 248–254.

Gul, F., and H. Sonnenschein. 1988. On delay in bargaining with one-sided uncertainty. *Econometrica* 56: 601–611.

Gul, F., H. Sonnenschein, and R. Wilson. 1986. Foundations of dynamic monopoly and the Coase conjecture. *Journal of Economic Theory* 39: 155–190.

Hart, O. 1989. Bargaining and strikes. *Quarterly Journal of Economics* 104: 25–44.

Hart, O., and J. Tirole. 1988. Contract renegotiation and Coasian dynamics. *Review of Economic Studies* 55: 509–540.

Kahn, C. 1986. The durable goods monopolist and consistency with increasing costs. *Econometrica* 54: 275–294.

Kennan, J., and R. Wilson. 1989. Bargaining with private information. *Journal of Economic Literature*, forthcoming.

Kennan, J., and R. Wilson. 1990. Theories of bargaining delays. Mimeo, Stanford Graduate School of Business.

Laffont, J.-J., and J. Tirole. 1990. Adverse selection and renegotiation in procurement. *Review of Economic Studies* 57: 597–626.

Maskin, E., and J. Tirole. 1989. Markov equilibrium. Mimeo, Harvard University and Massachusetts Institute of Technology.

Myerson, R., and A. Satterthwaite. 1983. Efficient mechanisms for bilateral trading. *Journal of Economic Theory* 28: 265–281.

Nalebuff, B. 1987. Credible pretrial negotiation. *Rand Journal of Economics* 18: 198–210.

Nash, J. F. 1950. The bargaining problem. *Econometrica* 18: 155–162.

Nash, J. F. 1953. Two-person cooperative games. *Econometrica* 21: 128–140.

Olsen, T. 1988. Durable goods monopoly, learning by doing and the Coase conjecture. CEPR publication 141, Stanford University.

Ordover, J., and A. Rubinstein. 1986. A sequential concession game with asymmetric information. *Quarterly Journal of Economics* 101: 879–888.

Perry, M. 1986. An example of price formation in bilateral situations: A bargaining model with incomplete information. *Econometrica* 54: 313–321.

Perry, M., and P. Reny. 1989. Bargaining without procedures. Mimeo, University of Jerusalem.

Reinganum, J., and L. Wilde. 1986. Settlement, litigation, and the allocation of litigation costs. *Rand Journal of Economics* 17: 557–566.

Rubinstein, A. 1982. Perfect equilibrium in a bargaining model. *Econometrica* 50: 97–109.

Rubinstein, A. 1985. A bargaining model with incomplete information about time preferences. *Econometrica* 53: 1151–1172.

Schmidt, K. 1990. Commitment through incomplete information in a simple repeated bargaining model. Discussion paper A-303, Universität Bonn.

Sobel, J. 1990. Durable goods monopoly with entry of new consumers. Mimeo, University of California, San Diego.

Sobel, J., and I. Takahashi. 1983. A multi-stage model of bargaining. *Review of Economic Studies* 50: 411–426.

Spier, K. 1989. The resolution of disputes: Enforcement in multiperiod bargaining models with asymmetric information. *Review of Economic Studies*, forthcoming.

Spulber, D. 1989. Contingent damages and settlement damages. Working paper, University of Southern California.

Stahl, D. 1990. Choice of walkout capacity in bargaining. Mimeo.

Ståhl, I. 1972. *Bargaining Theory*. Economics Research Institute, Stockholm School of Economics.

Stokey, N. 1981. Rational expectations and durable goods pricing. *Bell Journal of Economics* 12: 112–128.

Stokey, N., and R. Lucas, with E. Prescott. 1989. *Recursive Methods in Economic Dynamics*. Harvard University Press.

Vincent, D. R. 1989. Bargaining with common values. *Journal of Economic Theory* 48: 47–62.

Wilson, R. 1987a. Game-theoretic analyses of trading processes. In *Advances in Economic Theory*, ed. T. F. Bewley. Cambridge University Press.

Wilson, R. 1987b. Bilateral bargaining. Unpublished paper, Graduate School of Business, Stanford University.

V ADVANCED TOPICS

11 More Equilibrium Refinements: Stability, Forward Induction, and Iterated Weak Dominance

A succession of authors have argued for equilibrium refinements that capture some aspect of what is loosely called "forward induction." Forward induction plays a key role in the equilibrium refinements developed for signaling games, and it underlies the nonequilibrium notion of the iterated deletion of weakly dominated strategies. This chapter discusses these equilibrium refinements in some detail, and then presents the "burning money" game as a striking example of the power of iterated weak dominance. The chapter concludes by examining the argument that equilibrium refinements are too strong because they are not robust to certain changes in the information players have about one another's payoffs.

The idea of forward induction is that when an off-the-equilibrium-path information set is reached, the player on move at this information set should not suppose that it was reached by "accident" and then use the equilibrium strategies to look "down" the tree as in backward induction. Instead, the player should take into account what could have happened but did not in forming his beliefs about the nodes in the information set and about what is likely to happen next. Thus, players should reason "forward" from the beginning of the tree as well as backward from its end. For example, if you expected that with probability 1 your opponent in a bargaining game would accept your offer of $10 yet she refused it, and if you knew that her delay costs were positive, you might conclude that she expects to receive a higher offer in the future.[1] This idea is clearly at odds with the interpretation that deviations from equilibrium play are due to unintended errors, for if the refusal was an unintended error it would contain no information about your opponent's likely future play.[2]

11.1 Strategic Stability[†††]

The idea of forward induction is one of Kohlberg and Mertens' key reasons for introducing their notion of strategic stability. A second motivation is the idea that solution concepts should be invariant to "inessential" transformations of the extensive form.

As a first example of forward induction, consider the game in figure 11.1a, which is from van Damme 1989. Here player 1 has the choice of playing L, which ends the game, or playing R, in which case he and player 2 play a "battle of the sexes" subgame in which the players simultaneously choose between T ("tough") and W ("weak"). This subgame has three Nash equilibria: (T, W), (W, T), and a mixed-strategy equilibrium in which each player plays W with probability $\frac{1}{4}$.

---

1. Dekel (1990) studies the implications of strategic stability in a two-period simultaneous-move bargaining game.
2. For a similar reason, forward induction conflicts with the concept of Markov equilibrium discussed in chapter 13.

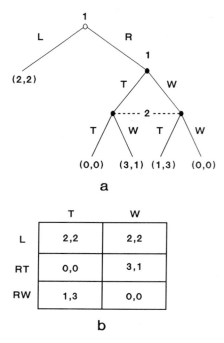

a

	T	W
L	2,2	2,2
RT	0,0	3,1
RW	1,3	0,0

b

Figure 11.1

The profile (LW, T) is a subgame-perfect equilibrium of this game. It is clearly a sequential-equilibrium profile, and it is also trembling-hand perfect in the agent strategic form. However, the equilibrium is not consistent with the following "forward induction" argument: There is no reason for player 1 to play RW, since this gives a payoff of at most 1 while L gives a payoff of 2. However, playing R followed by T would be rational if player 1 expected player 2 to play W. Hence, the argument goes, player 2 should expect that if player 1 plays R he will then play T and not W, so player 2 should play W. And player 1 should foresee this reasoning of player 2, so player 1 should play R instead of L. (This argument does not apply if we suppose that player 1 meant to play L and played R by "mistake," as is implicitly supposed by sequential equilibrium and perfect equilibrium.)

Figure 11.1b gives the reduced strategic form of this game. Note that the strategy RW is strictly dominated by L, and that if RW is removed the only trembling-hand perfect equilibrium is (RT, W), since if player 1 plays RT with positive probability (and RW is not included) then player 2 strictly prefers W to T. Note also that (L, T) is trembling-hand perfect in the strategic form with RW included: Even though RW is strictly dominated, player 1 might play it by "mistake," and this leads player 2 to play T if RW is as likely as RT. Note finally that if player 2 plays T, RW is better for player 1 than RT, so (L, T) is a proper equilibrium.

The link in similar examples between forward induction and removing strictly dominated strategies leads Kohlberg and Mertens to formulate the following requirement for their solution concept which they call "strategic stability." Unlike previous solution concepts, strategic stability is a set-valued concept. That is, instead of each solution being a single equilibrium profile, and the set of solutions being a set of equilibria, each solution is itself a "strategically stable set," and the set of solutions is the set of all such sets. As we will see, the reason Kohlberg and Mertens use a set-valued concept is that the conditions that they wish to impose, taken together, cannot be satisfied by a point-valued concept.

ID (Iterated Dominance)   Every strategically stable set of equilibria of a game G should contain a strategically stable set of equilibria of any game G′ obtained from G by deleting strictly dominated strategies.

Note that the example of figure 11.1 shows that proper equilibria do not satisfy condition ID: (L, T) is a proper equilibrium, but it is no longer proper once RW is deleted. Note also that condition ID indeed implies *iterated* strict dominance: If G′ is obtained by deleting dominated strategies from G, and G″ by deleting dominated strategies from G′, then the stable set of G contains a stable set of G′, which in turn contains a stable set of G″.

Although this definition does select the (RT, W) equilibrium in figure 11.1, it may not be clear whether it captures all that one might mean by "forward induction." We will return to this point after we have developed the Kohlberg-Mertens definition of strategic stability. Next, Kohlberg and Mertens wish their solution concept to satisfy the following condition:

A (Admissibility)   No mixed strategy in a strategically stable set can assign positive probability to any pure strategy that is weakly dominated.

Figure 11.2 shows that conditions ID and A are inconsistent with the existence of a point-valued solution concept. Here D is strictly dominated, so condition ID requires that the strategically stable set contain a stable set for the game illustrated in figure 11.3a. In that game L weakly dominates R for player 2, so condition A requires that the unique solution be (U, L). But in the original game M is strictly dominated as well, and deleting M instead of D yields the game in figure 11.3b, where (U, R) must be the unique solution from condition A. Hence, the solution to the original game must contain both (U, L) and (U, R), and so it cannot be single valued. Using this example, we can also see that it is not possible to strengthen condition ID to the requirement that a stable set be contained in a stable set of a game obtained by adding dominated strategies: In the game where player 1's only strategy is U, both (U, L) and (U, R) are stable; but if we then add the dominated strategy D, the unique stable solution is (U, R). An alternate way of saying this is that condition ID allows for the possibility that

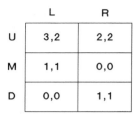

Figure 11.2

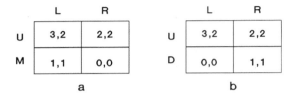

Figure 11.3

deleting dominated strategies can make stable profiles that were not stable previously.

Kohlberg and Mertens were the first to propose making sets of equilibria the objects of a theory of equilibrium refinements. Having a set of equilibria as the prediction of a theory is particularly troubling if different equilibria in the set involve different play along the equilibrium path. (Even if all the equilibria in the prediction agree on the path, one would like to know just what off-path play the players expect to see, but this concern may be less troubling.)

Here, Kohlberg and Mertens make use of the following result of Kreps and Wilson (see chapter 8 above):

Theorem 11.1 (Kreps and Wilson 1982)   In a fixed tree, for generic assignments of payoffs to terminal nodes, the set of Nash-equilibrium probability distributions over terminal nodes is finite.

Since the distribution over terminal nodes is a continuous function of the strategy profile, when there is only a finite number of equilibrium distributions, every equilibrium in the same connected component[3] must have the same probability distribution over endpoints, and hence the same play at every information set that is reached with positive probability. This avoids one potential drawback of using a set-valued solution concept.

The third main condition Kohlberg and Mertens want their solution concept to satisfy is invariance to certain transformations of the extensive

---

3. A topological space X is said to be connected if there do not exist two nonempty, disjoint open sets O_1 and O_2 such that $X \subset (O_1 \cup O_2)$, and $X \cap O_1$ and $X \cap O_2$ are both nonempty. The intuitive idea of a connected set is that one should be able to draw a path linking any two points of the set that itself lies in the set.

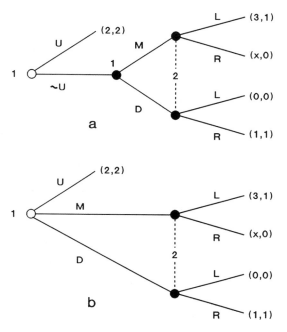

Figure 11.4

form. Consider the games illustrated in figure 11.4, where we set x to lie between 1 and 2. Kohlberg and Mertens argue that figures 11.4a and 11.4b are merely two different representations of the same game, because "the transformed tree is merely a different representation of the same decision problem." In a one-player game, the two alternate representations of player 1's decision are clearly equivalent: The choice of a best element from the set (U, M, D) is exactly the same as the two-stage choice where player 1 first decides whether he would choose M or D if he couldn't go U, and then decides whether he prefers U or the better of M and D.

Although the two games are arguably equivalent, the concept of sequential equilibrium gives different solutions for them, just as in the case of figures 8.6a and 8.6b. In figure 11.4b, (U, R) is sequential but not proper for $x \in (1, 2)$. In figure 11.4a, it is not even subgame perfect for player 1 to choose U: Player 1's second information set begins a proper subgame in which M strictly dominates D, so the unique Nash equilibrium in this subgame is (M, L). Given that player 2 will play L, player 1 will not play U, so sequential equilibrium does not satisfy the invariance property.

As we suggest in chapter 8, the analogy with decision theory is attractive but not fully convincing. Games and decisions differ in one key respect: Probability-0 events are both exogenous and irrelevant in decision problems, whereas what *would* happen if a player played differently in a game is both important and endogenously determined. This point is related to a possible difficulty with the Kohlberg-Mertens position that "the game under consideration fully describes the real situation," so that in particular "any probabilities of error have been modeled in the game tree." As we noted when we introduced backward induction, one can argue to the contrary that this "classical" point of view is not compatible with an attempt to refine Nash equilibrium by restricting the actions players take at information sets that they expect will never be reached, and that the "right" way to develop equilibrium refinements is in the context of a complete theory that provides an explanation (or several) for every conceivable sequence of observations. From the complete-theory viewpoint, every extensive form that does not explain every possible observation is simply a simplified representation of a more complex game in which all observations do have positive probability.

One such complete theory (not necessarily our favorite) is precisely the "trembles" story that players make "errors" with small probability. When this theory is used, the extensive forms in the two figures represent two distinct situations. In figure 11.4b, where player 1 chooses which of three actions to take, he plays either M or D only by mistake, and if the relative probability of mistakes is arbitrary (as in trembling-hand perfection) he may well play D more often than M. In figure 11.4a, if player 1 mistakenly fails to play U he has a chance to reconsider and choose which alternative action he most prefers, and so M will be much more likely than D.

This is not to say that one should be completely indifferent to the argument that a "good" theory would make the same predictions in the two cases. In fact, this example is troubling for economic applications of trembling-hand perfection, because the analyst will seldom know which of the two extensive forms is more descriptive. (One solution concept we develop at the end of this chapter, "c-perfection," allows (U, R) in both extensive forms.) Rather, the point is that arguing about which extensive forms are equivalent before deciding on a (complete) theory of how players behave may be putting the cart before the horse.

With these caveats in mind, we now state the invariance condition.

I (Invariance) The set of strategically stable equilibria should depend only on the reduced strategic form of the game; i.e., all extensive forms with the same reduced strategic form should have the same set of stable equilibria.

(Recall our definitions of the reduced strategic form in chapter 3, which identify "equivalent" strategic-form strategies. Kohlberg and Mertens use the stronger of the two definitions, which requires that a pure strategy s_i be deleted if there is a mixed strategy σ_i with support excluding s_i such that, for all s_{-i} and all players j, $u_j(s_i, s_{-i}) = u_j(\sigma_i, s_{-i})$.)

Kohlberg and Mertens appeal to the results of Thompson (1952) and Dalkey (1953) to argue that all extensive forms with the same reduced strategic form are equivalent. Those authors showed that if two extensive forms have the same reduced strategic form up to identifying equivalent pure strategies, then one of them may be transformed into the other by a sequence of four kinds of transformations:

coalescing and expanding information sets (as in the two extensive forms we considered),

interchange of simultaneous moves (recall that there are two ways of representing a two-player simultaneous-move game),

adding moves that are not revealed to the players and have no influence on payoffs, and

inflation and deflation of information sets (which we do not define, because it is irrelevant in perfect-recall games).

The Kohlberg-Mertens paper has renewed interest in the question of which extensive forms ought to be regarded as equivalent by a solution concept; see Elmes and Reny 1988.

In the 1986 paper, Kohlberg and Mertens go on to develop three increasingly restrictive definitions of sets of equilibria; Mertens (1988) has proposed further definitions. We will focus on the third and most restrictive definition of the 1986 paper, that of a stable set of equilibria. This concept requires that, for any trembles, there exists an equilibrium near the set.

Definition 11.1 (Kohlberg and Mertens 1986)   A closed set S of Nash equilibria is *stable* if it is minimal[4] with respect to the property that for each $\eta > 0$ there exists some $\varepsilon' > 0$ such that, for any $\varepsilon < \varepsilon'$ and any numbers

$$\{\varepsilon(s_i)\}_{\substack{i \in \mathscr{I} \\ s_i \in S_i}} \text{ with } 0 < \varepsilon(s_i) \leq \varepsilon,$$

the game where player i is constrained to play each s_i with probability at least $\varepsilon(s_i)$ has an equilibrium σ^ε that is within η (in the space of strategies) of some equilibrium in the set S. If every element of a stable component yields the same probability distribution over endpoints, this distribution is a *stable outcome*.

Comments

(1) From the standard upper-hemi-continuity argument, the set of all Nash equilibria has the property of containing an element that is close to an equilibrium of every perturbed game. Definition 11.1 gains its force from the minimality requirement.

(2) The perturbed game referred to in this definition is the same as that used to define trembling-hand perfection in the strategic form (see definition 8.5A). Thus, all equilibria in a stable set are trembling-hand perfect in the strategic form.

---

4. A set S is minimal with respect to property P if there does not exist a subset $S' \subset S$ such that S' satisfies P. (The symbol $\subset$ denotes strict inclusion.)

The key difference between trembling-hand perfection in the strategic form and stability is that perfection requires only that there exist a single sequence of perturbed games whose equilibria converge to σ, but a stable set must contain a limit point of the equilibria for every perturbed game. This is related to the fact that stability is defined in terms of sets and not single equilibria, as there may not be a single equilibrium that "works" for every allowed perturbation. If, however, there is a single equilibrium σ such that σ_i is a best response to every sequence $\sigma^n_{-i} \to \sigma_{-i}$, then that equilibrium is stable as a one-point set. Kohlberg (1981) calls such equilibria "truly perfect."[5] The game in figure 11.2 has no truly perfect equilibrium: In any perfect equilibrium σ, player 1 plays U with probability 1. If $\sigma^n_1(M) > \sigma^n_1(D)$, player 2's best response is L; if $\sigma^n_1(M) < \sigma^n_1(D)$, player 2's best response is R. (Although the payoffs in this strategic form are not generic—i.e., they involve ties—in the extensive-form example of figure 11.5 no truly perfect equilibrium exists for payoffs in the neighborhood of those payoffs that are shown.)

In the case of figure 11.1b, the reason stability eliminates the equilibrium (L, T) is that stability considers all perturbations. If player 1 trembles more onto RT than RW, player 2 responds with W, and this induces player 1 to deviate from the original equilibrium. The equilibrium (RT, W), in contrast, is truly perfect: No small trembles by player 1 will change player 2's optimal response.

(3) It is important that stability considers perturbations of the strategic form and not the agent-strategic form. In the agent-strategic form, a player's "mistakes" at different information sets are independent, and even considering all such independent trembles would not capture the forward-induction argument of figure 11.1a. In the agent-strategic form corresponding to this game, where player 1' chooses L or R, and player 1" chooses T or W, the profile where 1' plays L, 1" plays W, and 2 plays T is perfect for *any* independent trembles, since with independent trembles 1" is very likely to play W whether 1' plays L or R. Forward induction corresponds to the "correlated tremble" where, if 1' trembles onto R, 1" is more likely to play T than W. Stability has forward-induction-like properties because it does consider these correlated trembles.

Definition 11.2 A strategy profile s is a *strict equilibrium* if each s_i is a strict best response to s_{-i} in the *reduced* strategic form; that is, $u_i(s_i, s_{-i})$ is strictly greater than $u_i(s'_i, s_{-i})$ for $s'_i \neq s_i$.

Clearly any strict equilibrium is stable as a one-point set. But note that in order for s to be a strict equilibrium, it must assign positive probability to every information set of every player, since a player is

---

5. Okada [1981] independently introduced this concept, which he called "strictly perfect."

indifferent between actions at an information set whose probability is 0. This implies that strict equilibria are less likely to exist in dynamic games than in static games. Note also that mixed-strategy equilibria cannot be strict, and that a totally mixed equilibrium (where every strategy has a positive probability) is stable as a one-point set by definition. (The minimum probability constraints are not binding once they are less than the minimum weight the equilibrium gives any pure strategy.)

Theorem 11.2 (Kohlberg and Mertens 1986) There exists a stable set that is contained in a single connected component of the set of Nash equilibria, and every tree with generic payoffs has a stable payoff (i.e., a payoff that obtains for every equilibrium in a stable set). A stable set contains a stable set of any game obtained by deletion of a weakly dominated strategy, and a stable set contains a stable set of any game obtained by deleting any strategy that is not a weak best response to any of the opponents' strategy profiles in the set.

This last property, called "*never a weak best response*" (**NWBR**), embodies a version of forward induction not implied by the elimination of weakly dominated strategies: One can prune undominated strategies if they are not best responses to any of the opponents' strategy profiles in the component under consideration, even though they may be best responses to strategies outside the equilibria in the component. To see the force of this property, consider the game illustrated in figure 11.5. In this modification of the Kohlberg-Mertens example of figure 11.4, x has been set equal to $\frac{1}{4}$ and the payoffs $(2,2)$ if player 1 goes U have been replaced by a simultaneous-move coordination game, one of whose equilibria has payoffs $(2,2)$. Now the strategy D is not dominated for player 1, and neither is UA or UB. Thus, there are no weakly dominated strategies to be pruned. Yet,

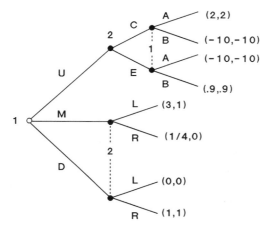

Figure 11.5

in any component of equilibria with outcome (UA, C), E is not a weak best response and can be pruned. Once E is pruned, D is weakly dominated and $(2, 2)$ is not a stable payoff. (Note that $(0.9, 0.9)$ is *not* removed by NWBR, as D is not dominated after C is deleted. One can show that $(0.9, 0.9)$ is a stable payoff.)

The NWBR property is frequently a useful way to show that some components of equilibria are not stable. For example, the refinements for signaling games that we develop below are shown to be weaker than stability by showing that all components they rule out can be ruled out by successive applications of NWBR. Exercise 11.5 gives another example.

One troubling property of stability, established by an example due to Gul, is that stable sets need not be connected and need not contain a sequential equilibrium. (Exercise 11.2 analyzes Gul's example.) This is one reason why alternative definitions of stability have been proposed. Hillas (1990) replaces the trembles in the definition of stability by perturbations of the player's best-response correspondences and says that a set is stable if it is the minimal closed set such that every profile of "nearby" best-response correspondences has a fixed point near the set. (Obviously this definition requires a topology on the space of best-response correspondences.) Hillas shows that stable sets under his definition are connected, and that they satisfy all of Kohlberg and Mertens' other conditions. Mertens (1989, 1990) retains the idea of using trembles as the perturbations but adds a topological requirement on the correspondence from perturbations to the stable set. This alternative definition also satisfies connectedness.

It is also interesting to note that stability and NWBR do not capture all that one might mean by "forward induction." We discuss van Damme's alternative definition of forward induction in section 11.3.

11.2   Signaling Games[†††]

Kohlberg and Mertens motivate their stability concept by referring to the properties they would like it to have and to its mathematical structure; they do not offer a behavioral argument that players "should" be expected to play as stability predicts. In papers on related equilibrium refinements, Cho and Kreps (1987) and Banks and Sobel (1987) do try to provide at least a heuristic behavioral foundation for stability-like ideas. One goal of these papers is to better understand stability through examining its implications in a simple class of games: the class of signaling games we introduced in chapter 8. (Recall the definition of a signaling game: The informed player, player 1, moves first and chooses an action a_1. Player 2 observes a_1 but not player 1's type θ, and chooses a_2, and then the game ends.)

A second purpose is to develop alternative equilibrium refinements that are weaker and easier to apply. A common theme of these refinements is to take as a behavioral axiom one aspect of the NWBR property described above: that of replacing the equilibrium path by its expected payoff. That is, the solutions suppose that the players are quite sure of the way their opponents will play along the equilibrium path, but that the players are less sure of the off-path play. Thus, if player 1 deviates from the equilibrium, player 2 tries to "explain" the deviation by asking which types of player 1 could do better by making this deviation, if it is met with some response that is "reasonable" in senses to be defined, than by sticking with the equilibrium strategy followed by the equilibrium response.

Before proceeding to the equilibrium concepts used in the above-mentioned papers, we first state two preliminary results which help relate the solution concepts in this section to stability.

Fact   In a signaling game, every stable set contains only sequential equilibria.

We observed above that this need not be true for general games. Signaling games have the special property that all strategic-form perfect equilibria are sequential (because each player moves only once, the agent-strategic form coincides with the strategic form), and stable sets contain only strategic-form perfect equilibria by definition. This fact means that if we begin with a stable set, and then, using NWBR, delete a strategy in which type θ plays action a_1, the resulting set must contain a stable component of the reduced game, and hence must contain a sequential equilibrium where beliefs assign probability 0 to type θ following action a_1. Thus, we can infer that sequential equilibria consistent with the deletions exist from the existence of stable components.

The idea of the Cho-Kreps paper is to use the concept of "equilibrium dominance" to argue that certain types should not be expected to use certain strategies. In contrast to stability, which follows Selten in looking at strategies and "trembles" in the strategic form, equilibrium dominance proceeds in the style of sequential equilibrium and develops a further restriction on the beliefs allowed in an extensive-form game. Recall that in a signaling game the only beliefs that aren't pinned down by an equilibrium are those of the receiver when he sees a "message" that has probability 0 according to the equilibrium strategies, and that, since the sender's message is perfectly observed, these beliefs are simply probability distributions over the sender's type. Fix an equilibrium outcome, and let $u_1^*(\theta)$ be type θ's expected payoff.

Definition 11.3 (Equilibrium Dominance)   Action a_1 can be eliminated for type θ by equilibrium dominance if

$$u_1^*(\theta) > \max_{a_2} u_1(a_1, a_2, \theta).$$

Note that this test can also be applied to a component of equilibria that all have the same equilibrium payoff. In particular, in generic signaling games, if at a fixed equilibrium the action a_1 is eliminated for type θ by equilibrium dominance, then at the connected component containing this equilibrium, all strategies σ_1 with $\sigma_1(a_1|\theta) > 0$ are eliminated by NWBR. It may seem reasonable to require that player 2 place probability 0 on type θ conditional on a_1 being played, and this restriction may reduce the set of a_2s which are best responses to a_1. Moreover, this restriction on the beliefs, if common knowledge, will lead to further restrictions on which types can reasonably be thought to choose strategy a_1, for now we can eliminate strategy a_1 for type θ' if $u_1^*(\theta')$, the equilibrium payoff of type θ', exceeds the best he could get with a_1, given that player 2 will not play a response that is only justified by beliefs that assign positive probability to type θ. This sort of argument leads to what Cho and Kreps call the "Intuitive Criterion" and the "Equilibrium Domination Test."

Defining these concepts requires some additional notation. For a nonempty subset T of Θ, let $\mathrm{BR}(T, a_1)$ be the set of all pure-strategy best responses for player 2 to action a_1 for beliefs $\mu(\cdot|a_1)$ such that $\mu(T|a_1) = 1$:

$$\mathrm{BR}(T, a_1) = \bigcup_{\mu: \mu(T|a_1)=1} \mathrm{BR}(\mu, a_1),$$

where

$$\mathrm{BR}(\mu, a_1) = \arg\max_{a_2} \sum_{\theta \in \Theta} \mu(\theta|a_1) u_2(a_1, a_2, \theta).$$

Let $\mathrm{MBR}(\mu, a_1)$ be the set of mixed best responses to a_1 given μ, that is, the set of all probability distributions over $\mathrm{BR}(\mu, a_1)$. Now let

$$\mathrm{MBR}(T, a_1) = \bigcup_{\mu: \mu(T|a_1)=1} \mathrm{MBR}(\mu, a_1).$$

(For $T = \varnothing$, set $\mathrm{BR}(\varnothing, a_1) = \mathrm{BR}(\Theta, a_1)$.) This is the set of all mixed best responses to some beliefs with support in T. It will be important in the following discussion that $\mathrm{MBR}(T, a_1)$ need not include every probability distribution over $\mathrm{BR}(T, a_1)$. As in figure 11.7 below, it may be that action a_2' is a best response for some beliefs about player 1, and a_2'' is a best response for other beliefs, but for *no* beliefs is it a best response to randomize between a_2' and a_2''. Let $T \backslash W$ denote the set-theoretic difference between T and W.

Definition 11.4 (Intuitive Criterion) Fix a vector of equilibrium payoffs $u_1^*(\cdot)$ for the sender. For each strategy a_1, let $J(a_1)$ be the set of all θ such that

$$u_1^*(\theta) > \max_{a_2 \in \mathrm{BR}(\Theta, a_1)} u_1(a_1, a_2, \theta).$$

If for some a_1 there exists a $\theta' \in \Theta$ such that

$$u_1^*(\theta') < \min_{a_2 \in \mathrm{BR}(\Theta \backslash J(a_1), a_1)} u_1(a_1, a_2, \theta'),$$

then the equilibrium fails the Intuitive Criterion.

In words, $J(a_1)$ is the set of types who get less than their equilibrium payoff by choosing a_1, provided the receiver plays an undominated strategy. The equilibrium fails the Intuitive Criterion if there exists a type who would necessarily do better by choosing a_1 than in equilibrium as long as the receiver's beliefs assign probability 0 to types in $J(a_1)$.

Cho and Kreps discuss the idea of iterating this criterion, which leads to a concept we call the Iterated Intuitive Criterion:

Definition 11.5 (Iterated Intuitive Criterion)   Fix a vector of equilibrium payoffs $u_1^*(\cdot)$ for the sender. Set $\Theta^0(a_1) = \Theta$ for all a_1. For each strategy a_1 and subset $\Theta^k(a_1)$ of types, let $J(\Theta^k(a_1), a_1)$ be the set of all $\theta \in \Theta^k(a_1)$ such that

(i) $u_1^*(\theta) > \displaystyle\max_{a_2 \in \mathrm{BR}(\Theta^k(a_1), a_1)} u_1(a_1, a_2, \theta).$

$J(\Theta^k(a_1), a_1)$ are the "types who are deleted for strategy a_1 at the kth round of iteration." Set

$$\Theta^{k+1}(a_1) = \Theta^k(a_1) \backslash J(\Theta^k(a_1), a_1).$$

This is the set of types who "could reasonably" choose strategy a_1 when they are certain of their equilibrium payoff, and who believe that player 2 will play some best response to beliefs concentrated on $\Theta^k(a_1)$. If for some a_1 there is a $\theta' \in \Theta^{k+1}(a_1)$ such that

(ii) $u_1^*(\theta') < \displaystyle\min_{a_2 \in \mathrm{BR}(\Theta^{k+1}(a_1), a_1)} u_1(a_1, a_2, \theta'),$

then the equilibrium is said to fail the $(k + 1)$st round of the Iterated Intuitive Criterion (note that if the equilibrium fails in round 1 then it fails the Intuitive Criterion). The equilibrium fails the Iterated Intuitive Criterion if it fails the $(k + 1)$st round for some k.

Cho and Kreps also offer a modified version of the Iterated Intuitive Criterion. The equilibrium-domination test is defined by replacing condition ii of definition 11.5 with the following:

(ii′)   If for some a_1 and all $a_2 \in \mathrm{BR}(\Theta^{k+1}(a_1), a_1)$ there is a $\theta' \in \Theta^{k+1}(a_1)$ such that

$$u_1^*(\theta') < u_1(a_1, a_2, \theta'),$$

then the equilibrium fails the $(k + 1)$st round of the equilibrium domination test.

The difference between the Intuitive Criterion and the equilibrium-domination test is in the order of quantifiers: Condition ii asks that there be a single type θ' who prefers a_1 for all responses in $\mathrm{BR}(\Theta^{k+1}(a_1), a_1)$; condition ii′ asks only that for each response in $\mathrm{BR}(\Theta^{k+1}(a_1), a_1)$ there be some type who prefers to deviate.

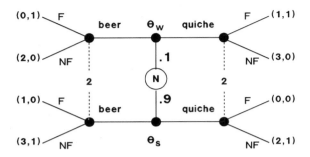

Figure 11.6

Cho and Kreps illustrate the Intuitive Criterion with the beer-quiche game, displayed in figure 11.6. Here, player 1 has two types: θ_w, who is "weak," and θ_s, who is "surly." The prior probability of weak is 0.1. Player 2 prefers to fight if she believes there is probability over $\frac{1}{2}$ that player 1 is weak, but she does not observe player 1's type. However, before deciding whether to fight, player 2 (who is very nosy) observes what player 1 has for breakfast. Player 1 has only two possible breakfasts, "beer" and "quiche"; the surly type prefers beer and the weak type likes quiche. However, regardless of their dietary preferences, either type would have either breakfast in order to avoid being fought. This game has two pooling equilibria, one in which both types have beer and another where they both have quiche; in both cases player 2 must fight with some probability when observing the out-of-equilibrium breakfast in order to make the mismatched type endure gastronomic horror. To support these outcomes as sequential equilibria, we specify that player 2's out-of-equilibrium beliefs are that if the unexpected breakfast is observed, there is probability at least $\frac{1}{2}$ that player 1 is weak.

Cho and Kreps argue that the pooling equilibrium where both types have quiche is not reasonable, and indeed it is eliminated by their Intuitive Criterion: In this equilibrium, the weak type is getting its highest possible payoff, and, so long as it is "convinced" that its equilibrium action will give the equilibrium payoff, it has no incentive to switch to drinking beer, regardless of how it expects player 2 to respond to beer. Once type θ_w is removed for $a_1 = $ beer, the set $\Theta^1(\text{beer}) = \Theta \setminus J(\Theta, \text{beer})$ is simply $\{\theta_s\}$, and player 2's unique best response to beliefs concentrated on θ_s is to not fight, which gives θ_s more than its equilibrium payoff.

Cho and Kreps offer the following heuristic justification of this process of elimination: Suppose that player 1 has the (unmodeled) chance to make a speech to player 2 at the same time he eats his breakfast. Then θ_s could say, "I'm having beer, and you should infer from this that I'm surly, for so long as it is common knowledge that you will not fight if I eat quiche, I would have no incentive to drink beer and make this speech if I were type θ_w."

This heuristic is suggestive but only partially compelling. One would prefer the communication stage to be explicitly modeled into the game, but then one runs into the difficulty that an equilibrium would have to specify how player 2 would respond to each possible speech, and also how player 2 would respond to the *absence* of a speech that would have been made had player 1's type been different. This implies that if only type θ_s is expected to make the speech, and the speech is not made, player 2 should infer that player 1 is type θ_w, which in turn would reduce type θ_w's incentive to remain silent. (Cho and Kreps attribute this reasoning to Stiglitz.) Once again, the problem arises from trying to refine the set of equilibria without specifying a complete theory of play, so that the discussion of beliefs involves considering counterfactuals.

The "both quiche" equilibrium can also be eliminated by applying iterated weak dominance to the strategic form of the corresponding two-player game, where the two types of player 1 are viewed as different information sets of the same player. (The first step is to show that the strategy "beer if weak, quiche if surly" is dominated. Any strategy for player 2 that makes beer optimal when player 1 is weak makes it optimal when he is surly. This kind of monotonicity property is discussed in chapter 6.) As we discuss in chapter 6 in the context of iterated strict dominance, whether the "right" strategic form has two or three players depends on whether we wish to assume that the different types of each player necessarily have the same beliefs about their opponents' strategies.[6]

Finally, Cho and Kreps study the implications of equilibrium refinements in Spence's model of job-market signaling. Cho and Kreps prove that when there are only two types, the only equilibrium not rejected by the Intuitive Criterion is the "Riley outcome," i.e., the separating equilibrium with the least amount of inefficient signaling. With more than two types, selecting the Riley outcome requires the stronger concept of "universal divinity," developed by Banks and Sobel; thus, we will introduce universal divinity before studying the Spence signaling game.

Universal divinity is defined with an iterative process like that in definition 11.5 above; the difference is that more type-strategy pairs may be deleted at each round. As above, we first fix an equilibrium, and let $u_1^*(\theta)$ be the equilibrium payoff of type θ. Define $D(\theta, T, a_1)$ to be the set of mixed-strategy best responses α_2 to action a_1 and beliefs concentrated on T that make type θ strictly prefer a_1 to his equilibrium strategy,[7]

---

6. Cho and Kreps note that using the "*ex ante*" strategic form that treats the different types as the same player has an additional implication: When one is computing the set of proper equilibria, a tremble by a low-probability type θ' has a low *ex ante* cost, and so is assigned a very small probability, even though the cost of that tremble conditional on the occurrence of type θ' may be quite high.

7. We use α_2 for a probability distribution over A_2 and $\sigma_2(\cdot \mid \cdot)$ for player 2's overall strategy. Thus, for a given a_1, $\sigma_2(\cdot \mid a_1)$ is some $\alpha_2 \in \Delta(A_2(a_1))$.

$$D(\theta, T, a_1) = \bigcup_{\mu:\,\mu(T|a_1)=1} \{\alpha_2 \in \mathrm{MBR}(\mu, a_1) \text{ s.t. } u_1^*(\theta) < u_1(a_1, \alpha_2, \theta)\},$$

and let $D^0(\theta, T, a_1)$ be the set of mixed best responses that make type θ exactly indifferent.[8]

Definition 11.6 A type θ is deleted for strategy a_1 under criterion D1 if there is a θ' such that

$$\{D(\theta, \Theta, a_1) \cup D^0(\theta, \Theta, a_1)\} \subset D(\theta', \Theta, a_1).$$

A type θ is deleted for strategy a_1 under criterion D2 if

$$\{D(\theta, \Theta, a_1) \cup D^0(\theta, \Theta, a_1)\} \subset \bigcup_{\theta' \neq \theta} D(\theta', \Theta, a_1).$$

(The symbol $\subset$ denotes strict inclusion.)

Obviously, having deleted a type for strategy a_1 under either of these conditions, we can impose further restrictions on player 2's responses to a_1; this leads to iterated versions of the two criteria that exactly parallel those of definition 11.5. Banks and Sobel call the iterated version of D2 universal divinity; their "divine equilibrium" results from the iterated application of a criterion slightly weaker than D1.[9]

Criterion D1 says that if the set of player 2's responses that make type θ willing to deviate to a_1 is strictly smaller than the set of responses that make type θ' willing to deviate, then player 2 should believe that type θ' is infinitely more likely to deviate to a_1 than type θ is. This is a strengthening of the Intuitive Criterion, as whenever type θ is removed by the Intuitive Criterion then the sets $D(\theta, \Theta, a_1)$ and $D^0(\theta, \Theta, a_1)$ are empty. Criterion D2 has roughly the same relation to D1 as the equilibrium domination test has to the Intuitive Criterion, as it replaces a single type θ' that deletes θ with the union over all other types.

Note that the "speeches" of Cho and Kreps do not serve to motivate D1 and D2, and Banks and Sobel do not provide a behavioral defense of these criteria. Note in particular that D1 and D2 test a deviation a_1 for a given type θ with respect to each particular mixed best response α_2 of player 2. Type θ is not allowed to be uncertain about which element of $\mathrm{MBR}(T, a_1)$ player 2 will choose, which would correspond to considering all α_2 in the convex hull of the set of best responses.[10] To see the difference this makes,

8. The reader may wonder why the equilibrium-dominance criteria use best responses and the divinity criteria use mixed best responses. Note first that introducing mixed best responses in the equilibrium-dominance criteria would not change those criteria, because the max in condition i and the min in condition ii would be unaffected. Note also that replacing MBR by BR in the divinity criteria would reduce their bite, except in the subclass of signaling games studied in theorem 11.3, where player 2's best response is a singleton. Another possibility is to replace MBR by the convex hull of best responses, as in the discussion of figure 11.7.
9. Divinity does not strike a type completely when it fails D1, but instead requires that if type θ fails D1 for strategy a_1 then the probability of type θ should not increase when a_1 is observed.

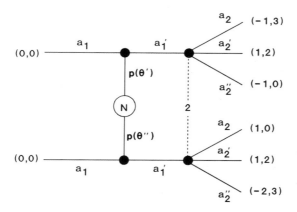

Figure 11.7

consider the game illustrated in figure 11.7. Here the set $\text{MBR}(\mu, a_1')$ of player 2's mixed best responses to a_1' is a_2 if $\mu(\theta'|a_1') > \frac{2}{3}$, any mixture between a_2 and a_2' if $\mu(\theta'|a_1') = \frac{2}{3}$, a_2' if $\frac{1}{3} < \mu(\theta'|a_1') < \frac{2}{3}$, any mixture between a_2' and a_2'' if $\mu(\theta'|a_1') = \frac{1}{3}$, and a_2'' if $\mu(\theta'|a_1') < \frac{1}{3}$. Thus, although a_2 and a_2'' are both best responses to a_1' for some beliefs, there are no beliefs for which a mixture between a_2 and a_2'' is a best response.

This game has a pooling equilibrium where both types of player 1 play a_1, and player 2 plays a_2' in response to a_1', supported by the beliefs $\mu(\theta'|a_1') = \frac{1}{2}$. This equilibrium satisfies the Cho-Kreps conditions, since both types of player 1 would do better with a_1' followed by (undominated response) a_2'' than they do in the equilibrium. Let us check whether the pooling equilibrium satisfies D1 and D2.

To compute the sets $D(\theta', \Theta, a_1')$ and $D(\theta'', \Theta, a_1')$ we first compute which responses $\alpha_2 = \sigma_2(\cdot | a_1')$ would make each type prefer a_1', and then take the intersection with the set of mixed best responses. Simple algebra shows that type θ' prefers a_1' to the equilibrium a_1 if $\alpha_2(a_2'') > \frac{1}{3}$, whereas type θ'' prefers a_1' if $3\alpha_2(a_2'') > 3\alpha_2(a_2) + \alpha_2(a_2')$. Figure 11.8 displays these deviation regions on the probability simplex corresponding to α_2, where the bold edges of the probability simplex correspond to α_2s that put probability 0 on either a_2'' or a_2 and are therefore mixed best responses for some beliefs about player 1.

Inspection of figure 11.8 shows that $D(\theta'', \Theta, a_1')$ strictly contains $D(\theta', \Theta, a_1')$, so that by criterion D1 a deviation to a_1' must be interpreted as coming from θ''. This causes player 2 to respond with a_2'', which induces both types to deviate. However, the deviation regions are not nested when one considers all mixtures over best responses: The mixed strategy

10. This point was made by van Damme (1987) and developed further by Fudenberg and Kreps (1988) and by Sobel, Stole, and Zapater (1990). Substituting the convex hull of (BR) for MBR yields "codivinity," which is not implied by stability in some nongeneric games.

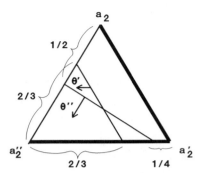

Figure 11.8

$\{\alpha_2(a_2) = \frac{3}{5}, \alpha_2(a_2'') = \frac{2}{5}\}$ would induce type θ' to deviate, but not type θ'', and therefore no type can be eliminated.

In practice, instead of checking condition D2 directly it is often easier to check the following stronger condition. Cho and Kreps call it NWBR, and indeed it is closely related to the NWBR property of Kohlberg and Mertens (i.e., that a stable component remains stable after the deletion of any strategy that is not a weak best response in any of the equilibria in that component). Since the Cho-Kreps version of NWBR is not precisely the same as that of Kohlberg and Mertens, we will call it *NWBR in signaling games*. A type-action pair can be deleted under this criterion if

$$D^0(\theta, \Theta, a_1) \subset \bigcup_{\theta' \neq \theta} D(\theta', \Theta, a_1).$$

Note that any type that is deleted for a_1 under D2 is deleted under NWBR in signaling games.

Since in generic games each stable component consists of equilibria with the same distribution on endpoints, stability implies NWBR in signaling games for generic payoffs. More precisely, fix a signaling game in which each stable component is identified with a stable outcome, and suppose that in one of the equilibria in this component NWBR in signaling games deletes type θ using action a_1. Deleting strategies where θ uses a_1 is consistent with stability if a_1 is not a weak best response for θ in any of the equilibria in the component. If a_1 were a weak best response for θ in some equilibrium in this component, then in that equilibrium player 2's response $\alpha_2(a_1)$ would lie in $D^0(\theta, \Theta, a_1)$. NWBR in signaling games then implies that player 2's response would be in $D(\theta', \Theta, a_1)$ for some other type θ', so that θ' would strictly prefer to deviate and we would not have an equilibrium after all. Thus, deleting equilibria according to NWBR in signaling games, or the weaker D2 or the still weaker condition of equilibrium dominance, does not eliminate any stable outcomes. Since stable outcomes exist in generic signaling games, universally divine (and hence "Intuitive") equilibria exist in the same generic class.

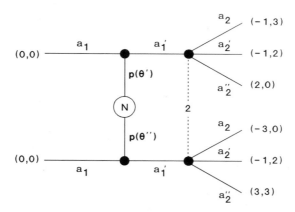

Figure 11.9

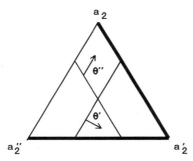

Figure 11.10

To see why NWBR in signaling games is stronger than D2, consider the example illustrated in figure 11.9, which is taken from Cho and Kreps. Player 2's payoffs in this figure are exactly as in figure 11.7; all that has been changed is the payoffs of player 1 when choosing a_1'. In this game, type θ'' strictly prefers a_1' to the equilibrium a_1 if $\alpha_2(a_2'') < \frac{1}{3}$, and type θ' strictly prefers a_1' if $\alpha_2(a_2') > \frac{1}{2}$.

Figure 11.10 displays the intersection of these deviation regions with the set of mixed best responses. (Recall that the mixed best responses for player 2 are the two bold edges in figure 11.10.) Since neither one contains the other, D2 has no bite. But the unique mixed best response $D^0(\theta'', \Theta, a_1')$ that makes θ'' exactly indifferent between the equilibrium and a_1' is contained in $D(\theta', \Theta, a_1')$, so that type θ'' choosing a_1' can be deleted by NWBR in signaling games. This deletion implies that player 2 must respond to a_1' with a_2, which is not an equilibrium response in the original game, so that the equilibrium where both types choose a_1 fails NWBR in signaling games.

Cho and Sobel (1990) have shown that eliminating equilibria that fail D1 is equivalent to stability in the class of "monotonic signaling games."

Definition 11.7 A *monotonic signaling game* has payoffs such that, for all a_1, and for all mixed strategies α_2 and α_2' in the set $\mathrm{MBR}(\Theta, a_1)$ of mixed best responses to a_1, if for some $\theta \in \Theta$

$$u_1(a_1, \alpha_2, \theta) > u_1(a_1, \alpha_2', \theta),$$

then for all $\theta' \in \Theta$

$$u_1(a_1, \alpha_2, \theta') > u_1(a_1, \alpha_2', \theta').$$

Many signaling games in the literature are monotonic. For example, if a_2 is a monetary payment to player 1 and player 1 is risk neutral, then monotonicity follows from the fact that all types of player 1 prefer the α_2 with the highest expected value. This is the case, for example, in Spence's model of job-market signaling. (With a risk-averse player 1, monotonicity is more restrictive.) The Cho-Sobel proof relies on a complex characterization of stability in signaling games due to Banks and Sobel. However, one implication of the monotonicity assumption is easy to obtain:

Lemma 11.1 (Cho and Sobel 1990)   In monotonic signaling games, criterion D1 is equivalent to NWBR.

Proof   Exercise 11.4.

Example 11.1: Spence's Job-Market Signaling

As an illustration of refinements in signaling games, we now consider a variant of the Spence model we studied in example 8.2. Suppose that there are three types of player 1: θ', θ'', and θ'''. Player 1 moves first, selecting a level of education a_1 from the set $[0, \infty)$. (We work with a continuum of education levels to simplify the analysis, but this does involve some loss of rigor. Note that stability is defined only for finite games, but the Intuitive Criterion and universal divinity can be applied to games with a continuum of actions.) Player 2, the firm, wants to minimize the quadratic difference between the wage a_2 offered to player 1 and player 1's productivity, which is $a_1 \theta$, with $\theta' = 2$, $\theta'' = 3$, and $\theta''' = 4$. (This quadratic loss is meant to stand in for a situation of Bertrand competition among several competing firms; allowing several firms would take us out of the signaling-game model.) Player 1's utility is the difference between his wage, a_2, and his disutility of education, which is $a_1{}^2/\theta$. The key aspect of these preferences is that they satisfy the Spence-Mirrlees or sorting or single-crossing condition: The marginal disutility of education is decreasing in player 1's type. This is why there are equilibria where the level of education chosen increases in player 1's productivity: The high-productivity types will be willing to choose higher levels of education than the low-productivity types for a given increment in wages.

As in chapter 8, this game can have a great multiplicity of sequential equilibria. For some parameter values there is a pooling equilibrium, where all three types choose the same education level, supported by the beliefs that the observation of any other education level implies that player 1 is the low-productivity type θ'. There is typically a continuum of different separating equilibria, where each type chooses a different education level.[11] And there are all sorts of "semi-separating" equilibria, where the supports of the education levels chosen by the different types intersect but do not coincide.[12]

Riley (1979) argued that the following equilibrium is the most reasonable one: The lowest-productivity type chooses the level of education that maximizes his utility on the assumption that his type will be fully revealed, so that his wage will equal his productivity. Call this level $a_1^*(\theta')$; with our parameter values, $a_1^*(\theta') = 2$ and $u_1^*(\theta') = 2$. The next-most-able type, θ'', chooses the level of education, $a_1^*(\theta'')$, that maximizes his utility when he is paid his productivity, $3a_1 - a_1^2/3$, subject to the constraint that type θ' should not strictly prefer the combination a_1 and a wage of $3a_1$ to his "own" education level and wage. Thus, $a_1^*(\theta'')$ must satisfy

$$3a_1^*(\theta'') - \frac{(a_1^*(\theta''))^2}{2} \leq 2 \Rightarrow a_1^*(\theta'') \simeq 5.2.$$

For $\theta = \theta'''$ (and subsequent types, if any), $a_1^*(\theta)$ is defined as the minimum of the education level that type would choose under perfect information and the education level required to prevent the next-lowest type from "pretending" to be type θ. (One can show that these adjacent incentive constraints are the binding ones: If no type wishes to pretend that it is the next-highest type, then no type prefers any deviation.) The Riley outcome is Pareto efficient in the class of separating equilibria, as it involves the minimum amount of education needed for separation. However, other equilibria can be more efficient in terms of the players' *ex ante* payoffs.

Cho and Kreps show that the Intuitive Criterion selects the Riley outcome if there are only two possible types of player 1: Suppose that in equilibrium types θ' and θ'' both assign positive probability to action $\hat{a}_1$, and let $\tilde{a}_1$ be the highest education level such that type θ' at least weakly prefers education $\tilde{a}_1$ and wage $\tilde{a}_1\theta''$ to his equilibrium action. Since there are only two types, the wage $a_2(\hat{a}_1)$ paid to the pooling action is at most $\hat{a}_1\theta''$, and $\tilde{a}_1 \geq \hat{a}_1$. From single crossing, type θ'' will then strictly prefer actions a_1 just above $\tilde{a}_1$ and wage $a_1\theta''$ to his equilibrium choice of $\hat{a}_1$. Since wage offers greater than $a_1\theta''$ are weakly equilibrium dominated for player 2, the Intuitive Criterion requires that player 2 assign probability 0 to type θ' after all actions just above $\tilde{a}_1$, so that the wage paid for these education levels must be $a_1\theta''$, and player θ'' will strictly prefer to deviate from the equilibrium because of single crossing.

11. With a continuum of types, there is a unique separating equilibrium; see Mailath 1987.
12. The reader will check that the supports cannot coincide because of the sorting condition.

An interesting feature of the Riley outcome selected by the Intuitive Criterion is that this outcome is independent of player 2's beliefs about player 1 as long as the support of the distribution is kept constant, but varies discontinuously when a type is added or deleted. To illustrate this, suppose that initially there is only one possible type: θ''. Type θ'' chooses $a_1(\theta'')$ so as to maximize $3a_1 - a_1{}^2/3$, so $a_1(\theta'') = 4.5$. Suppose now that player 1 has type θ'' with probability $1 - \varepsilon$ and type θ' with probability ε, where ε is small. The Riley outcome predicts that type θ'' chooses $a_1^*(\theta'') \simeq 5.2$. It seems extreme that allocations would be so sensitive to beliefs. Indeed, in the case of a small ε a pooling allocation at a_1 close to 4.5 seems more reasonable.

The Cho-Kreps argument fails when there are three or more types, for in order to delete action a_1 for type θ', a_1 must be large enough that type θ' would not gain by choosing it even if he were paid the wage of type θ''' whose productivity is two steps higher. If type θ'' picks this high an a_1, he is sure to get $a_1\theta''$ or more, but this will no longer guarantee that type θ'' gets more from the deviation than he did from the equilibrium. The point is that, to rule out type θ', the Intuitive Criterion asks us to consider the best possible response, which is $a_1\theta'''$, whereas to conclude that type θ'' would deviate once θ' is ruled out we must allow for the possibility that when type θ'' deviates he is paid his own productivity. With only two types, the best possible response θ' could hope for and the response θ'' could guarantee once θ' is ruled out are the same, which is why in general the Intuitive Criterion has more power in this case. In fact, with only two types it selects the Riley outcome.

Cho and Kreps observe that D1 picks out the Riley outcome with three types.

Cho and Sobel characterize the implications of D1 in a larger class of signaling games: Let $A_1 = [0, 1]^N$ for some N and $A_2 = [0, 1]$, and suppose the set of types Θ is the set of integers from 1 to $\#\Theta$.

Theorem 11.3 (Cho and Sobel 1990) Suppose that a signaling game satisfies the following conditions:
(i) (Monotonicity) If $a_2' > a_2$, then all types θ prefer a_2' to a_2.
(ii) For each $\mu \in \Delta(\Theta)$, $\text{MBR}(\mu, a_1)$ is a single point; MBR is continuous in μ, and if μ' is greater than μ in the sense of first-order stochastic dominance, then $\text{MBR}(\mu', a_1) > \text{MBR}(\mu, a_1)$, so that player 2's response is more favorable to player 1 when player 2 thinks player 1's type is higher.
(iii) Player 1's utility function is differentiable and satisfies the Spence-Mirrlees sorting condition: $-(\partial u_1/\partial a_{1j})/(\partial u_1/\partial a_2)$ is decreasing in θ for each component a_{1j} of a_1.
Then there exists a unique equilibrium satisfying D1.

Since Cho and Sobel require player 1's action space to be bounded above, one possible equilibrium configuration has a set of types pooling at the highest possible action. Cho and Sobel show that this is the only possible kind of pooling, so if no type chooses to send the highest action then the equilibrium must be fully separating, and corresponds to a generalized version of the Riley outcome. Say that $a_1'' > a_1'$ if every component of a_1'' is at least as large as the corresponding component of a_1', and a_1'' is strictly larger in at least one component. A key step in the proof is the following lemma:

Lemma 11.2 Under the hypotheses of theorem 11.3, if type θ'' chooses action a_1' with positive probability in equilibrium, then D1 implies that $\mu(\theta'|a_1'') = 0$ if $\theta'' > \theta'$ and $a_1'' > a_1'$.

Proof Fix an equilibrium (σ_1^*, σ_2^*) such that type θ'' chooses a_1' with positive probability. Let $a_2^*(a_1)$ be the action prescribed by $\sigma_2^*(\cdot|a_1)$. For each $a_1'' > a_1'$ and θ, let $\hat{a}_2(\theta) \in BR(\Theta, a_1'')$ satisfy $u_1(a_1'', \hat{a}_2(\theta), \theta) = u_1^*(\theta)$; if no such $\hat{a}_2$ exists, set $\hat{a}_2(\theta) = +\infty$. We claim that single crossing implies that $\hat{a}_2(\theta') > \hat{a}_2(\theta'')$. To see this, consider figure 11.11, which displays the situation $\hat{a}_2(\theta') \leq \hat{a}_2(\theta'')$ for the case of a single-dimensional a_1. By definition, type θ'' is indifferent between $A = (a_1', a_2^*(a_1'))$ and $B = (a_1'', \hat{a}_2(\theta''))$. But single crossing means that in the (a_1, a_2) space the indifference curve of type θ' is steeper than that of type θ'' at any point, and therefore the two indifference curves depicted in figure 11.11 cannot intersect at any $a_1 < a_1''$. Therefore, type θ' strictly prefers $(a_1', a_2^*(a_1'))$ to his equilibrium strategy—a contradiction. We leave it to the reader to provide the algebraic proof for a multi-dimensional a_1 (see also chapter 7).

Because $\hat{a}_2(\theta') > \hat{a}_2(\theta'')$,

$$D(\theta', \Theta, a_1'') \cup D^0(\theta', \Theta, a_1'')$$

$$= \{a_2 \in BR(\Theta, a_1'') \text{ such that } a_2 \geq \hat{a}_2(\theta')\} \subset D(\theta'', \Theta, a_1'')$$

$$= \{a_2 \in BR(\Theta, a_1'') \text{ such that } a_2 > \hat{a}_2(\theta'')\}.$$

Hence, θ' is eliminated by D1 for a_1''. ∎

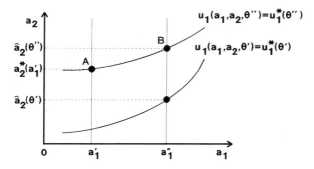

Figure 11.11

With this lemma it is easy to see that any equilibrium where two or more types assign positive probability to the same a_1^* must fail criterion D1. Let θ^* be the highest-productivity type that plays a_1^*, and let a_2^* be player 2's equilibrium response to a_1^*. From the sorting condition, for each a_1' greater than but sufficiently close to a_1^*, there are undominated responses that make type θ^* prefer to deviate to a_1' but do not tempt the lower-productivity types. Thus, type θ^*'s "deviation region" (in the sense of definition 11.6) strictly includes the deviation regions of the other types who play a_1^*, and so, if player 1 chooses any education level "just above" a_1^*, player 2 must assign probability 0 to all types with productivity less than θ^*'s. Then, since player 2's equilibrium response is continuously increasing in the beliefs about player 1, type θ^* can induce a nonnegligible increase in a_2 by an infinitesimal increase in a_1.

Not only does D1 select the Riley outcome, it imposes the following restrictions on the beliefs which are used to support it: Player 2 must assign probability 1 to type θ' after any action in the interval $[a_1^*(\theta'), a_1^*(\theta''))$, probability 1 to type θ'' after any action in $[a_1^*(\theta''), a_1^*(\theta'''))$, etc. Since the motivation for D1 is to refine the set of equilibria using "reasonable" restrictions on beliefs, to the extent that the above 0-1 restrictions are implausible they may cast doubt on D1 as an equilibrium concept.

11.3   Forward Induction, Iterated Weak Dominance, and "Burning Money"[†††]

Just as iterated strict dominance and rationalizability can be used to narrow down the set of predictions without invoking equilibrium refinements by using rationality arguments alone, the concept of iterated weak dominance (IWD) can be used to capture some of the force of forward and backward induction without assuming that players will coordinate their expectations on a particular equilibrium.[13] Since the idea of forward induction is that players interpret a deviation as a signal of how their opponent intends to player in the future, forward induction seems more compatible with a situation of considerable strategic uncertainty—i.e., a nonequilibrium situation—than with a situation where the strategic uncertainty has been resolved and all players are certain that they know their opponents' strategies. (This is another version of our argument that probability-0 events are best thought of as events whose probability is low.)

One difficulty with iterated weak dominance, as opposed to iterated strict dominance (see section 2.1), is that different orders of deletion can give different solutions, as is shown by the game in figure 11.12. Here, if we first eliminate player 1's weakly dominated strategy D at the first round, the

---

13. Pearce's (1984) extensive-form rationalizability and the notion of iterated conditional dominance are other ways to make "refined" predictions in a nonequilibrium context.

	L	R
U	1,0	0,1
D	0,0	0,2

Figure 11.12

solution is (U, R) because L is dominated for player 2; if we eliminate L at the first round, player 1 becomes indifferent between U and D, and so the solution set is both (U, R) and (D, R). The standard response to this problem is to specify the maximal amount of deletion at each round, i.e., that at each round all weakly dominated strategies of all players are deleted.[14]

Iterated weak dominance incorporates backward induction in games of perfect information: The suboptimal choices at the last information sets are weakly dominated; once these are removed, all subgame-imperfect choices at the next-to-last information sets are removed at the next round of iteration; and so on. Iterated weak dominance also captures part of the forward-induction notions implicit in stability, as a stable component contains a stable component of the game obtained by deleting a weakly dominated strategy. For example, the stable outcome of the Kohlberg-Mertens example in figure 11.1 can be obtained by iterated weak dominance: The play of RW for player 1 is strictly dominated, and player 2's playing T is weakly dominated once RW has been removed.

The most striking example we have seen of the power of iterated weak dominance is Ben-Porath and Dekel's (1988) study of the following class of

---

14. Rochet (1980) provides a partial answer to the question of when every order of deleting weakly dominated strategies gives the same solution. His answer is partial in two respects: First, instead of considering weak dominance as we have defined it, he looks only at "pure-strategy dominance"—the process he considers does not delete all weakly dominated strategies, only those that are dominated by another *pure* strategy. (See chapter 1 above for an example in which a mixed strategy strictly dominates a pure strategy that is not (pure-strategy) dominated.) Second, Rochet considers only games in which *some* order of deletion yields a unique prediction. He shows that, if any order of iterated pure-strategy weak dominance yields a unique solution, then any order of deletion yields this same unique solution, under the following assumption: If for some player i and strategy profiles s and s'

$$u_i(s) = u_i(s'),$$

then

$$u_j(s) = u_j(s') \text{ for all } j.$$

(Note that this condition is not satisfied in the strategic form in figure 11.12: (D, L) and (D, R) yield the same payoff for player 1, but not for player 2. In an extensive-form game, a sufficient condition for Rochet's assumption to be satisfied is that there not exist a player i and two terminal nodes z and z' such that $u_i(z) = u_i(z')$. This sufficient condition is satisfied in generic extensive-form games.) Moulin (1986) uses Rochet's theorem to show that backward induction and (any order of) iterated weak dominance give the same unique solution for generic payoffs in finite games of perfect information. Iterated weak dominance can be stronger than backward induction if some player has the same payoff at two distinct terminal nodes.

games[15]: Players 1 and 2 are going to play a simultaneous-move game of coordination, which has several pure-strategy equilibria, all of which are better than not coordinating (so that the mixed equilibria are Pareto dominated) and one of which gives player 1 his highest possible payoff. Before they play, however, player 1 has the option of publicly "burning" a small amount of utility. If the maximum amount player 1 can burn is sufficiently large, and the amount to be burned can be specified sufficiently finely, then the unique outcome according to iterated weak dominance is that player 1 burns no utility, and then the players play the stage-game equilibrium that gives player 1 his highest payoff. This strong conclusion can be viewed either as an argument about how players can arrange to coordinate in a particular way or as evidence that iterated weak dominance (and hence stability) is too restrictive.

Rather than state the theorem of Ben-Porath and Dekel formally, we give an illustrative example: In the first period, player 1 can either "not burn" or "burn" 2.5 utils. After this choice is observed, he and player 2 will play the simultaneous-move game at the top of figure 11.13. Note that without the possibility of burning there is no way to distinguish between the equilibria (U, L) and (D, R); player 1 prefers the first equilibrium and

	L	R
U	9,6	0,4
D	4,0	6,9

a

	L,L	L,R	R,L	R,R
Burn, U	6.5,6	6.5,6	−2.5,4	−2.5,4
Burn, D	1.5,0	1.5,0	3.5,9	3.5,9
Not Burn, U	9,6	0,4	9,6	0,4
Not Burn, D	4,0	6,9	4,0	6,9

b

Figure 11.13

15. Van Damme (1989) independently discovered the power of forward induction in a game in which players can "burn utility." He develops other examples of the power of forward induction and of its relation to stability. See exercise 11.5.

player 2 prefers the second. This creates the game depicted in the bottom of the figure, where the first component of player 2's strategy is how to play if player 1 burns and the second component is how to play if player 1 doesn't burn. In this extended game, the strategy (Burn, D) is strictly dominated for player 1 by (Not burn, D), and so at the next round of iteration player 2's best response to Burn is to play L. (That is, once (Burn, D) is deleted, any strategy s_2 with $s_2(\text{Burn}) = \text{R}$ is weakly dominated for player 2 by the strategy $\hat{s}_2$, where $\hat{s}_2(\text{Not burn}) = s_2(\text{Not burn})$ and $\hat{s}_2(\text{Burn}) = \text{L}$.) Therefore, after two rounds of iteration (Burn, U) guarantees player 1 a payoff of 6.5 and strictly dominates (Not burn, D). Hence, after three rounds of iteration, player 2 should conclude that even if player 1 does not burn he is certain to play U, and so player 1 can use the strategy (Not burn, U) and be sure of a payoff of 9! That is, the mere fact that player 1 could have chosen to burn utility but did not do so ensures that he obtains the equilibrium he most prefers.

Even for a 2×2 second-stage game, for general payoffs the result requires that player 1 have a number of different possible levels of burning. To see this, suppose that the second-stage payoffs are $(90, 90)$ to (U, L), $(72, 72)$ to (D, R), and $(0, 0)$ otherwise, and denote the cost of burning by b. Player 1's maximin strategy in the second stage is $(\frac{4}{9}\text{U}, \frac{5}{9}\text{D})$, which guarantees a payoff of 40; this is also player 1's minmax payoff. If $b \geq 50$, then the best player 1 can obtain by burning is less than 40, so (Burn, U) and (Burn, D) are both weakly dominated by not burning (followed by playing $(\frac{4}{5}\text{U}, \frac{5}{9}\text{D})$) and after one round of elimination the game with burning reduces to the original game. If $b < 32$, then (Burn, D) is a best response to the strategy of player 2, "minmax if player 1 does not burn, and play R if he burns," so no strategies are even weakly dominated. If $b \in [32, 50]$, then only (Burn, D) is weakly dominated (by Not burn, $(\frac{4}{5}\text{U}, \frac{5}{9}\text{D})$).) Once this strategy is removed, all of player 2's strategies that play R after Burn are weakly dominated; but this is as far as the iteration goes. (Burn, U) gives player 1 a payoff of $90 - b \leq 58$; (Not burn, D) could give as much as 72. In this game, IWD is not powerful. The point of the Ben-Porath–Dekel paper is that when there is a sufficiently fine grid of burning levels, player 1 can ensure his most preferred equilibrium *without* burning.[16]

When there is a sufficiently fine grid, the result of Ben-Porath and Dekel is surprisingly strong. Ben-Porath and Dekel respond to the unease that

16. The argument fails when the amount to be burned is chosen from an interval. Thus, if we regard the case of infinitely divisible money as the limit of increasingly fine discrete grids, the set of profiles satisfying iterated weak dominance fails to be lower hemi-continuous. This is related to the familiar observation that the reaction and equilibrium correspondences need not be lower hemi-continuous. From the literature on ε-equilibrium, one suspects that there are various ways of perturbing the game or the solution concept to obtain the solutions of the continuum case in the model with a fine discrete grid, so that the result of Ben-Porath and Dekel may be least compelling when a very fine grid is required.

this strength may inspire by suggesting that attention be restricted to games of "common interest," where a single equilibrium of the stage game gives both players their most preferred payoff, so that player 2 is not disadvantaged when player 1 is the only one to burn. They argue that if the stage game is not one of common interest, then both players would "try to be able to burn," and once players can burn or not simultaneously, iterated weak dominance gives much weaker predictions. But we believe that it is still instructive to look at the power of IWD in games like our example where only player 1 burns and the game is not of common interest. *If* one did one's best to ensure that the "real" extensive form was as close as possible to that of the burning-utility game, what outcome should one expect? The process of iteration in the example is sufficiently involved that we have little confidence that the outcome would be as predicted. In part this is because the iteration requires a large number of steps (four), and, like chains of backward induction, chains of forward induction become more suspicious as they grow longer. One way of formalizing this suspicion is to recall that, as with backward induction, each step of the forward-induction process requires another level of the assumption that "player 1 knows that player 2 knows ... that no one will play a strategy that is weakly dominated if the payoffs are as specified." Another way to justify the suspicion is to ask if even the second step of the induction is plausible: If player 2 sees that player 1 "burned utility," will she reason in the style of forward induction that this is a rational decision by a player 1 whose payoffs are exactly as originally believed? Or will player 2 decide that player 1 is "crazy" and derives positive utility from what was supposed to be a costly act? This latter explanation is at the heart of the Fudenberg-Kreps-Levine (1988) and Dekel-Fudenberg (1990) papers on the robustness of refined predictions to the possibility that players always assign a small but nonzero probability to their opponents' payoffs being very different than originally supposed.

Before developing these papers, we would like to discuss an alternative viewpoint. Van Damme (1989) argues that stability is too weak to capture all the implications of forward induction in an equilibrium context. He proposes that one implication of forward induction ought to be the following:

Definition 11.8 (van Damme 1989)   A solution concept S is *consistent with forward induction in the class of generic two-person extensive forms* if there is no equilibrium in S such that some player i, by deviating at a node along the equilibrium path, can ensure (with probability 1) that a proper subgame Γ is reached where (according to S) all solutions but one give the player strictly less than the equilibrium, and where exactly one solution gives the player strictly more.

This definition combines a sort of backward-induction notion with the idea that deviations should be interpreted as a signal of how the deviator intends to play in the future. If player i does deviate in a way that ensures that a proper subgame Γ satisfying the definition is reached, and if it is common knowledge that S gives the set of expected solutions in each subgame, then player i's opponent "should" conclude that player i will play in Γ according to the unique solution that gives him a higher payoff than the equilibrium path. (This idea that players' actions can signal which of several equilibria they expect was first proposed by McLennan 1985, who developed it in a different way.) The reason that the definition covers only two-player games is to ensure that player i has a nontrivial choice in the subgame Γ: In a three-player game, if player 1 deviates but will not play again, there is no particular reason to expect players 2 and 3 to choose the equilibrium that player 1 most prefers. (If only player j moves in Γ, a generic Γ will have a unique solution from backward induction.) An alternative definition would say that if player i deviates in a way that satisfies the definition, then all other players should expect player i to play according to the unique solution that justifies the deviation; this would impose no restrictions if player i did not move again.

Van Damme uses the "outside-option" game illustrated in figure 11.14 to show that stability does not satisfy his definition of forward induction.

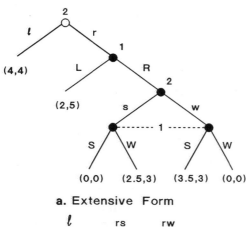

a. Extensive Form

ℓ		rs	rw
L	4,4	2,5	2,5
RS	4,4	0,0	3.5,3
RW	4,4	2.5,3	0,0

b. Reduced Strategic Form

Figure 11.14

We will go through his argument both for its own sake and to help illustrate the mechanics of checking for stability.

To begin, let us analyze the stable equilibria of the subgame $\Gamma(r)$ after player 2's choice of r. Here player 1 can either choose his "outside option" of L, resulting in payoffs $(2,5)$, or play a "battle of the sexes" game with player 2. There are three subgame-perfect equilibrium outcomes: (RS, w), (RW, s), and L, with payoffs $(3.5, 3)$, $(2.5, 3)$, and $(2, 5)$; the last outcome is supported by the mixed-strategy equilibrium if the "battle of the sexes" game is reached. Since both pure equilibria of this game give player 1 more than his outside option, the fact that player 1 didn't choose L does not "signal" his intentions, and we would expect that stability would not reduce the set of subgame-perfect equilibria. (Note the contrast with the burning-utility game, where stability picked out a unique outcome.)

To verify this intuition, we must identify a component of equilibria of $\Gamma(r)$ with outcome L such that for every perturbation of the game there is an equilibrium near some element of the component. In figure 11.14b, the game $\Gamma(r)$ simply corresponds to deleting player 2's strategy ℓ.

Let q denote the probability that player 2 plays rs, and consider the component $\{(L, (q, (1 - q)))\}$, when $q \in \{\frac{3}{7}, \frac{4}{5}\}$. (Note that this component is not connected, but that both equilibria do have the same outcome.) For either q, player 1 at least weakly prefers L to RS and RW, so both profiles are Nash equilibria. Now perturb $\Gamma(r)$ by requiring that player i place probability at least $\varepsilon(s_i)$ on strategy s_i. If $\varepsilon(RS) \geq \varepsilon(RW)$, an equilibrium of the perturbed game is for player 1 to play L with probability $1 - 2\varepsilon(RS)$, and play RS and RW with equal probability $\varepsilon(RS)$, and for player 2 to play rs with probability $q = \frac{4}{5}$. Given that $q = \frac{4}{5}$, player 1 is indifferent between L and RW, and so is willing to give RW more than the minimum required probability. Given that RS and RW are equally likely, player 2 is willing to randomize; for small εs, player 2's strategy clearly meets the minimum-probability constraint. As $\varepsilon_1(\cdot) \to 0$, this profile converges to $(L, (\frac{4}{5}, \frac{1}{5}))$, which belongs to the component we constructed. If $\varepsilon(RS) \leq \varepsilon(RW)$, an equilibrium is for player 1 to give both RS and RW probability $\varepsilon(RW)$ and for player 2 to set $q = \frac{3}{7}$.

Note that the stable component does not include the sequential-equilibrium strategies $(L, (\frac{1}{2}, \frac{1}{2}))$, for these strategies do not make player 1 indifferent between L and R. As we saw above, it is important that player 1 be indifferent, so that if (for example) the perturbations make him tremble more onto RS than RW, he is willing to play RW with greater than the minimum probability in order to restore player 2's indifference between rs and rw.

Next we claim that, in the overall game, player 2 playing ℓ is a stable outcome. Exercise 11.7 asks you to prove this; the starting point is to consider the component

$\{((\frac{3}{5}\,L, \frac{2}{5}\,RS, 0\,RW), \ell), ((\frac{1}{3}\,L, 0\,RS, \frac{2}{3}\,RW), \ell)\}.$

In both of the equilibria in this component, player 2 is indifferent between ℓ and an alternative. In the first equilibrium the alternative is rw; in the second it is rs. In any perturbed game, player 2 will be willing to play either rs or rw with enough probability to make player 1 indifferent between L and R.

The stable outcome ℓ shows that stability does not satisfy definition 11.8. Fix only equilibria with outcome ℓ, and suppose that player 2 deviates to r. Since $\Gamma(r)$ has a unique stable outcome that makes playing r rational for player 2, namely "1 plays L," definition 11.8 requires that if player 2 plays $\imath$ she receives payoff 5, which eliminates the equilibrium where 2 plays ℓ.

11.4   Robust Predictions under Payoff Uncertainty[†††]

Fudenberg, Kreps, and Levine (1988) and Dekel and Fudenberg (1990) discuss what kinds of refined predictions are possible if the main story the players use to explain unexpected deviations is that payoffs are different from what had been originally supposed. The typical game-theoretic assumption is that the payoffs (as functions of the terminal nodes) are correctly specified and indeed are common knowledge. Fudenberg, Kreps, and Levine suggest that this assumption is best viewed as an approximation, as neither the game theorist analyzing the game nor the players in it should be completely certain that the payoffs are as in the "most likely" case depicted by the extensive form.

Allowing for even a small probability of different payoffs has very strong implications, because a small *ex ante* probability can become quite large if there is an unexpected observation. We already observed this in our discussion of Spence's model of job-market signaling model, in section 11.2. Let us give two further examples of this before developing the formal results.

First consider the game illustrated in figure 11.15a. Here the unique subgame-perfect equilibrium is for 1 to play (D, u) and 2 to play R; the profile (U, L) is Nash but not subgame perfect. However, if player 2 expects player 1 to usually play U, and interprets D as a signal that player 1's payoffs are such that player 1 would play d at his second information set, then player 2 is justified in playing L. The extensive form that goes with this story is shown in figure 11.15b: Here player 1 has two possible types, θ and θ'; type θ's payoffs are as in figure 11.15a and those of type θ' make $D_2 d_2$ a weakly dominant strategy. In this game, regardless of the prior probability of θ', the profiles where θ plays $U_1 u_1$, θ' plays $D_2 d_2$, and player 2 plays L are sequential, and indeed are stable as a singleton set, as each player's strategy is a strict best response to the strategy of his opponent. Thus, a "small amount" of payoff uncertainty—i.e., a small probability that a player's payoffs are very different than had been supposed—can justify

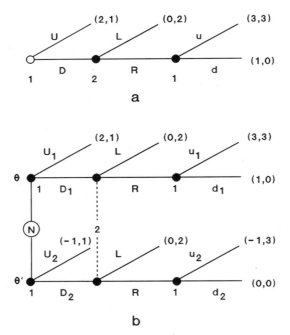

Figure 11.15

subgame-imperfect outcomes. Note that the imperfect equilibrium (Ud, L) of figure 11.15a is trembling-hand perfect in the associated strategic form: If player 1 mostly plays U and "trembles" onto Dd much more than onto Du, then player 2's choice of L is optimal. As discussed in chapter 8 above, Selten introduced the agent-strategic form precisely to rule out this sort of "correlation" in player 1's deviations. If deviations are due to different payoffs rather than to trembles, the argument that a player's deviations should be independent is less convincing.

The game depicted in figure 11.16 illustrates another implication of interpreting deviations as due to payoff uncertainty. In the subgame in which players 1 and 3 have played R and r, the payoffs of players 1 and 2 are independent of player 3's choice between A and B, and they play a "matching pennies" game between themselves. Thus, any Nash equilibrium of the subgame has players 1 and 2 randomize $\frac{1}{2}$-$\frac{1}{2}$ between their two actions. Player 3 thus gets more than 0 in the subgame, and must play r. There is also an imperfect Nash equilibrium where player 1 chooses L and player 3 chooses d. This choice by player 3 can be "justified" if he interprets a deviation by player 1 to R as meaning that players 1 and 2 are going to *correlate* their play in the simultaneous-move subgame, particularly if player 3 expects to face the joint distribution $(\frac{1}{2}(H, h), \frac{1}{2}(T, t))$. This would be the case if player 3 attached a small but positive *ex ante* probability to both of his opponents' having different payoffs, and if he believed that

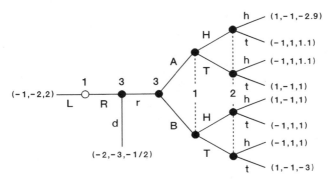

Figure 11.16

these probabilities were correlated, so that player 1's deviation could signal information about player 2's future play. For instance, there could be three states of the world. In state ω_1 (which has probability close to 1), the payoffs are those of figure 11.16. In state ω_2, the payoffs for players 1 and 2 in the subgame following Rr are such that they both have a dominant strategy: to play H and h respectively and get 1. In state ω_3, which is as likely as ω_2, the payoffs for players 1 and 2 in the subgame following Rr are such that T and t are dominant strategies for players 1 and 2, who get 1. Player 3 does not know which state prevails, but players 1 and 2 do. The introduction of states ω_2 and ω_3 models the above correlation of the strategies of players 1 and 2 in the subgame. Such correlations may not seem reasonable in the case where deviations are "trembles"; they seem more natural when deviations are due to payoff uncertainty.

As a special case of correlated types introducing correlated trembles, suppose that the opponents' payoffs do have probability 1 of being as originally supposed, but that there is a small chance the opponents have access to a correlating device. Then, although along the equilibrium path the players will proceed as if no correlating devices were available, they may interpret some unexpected observations as a signal that correlating devices were available after all. Jean-François Mertens has created an example in which this seems particularly apt: Suppose that players 1 and 2, who cannot communicate, have the option of whether or not to play a "battle of the sexes" game in which the payoffs to coordinating are $(1, 2)$ and $(2, 1)$, the payoffs to not coordinating are $(-10, -10)$, and the payoffs for not playing are $(0, 0)$. Player 3 believes there are no correlating devices available, and thus might well predict that players 1 and 2 would choose not to play. However, if, contrary to expectations, players 1 and 2 do agree to play the game, then player 3 might well conclude that a correlating device was available after all.

The examples above show that allowing for small payoff uncertainty can have a "large" effect. (Van Damme (1983) and Myerson (1986) give other examples.) Fudenberg, Kreps, and Levine (1988) give precise definitions of various kinds of "small" uncertainties and characterize the implications of each. Fudenberg et al. consider only the case of payoff uncertainty. That is, they assume that there is no doubt at all about the physical rules of the game (who moves when, what their choices are, and which previous actions they have observed), and that the only "additional" uncertainty concerns other players' payoffs. This leads to the notion of an *elaboration* $\tilde{E}$ of an extensive-form game E.

Definition 11.9 An *elaboration* $\tilde{E}$ of an extensive-form game E is formed as follows. An integer N is given, along with a probability distribution μ on $\mathcal{N} = \{1, 2, \ldots, N\}$. The game tree $\tilde{T}$ of $\tilde{E}$ is an N-fold replication of the tree T of E: $\tilde{T} = T \times \mathcal{N}$; each $n \in \mathcal{N}$ corresponds to a "version" of the game. If player i moves at x in T, he moves at (x, n) in $\tilde{T}$ for all n; the probability distribution over initial nodes $\tilde{w} = (w, n)$ of $\tilde{T}$ is $\rho(w)\mu(n)$, where ρ is the distribution over initial nodes of T. Each player i has a partition $P_i(n)$ over n, and information sets of $\tilde{E}$ take the form $h(x) \times P_i(n)$, where $h(x)$ is the information set of E containing x. Actions at information sets are inherited in the obvious way. Finally, the payoff to player i at terminal node (z, n) is $u_i(z, n)$. The elaboration has *personal types* if player i's payoff is a constant over all $n \in P_i(n)$, so that each player's payoff depends only on z and his own information about nature's choice of version. In this case we identify $P_i(n)$ with player i's "type."[17]

The incomplete-information game depicted in figure 11.15b is an elaboration with personal types of the game in figure 11.15a. Note that the definition of personal types does not require that the distributions over types be independent.

The next step is to specify when an elaboration is "close" to the game upon which it is based.

Definition 11.10 (convergence criterion)   The following are sufficient conditions for a sequence of elaborations $\tilde{E}^k$ of an extensive-form game E to *approach E:*
(i) there is a uniform bound on the absolute values of the payoffs in each version and on the number of versions per elaboration,
(ii) there is a single version 1 such that $\lim_{k \to \infty} \mu^k(1) = 1$, and
(iii) for each i and z, $\lim_{k \to \infty} \mu_i^k(z, 1) = u_i(z)$.

Conditions ii and iii require that there be a single version whose probability tends to 1 and in which the payoffs converge to those of the original game. If one replaced these conditions by the requirement that the total probability of all versions with payoffs close to the original game converge

---

17. The original game E can be a game of incomplete information, so that $P_i(n)$ is not a full description of player i's type in the usual sense—it is his "meta-type."

to 1, then the definition would allow a sequence of games with a correlating device to approach a game in which the device is not present (correlated equilibrium corresponds to elaborations where all versions have the same payoffs as the original game). This seems too loose a notion of "closeness"; we argued above that a game with a small probability of a correlating device should be close to a game without one, but a game in which correlating devices are certain to be present is a different matter.

The first part of condition i ensures that the small probability of different payoffs makes only a small difference in *ex ante* payoffs. Without the uniform bound, the payoffs in other versions can inflate as their probability decreases, so that the limiting values of the *ex ante* payoffs can be quite different than in the original game. Fudenberg, Kreps, and Levine assert that the bound on the number of versions is "probably unnecessary to support a notion of closeness." (Note that the definition gives sufficient conditions for convergence, but not necessary conditions. This is because the definition as stated does not generate a topology on the space of elaborations of a given extensive form.) To characterize the implications of allowing "small" perturbations of the kind defined in the convergence criterion, one must specify an equilibrium refinement to be used in the perturbed games. Fudenberg, Kreps, and Levine use the concept of strict equilibrium, which is very demanding: Any strict equilibrium is stable as a singleton set. They use such a strong concept because their critique of refinements such as stability is most forceful when the equilibria rejected by the refinement can be shown to satisfy a strong version of the refinement in the perturbed game.

Definition 11.11   An equilibrium σ of the strategic form corresponding to an extensive-form game E is *near-strict with personal types* if there is a sequence of elaborations with personal types $\tilde{E}^k$ of E that approaches E in the sense of the convergence criterion, and a sequence of strict equilibria $\tilde{\sigma}^k$ of the reduced strategic forms corresponding to $\tilde{E}^k$, such that the behavior prescribed by $\tilde{\sigma}^k$ at all nodes $(x, 1)$ converges to the behavior prescribed by σ at x.

Recall that the definition of personal elaborations allows the distributions over types to be correlated, which we argued was natural. The equilibrium concept that characterizes the set of near-strict equilibria thus also involves correlation.

Definition 11.12   A strategy profile σ of a strategic form is *c-perfect* if for each player i there is a sequence ϕ_{-i}^k of totally mixed probability distributions over S_{-i} such that $\phi_{-i}^k \to \sigma_{-i}$ and σ_i is a best response to each ϕ_{-i}^k.

C-perfection weakens trembling-hand perfection in the strategic form in two ways. First, instead of using a common $\sigma^k \to \sigma$, each player is allowed to have his own beliefs about the "trembles" of his opponents. If trembles are thought of as occurring very rarely, it seems plausible that the players could have such differing beliefs. Second, the beliefs about the opponents' trembles need not take the form of a mixed strategy, but can be any probability distribution over joint actions by the opponents, so that correlated trembles are allowed. (The c in the term c-perfection is meant to represent this correlation.) Both of these considerations are irrelevant in two-player games, where c-perfection reduces to trembling-hand perfection in the strategic form.

Theorem 11.4 (Fudenberg, Kreps, and Levine 1988)[18]   A pure-strategy profile s of an extensive-form game E with payoffs u is near-strict with respect to personal types if and only if there is a sequence $u^k \to u$ such that s is c-perfect in the corresponding strategic-form games.

Theorem 11.4 implies that any c-perfect pure-strategy[19] equilibrium is near-strict; the set of near-strict equilibria also includes equilibria that are c-perfect in games where the payoffs are slightly different. That is, to obtain an "if and only if" characterization, one takes the closure of the c-perfect set with respect to payoff perturbations in the extensive form. Note that these payoff perturbations, where payoffs are certain to be close to those of the original game, are much more restrictive than the perturbations considered in the definition of near-strict.

Fudenberg, Kreps, and Levine argue that when payoff uncertainty is the dominant explanation for deviations, one should not use concepts more restrictive than c-perfection unless one is prepared to argue which forms of payoff uncertainty—that is, which versions of the game—are viewed as more likely. In our view there is always payoff uncertainty: There are no economically interesting situations where the players are completely sure of their opponents' payoffs, and it may not even be reasonable to suppose that this is true as a thought experiment. However, this does not mean that we view the results of Fudenberg, Kreps, and Levine as relevant to all situations. They analyze the effects of small payoff uncertainties, ignoring all other explanations for deviations such as mistakes and experimentation. Thus, their results describe situations where payoff uncertainty is "large" relative to these other explanations. The right model for a given situation depends on which explanation(s) is most likely, and this information is not captured by the usual extensive form. Thus, we fear that it may not be possible to have a single theory of refinements that is appropriate for all extensive-form games.

---

18. Fudenberg et al. (1988) state the theorem as an equivalence between the set of near-strict equilibria and those which are "quasi c-perfect." As Dekel and Fudenberg (1990) explain, the theorem can be stated in the simpler form given here.
19. A mixed-strategy equilibrium can be made near-strict by first transforming it to a pure-strategy equilibrium with private information à la Harsanyi (see chapter 6 above).

Exercises

Exercise 11.1** Apply the Intuitive Criterion, the Iterated Intuitive Criterion, and universal divinity to the games illustrated in Figure 11.17.

Exercise 11.2*** Show that the game depicted in figure 11.18 has a stable component that does not contain a sequential equilibrium.

Exercise 11.3** Apply iterated weak dominance to the two-player strategic form of the beer-quiche game depicted in figure 11.6.

Exercise 11.4*** Show that D1 and NWBR are equivalent in monotonic signaling games. Hint: Suppose that θ is deleted for a_1 under NWBR in signaling games, so that every response that makes θ indifferent between a_1 and his equilibrium action makes some other types strictly prefer to deviate. Then, since all types have the "same" preferences over responses to a given a_1, all responses that make θ strictly prefer a_1 must make the other types strictly prefer a_1 as well. (This sketch has to be sharpened a bit to get the strict inclusion required by D1.)

Exercise 11.5** Use NWBR to find the stable components of the following game (taken from van Damme 1989). In period 1, players 1 and 2 simultaneously decide whether to burn 0 or 1 util; in period 2, they play the "battle of the sexes" game illustrated in figure 11.19.

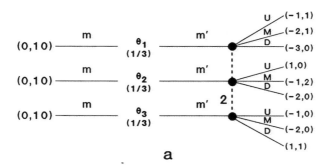

a

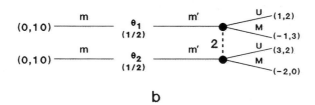

b

Figure 11.17

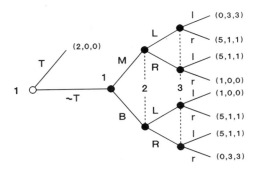

Figure 11.18

	S	W
S	0,0	3,1
W	1,3	0,0

Figure 11.19

(a) Show that strategic stability requires both players to burn utility, so that all stable equilibria are inefficient.

(b) Now suppose that the cost of burning is $1 + \varepsilon$, and that there is a publicly observed random variable ω at the beginning of period 2, where ω has the uniform distribution on the first 100 positive integers. Does the NWBR argument still work, and (***) are there stable components where the players do not burn utility?

Exercise 11.6**

(a) Show that there is a stable component where player 1 plays U with probability 1 in Figure 11.20.

(b) This strategic form is consistent with the extensive form depicted in figure 11.21. In this extensive form, there are three equilibria in the subgame following $\sim$ U, only one of which, (M, L), gives player 1 a payoff greater than the 2 he gets from choosing U. What do you think "forward induction" should imply here?

Exercise 11.7** Verify that $\{((\frac{3}{5}L, \frac{2}{5}RS, 0\,RW), \ell), ((\frac{1}{3}L, 0\,RS, \frac{2}{3}RW), \ell)\}$ is a stable component of the game in figure 11.14.

Exercise 11.8** Consider the twice-repeated version of the "battle of the sexes" stage game shown in figure 11.19, where the players maximize the sum of their per-period payoffs.

(a) Show that, although the path where players choose (S, W) in both periods can be supported by a subgame-perfect equilibrium, the component

	L	M	R
U	2,2	2,2	2,2
M	3,3	0,2	3,0
D	0,0	3,2	0,3

Figure 11.20

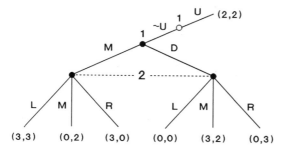

Figure 11.21

where (S, W) is played in both periods does not contain a stable set. (Hint: Use iterated applications of NWBR on the game's reduced strategic form.)

(b) Characterize the stable paths that are in pure strategies.

(c) Construct a stable component where both players randomize $(\frac{1}{2}, \frac{1}{2})$ in the first stage. (Van Damme (1989) and Osborne (1987) discuss forward induction in repeated games.)

Exercise 11.9*** Show that an essential equilibrium (see chapter 12) is "truly perfect."

Exercise 11.10** Consider Farqharson's (1969) model of plurality voting with ties (example 2, p. 73, in Moulin 1986). An election has three candidates or policies, A, B, and C, and three voters, $i = 1, 2, 3$. The voting rule is plurality voting, and player 1 breaks ties: The elected candidate is the one chosen by voters 2 and 3 if they vote for the same candidate, and the one chosen by voter 1 otherwise. Suppose that $u_1(A) > u_1(B) > u_1(C)$, $u_2(C) > u_2(A) > u_2(B)$, and $u_3(B) > u_3(C) > u_3(A)$. Show that iterated weak dominance predicts that candidate C will be elected! Comment.

References

Banks, J., and J. Sobel. 1987. Equilibrium selection in signalling games. *Econometrica* 55: 647–662.

Ben-Porath, E., and E. Dekel. 1988. Coordination and the potential for self-sacrifice. Mimeo.

Cho, I. K., and D. M. Kreps. 1987. Signalling games and stable equilibria. *Quarterly Journal of Economics* 102: 179–221.

Cho, I. K., and J. Sobel. 1990. Strategic stability and uniqueness in signalling games. *Journal of Economic Theory* 50: 381–413.

Dalkey, N. 1953. Equivalence of information patterns and essentially determinate games. In *Contributions to the Theory of Games II*, ed. H. Kuhn and A. Tucker. Princeton University Press.

Dekel, E. 1990. Simultaneous offers and the inefficiency of bargaining: A two-period example. *Journal of Economic Theory* 50: 300–308.

Dekel, E., and D. Fudenberg. 1990. Rational play under payoff uncertainty. *Journal of Economic Theory* 52: 243–267.

Elmes, S., and P. Reny. 1988. On the equivalence of extensive form games. Mimeo, Columbia University and University of Western Ontario.

Farqharson, R. 1969. *Theory of Voting*. Yale University Press.

Fudenberg, D., and D. Kreps. 1988. A theory of learning, experimentation, and equilibrium in games. Mimeo, Massachusetts Institute of Technology.

Fudenberg, D., D. M. Kreps, and D. K. Levine. 1988. On the robustness of equilibrium refinements. *Journal of Economic Theory* 44: 354–380.

Hillas, J. 1990. On the definition of the strategic stability of equilibria. *Econometrica* 58: 1365–1390.

Kohlberg, E. 1981. Some problems with the concept of perfect equilibrium. Rapporteurs' report of the NBER conference on the Theory of General Economic Equilibrium by Karl Dunz and Nirvikar Singh, University of California, Berkeley.

Kohlberg, E., and J.-F. Mertens. 1986. On the strategic stability of equilibria. *Econometrica* 54: 1003–1038.

Kreps, D., and R. Wilson. 1982. Sequential equilibria. *Econometrica* 50: 863–894.

Mailath, G. 1987. Incentive compatibility in signaling games with a continuum of types. *Econometrica* 55: 1349–1365.

McLennan, A. 1985. Justifiable beliefs in sequential equilibrium. *Econometrica* 53: 889–904.

Mertens, J.-F. 1989. Stable equilibria—a reformulation. I. Definition and basic properties. *Mathematics of Operations Research* 14: 575–624.

Mertens, J.-F. 1990. Stable equilibria—a reformulation. II. *Mathematics of Operations Research*, forthcoming.

Moulin, H. 1986. *Game Theory for the Social Sciences* (second edition, revised). New York University Press.

Myerson, R. (1986). Multi-stage games with communication. *Econometrica* 54: 323–358.

Okada, A. 1981. On the stability of perfect equilibrium points. *International Journal of Game Theory* 10: 67–73.

Osborne, M. 1987. Signaling, forward induction and stability in finitely repeated games. Mimeo, Department of Economics, McMaster University.

Pearce, D. 1984. Rationalizable strategic behavior and the problem of perfection. *Econometrica* 52: 1029–1050.

Riley, J. 1979. Informational equilibrium. *Econometrica* 47: 331–359.

Rochet, J.-C. 1980. Selection of a unique equilibrium payoff for extensive games with perfect information. Mimeo, Université de Paris IX.

Selten, R. 1965. Re-examination of the perfectness concept for equilibrium points in extensive games. *International Journal of Game Theory* 4: 25–55.

Sobel, J., L. Stole, and I. Zapater. 1990. Fixed-equilibrium rationalizability in signalling games. *Journal of Economic Theory* 52: 304–331.

Thompson, F. 1952. Equivalence of games in extensive form. Report RN 759, Rand Corporation.

van Damme, E. 1983. *Refinements of the Nash Equilibrium Concept.* Springer-Verlag.

van Damme, E. 1987. *Stability and Perfection of Nash Equilibria.* Springer-Verlag.

van Damme, E. 1989. Stable equilibria and forward induction. *Journal of Economic Theory* 48: 476–496.

12 Advanced Topics in Strategic-Form Games

This chapter collects several classes of results on strategic-form games that require more apparatus than those in chapter 1. Readers lacking mathematical training are advised to skim through the issues, ideas, and results and ignore technical details. Some of the results are stated only for reference, with no attempt made at explaining the proofs.

Section 12.1 develops properties of finite strategic-form games that hold for generic strategic forms. Generically, strategic forms have a finite and odd number of equilibria, and these equilibria are robust in the sense that any perturbed game with nearby payoffs has equilibria that are nearby.

Section 12.2 extends the existence analysis of subsection 1.3.3 to games with "continuous" action spaces (i.e., convex subsets of $\mathbb{R}^n$) and discontinuous payoff functions.

Section 12.3 analyzes the properties of "supermodular games." Roughly speaking, in supermodular games each player's strategies are ordered, and each player's best response is increasing in his opponents' strategies. Supermodular games have pure-strategy Nash equilibria even if the payoffs are neither quasi-concave nor continuous. The sets of Nash-equilibrium strategies and of rationalizable strategies have upper and lower bounds, which furthermore coincide. Also, the learning and comparative-statics properties of supermodular games are straightforward.

12.1 Generic Properties of Nash Equilibria[†††]

Although Nash equilibria exist in every finite strategic-form game, some other interesting properties of the Nash concept hold only for "almost all finite strategic forms." This section examines two such properties: the finiteness and oddness of the number of equilibria, and the robustness of equilibria to small perturbations of the payoffs.

By "almost all" we mean the following: A finite game with I players in which each player i has $\#S_i$ strategies can be seen as a payoff vector $\{u_i(s)\}_{i \in \mathscr{I}, s \in S}$ in the Euclidean space of dimension $I \cdot \Pi_{i=1}^I \#S_i$. For a fixed set of I strategy spaces, "game u" is the game with the fixed strategy spaces and payoff vector u. "Almost all games" satisfy a property if the set of games (i.e., payoff vectors, with the number of players and the strategy spaces kept fixed) that satisfy this property is open and dense in the above Euclidean space. A property is satisfied for "generic games" if it is satisfied for "almost all games."

12.1.1 Number of Nash Equilibria

As Debreu (1970) showed (see Mas-Colell 1985), competitive economies "in general" have a finite and odd number of Walrasian equilibria. "In general" refers to the fact that oddness does not hold for any economy, but rather for almost all of them (more precisely, for an open and dense set of

	L	R
U	1,1	0,0
D	0,0	0,0

Figure 12.1

economies). We may wonder whether a similar result holds for the set of
Nash equilibria of a game. It is easy to find games with an even number of
Nash equilibria. For example, the game illustrated in figure 12.1 has two
equilibria: the pure-strategy profiles (U, L) and (D, R). Wilson (1971) has
shown that this game is "exceptional"[1]:

Theorem 12.1 (Wilson's (1971) Oddness Theorem)   Almost all finite games
have a finite and odd number of equilibria.

12.1.2   Robustness of Equilibria to Payoff Perturbations

In practice it is unlikely that the modeler will have specified payoff functions
that are exactly correct. The issue is then whether the Nash predictions of
the original game with payoffs u are approximate Nash predictions of the
real game with nearby payoffs $\tilde{u}$.

The issue of robustness has many facets. In this subsection, we fix the
strategic form (the set of players and their strategy spaces) and relax the
assumption that the modeler has specified the correct payoffs, but we
maintain the hypothesis that the payoffs are common knowledge among
the players themselves. In chapters 11 and 14 we discuss other robustness
issues by relaxing the assumption that payoffs are common knowledge
among the players.

To define the notion of proximity in finite games, we introduce distances
between payoff vectors and between strategy profiles. Let

$$u = \{u_i(s)\}_{i \in \mathscr{I}, s \in S}$$

and

$$\tilde{u} = \{\tilde{u}_i(s)\}_{i \in \mathscr{I}, s \in S}$$

denote two payoff vectors, and let

$$\sigma = \{\sigma_i(s_i)\}_{i \in \mathscr{I}, s_i \in S_i}$$

and

$$\tilde{\sigma} = \{\tilde{\sigma}_i(s_i)\}_{i \in \mathscr{I}, s_i \in S_i}$$

denote two mixed strategy profiles. Let

1. For further odd-number theorems see Eaves 1971, 1973, 1976 and Harsanyi 1973.

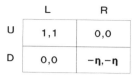

	L	R
U	1,1	0,0
D	0,0	$-\eta,-\eta$

Figure 12.2

$$D(u, \tilde{u}) = \max_{i \in \mathscr{I}, s \in S} |u_i(s) - \tilde{u}_i(s)|$$

and

$$d(\sigma, \tilde{\sigma}) = \max_{i \in \mathscr{I}, s_i \in S_i} |\sigma_i(s_i) - \tilde{\sigma}_i(s_i)|.$$

A Nash equilibrium of a game is "essential" or "robust" if there exists a nearby Nash equilibrium for any nearby game:

Definition 12.1 A Nash equilibrium σ of game u is *essential* or *robust* if for any $\varepsilon > 0$ there exists $\eta > 0$, such that for any $\tilde{u}$ such that $D(u, \tilde{u}) < \eta$ there exists a Nash equilibrium $\tilde{\sigma}$ of game $\tilde{u}$ such that $d(\sigma, \tilde{\sigma}) < \varepsilon$. A game u is essential if all its equilibrium points are essential.

Figure 12.1 gives an example of a nonessential game. The strategy profile $\sigma = (D, R)$ is a Nash equilibrium of game u in that figure. However, the only Nash equilibrium of the slightly perturbed game $\tilde{u}$ in figure 12.2 is $\tilde{\sigma} = (U, L)$. Note that $D(u, \tilde{u}) = \eta$ and $d(\sigma, \tilde{\sigma}) = 1$, and so one of the Nash equilibria of game u, viz. $\sigma = (D, R)$, is far away from the nearest (and only) Nash equilibrium, $\tilde{\sigma} = (U, L)$, of game $\tilde{u}$. Again, the game depicted in figure 12.1 is exceptional, as the following theorem demonstrates.

Theorem 12.2 (Wu and Jiang 1962) Almost all finite strategic-form games are essential.

The proof of this theorem relies on the essential fixed-point theorem of Fort (1950). Consider a compact metric space Σ with distance d. A fixed point σ of a continuous mapping f from Σ into itself is essential if for any $\varepsilon > 0$ there exists $\eta > 0$ such that, for any continuous mapping $\tilde{f}$ such that $d(f, \tilde{f}) = \max_{\sigma \in \Sigma} d(f(\sigma), \tilde{f}(\sigma)) < \eta$, there exists a fixed point $\tilde{\sigma}$ of $\tilde{f}$ with $d(\sigma, \tilde{\sigma}) < \varepsilon$. A mapping is essential if all its fixed points are essential. Fort's essential fixed-point theorem asserts that the set of essential mappings is dense in the set of continuous mappings. (This set is also open from its definition.)

Fort's theorem compares fixed points of nearby mappings. It does not quite yield a comparison of equilibria of nearby games. Recalling that Nash equilibria can be obtained as fixed points of certain continuous mappings, Wu and Jiang identify a game u with the "Nash mapping associated with game u." This Nash mapping is the function f_u from Σ into itself, where

$$f_u = \{f_u^{s_i}\}_{i \in \mathcal{I}, s_i \in S_i}$$

and

$$f_u^{s_i}(\sigma) = \frac{\sigma_i(s_i) + \max\{0, u_i(s_i, \sigma_{-i}) - u_i(\sigma)\}}{1 + \sum_{s_i' \in S_i} \max\{0, u_i(s_i', \sigma_{-i}) - u_i(\sigma)\}}.$$

f_u is continuous in σ, and σ is a fixed point of f_u if and only if σ is a Nash equilibrium of game u.[2]

The correspondence from payoffs u to Nash mappings f_u is not one-to-one. If u is replaced by $\tilde{u}$ such that $\tilde{u}_i(s) \equiv u_i(s) + v_i(s_{-i})$ for all i and s, where v_i is an arbitrary function from S_{-i} into $\mathbb{R}$, then $f_u = f_{\tilde{u}}$. More generally, one would like to consider equivalence classes of games. Two games u and $\tilde{u}$ having the same von Neumann-Morgenstern utility functions for all players (i.e., satisfying $\tilde{u}_i(s) = \lambda_i u_i(s) + v_i(s_{-i})$ for some $\lambda_i > 0$ and all i and s) are equivalent. It is easily seen that two equivalent games have not only the same set of Nash equilibria but also the same set of essential equilibria. To identify equivalent games, Wu and Jiang normalize games by requiring that

(i) $\sum_{s_i \in S_i} u_i(s_i, s_{-i}) = 0$ for all s_{-i}

and

(ii) $\sum_{\substack{s_{-i} \in S_{-i} \\ s_i, s_i' \in S_i}} |u_i(s_i, s_{-i}) - u_i(s_i', s_{-i})| =$ either 0 or 1.

(The 0 on the right-hand side of constraint ii is meant to accommodate the payoff function that is constant in player i's strategy; this payoff function is nongeneric anyway. Any payoff function that is not constant in player

---

2. That a Nash equilibrium is a fixed point of f_u is trivial. Conversely, a fixed point of f_u must satisfy, for all $s_i \in S_i$,

$$\sigma_i(s_i)\left(\sum_{s_i' \in S_i} \max\{0, u_i(s_i', \sigma_{-i}) - u_i(\sigma)\}\right) = \max\{0, u_i(s_i, \sigma_{-i}) - u_i(\sigma)\}.$$

Let $\tilde{S}_i \subseteq S_i$ denote the support of σ_i. Because $\sum_{s_i \in \tilde{S}_i} \sigma_i(s_i) = 1$,

$$\sum_{s_i' \in S_i} \max\{0, u_i(s_i', \sigma_{-i}) - u_i(\sigma)\} = \sum_{s_i' \in \tilde{S}_i} \max\{0, u_i(s_i', \sigma_{-i}) - u_i(\sigma)\},$$

which implies that, for all $s_i' \notin \tilde{S}_i$,

$$u_i(s_i', \sigma_{-i}) \le u_i(\sigma).$$

If $\sigma_i(s_i) > 0$, then either

$$u_i(s_i, \sigma_{-i}) \le u_i(\sigma) \text{ for all } s_i \in S_i,$$

and then σ_i is a best response to σ_{-i}, or

$$u_i(s_i, \sigma_{-i}) > u_i(\sigma) \text{ for all } s_i \in \tilde{S}_i,$$

which is impossible.

i's strategy can be scaled up or down so that the sum in constraint ii is equal to 1.) Constraints i and ii eliminate the $1 + \Pi_{j \neq i}(\# S_j)$ degrees of freedom left in the specification of player i's preferences before normalization for each i.[3] It is straightforward to show that, in the compact metric space of normalized games, two games are equivalent if and only if they are identical. Wu and Jiang then apply Fort's theorem to the subspace of all Nash mappings corresponding to normalized games.

Remark The Wu-Jiang theorem states that the Nash equilibria of generic strategic-form games are robust to perturbations of the payoffs. This is of interest only for simultaneous-move games. To see this, consider the extensive form in figure 12.3a. It depicts a sequential-move game in which player 1 first chooses between L_1 and R_1. If player 1 chooses R_1, the game stops and the payoffs are (a, b). If he chooses L_1, player 2 gets to play. The payoffs are (c, d) if player 2 chooses L_2, and (e, f) if he chooses R_2. Figure 12.3b depicts the associated strategic form, which prescribes payoffs for any pair of strategies. (Strategies L_2 and R_2 for player 2 are shorthand for "L_2 if L_1" and "R_2 if L_1"). It is clear that *genericity in the tree is not equivalent to genericity in the strategic form*: For a given game tree (extensive form), generic extensive-form payoffs can lead to nongeneric payoffs in the strategic form. In figure 12.3b, the payoffs corresponding to the lower row of the strategic form cannot be perturbed independently. That is, for a given game tree, the payoffs are constrained to belong to a subspace of the space of

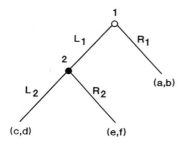

a. Extensive form

	L_2	R_2
L_1	c,d	e,f
R_1	a,b	a,b

b. Strategic form

Figure 12.3

3. There are $\Pi_{j \neq i}(\# S_j)$ equations in constraint i, and 1 in constraint ii.

strategic-form payoffs $\mathbb{R}^{I \cdot \Pi_i(\#S_i)}$ which in general has measure 0 in that space. (For instance, in the game of figure 12.3, the (unnormalized) payoffs in the extensive form belong to $\mathbb{R}^6$ and the set of (unnormalized) payoffs in the strategic form is $\mathbb{R}^8$.) Generic results in the set of strategic-form payoffs are then meaningless.

12.2   Existence of Nash Equilibrium in Games with Continuous Action Spaces and Discontinuous Payoffs[†††]

A number of games in the economic literature have discontinuous and/or non-quasi-concave payoff functions. Economic models often have payoffs that are not quasi-concave. On the other hand, one might argue that discontinuities are sometimes built in by the modeler and that small perturbations of the game "smooth" the payoff functions. For instance, in the Hotelling model discussed below, which assumes that products are differentiated only by "location," at some price profiles a small cut in price allows one firm to corner the other firm's "backyard market." When another parameter of differentiation—such as different tastes for quality (if qualities differ) or different transportation costs among consumers—is introduced, the discontinuity may disappear (see, e.g., De Palma et al. 1985). Yet discontinuous games are sometimes of interest. First, smoothing usually requires a more complex model. Second, mechanism design (such as the design of an optimal auction) leads to the consideration of discontinuous games. For instance, a seller with an object for sale may want to offer it to the highest bidder, creating a discontinuous game for the buyers.

Consider the Hotelling model of competition on the line developed in example 1.4. Consumers are distributed uniformly along the interval $[0, 1]$ and have unit transportation cost t. Suppose, in contrast with example 1.4, that firms are located in the interior of the interval, firm 1 at $x = \frac{1}{3}$ and firm 2 at $x = \frac{2}{3}$. Again, we assume that the buyers' valuation for the good supplied by the firms is sufficiently large that we do not have to worry about the buyers' not purchasing in the relevant price range. Consider a consumer located at $x \leq \frac{1}{3}$. This consumer belongs to firm 1's "back yard." His choice between the two firms is determined by the comparison between the generalized prices $p_1 + t(\frac{1}{3} - x)$ and $p_2 + t(\frac{2}{3} - x)$, i.e., between p_1 and $p_2 + t/3$. Thus, all consumers located to the left of firm 1 always make the same brand choice as the consumer located at $x = \frac{1}{3}$. The firms' demands are thus discontinuous at $p_2 = p_1 - t/3$. Figure 12.4 depicts firm 2's profit function u_2 for $p_1 \in (c + t/3, c + 5t/3)$, which is both discontinuous and non-quasi-concave.[4] D'Aspremont, Gabszewicz, and Thisse (1979) showed that there exists no pure-strategy equilibrium for this game.

---

4. Figure 12.4 assumes that when $p_2 = p_1 - t/3$ or $p_2 = p_1 + t/3$, so that the consumers in the back yard of one of the firms are indifferent between the two firms, these consumers go to the nearer firm. Of course, alternative conventions could be made.

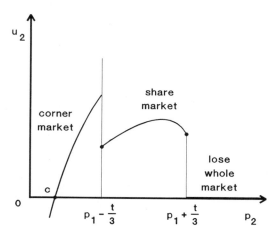

Figure 12.4

Dasgupta and Maskin (1986a) supply two existence theorems for discontinuous games. First, assuming quasi-concavity, they provide conditions (upper semi-continuity and continuous maximum) that are weaker than continuity and allow the use of Kakutani's theorem to guarantee the existence of a pure-strategy equilibrium. Second, they provide conditions for the existence of mixed-strategy equilibria in games without quasi-concave payoffs.

12.2.1   Existence of a Pure-Strategy Equilibrium

With discontinuous payoffs, a compact strategy space no longer ensures that a player's optimal reaction to his opponents' strategies exists. To guarantee existence, we assume that payoff functions are upper semi-continuous. An upper semi-continuous function is a function that has no jumps down.

Definition 12.2   A function $u_i(\cdot)$ on S is *upper semi-continuous* at s, if, for any sequence s^n converging to s,

$$\limsup_{n \to +\infty} u_i(s^n) \le u_i(s).[5]$$

Note that the function u_2 depicted in figure 12.4 fails to be upper semi-continuous at $p_2 = p_1 - t/3$. (We will show that this example has a mixed-strategy equilibrium as it satisfies a weak form of upper semi-continuity; however, there is no point in trying to "patch" upper semi-continuity to prove the existence of a pure-strategy equilibrium for this game, because payoffs are not quasi-concave.)

5. The limit superior or "lim sup" of a sequence $x^n \in \mathbb{R}$ is the smallest x such that, for all $\varepsilon > 0$, there is an N such that $x^n \le x + \varepsilon$ for all $n > N$. Similarly, the limit inferior or "lim inf" of a sequence $x^n \in \mathbb{R}$ is the largest x such that, for all $\varepsilon > 0$, there is an N such that $x^n \ge x - \varepsilon$ for all $n > N$.

Let

$$r_i^*(s_{-i}) \equiv \left\{ s_i \in S_i \,\middle|\, u_i(s_i, s_{-i}) = \max_{s_i' \in S_i} u_i(s_i', s_{-i}) \right\}$$

denote the set of player i's optimal *pure-strategy reactions* to pure strategies s_{-i}. If the strategy set S_i is compact, a maximand exists and $r_i^*(s_{-i})$ is indeed non-empty-valued for all s_{-i}. (To see this, consider a sequence s_i^n such that $\lim_{n \to \infty} u_i(s_i^n, s_{-i}) = \sup_{s_i' \in S_i} u_i(s_i', s_{-i})$. Because S_i is compact, s_i^n has a converging subsequence, with limit $\bar{s}_i \in S_i$, say. But upper semi-continuity of u_i implies that $u_i(\bar{s}_i, s_{-i}) \geq \sup_{s_i' \in S_i} u_i(s_i', s_{-i})$, and thus an optimal reaction exists.) Note that r_i^* differs from the best response or *reaction correspondence*, r_i, which for a given s_{-i} is the convex hull of the points in r_i^* (i.e., includes the mixed best responses).

To prove the existence of a pure-strategy equilibrium, we follow the method of theorem 1.2. That is, we use Kakutani's theorem to prove that the pure-strategy reaction correspondence $r^*: S \rightrightarrows S$ (defined by $[r^*(s)]_i = r_i^*(s)$) has a fixed point. We will assume that the strategy spaces are compact, convex, nonempty subsets of finite Euclidean spaces. Because r^* is thus nonempty (as noted above), it remains to make assumptions that ensure that r^* is convex valued and has a closed graph.

As in theorem 1.2, to guarantee convex-valuedness we require the payoff functions to be quasi-concave in their own strategy. That is, for all s_{-i}, the set of s_i such that $u_i(s_i, s_{-i}) \geq k$ is convex for all k and thus, in particular, is convex for $k = \max_{s_i \in S_i} u_i(s_i, s_{-i})$.

To ensure that the pure-strategy reaction correspondence has a closed graph (that is, if

$$(s_i^n, s_{-i}^n) \xrightarrow[n \to \infty]{} (s_i, s_{-i})$$

and

$$s_i^n \in r_i^*(s_{-i}^n) \text{ for all } n,$$

then $s_i \in r_i^*(s_{-i})$), we need an assumption that, together with upper semi-continuity of payoffs, ensures closed graph:

Definition 12.3 A function u_i has a *continuous maximum*[6] if $u_i^*(s_{-i}) \equiv \max_{s_i} u_i(s_i, s_{-i})$ is continuous in s_{-i}.

It is easy to see that a continuous maximum and upper semi-continuity imply that r_i^* has a closed graph. If not, there is a sequence $(s_i^n, s_{-i}^n) \to (\bar{s}_i, \bar{s}_{-i})$ with $s_i^n \in r_i^*(s_{-i}^n)$, but $\bar{s}_i \notin r_i^*(\bar{s}_{-i})$. Then

---

6. Dasgupta and Maskin impose instead the stronger condition of "graph continuity." A function u_i is graph continuous if, for all $\bar{s}$, there exists a function $f_i: S_{-i} \to S_i$ such that $\bar{s}_i = f_i(\bar{s}_{-i})$ and such that $u_i(f_i(s_{-i}), s_{-i})$ is continuous at $s_{-i} = \bar{s}_{-i}$.

$$\max_{s_i} u_i(s_i, \bar{s}_{-i}) > u_i(\bar{s}_i, \bar{s}_{-i})$$

$$\geq \limsup_{n \to \infty} u_i(s_i^n, s_{-i}^n) = \limsup_{n \to \infty} \left[\max_{s_i} u_i(s_i, s_{-i}^n) \right],$$

contradicting the assumption of a continuous maximum. We have therefore proved the following:

Theorem 12.3 (Dasgupta and Maskin 1986a)   Let S_i be a nonempty, convex, and compact subset of a finite-dimensional Euclidean space, for all i. If, for all i, u_i is quasi-concave in s_i, is upper semi-continuous in s, and has a continuous maximum, there exists a pure-strategy Nash equilibrium.

12.2.2   Existence of a Mixed-Strategy Equilibrium

The idea of the Dasgupta-Maskin result on the existence of a mixed-strategy equilibrium is to approximate the strategy spaces (which are closed intervals of $\mathbb{R}$) by finite grids, and to provide conditions ensuring that the limits of the Nash equilibria of the discretized games do not have "atoms" (nonnegligible probability) on any of the discontinuity points of the payoff functions.[7]

Consider a sequence of finite approximations S_i^n of S_i converging to S_i for all i. By Nash's existence theorem, each discretized game with strategy sets $\times_i S_i^n$ has a mixed-strategy equilibrium $\sigma^n \equiv (\sigma_1^n, \ldots, \sigma_I^n)$; that is,

$$u_i(\sigma_i^n, \sigma_{-i}^n) \geq u_i(s_i, \sigma_{-i}^n) \text{ for all } s_i \in S_i^n \text{ and for all } i. \tag{12.1}$$

Because the space of probability measures on S_i is compact under the topology of weak convergence, there is a subsequence of Nash-equilibrium mixed-strategy profiles, which without loss of generality can be taken to be the sequence itself, that converges to some mixed strategy σ^* on S. Now, if payoffs were continuous, we would be finished: $u_i(\sigma_i^n, \sigma_{-i}^n)$ and $u_i(s_i, \sigma_{-i}^n)$ would converge to $u_i(\sigma_i^*, \sigma_{-i}^*)$ and $u_i(s_i, \sigma_{-i}^*)$, respectively, and the limit strategies σ^* would form a Nash equilibrium of the limit game (this is the essence of theorem 1.3). More generally, if the equilibria σ^n put vanishingly small probability on the discontinuity points of the payoff functions, σ^* would be a Nash equilibrium. Thus, the challenge is to find conditions that ensure that discontinuity points do not matter in the limit game.

Dasgupta and Maskin introduce two assumptions. First, they require that the *sum* of payoffs ($\sum_i u_i$) be upper semi-continuous.[8] (This assumption

---

7. Simon (1987) relaxes this condition by requiring only that at least one limit has this no-atom property, instead of all of them.

8. Dasgupta and Maskin give the following example of a game that does not satisfy upper semi-continuity of the sum of the payoffs (but satisfies the other assumptions of theorem 12.4) and does not have a mixed-strategy equilibrium: Let $I = 2$, $S_i = [0, 1]$, and $u_i(s_1, s_2) = 0$ if $s_1 = s_2 = 1$ and $= s_i$ otherwise. If one player puts positive weight on 1, then the other player has no optimal strategy, as he wants to play as close as possible to 1 but doesn't want to play 1. And if both players put zero weight on 1, then each wants to play 1—a contradiction.

is satisfied in the Hotelling game.) In particular, they require that this sum not jump down in the limit of the equilibrium strategies

$$\limsup_{n \to \infty} \sum_{i=1}^{I} u_i(\sigma^n) \le \sum_{i=1}^{I} u_i(\sigma^*). \tag{12.2}$$

Next, they make the assumption of weakly lower semi-continuous payoffs: Let $S^{**}(i)$ denote the set of s such that u_i is discontinuous at s and

$$S_{-i}^{**}(s_i) \equiv \{s_{-i} \in S_{-i} | (s_i, s_{-i}) \in S^{**}(i)\}.$$

Assume that discontinuities occur only on a subset (of measure 0) in which a player's strategy is "related" to another player's. That is, for any two players i and j, there exist a finite number of functions $f_{ij}^d \colon S_i \to S_j$, where d is an index, that are one-to-one and continuous[9] such that, for each i,

$$S^{**}(i) \subseteq S^*(i) = \{s \in S | \exists j \ne i, \exists d \text{ such that } s_j = f_{ij}^d(s_i)\}.$$

In the Hotelling example above, discontinuities arose when $p_1 = p_2 - t/3$ or $p_1 = p_2 + t/3$.

$u_i(s)$ is *weakly lower semi-continuous* in s_i if for all s_i there exists $\lambda \in [0, 1]$ such that, for all $s_{-i} \in S_{-i}^{**}(s_i)$,

$$\lambda \liminf_{s_i' \uparrow s_i} u_i(s_i', s_{-i}) + (1 - \lambda) \liminf_{s_i' \downarrow s_i} u_i(s_i', s_{-i}) \ge u_i(s_i, s_{-i}).$$

In a sense, this says that u_i does not jump up when s_i' tends to s_i either from the left, or from the right, or both. Roughly, this assumption implies that player i can do about as well with strategies near s_i as with s_i, even if player i's rivals' strategies put weight on the discontinuity points of u_i.[10] Weak lower semi-continuity holds in the Hotelling game.[11] The proof of existence of a mixed-strategy equilibrium under these assumptions is involved, and we refer the reader to the original paper.

---

9. The assumption that these functions are one-to-one prevents discontinuity curves that are "vertical" or "horizontal" in the Cartesian product space of strategies. The following example (inspired by example 4 of Dasgupta and Maskin) demonstrates the possibility of nonexistence if this assumption is not satisfied. Let $I = 2$ and $S_i = [0, 1]$. Let $u_i(s_1, s_2) = -(s_i - \frac{1}{2})^2$ if $s_i, s_j \ne \frac{1}{2}$; $= -1$ if $s_i = \frac{1}{2}$ and $s_j \ne \frac{1}{2}$; $= +1$ if $s_i \ne \frac{1}{2}$ and $s_j = \frac{1}{2}$; $= 0$ if $s_i = s_j = \frac{1}{2}$. That is, each player wants to be as close as possible to $\frac{1}{2}$, but not to play $\frac{1}{2}$ (which would transfer payoff to his rival). The sum of the payoffs is upper semi-continuous: If anything, Σu_i jumps up at the horizontal and vertical lines corresponding to $s_1 = \frac{1}{2}$ and $s_2 = \frac{1}{2}$. Furthermore, u_i is weakly lower semi-continuous; indeed, for any s_j, player i does better by playing close to $\frac{1}{2}$ than by playing $\frac{1}{2}$. But there is no mixed-strategy equilibrium: For any mixed strategy of his rival, a player wants to play as close as possible to $\frac{1}{2}$, but not to play $\frac{1}{2}$.

10. That weak lower semi-continuity is needed for existence is demonstrated in exercise 12.2.

11. Note that $S_1^{**}(p_2) = \{p_2 - t/3, p_2 + t/3\}$, and, for all $p_1 \in S_1^{**}(p_2)$,

$$\liminf_{p_2' \uparrow p_2} u_2(p_2', p_1) \ge u_2(p_2, p_1)$$

(because firm 2, by undercutting a little, can make sure that it sells to its own back yard or that it invades the other's, depending on the case).

Theorem 12.4 (Dasgupta and Maskin 1986a)   Let S_i be a closed interval of $\mathbb{R}$. Suppose that u_i is continuous except on a subset $S^{**}(i)$ of $S^*(i)$, where $S^*(i)$ is defined above; that $\sum_{i=1}^{I} u_i(s)$ is upper semi-continuous; and that $u_i(s_i, s_{-i})$ is bounded and weakly lower semi-continuous in s_i. Then the game has a mixed-strategy equilibrium.

Dasgupta and Maskin prove other existence theorems, in particular for the case in which a discontinuity in a player's payoff occurs independent of discontinuities in the other players' payoffs (as is the case when firms must incur a fixed cost to be in the market). They also prove that symmetric games satisfying the assumptions of theorem 12.4 possess a symmetric mixed-strategy equilibrium. They apply theorem 12.4 to examples such as the above Hotelling game, price competition with capacity constraints, and the insurance market with adverse selection (Dasgupta and Maskin 1986b).

12.3   Supermodular Games[†††]

Supermodular games, developed by Topkis (1979), were applied to economic problems first by Vives (1990) and then by Milgrom and Roberts (1990). Roughly, they are games in which each player's marginal utility of increasing his strategy rises with increases in his rivals' strategies. In such games the best response correspondences are increasing, so that the players' strategies are "strategic complements." When there are two players, a change in variables allows this framework to also accommodate the case of decreasing best responses (that is, "strategic substitutes").[12]

Supermodular games are particularly well behaved. They have pure-strategy Nash equilibria. The upper bound (defined below) of player i's Nash-equilibrium strategies exists (which is not trivial if the strategy sets are not one-dimensional) and is a best response to the upper bounds of his rivals' sets of Nash-equilibrium strategies, and similarly for the lower bounds. Furthermore, the upper and lower bounds of the sets of Nash equilibria and rationalizable strategies coincide.

The simplicity of supermodular games makes convexity and differentiability assumptions unnecessary, although they are satisfied in most applications. What is needed for the theory is an order structure on strategy spaces and a weak continuity requirement on payoffs, in addition to the above-mentioned property that the marginal utility of each player's strategy is monotonic in the strategies of his rivals, and a "supermodularity requirement."

Suppose that each player i's strategy set S_i is a subset (not necessarily compact and convex) of a finite-dimensional Euclidean space $\mathbb{R}^{m_i}$. Then

---

12. See Bulow et al. 1985 and Fudenberg and Tirole (1984) for discussions of the use of these concepts in industrial organization. (Bulow et al. coined the strategic complements/substitutes terminology.)

$S \equiv \times_{i=1}^{I} S_i$ is a subset of $\mathbb{R}^m$, where $m \equiv \sum_{i=1}^{I} m_i$. Let x and y denote two vectors in some Euclidean space $\mathbb{R}^K$. Let $x \geq y$ if $x_k \geq y_k$ for all $k = 1, \ldots, K$, and let $x > y$ if $x \geq y$ and there exists k such that $x_k > y_k$. The order $\geq$ is only a partial order: If a vector dominates another in one component but is dominated in another component, the vectors cannot be compared. Next we define the "meet" $x \wedge y$ and the "join" $x \vee y$ of x and y:

$$x \wedge y \equiv (\min(x_1, y_1), \ldots, \min(x_K, y_K)),$$

$$x \vee y \equiv (\max(x_1, y_1), \ldots, \max(x_K, y_K)).$$

S is a *sublattice* of $\mathbb{R}^m$ if $s \in S$ and $\tilde{s} \in S$ imply that $s \wedge \tilde{s} \in S$ and $s \vee \tilde{s} \in S$.[13]

A set S has a *greatest* element $\bar{s}$ (respectively, a *least* element $\underline{s}$) if $s \leq \bar{s}$ (respectively, $s \geq \underline{s}$) for all $s \in S$. A topological result of Birkhoff (1967) says that if S is a nonempty, compact sublattice of $\mathbb{R}^m$, it has a greatest element and a least element.

The following notion formalizes the notion of strategic complementarity:

Definition 12.4 $u_i(s_i, s_{-i})$ has *increasing differences in* (s_i, s_{-i}) if, for all $(s_i, \tilde{s}_i) \in S_i^2$ and $(s_{-i}, \tilde{s}_{-i}) \in S_{-i}^2$ such that $s_i \geq \tilde{s}_i$ and $s_{-i} \geq \tilde{s}_{-i}$,

$$u_i(s_i, s_{-i}) - u_i(\tilde{s}_i, s_{-i}) \geq u_i(s_i, \tilde{s}_{-i}) - u_i(\tilde{s}_i, \tilde{s}_{-i}).$$

$u_i(s_i, s_{-i})$ has *strictly increasing differences in* (s_i, s_{-i}) if, for all $(s_i, \tilde{s}_i) \in S_i^2$ and $(s_{-i}, \tilde{s}_{-i}) \in S_{-i}^2$ such that $s_i > \tilde{s}_i$ and $s_{-i} > \tilde{s}_{-i}$,

$$u_i(s_i, s_{-i}) - u_i(\tilde{s}_i, s_{-i}) > u_i(s_i, \tilde{s}_{-i}) - u_i(\tilde{s}_i, \tilde{s}_{-i}).$$

Increasing differences says that an increase in the strategies of player i's rivals raises the desirability of playing a high strategy for player i (see theorem 12.7).

Definition 12.5 $u_i(s_i, s_{-i})$ is *supermodular* in s_i if for each s_{-i}

$$u_i(s_i, s_{-i}) + u_i(\tilde{s}_i, s_{-i}) \leq u_i(s_i \wedge \tilde{s}_i, s_{-i}) + u_i(s_i \vee \tilde{s}_i, s_{-i})$$

for all $(s_i, \tilde{s}_i) \in S_i^2$. u_i is *strictly supermodular* in s_i if this inequality is strict whenever s_i and $\tilde{s}_i$ cannot be compared with respect to $\geq$.

Note that supermodularity is automatically satisfied if S_i is single-dimensional. We will need supermodularity in the case of multi-dimensional strategy spaces to prove that each player's best responses are increasing with his rivals' strategies. To see why, suppose that $m_i = 2$. From increasing differences, if s_{-i} increases, the optimal $s_{i,1}$ for a given $s_{i,2}$ increases and so does the optimal $s_{i,2}$ for a given $s_{i,1}$. However, if $\partial^2 u_i / \partial s_{i,1} \partial s_{i,2} < 0$ (with u_i assumed differentiable), a higher $s_{i,2}$ makes a lower $s_{i,1}$ desirable, and

---

13. Note that $\mathbb{R}^m$ is a lattice in that any two vectors x and y have a meet and a join in $\mathbb{R}^m$.

conversely. This indirect effect of an increase in s_{-i} may outweigh the direct effect, which means that the effect of an increase in s_{-i} on $s_{i,k}$ ($k = 1, 2$) is ambiguous; all we can say is that $s_{i,1}$ and $s_{i,2}$ cannot both decrease, because this would contradict increasing differences. The supermodularity assumption is thus an assumption of complementarity among the components of a player's strategies; it ensures that these components move together when the rivals' strategies (or the exogenous environment) change.

As Topkis has shown, if $S_i = \mathbb{R}^{m_i}$ and if u_i is twice continuously differentiable in s_i, then u_i is supermodular in s_i if and only if, for any two components s_{ik} and $s_{i\ell}$ of s_i (with $k \neq \ell$), $\partial^2 u_i / \partial s_{ik} \partial s_{i\ell} \geq 0$.

Definition 12.6 A *supermodular game* (respectively, a *strictly supermodular game*) is such that, for each i, S_i is a sublattice of $\mathbb{R}^{m_i}$, u_i has increasing differences (strictly increasing differences) in (s_i, s_{-i}), and u_i is supermodular (strictly supermodular) in s_i.

Remark Increasing differences in (s_i, s_{-i}) and supermodularity in s_i are both implied by *supermodularity in s*, which requires that, for all s and $\tilde{s}$,

$$u_i(s \vee \tilde{s}) + u_i(s \wedge \tilde{s}) \geq u_i(s) + u_i(\tilde{s}).$$

Supermodularity in s clearly implies supermodularity in s_i (take s and $\tilde{s}$ to differ only in s_i in the above definition). One can see that it implies increasing differences by considering $s_i \geq \tilde{s}_i$ and $s_{-i} \geq \tilde{s}_{-i}$ and letting $u \equiv (\tilde{s}_i, s_{-i})$ and $v \equiv (s_i, \tilde{s}_{-i})$. Then $u \vee v = (s_i, s_{-i})$ and $u \wedge v = (\tilde{s}_i, \tilde{s}_{-i})$. Applying the supermodularity definition to u and v yields increasing differences. In practice, it is often easier to recognize increasing differences than to recognize supermodularity.

If u_i is twice continuously differentiable, u_i is supermodular if and only if, for any two components s_ℓ and s_k of s, $\partial^2 u_i / \partial s_\ell \partial s_k \geq 0$.[14]

Examples[15]

Bertrand game Consider an oligopoly with demand functions

$$D_i(p_i, p_{-i}) = a_i - b_i p_i + \sum_{j \neq i} d_{ij} p_j,$$

---

14. To see this, let e_ℓ be the vector equal to 1 for the ℓth component and 0 for the other components. Let ε and η be the two positive infinitesimals. Then supermodularity means that, for all s,

$$u_i(s + e_\ell \varepsilon) + u_i(s + e_k \eta) \leq u_i(s) + u_i(s + e_\ell \varepsilon + e_k \eta),$$

or

$$(\varepsilon \eta) \frac{\partial^2 u_i}{\partial s_\ell \partial s_k} \geq 0.$$

The proof of the converse is omitted.
15. See Topkis 1979, Vives 1990, and Milgrom and Roberts 1990 for other applications.

where $b_i > 0$ and $d_{ij} > 0$. Let

$$u_i(p_i, p_{-i}) = (p_i - c_i)D_i(p_i, p_{-i}).$$

Then $\partial^2 u_i/\partial p_i \partial p_j > 0$ for all $i, j \neq i$, so the game in which firms choose prices simultaneously has increasing differences. But many Bertrand games are not supermodular. For instance, the Hotelling game described in section 12.2 does not have increasing differences: Though firm i's best response to p_j is increasing in p_j as long as it is optimal for firm i to share the market (the demand function has the above linear form in the range of prices for which both firms have positive market share), an increase in p_j may make it more attractive for firm i to corner the whole market—i.e., to lower its price to $(p_j - t/3)$.[16]

Cournot game   Consider a duopoly. Firm i ($i \in \{1, 2\}$) chooses a quantity $q_i \in [0, \bar{q}_i]$. Suppose that the inverse demand functions $P_i(q_i, q_j)$ are twice continuously differentiable, and that P_i and firm i's marginal revenue (i.e., $P_i + q_i \partial P_i/\partial q_i$) are decreasing in q_j. Firm i's cost, $C_i(q_i)$, is assumed differentiable. The payoffs are

$$u_i(q_i, q_j) = q_i P_i(q_i, q_j) - C_i(q_i).$$

If $s_1 \equiv q_1$ and $s_2 \equiv -q_2$, the transformed payoffs satisfy $\partial^2 u_i/\partial s_i \partial s_j \geq 0$ for all $i \neq j$ (note that this transformation works only for $I = 2$). Thus, the game is supermodular.

Aggregate-demand externalities   The stag-hunt game of chapter 1 is supermodular. Let "hunt the hare" be action 1 and "hunt the stag" be action 2. The game exhibits increasing differences in that, if a player hunts the stag instead of the hare, hunting the stag becomes more attractive to the other players.

As we note in chapter 1, aggregate-demand-externality models in macroeconomics have a similar flavor. For example, a simple search model *à la* Diamond (1982) has payoff functions

$$u_i(s) = \alpha s_i \sum_{j \neq i} s_j - c(s_i),$$

where s_i is player i's search intensity, $c(s_i)$ is the cost of search, $s_i \sum_{j \neq i} s_j$ is the probability of finding a trading partner, and α is the gain when a partner is found. Note that $\partial^2 u_i/\partial s_i \partial s_j = \alpha > 0$ for $j \neq i$. The game is supermodular (and in general has multiple equilibria, some with high search activity and some with low search activity). See also exercise 12.3.

---

16. For instance, for $c_2 = 0$, firm 2 charges $p_2 = (p_1 + t)/2$ for $p_1 \in [0, (3 - 4/\sqrt{3})t)$ and undercuts to $p_1 - t/3 < (p_1 + t)/2$ for p_1 a bit above $(3 - 4/\sqrt{3})t$.

Remark Vives (1990, section 6) noted that the theory of supermodular games can also be applied to games in which players have private information. We invite the reader to think about why this is so.[17]

From the point of view of existence of a pure-strategy Nash equilibrium, supermodular games derive their interest from the following result:

Theorem 12.5 (Tarski 1955)   If S is a nonempty, compact sublattice of $\mathbb{R}^m$ and $f: S \to S$ is increasing ($f(x) \le f(y)$ if $x \le y$), f has a fixed point in S.

To obtain intuition about this fixed-point theorem, consider the single-dimensional case $S = [0, 1]$, which is depicted in figure 12.5. In order not to have a fixed point, the function f in figure 12.5 would need to "escape" the area above the diagonal and "jump into" the area below the diagonal; but increasing functions do not jump down. In the multi-dimensional case the intuition is the same, as no component of $f(x)$ jumps down when an arbitrary component of x increases.[18]

Tarski's theorem is relevant here because the set $r_i^*(s_{-i})$ of s_i that maximize $u_i(\cdot, s_{-i})$ turns out to be a sublattice and to "increase" with s_{-i}, as we now show.

If S_i is compact and u_i is upper semi-continuous in s_i, r_i^* is nonempty since $u_i(s_i, s_{-i})$ attains a maximum in s_i on S_i. (Consider a sequence s_i^n such

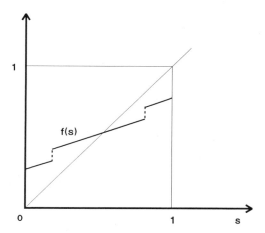

Figure 12.5

17. For an interesting application, see, in Milgrom and Roberts 1990, the Hendricks-Kovenock (1989) oil-drilling game, in which each firm would like the other to explore the other's tract in order to learn about the profitability of its own tract.

18. A related result of Tarski (in the single-dimensional case) is that a function f from $[0, 1]$ to $[0, 1]$ which has no downward jump has a fixed point, even if the function is not everywhere nondecreasing. (Figure 12.5 again yields the intuition for this result.) Vives (1990) uses this second result of Tarski to give a simple proof of the result (originally due to McManus (1964) and Roberts and Sonnenschein (1977)) that a (symmetric) pure-strategy equilibrium exists in symmetric homogeneous-good Cournot games with convex cost functions.

that $\sup_{s_i \in S_i} u_i(s_i', s_{-i}) = \lim_{n \to +\infty} u_i(s_i^n, s_{-i})$. Compactness implies that there exists a converging subsequence $s_i^n \to s_i$. Upper semi-continuity implies that $u_i(s_i, s_{-i}) \geq \limsup_{n \to +\infty} u_i(s_i^n, s_{-i})$, so that s_i is indeed a best response to s_{-i}.)

To show that $r_i^*(s_{-i})$ is a sublattice for each s_{-i}, suppose that s_i and $\tilde{s}_i$ are both elements of $r_i^*(s_{-i})$ and that $u_i(s_i \wedge \tilde{s}_i, s_{-i}) < u_i(s_i, s_{-i}) = u_i(\tilde{s}_i, s_{-i})$. Supermodularity of $u_i(\cdot, s_{-i})$ then implies that $u_i(s_i \vee \tilde{s}_i, s_{-i}) > u_i(s_i, s_{-i}) = u_i(\tilde{s}_i, s_{-i})$, which contradicts the assumption that s_i and $\tilde{s}_i$ are best responses to s_{-i}. The same reasoning applies to the join.

Because $r_i^*(s_{-i})$ is a nonempty compact sublattice of $\mathbb{R}^{m_i}$, it has a greatest element $\bar{s}_i(s_{-i})$. We leave it to the reader to check that increasing differences of u_i implies that $\bar{s}_i(\cdot)$ is nondecreasing:

$$s_{-i} \geq \tilde{s}_{-i} \Rightarrow \bar{s}_i(s_{-i}) \geq \bar{s}_i(\tilde{s}_{-i}).$$

We can now apply Tarski's theorem to $f(\tilde{s}) = (\bar{s}_1(\tilde{s}), \ldots, \bar{s}_I(\tilde{s}))$. By construction, a fixed point $\bar{s}$ of f (which exists) is a pure-strategy Nash equilibrium. It can be shown (see also the proof of theorem 12.8 below) that $\bar{s}$ is the greatest element in the set of Nash equilibria. The intuition is again that a higher strategy triggers a higher best response. Last, by symmetry, the analysis applies to lower bounds as well. This proves part a of theorem 12.6:

Theorem 12.6

(a) (Topkis 1979) If, for each i, S_i is compact and u_i is upper semi-continuous in s_i for each s_{-i}, and if the game is supermodular, the set of pure-strategy Nash equilibria is nonempty and possesses greatest and least equilibrium points $\bar{s}$ and $\underline{s}$.

(b) (Vives 1990) If furthermore the game is *strictly* supermodular, the set of Nash equilibria is a nonempty complete sublattice. ("Complete" means that the sup and the inf of any subset belongs to the set.)

The proof of part a of theorem 12.6 relies on the monotonicity of the upper bound on the reaction correspondence for supermodular games. For supermodular games with *strictly* increasing differences, one can prove monotonicity of the entire reaction correspondence—a fact of considerable economic interest:

Theorem 12.7 (Topkis 1979)   Consider a supermodular game with strictly increasing differences. If $s_i \in r_i^*(s_{-i})$, $\tilde{s}_i \in r_i^*(\tilde{s}_{-i})$, and $s_{-i} \geq \tilde{s}_{-i}$, then $s_i \geq \tilde{s}_i$.

Proof   Theorem 12.7 results from the following chain of inequalities:

$$0 \leq u_i(s_i, s_{-i}) - u_i(s_i \vee \tilde{s}_i, s_{-i})$$

$$\leq u_i(s_i, \tilde{s}_{-i}) - u_i(s_i \vee \tilde{s}_i, \tilde{s}_{-i})$$

$$\leq u_i(s_i \wedge \tilde{s}_i, \tilde{s}_{-i}) - u_i(\tilde{s}_i, \tilde{s}_{-i}) \leq 0.$$

The first and fourth inequalities result from optimality of s_i against s_{-i} and $\tilde{s}_i$ against $\tilde{s}_{-i}$, the second from increasing differences and $s_i \leq s_i \vee \tilde{s}_i$, and the third from supermodularity in player i's strategy. Last, note that if $s_i \not\geq \tilde{s}_i$ then $s_i < s_i \vee \tilde{s}_i$, and strictly increasing differences implies that the second inequality is strict.   ∎

After this investigation of the Nash set, we study iterated strict dominance and learning processes of supermodular games. Vives (1990) notes that such games have nice stability properties. He analyzes Cournot tâtonnement (the sequence of strategies starting from some arbitrary pure-strategy profile s^0 is given by $s^n \in r^*(s^{n-1})$, as explained in section 1.2) and makes the convention that if player i's rivals choose the same strategies at steps n and $n + 1$ then player i also chooses the same strategy at steps $n + 1$ and $n + 2$ (that is, $s_{-i}^{n+1} = s_{-i}^n \Rightarrow s_i^{n+2} = s_i^{n+1}$). He shows that the tâtonnement process converges monotonically to an equilibrium point of the game when the starting point, s^0, is "below" or "above" the best reply correspondences of the players. A related result is proved by Milgrom and Roberts (1990). Consider a model of learning in which players repeatedly play the same game, play myopically (maximize current payoff at each stage), and form expectations about their rivals' play on the basis of their previous behavior in such a way that they assign small probabilities to strategies that they have not observed for a long time (see section 1.2). In a stage game with simultaneous moves, models in which players learn their opponents' strategies predict that the players will not use strictly dominated strategies, that their opponents will learn this, and so on. Thus, strategic learning therefore rules out all strategies ruled out by iterated strict dominance.

Milgrom and Roberts also use the monotonic sequences of strategies studied in learning processes to analyze rationalizable strategies. They show that the greatest and the least Nash equilibria, $\bar{s}$ and $\underline{s}$ (whose existence was ascertained in theorem 12.6a), are also the greatest and the least elements in the set of strategies that survive iterated deletion of strictly dominated strategies:

Theorem 12.8 (Milgrom and Roberts 1990)   Consider a supermodular game such that, for each i, S_i is a complete sublattice and is bounded, and such that u_i is continuous and is bounded above. Then the iterated deletion of strictly dominated strategies yields a set of strategies in which the greatest and the least elements are the Nash equilibrium $\bar{s}$ and $\underline{s}$.

Proof   The proof of theorem 12.8 is both simple and instructive. Start from the upper bounds $s^0 = (s_1^0, \ldots, s_I^0)$ on the strategy sets. Let s_i and s_i' denote two elements of $r_i^*(s_{-i}^0)$ such that there exists no s_i'' in $r_i^*(s_{-i}^0)$ such that either $s_i'' > s_i$ or $s_i'' > s_i'$. Suppose that $s_i \neq s_i'$ (that is, s_i exceeds s_i' along some dimension and s_i' exceeds s_i along another). We claim that $s_i \wedge s_i'$ is a strictly better response to s^0_{-i} than s_i, a contradiction: Supermodularity implies that

$$u_i(s_i, s^0_{-i}) - u_i(s_i \wedge s'_i, s^0_{-i}) \leq u_i(s_i \vee s'_i, s^0_{-i}) - u_i(s'_i, s^0_{-i}) < 0,$$

where the strict inequality results from the fact that $s_i \vee s'_i > s'_i$ and therefore cannot be a best response to s^0_{-i}. We thus conclude that $r^*_i(s^0_{-i})$ has a greatest element s^1_i. Let $s^1 \equiv (s^1_1, \ldots, s^1_I) \leq s^0$. One then defines s^n by induction; any s_i that does not satisfy $s_i \leq s^n_i$ is strictly dominated by $s_i \wedge s^n_i < s_i$: Because all strategies s_{-i} remaining after $n-1$ rounds of deletion satisfy $s_{-i} < s^{n-1}_{-i}$ from the induction hypothesis,

$$u_i(s_i, s_{-i}) - u_i(s_i \wedge s^n_i, s_{-i}) \leq u_i(s_i, s^{n-1}_{-i}) - u_i(s_i \wedge s^n_i, s^{n-1}_{-i})$$

$$\leq u_i(s_i \vee s^n_i, s^{n-1}_{-i}) - u_i(s^n_i, s^{n-1}_{-i})$$

$$< 0,$$

where the first inequality results from increasing differences, the second from supermodularity, and the third from the facts that s^n_i is the greatest best response to s^{n-1}_{-i} and that $s_i \vee s^n_i > s_i$.

Because the sequence s^n is bounded below and decreasing, it converges to some $\bar{s}$. To show that $\bar{s}$ is a Nash equilibrium, fix an arbitrary s_i; by optimality of s^{n+1}_i against s^n_{-i},

$$u_i(s^{n+1}_i, s^n_{-i}) \geq u_i(s_i, s^n_{-i});$$

by continuity, $u_i(s^{n+1}_i, s^n_{-i})$ converges to $u_i(\bar{s}_i, \bar{s}_{-i})$ and $u_i(s_i, s^n_{-i})$ converges to $u_i(s_i, \bar{s}_{-i})$. Because weak inequalities are preserved in the limit, $\bar{s}_i$ is a better response to $\bar{s}_{-i}$ than s_i, for each s_i.

By symmetry, the least Nash equilibrium $\underline{s}$ is also the lower bound on the set of strategies that survive iterated deletion of strictly dominated strategies.   ∎

Note that theorem 12.8 implies that, if there exists a unique Nash equilibrium, the game is solvable by iterated strict dominance.

Supermodular games have several other convenient features. First, comparative-statics exercises are straightforward. Suppose that payoffs are indexed by a parameter α, $u_i(s_i, s_{-i}, \alpha)$, and that u_i has increasing differences in $((s_i, \alpha), s_{-i})$. Then the greatest and least Nash strategies, $\bar{s}_i(\alpha)$ and $\underline{s}_i(\alpha)$, are nondecreasing functions of α (Milgrom and Roberts 1990, theorem 6). (This result is particularly useful when there is a unique Nash equilibrium. Special versions of it have been used in the papers mentioned in footnote 12.)

Second, one can compare payoffs in two Nash equilibria s and $\tilde{s}$ that satisfy $s \geq \tilde{s}$ (Milgrom and Roberts 1990, theorem 7). If $u_i(s_i, s_{-i})$ is increasing in s_{-i}, then $u_i(s) \geq u_i(\tilde{s})$ (for instance, Bertrand oligopolists prefer an equilibrium with high prices for all firms). If $u_i(s_i, s_{-i})$ is decreasing in s_{-i}, then $u_i(s) \le u_i(\tilde{s})$ (for instance, in Cournot duopoly, a firm prefers the equilibrium in which it produces the highest output—i.e., in which its rival produces the lowest output).

Exercises

Exercise 12.1* Are the games in figure 1.10a (matching pennies), and figure 1.18 for $\lambda = 0$, essential?

Exercise 12.2*** The Sion-Wolfe (1957) example of nonexistence of a mixed-strategy equilibrium is a two-person, zero-sum game with strategy sets $S_1 = S_2 = [0, 1]$ and with the following payoffs:

$$u_1(s_1, s_2) = \begin{cases} -1 \text{ if } s_1 < s_2 < s_1 + \frac{1}{2} \\ 0 \text{ if } s_1 = s_2 \text{ or } s_2 = s_1 + \frac{1}{2} \\ 1 \text{ otherwise.} \end{cases}$$

(a) Show that

$$\sup_{\sigma_1} \inf_{\sigma_2} u_1(\sigma_1, \sigma_2) = \frac{1}{3} < \inf_{\sigma_2} \sup_{\sigma_1} u_1(\sigma_1, \sigma_2) = \frac{3}{7},$$

where the sup inf is obtained by player 1 putting weight on 0, $\frac{1}{2}$, and 1 and the inf sup is obtained by player 2 putting weight on $\frac{1}{4}$, $\frac{1}{2}$, and 1).

(b) Conclude that there exists no Nash equilibrium.

(c) What assumption in theorem 12.4 is violated here?

Exercise 12.3* Games with strategic complementarities can often be studied simply by using standard techniques, although the theory of super-modularity offers a more elegant and general method. Bulow et al. (1985) and Fudenberg and Tirole (1984) offer industrial-organization examples. This exercise develops a macroeconomic example. As Cooper and John (1988) argue, many models with "aggregate demand externalities," "spill-overs," or "Keynesian effects" have a common structure. Consider a symmetric I-player game in which player i's payoff is $u(s_i, s_{-i})$, where $s_i \ge 0$. Assume that $u(\cdot, s_{-i})$ is strictly concave in s_i. When player i's opponents choose the same action s^*, player i's payoff is written

$$U(s_i, s^*) \equiv u(s_i, (s^*, \dots, s^*)).$$

Since all players have the same payoff function, they have the same reaction function. Let $r^*(s^*)$ be the optimal response (for any player) to the profile in which all of his opponents play s^*. Assume that $\partial U/\partial s^* > 0$ (that is, that the game exhibits "positive spillovers," in the terminology of Cooper and John), and that $\partial^2 U/\partial s^{*2} < 0$. We will focus on *symmetric* Nash equilibria.

(a) Show on a diagram why there may exist multiple symmetric equilibria.

(b) Show that any symmetric Nash equilibrium involves action $s^* < \hat{s}$, where $\hat{s}$ is the optimal symmetric level of activity. Show that the symmetric Nash equilibria are Pareto ranked.

(c) Let $r^{*\prime} \equiv -(\partial^2 U/\partial s_i \partial s^*)/(\partial^2 U/\partial s_i^2)$ denote the slope of the reaction function. A "stable" equilibrium (see chapter 1) is such that $r^{*\prime} < 1$. Index the utility functions by a parameter γ_i, $U(s_i, s^*, \gamma_i)$, such that $\partial^2 U/\partial s_i \partial \gamma_i > 0$. Show that a symmetric stable equilibrium (corresponding to $\gamma_i = \gamma$ for all i) exhibits "multiplier effects":

$$\frac{d(\sum s_j^*)}{d\gamma_i} > \frac{\partial r_i^*}{\partial \gamma_i}.$$

(d) Think of reasons why spillovers may be relevant. (Hint: Consider monopolistic competition, search, learning spillovers, and so on.)

References

Birkhoff, G. 1967. *Lattice Theory*, third edition. Colloquium Publications.

Bulow, J., J. Geanakoplos, and P. Klemperer. 1985. Multimarket oligopoly: Strategic substitutes and complements. *Journal of Political Economy* 93: 488–511.

Cooper, R., and A. John. 1988. Coordinating coordination failures in Keynesian models. *Quarterly Journal of Economics* 102: 441–464.

Dasgupta, P., and E. Maskin. 1986a. The existence of equilibrium in discontinuous economic games. 1: Theory. *Review of Economic Studies* 53: 1–26.

Dasgupta, P., and E. Maskin. 1986b. The existence of equilibrium in discontinuous economic games. 2: Applications. *Review of Economic Studies* 53: 27–42.

d'Aspremont, C., J. Gabszewicz, and J. Thisse. 1979. On Hotelling's stability in competition. *Econometrica* 47: 1145–1150.

Debreu, G. 1970. Economies with a finite set of equilibria. *Econometrica* 38: 387–392.

de Palma, A., V. Ginsburgh, Y. Panageorgiou, and J. F. Thisse. 1985. The principle of minimum differentiation holds under sufficient heterogeneity. *Econometrica* 53: 767–782.

Diamond, P. 1982. Aggregate demand management in search equilibrium. *Journal of Political Economy* 90: 881–894.

Eaves, C. 1971. The linear complementarity problem. *Management Science* 17: 612–634.

Eaves, C. 1973. Polymatrix games with joint constraints. *SIAM Journal of Applied Mathematics* 24: 418–423.

Eaves, C. 1976. A short course in solving equations with PL homotopies. *SIAM-AMS Proceedings* 9: 73–143.

Fort, M. 1950. Essential and non-essential fixed points. *American Journal of Mathematics* 72: 315–322.

Fudenberg, D., and J. Tirole. 1984. The fat-cat effect, the puppy-dog ploy and the lean and hungry look. *American Economic Review: Papers and Proceedings* 74: 361–368.

Harsanyi, J. 1973. Oddness of the number of equilibrium points: A new proof. *International Journal of Game Theory* 2: 235–250.

Hendricks, K., and D. Kovenock. 1989. Asymmetric information, information externalities, and efficiency: The case of oil exploration. *Rand Journal of Economics* 20: 164–182.

Mas-Colell, A. 1985. *The Theory of General Economic Equilibrium: A Differentiable Approach.* Cambridge University Press.

McManus, M. 1964. Equilibrium, numbers and size in cournot oligopoly. *Yorkshire Bull Economic Society Res.* 68–75.

Milgrom, P., and J. Roberts. 1990. Rationalizability, learning and equilibrium in games with strategic complementarities. *Economica* 58: 1255–1278.

Roberts, J., and H. Sonnenschein. 1977. On the foundations of the theory of monopolistic competition. *Econometrica* 45: 101–114.

Simon, L. 1987. Games with discontinuous payoffs. *Review of Economic Studies* 54: 569–597.

Sion, M., and P. Wolfe. 1957. On a game without a value. *Annals of Mathematical Studies* 39: 299–306.

Tarski, A. 1955. A lattice-theoretical fixpoint theorem and its applications. *Pacific Journal of Mathematics* 5: 285–309.

Topkis, D. 1979. Equilibrium points in nonzero-sum n-person submodular games. *SIAM Journal of Control and Optimization* 17: 773–787.

Vives, X. 1990. Nash equilibrium with strategic complementarities. *Journal of Mathematical Economics* 19: 305–321.

Wilson, R. 1971. Computing equilibria of n-person games. *SIAM Journal of Applied Mathematics* 21: 80–87.

Wu, Wen-Tsün, and Jiang Jia-He. 1962. Essential equilibrium points of n-person non-cooperative games. *Scientia Sinica* 11: 1307–1322.

Chapter 5 addressed repeated games, in which the "physical environment" is the same in every period. Now we study environments in which the past has a direct influence on current opportunities, say by determining the level of installed capacity or the quantity of discovered but unexploited natural resources. Such environments can be modeled as discrete-time games (sections 13.1 and 13.2) or with their continuous-time analogue of differential games (section 13.3).

In studying repeated games, we considered strategies in which past play influences current and future strategies, not because it had a direct effect on the environment, but rather because all players believe that the past play matters. When studying more complex environments, economists often focus attention on equilibria in a smaller class of "Markov" or "state-space" strategies in which the past influences current play only through its effect on a state variable that summarizes the direct effect of the past on the current environment. A *Markov perfect equilibrium* (MPE) is a profile of Markov strategies that yields a Nash equilibrium in every proper subgame. Since the state captures the influence of past play on the strategies and payoff functions for each subgame, if a player's opponents use Markov strategies, that player has a best response that is Markov as well. Thus, a Markov perfect equilibrium is still a perfect equilibrium when the Markov restriction is not imposed. However, there can be many other equilibria. The simplest example is that of infinitely repeated games, where the state variable is null, so the only Markov equilibria correspond to the infinite repetition of one of the equilibria of the stage game. For instance, the only Markov equilibrium of the infinitely repeated prisoner's dilemma (figure 4.1) has both players defect in every period.[1] Section 13.1 presents less trivial examples in which the Markov restriction has bite, and studies MPE in specific classes of games.

---

1. We do not assert that this is the most likely outcome when players are patient. The inadequacy of the Markov equilibrium here can be interpreted either as a critique of the Markov assumption or as a sign that the complete-information model omits some important features. Chapter 9, on reputation effects, shows how small amounts of certain kinds of incomplete information lead to Markov equilibria with more intuitive outcomes.

Chapter 9 perturbs the information structure of the game. Another way of avoiding the strong implication of Markov perfection in repeated games is to relax the requirement that players always move simultaneously. Maskin and Tirole (1988b), using the Markov restriction, obtain collusion in a repeated price game in which prices are locked in for two periods. They argue that what is meant by "reaction" is often an attempt by firms to react to a state that affects their *current* profits; for instance, when facing a low price by their opponents, they may want to regain market share. In the classic repeated-game model, firms move simultaneously, and there is no physical state to react to. If, however, one allows firms to alternate moves, they can react to their opponent's price. (Maskin and Tirole derive asynchronicity as the (equilibrium) result of the two-period commitments.) Gertner (1986) formalizes collusion with Markov strategies when commitment (inertia) takes the form of a fixed cost of changing prices. Halperin (1990) obtains an elegant characterization of the set of MPE when firms face such "menu costs" with or without inflation. He shows the existence of staggered and synchronized price cycles, some of which follow an (S, s) rule.

The MPE restriction pushes the notion that "bygones are bygones" further than perfect equilibrium does. It also runs counter to the equilibrium restrictions based on forward induction that we developed in chapter 11: The idea of forward induction is that past actions will be interpreted as signals of future intentions even though those actions may not influence payoffs in the continuation game.

Although some dynamic games are not presented with an explicit state variable, the notion of state is implicit in the way that the past influences the present. Subsection 13.2.1 shows how to construct an explicit state variable and thus extend the definition of MPE to general games of complete information.

Subsection 13.2.3 considers the robustness of MPE (in the sense of lower hemi-continuity) to small changes in the payoff functions. That is, start from an "original game" where the state space is small and so the MPE has bite; for example, one might think that the payoff-relevant state in an investment game is the current capacities of the players. Now consider a perturbation of the payoff functions that makes more aspects of the past relevant to the payoff; for example, learning by doing might imply that the exact time sequence of investments has a small effect as well. Lower hemi-continuity means that, for each MPE of the original game, there is an MPE of the perturbed game that is close to it, in that strategies depend "mostly" on the state variables of the original game. Lower hemi-continuity obtains for generic games. This is reassuring, as it is difficult to be certain that some past variables have absolutely no effect on the environment.

Despite this lower hemi-continuity, the set of Markov perfect equilibria can change discontinuously when the payoffs are perturbed. This is because MPE allows strategies to depend "a lot" on any state variable, even one whose influence on payoffs is small, but requires that strategies not depend at all on variables that have exactly zero influence. For example, if we modify the prisoner's dilemma by adding a "state variable" that keeps track of the number of times a player has defected, and allow this variable to have an arbitrarily small influence on payoffs, then "always cooperate" becomes the outcome of an MPE. This shows that the set of MPE is not upper hemi-continuous. (In the spirit of Markov perfection, it is then natural to select MPE in which variables that have a small effect on payoffs also have a small effect on strategies. The lower-hemi-continuity result guarantees that this can be done generically.)

Section 13.3 analyzes the class of differential games, which are the continuous-time analogues of stochastic games in which the state evolves according to a (deterministic) differential equation.[2] Since the optimal

2. There is also a literature on stochastic differential games, in which the state follows a stochastic differential equation.

solution to a one-player differential game (i.e., a control problem) can be chosen to be Markov, the MPE of differential games may correspond to the multi-player versions of control optima.[3] Rightly or wrongly, the analogy with control optima has led control theorists to study a class of MPE in differential games. Case (1969) and Starr and Ho (1969) have analyzed smooth perfect equilibria in Markov strategies in differential games. Section 13.4 treats the capital-accumulation game as an example of a differential game.

Because we develop few applications of the Markov concept in this chapter, we refer to other contributions for further examples.[4]

13.1 Markov Equilibria in Specific Classes of Games[†††]

13.1.1 Stochastic Games: Definition and Existence of MPE

Our first application of the Markov concept is to stochastic games.[5] The idea behind a stochastic game is that the history at each period can be summarized by a "state" (e.g., capital levels, goodwills). Current payoffs depend on this state and on current actions (e.g., investments, prices, advertising levels). The state follows a Markov process; that is, the probability distribution on tomorrow's state is determined by today's state and actions.

A stochastic game is defined by state variables $k \in K$, action spaces $A_i(k)$ with mixed actions $\mathscr{A}_i(k)$, a transition function $q(k^{t+1}|k^t, a^t)$ which gives the probability that the next period's state is k^{t+1} conditional on its being k^t at date t and on the playing of action a^t, and payoff functions $u_i = \sum_{t=0}^{\infty} \delta^t g_i(k^t, a^t)$. (Note that we are abusing notation when we define action spaces. Formally, the sets A_i and $\mathscr{A}_i$ of pure and mixed stage-game strategies, respectively, are functions of the whole history h^t. However, if, as we assume, they are measurable with respect to the state, it is notationally simpler to make them functions of the state only.) The game starts in some

3. In a control problem, there is a Markov optimum whenever an optimum exists. However, in a game with a continuum of actions, the conditions required for existence of a *Markov* perfect equilibrium are stronger than those for the existence of a perfect equilibrium. Harris (1990) discusses this. See also subsection 13.2.2.
4. See the literatures on resource extraction (Amir 1989; Amit and Halperin 1989; Dutta and Sundaram 1988; Lancaster 1973; Levhari and Mirman 1980; Loury 1990; Sundaram 1989), on bequest equilibria (Bernheim and Ray 1989; Harris 1985; Kohlberg 1976; Leininger 1986), on research and development (Harris and Vickers 1987), and on dynamic monopoly or oligopoly (Bénabou 1989; Dana and Montrucchio 1987; Eaton and Engers 1990; Gertner 1986; Harris 1988; Judd 1990; Kirman and Sobel 1974; Maskin and Tirole 1987, 1988a,b; Villas-Boas 1990). The Markov concept is also often used in games of incomplete information.
5. For notational simplicity, the definition of stochastic games in this section is more restrictive than that in much of the literature. For more on stochastic games, see Friedman 1986, Shapley 1953, and Sobel 1971.

state k^0 at date 0. The players know the entire history of play, $h^t = (k^0, a^0, k^1, a^1, \ldots, k^{t-1}, a^{t-1}, k^t)$, when they choose their period-t actions. An oligopoly example of a stochastic game is found in the paper by Kirman and Sobel (1974), where k^t is a vector of individual states k_i^t which represent firm i's stock of goodwill, and a_i^t is firm i's choices of price and advertising level at date t. A perfect equilibrium of this game allows the strategies $\sigma_i(h^t)$ to be functions of the entire history h^t. MPE requires that, for each player i and time t, $\sigma_i(h^t) = \sigma_i(\hat{h}^t)$ if the two histories have the same value of the state variable k^t. Another way of putting this is that the set M_i of Markov strategies for player i can be identified with the set of all maps σ_i: with $\sigma_i(k) \in \mathscr{A}_i(k)$. (Again, we abuse notation by defining strategies as functions of the state.)

Theorem 13.1 Markov perfect equilibria exist in stochastic games with a finite number of states and actions.

Proof
In the spirit of the agent strategic form of section 8.4, construct a *Markov strategic form* in which each agent (i, k) chooses a mixed action in $\mathscr{A}_i(k)$, and each agent (i, k) has the payoff function of player i at state K. Since there are finitely many states, the Markov strategic form has a finite number of players, each of whom has a finite number of pure actions. Thus theorem 1.1 implies the game has a Nash equilibrium. Moreover, σ^* is a Markov profile of the original game, and for each player i, σ_i^* is a best response in Markov strategies to σ_{-i}^*. Hence, as we remarked above, σ^* is a Nash equilibrium. Finally, this Markov equilibrium is perfect because player i optimizes in each state by construction.

This existence theorem has been extended to countable state spaces (see e.g., Parthasarathy 1982 and Rieder 1979). Existence theorems for uncountable (e.g., continuous) state spaces are much harder to obtain. Whitt (1980) proves the existence of an ε-equilibrium; Duffie et al. (1988) extend the basic game by adding a sequence of independently and identically distributed non-payoff-relevant public randomizations, and prove the existence of an extended concept of MPE in which the strategies in any given period depend (at least) on the current state, on the continuation payoff that players in the preceding stage anticipated would obtain if that state were realized, and on the current public signal. Mertens and Parthasarathy (1987) prove the existence of an extended concept of MPE in which strategies in any given period depend on the current state and on the anticipated continuation payoff (but not on any public signal). Harris (1990) proves the existence of an extended concept of MPE in which strategies in any given period depend on the current state and on the current public signal (but not on any continuation payoff). Thus, Harris' extended concept coincides with MPE, except for the dependence on the payoff-irrelevant public randomization at the beginning of the period.

13.1.2 Separable Sequential Games

A class of games for which one can obtain a general characterization of Markov-equilibrium strategies is the class of games of perfect information with separable payoffs.

Definition 13.1 A *separable sequential game* is defined by the following:
(i) a countable set of players, $t = 0, 1, \ldots$,
(ii) a state variable $k^t \in K \subseteq \mathbb{R}$ with evolution equation $k^{t+1} = f_{t+1}(a^t)$,
(iii) a sequence of action spaces $A^t(k^t) \subseteq \mathbb{R}$,
(iv) an objective function for each player of the form

$$u_t = g_t(k^t, a^t) + w_t(k^{t+1}, a^{t+1}, a^{t+2}, \ldots),$$

(v) perfect information (player t knows $h^t = (a^0, \ldots, a^{t-1})$ before choosing action a^t).

This class is more general than it might appear. First, the evolution equation could depend on both a^t and k^t; it suffices to relabel a^t to identify it with k^{t+1}.[6] Second, finite-horizon games belong to this class. Third, a given player may play in different periods; it suffices to distinguish his various "incarnations." Thus, player i playing at date t and player i playing at date t' can be formalized as two distinct players whose objective functions are derived from the same preferences. For instance, the class includes alternating-move duopoly games in which each firm's action is committed for two periods and players take turns.[7] If firm 1 plays in odd periods and has current payoff function $g_1(a_1, a_2)$, its objective function is

$$u_1 = g_1(a^{-1}, a^0) + \delta g_1(a^1, a^0) + \cdots + \delta^{2k+1} g_1(a^{2k+1}, a^{2k})$$
$$+ \delta^{2k+2} g_1(a^{2k+1}, a^{2k+2}) + \cdots.$$

We leave it to the reader to transform the alternating-move duopoly game into a separable sequential game.

A (pure) Markov strategy in a separable sequential game is a strategy $s: K \times T \to A$ (where T and A are the time and action spaces). If the functions g_t and w_t and the action spaces A^t are time independent, a (pure) Markov strategy is a time-invariant map, $s: K \to A$. The strategy can be interpreted as a "reaction function." (In contrast with the reaction

---

6. For instance, suppose $k^{t+1} = f(k^t, a^t)$ is the date-$(t + 1)$ capital given date-t capital and savings, $g_t(k^t, a^t) = g(k^t - a^t)$, and $\gamma(k^t, k^{t+1})$ is the amount of savings needed to obtain capital k^{t+1} from capital k^t. Then, redefine the action as the choice of tomorrow's capital, $a^t = k^{t+1}$, the transition equation as the identity, and the current-payoffs function as $g(k^t - \gamma(k^t, a^t))$.
7. See Cyert and DeGroot 1970; Dana and Montrucchio 1987; Eaton and Engers 1990; Gertner 1986; Maskin and Tirole 1987, 1988a,b.

functions defined in chapter 1, these are "real" reaction functions, i.e., parts of equilibrium strategies.)

A nice feature of separable sequential games is that the reaction functions are monotonic under the standard sorting condition.[8]

Definition 13.2 The function g_t satisfies the *sorting condition* if it is twice differentiable and either

$$(\text{CS}^+) \; \frac{\partial^2 g_t}{\partial k^t \partial a^t} \geq 0$$

or

$$(\text{CS}^-) \; \frac{\partial^2 g_t}{\partial k^t \partial a^t} \leq 0.$$

The sorting condition CS^+ means that a higher state variable makes a higher action more desirable. And conversely for CS^-. We now borrow from the mechanism-design literature (see chapter 7) the simple proof of the following proposition:

Theorem 13.2 Consider a separable sequential game satisfying the sorting condition. Suppose that the action spaces are state independent. Then the equilibrium Markov strategies $s^t(k^t)$ are nondecreasing (under CS^+) or nonincreasing (under CS^-).

Proof Fix Markov strategies for players $(t + 1), (t + 2), \ldots,$ and let

$$v_t(k^{t+1}) \equiv w_t(k^{t+1}, s^{t+1}(k^{t+1}), s^{t+2}(f_{t+2}(s^{t+1}(k^{t+1}))), \ldots)$$

denote the continuation valuation of player t for state variable k^{t+1}. Now consider two possible states, k and $\tilde{k}$, at date t, and let $a = s^t(k)$ and $\tilde{a} = s^t(\tilde{k})$. By definition of equilibrium, player t prefers action a to action $\tilde{a}$ when the state is k:

$$g_t(k, a) + v_t(f_{t+1}(a)) \geq g_t(k, \tilde{a}) + v_t(f_{t+1}(\tilde{a})).$$

Similarly, in state $\tilde{k}$, player t prefers action $\tilde{a}$ to action a:

$$g_t(\tilde{k}, \tilde{a}) + v_t(f_{t+1}(\tilde{a})) \geq g_t(\tilde{k}, a) + v_t(f_{t+1}(a)).$$

These inequalities are called the *incentive-compatibility constraints*. Adding them up eliminates the continuation valuations:

$$g_t(k, a) + g_t(\tilde{k}, \tilde{a}) - g_t(k, \tilde{a}) - g_t(\tilde{k}, a) \geq 0,$$

which can be rewritten as

---

8. This condition is also called "single crossing condition" or "Spence-Mirrlees condition" or "constant sign partial derivatives" (Guesnerie and Laffont 1984).

$$\int_a^{\tilde{a}} \int_k^{\tilde{k}} \frac{\partial^2 g_t}{\partial x \partial y} dx dy \geq 0.$$

Hence, if $\tilde{k} > k$, then $\tilde{a} \geq a$ if CS^+ holds and $\tilde{a} \leq a$ if CS^- holds. ∎

Theorem 13.2 extends trivially to the case in which the players play mixed (Markov) strategies. The result is then that the supports of the mixed strategies are ordered; for instance, under CS^+, if $k^t > \tilde{k}^t$, then

$$\min\{a \mid \sigma^t(a \mid k^t) > 0\} \geq \max\{a \mid \sigma^t(a \mid \tilde{k}^t) > 0\}.$$

Also, it is easy to see where the proof of monotonicity breaks down if preferences are nonseparable or if players $(t + 1), (t + 2), \ldots$ use non-Markov strategies. For instance, in the latter case, the continuation valuation v_t may depend on a^t and k^t (in a nonseparable way), so that the addition of the two incentive-compatibility constraints does not eliminate the continuation valuations.

Maskin and Tirole (1987, 1988a) make heavy use of the monotonicity property in alternating-move Cournot duopoly games. If the cross-partial derivatives of the firms' per-period payoffs with respect to the two outputs are negative (the two outputs are strategic substitutes in the terminology of subsection 12.3), then the Markov strategies or reaction curves are downward sloping, like the (fictitious) reaction curves of the static Cournot game (see chapter 1).

13.1.3 Examples from Economics

We now give two economic applications of Markov equilibrium. These applications differ slightly from the framework of subsection 13.1.1 in that strategy and state spaces are continuous rather than finite and in that the number of players is infinite in the first application. Furthermore, the transition function for the state is deterministic in both applications. These applications are developed in some detail, and readers familiar with examples of stochastic games may want to skip them.

Example 1: Bequest Games
Intergenerational family transfers give rise to games among successive generations. Suppose that each generation cares about the consumption of the next generation, which cares about the consumption of the following generation, and so on. This succession of generations does not behave like a single decision maker: A generation wants to leave a bequest to the next generation, but the two generations have different preferences about what to do with the bequest. A simple class of bequest games that has been much studied in the literature is the following:

There is a single good, used both for consumption and as productive capital. Generation t ($t = 0, 1, \ldots$) lives for a single period (period t) and inherits an amount of good $k^t \geq 0$ from generation $t - 1$. Generation t's utility depends on its own consumption, c^t, and on the next generation's consumption, c^{t+1}:

$$u_t = u(c^t, c^{t+1}).$$

Generation t saves $a^t \in [0, k^t]$ and consumes $c^t = k^t - a^t$. The next generation's inheritance or capital is $k^{t+1} = f(a^t)$, where f is an increasing function with $f(0) = 0$. A pure-strategy MPE of the bequest game is a strategy $s(k)$ such that

$$s(k) \in \arg\max_{x \in [0,k]} u(k - x, f(x) - s(f(x))).$$

We first derive an MPE for a parametric example. We then investigate general properties of MPE and study existence.

A Parametric Example Suppose that

$$u(c^t, c^{t+1}) = \ln c^t + \delta \ln c^{t+1}$$

and

$$f(a) = a^\alpha,$$

where δ and α belong to $(0, 1)$. Let us look for an MPE in which the stationary savings strategy $s(\cdot)$ is differentiable. The first-order condition for the program

$$\max_{x \in [0,k]} \{\ln(k - x) + \delta \ln[x^\alpha - s(x^\alpha)]\}$$

is

$$\frac{1}{k - x} = \frac{\delta \alpha x^{\alpha-1}[1 - s'(x^\alpha)]}{x^\alpha - s(x^\alpha)}.$$

This suggests looking for a linear strategy, $s(k) = sk$, where $s \in (0, 1)$, so

$$(1 - s)x = \delta\alpha(1 - s)(k - x)$$

or

$$x = \frac{\delta\alpha}{1 + \delta\alpha}k, \text{ so that } s \equiv \frac{\delta\alpha}{1 + \delta\alpha}.$$

There thus exists an MPE with a constant savings ratio; this savings ratio grows with the discount factor and with the productivity of savings.

Characterization of Equilibrium Markov Strategies When the utility function is separable, it is possible to characterize the slope of any equilibrium strategy $s(\cdot)$ as in subsection 13.1.2.

Theorem 13.2 as stated requires action spaces to be state independent, but its result extends trivially to some situations where action spaces are state dependent (such as separable bequest games). In these games,

$$u_t = u(c^t) + z(c^{t+1});$$

using the general notation,

$$g_t(k^t, a^t) = u(k^t - a^t)$$

and

$$w_t(k^{t+1}, a^{t+1}, a^{t+2}, \ldots) = z(k^{t+1} - a^{t+1}).$$

The action space $A^t(k^t) = [0, k^t]$ is state dependent. Thus, if $k > \tilde{k}$, $s(k) \in [0, k]$ may not be feasible for state $\tilde{k}$ (although $s(\tilde{k}) \in [0, \tilde{k}]$ is feasible for state k). But if $s(k) > \tilde{k}$, then $s(k) > s(\tilde{k})$ and hence monotonicity holds anyway. If $s(k) \leq \tilde{k}$, the proof of theorem 13.2 shows that $s(k) \geq s(\tilde{k})$ if u is concave (so that $\partial^2 g_t / \partial k^t \partial a^t \geq 0$). We conclude that equilibrium strategies are nondecreasing in the bequest game.

Existence Proving the existence of a pure-strategy MPE in the bequest game is much harder than showing monotonicity. Bernheim and Ray (1989) and Leininger (1986) have obtained existence results.

It is instructive to understand why the "natural method" of proving existence of an MPE does not work. Suppose that a generation inherits k and chooses $x = s(k)$ so as to maximize $u(k - s, f(x) - \bar{s}(f(x)))$, where $\bar{s}(\cdot)$ is the strategy of the following generation. If $\bar{s}(\cdot)$ is continuous, a maximum exists. One can thus map the continuous function $\bar{s}(\cdot)$ into a function $s(\cdot)$. A fixed point of this mapping is an MPE. However, $s(\cdot)$ need not be continuous, so one cannot use a fixed-point theorem on the space of continuous functions on some bounded interval $[0, \bar{k}]$.

Leininger shows that if f is continuous and increasing and u belongs to some class of utility functions (containing the separable function $u(c^t, c^{t+1}) = v(c^t) + \delta v(c^{t+1})$, where v is strictly concave and increasing),[9] there exists a monotonic, pure-strategy equilibrium in the finite-horizon version of the bequest game ($s(\cdot)$ nondecreasing). Similarly, Bernheim and Ray and Leininger show the existence of a monotonic MPE in the infinite-horizon version.

9. More precisely, consider the optimal-choice correspondence $\Phi(k|\bar{s})$, that is, the set of maximizers of $u(k - x, f(x) - \bar{s}(f(x)))$. (This correspondence is nonempty, compact valued, and upper hemi-continuous if u is continuous.) It is required that u be continuous and increasing in both variables, and that any selection $s^*(k)$ in $\Phi(k|\bar{s})$ be such that s^* is nondecreasing. Bernheim and Ray show that continuous and increasing utility functions that satisfy, for all $c^t \geq \tilde{c}^t \geq 0$ and $c^{t+1} \geq \tilde{c}^{t+1} \geq 0$,

$$u(c^t, c^{t+1}) + u(\tilde{c}^t, \tilde{c}^{t+1}) - u(\tilde{c}^t, c^{t+1}) - u(c^t, \tilde{c}^{t+1}) \geq 0,$$

a weaker form of the sorting condition, belong to the class of utility functions defined by Leininger.

Example 2: Extraction of a Common Resource

Several economists have looked at the extraction of a renewable resource by competing players. The classical fishing game (Lancaster 1973; Levhari and Mirman 1980) has two commercial fisheries at each period, fishing simultaneously in the same pool. Because the number of fish in the pool in the next fishing season depends on how many are left by the fisheries at the end of the current season, the fisheries exert a negative externality on each other that usually prevents a socially efficient fishing policy.

Consider the following model: Let $k^t \geq 0$ denote the current stock of the common resource. At date t, players 1 and 2 simultaneously choose how much to extract ($a_1^t \geq 0$ and $a_2^t \geq 0$). If $k^t \geq a_1^t + a_2^t$, player i gets instantaneous payoff $g_i(a_i^t)$, and the stock at the beginning of date $t + 1$ is $k^{t+1} = f(k^t - a_1^t - a_2^t)$, where f is the transition or reproduction function. (Note that f is assumed to be deterministic, and that it depends on the state variable k^t and the actions a^t in a specific way.) If $k^t < a_1^t + a_2^t$, some rule allocates the limited stock to the players; let us assume for instance that each player gets $k^t/2$, which yields payoff $g_i(k^t/2)$ and that $k^{t+1} = f(0) = 0$. We take the state space and the action spaces to be the intervals $[0, \bar{k}]$ and $[0, \bar{a}]$, respectively, where $\bar{k}$ and $\bar{a}$ are "sufficiently large" (see Dutta and Sundaram 1988 for details). We thus work with continuous spaces, rather than with discrete spaces as in the above existence proof. Assume further that $g_i(\cdot)$ is continuously differentiable and strictly concave with $\lim_{x \to 0} g_i'(x) = +\infty$ (this assumption prevents a corner solution at zero extraction when $k^t > 0$), and that f is continuously differentiable, strictly increasing, and strictly concave with $f'(0) > 1/\delta$ and $f'(+\infty) < 1$.

For a profile $(s_1(\cdot), s_2(\cdot))$ of pure Markov strategies, let $\psi(k) = k - s_1(k) - s_2(k)$ denote the remaining stock at the end of the period. Note that, without loss of generality, one can restrict s_i to belong to $[0, k]$.

A natural research procedure is to look for continuously differentiable strategies $s_i(\cdot)$ (assuming that a differentiable MPE exists). An MPE must then satisfy the Bellman equation : For all k,

$$g_i'(s_i(k)) = \delta g_i'(s_i(f(\psi(k))))f'(\psi(k))[1 - s_j'(f(\psi(k)))]. \tag{13.1}$$

To obtain equation 13.1, suppose that player i extracts one more unit of the common resource at date t when the stock is k. He increases his current payoff by $g_i'(s_i(k))$. Assume further that he also modifies his date-$(t + 1)$ strategy so as to get back to the original, equilibrium stock k^{t+2} at date $t + 2$. The reduction of the stock at date $t + 1$ resulting from a unit increase in the date-t extraction is equal to $f'(\psi(k))$. Because player j's extraction at $t + 1$ depends on k^{t+1}, the total reduction in stock at the end of $t + 1$ is

$$f'(\psi(k))[1 - s_j'(f(\psi(k)))].$$

Player i must reduce his date-$(t + 1)$ extraction by this amount. Equation 13.1 simply asserts that the increase in a_i^t and the decrease in a_i^{t+1} (or the converse) does not affect player i's intertemporal welfare. (Of course, the above reasoning is valid only if one can reduce extraction slightly in each period whenever $k \neq 0$. The assumption on $g_i'(0)$ is meant to ensure that this is feasible.)

Dutta and Sundaram (1988) note that equation 13.1 has strong implications. First, tomorrow's stock is a strictly monotonic function of today's stock (and is thus strictly increasing as $\psi(0) = 0$): Because $\psi(\cdot)$ is continuous, nonmonotonicity would imply the existence of a pair $(k, \tilde{k})$, with $k \neq \tilde{k}$, such that $\psi(k) = \psi(\tilde{k})$. Equation 13.1 would then imply that $s_i(k) = s_i(\tilde{k})$ for $i = 1, 2$. But because

$$k - s_1(k) - s_2(k) = \tilde{k} - s_1(\tilde{k}) - s_2(\tilde{k}),$$

$k = \tilde{k}$ after all. Second, let us compare a steady state of the game and the steady state of a centrally planned economy, in which a social planner would choose extraction rates so as to maximize a weighted sum of the two players' intertemporal utilities. (It can be shown that in both the game and the centrally planned situation the stock converges monotonically to a steady state. In the case of the game, this results from the fact that tomorrow's equilibrium stock is an increasing function of today's stock. Because strategies are continuous in the current stock, from any initial level the stock converges monotonically to a steady state level. The proof for a centrally planned economy is similar.) In a steady state $\hat{k}$ of the game, $\hat{k} = f(\psi(\hat{k}))$. Then equation 13.1 implies that for all i

$$\delta f'(\psi(\hat{k}))[1 - s_i'(\hat{k})] = 1.$$

Consider now a *stable* steady state $\hat{k}$ (that is, the stock converges to $\hat{k}$ for any initial level in a neighborhood of $\hat{k}$). Stability implies that

$$\left. \frac{d(k - f(\psi(k)))}{dk} \right|_{k = \hat{k}} > 0,$$

or

$$f'(\psi(\hat{k}))(1 - s_1'(\hat{k}) - s_2'(\hat{k})) < 1.$$

Using the Bellman equation at $\hat{k}$ then yields, for all i,

$$\delta[1 - s_i'(\hat{k})] > 1 - s_1'(\hat{k}) - s_2'(\hat{k}).$$

We thus conclude that $s_1'(\hat{k}) = s_2'(\hat{k}) > 0$, so that

$$\delta f'(\psi(\hat{k})) > 1.$$

In contrast, the steady state of the centrally planned economy, k^*, must be the "golden rule" level, $\delta f'(\psi^*(k^*)) = 1$, because the central planner must be indifferent between sacrificing one unit of consumption today and having $f'(\psi^*(k^*))$ more tomorrow. ($\psi^*(\cdot)$ is the central planner's savings function.) Because f is strictly concave, $\psi^*(k^*) > \psi(\hat{k})$, and therefore $\hat{k} = f(\psi(\hat{k})) < k^* = f(\psi^*(k^*))$. There is always a "tragedy of the commons" in a stable steady state of a differentiable MPE. (Dutta and Sundaram (1990b) give examples of *over*-accumulation in MPEs not satisfying the conditions above.)

Levhari and Mirman (1980) found that a differentiable MPE exists for the specification $f(k) = k^\alpha$ ($0 < \alpha < 1$) and $g_i(x) = \ln x$. If a linear solution ($s_i(k) = sk$) is postulated, equation 13.1 yields

$$s_i(k) = \frac{1 - \alpha\delta}{2 - \alpha\delta} k$$

and

$$\psi(k) = \frac{\alpha\delta}{2 - \alpha\delta} k,$$

so that

$$\hat{k} = \left(\frac{\alpha\delta}{2 - \alpha\delta}\right)^{\alpha/(1-\alpha)} < k^* = (\alpha\delta)^{\alpha/(1-\alpha)}.$$

(Levhari and Mirman actually use a different method to derive this infinite-horizon equilibrium. They compute the finite-horizon equilibrium and take its limit when the horizon tends to infinity.)

Dutta and Sundaram (1988) also present a more general analysis of this problem in which it is not assumed that the Markov strategies are differentiable. They consider the broader class of pure Markov strategies that (a) satisfy $\sum_i s_i(k) \le k$ for all k and (b) have $s_i(\cdot)$ lower semi-continuous in k.[10] Rather than studying existence (Sundaram (1989) shows that if $g_1 = g_2$ there exists a symmetric MPE in this class, and Dutta and Sundaram (1990a) generalize this result to stochastic reproduction functions), Dutta and Sundaram (1988) show that, if an equilibrium in this class exists, the path of k^t is monotonic and the steady state can be below or above the golden rule.

Amir (1989) provides a general existence theorem using a lattice-theoretic approach. (To the best of our knowledge, this is the first application to dynamic games of the theory of supermodular games developed in section 12.3 above.) Under the assumptions described above (duopoly, increasing and concave production function, compact action space), Amir shows that

---

10. $s_i(\cdot)$ is lower semi-continuous at point k if, for any sequence $k^n \to k$, $\lim\inf_{n\to\infty} s_i(k^n) \ge s_i(k)$. If $s_j(\cdot)$ is not lower semi-continuous, player i's best response may not be well defined. Suppose that, given k^t, player j plays $s_j(k^t)$. Suppose further that $s_j(\cdot)$ jumps up at $\tilde{k}$; then player i's payoff jumps down at $\tilde{s}_i^t = k^t - s_j(k^t) - f^{-1}(\tilde{k})$, because player j's date-$(t + 1)$ reaction jumps up. If player i's intertemporal objective function increases to the left of $\tilde{s}_i^t$, player i may face an "openness problem," so that his best response may not exist. Sundaram (1989) shows that, if the strategies are lower semi-continuous (and belong to $[0, k]$), each player has an optimal Markov strategy and his value function is an upper semi-continuous function of the state.

there exist MPE equilibrium strategies satisfying conditions a and b above and also (c) having Lipschitz constant equal to 1 (for all i, k, and $\tilde{k}$, $|s_i(\tilde{k}) - s_i(k)| \leq |\tilde{k} - k|$.)

Amit and Halperin (1989) study the I-player continuous-time version of this game (see section 13.3 for the definition of differential games). They demonstrate the existence of a family of MPE in the class of strategies that are continuous and continuously differentiable (except perhaps at isolated levels) functions of k. One of these MPE Pareto dominates the others.

13.2 Markov Perfect Equilibrium in General Games: Definition and Properties[†††]

The definition and the characterization of MPE follow those of Maskin and Tirole (1989).

13.2.1 Definition

Whereas the payoff-relevant variable is usually an explicit part of models that use the MPE concept, MPE can be defined starting from any extensive game without an explicit state variable. This subsection shows how to extend MPE to general multi-stage games with observed actions, in which at each date t all players know all actions chosen before date t. There are $T + 1$ periods ($t = 0, \ldots, T$) where T can be finite or infinite. At date t, player i ($i = 1, \ldots, I$) knows the *history* $h^t = (a^0, \ldots, a^{t-1})$ (where $a^{\tau} \equiv (a_1^{\tau}, \ldots, a_I^{\tau})$) and chooses an action a_i^t in a finite action set $A_i^t(h^t)$. (This formalism allows stochastic games (by letting one of the players be nature) as well as perfect-information or sequential games (in which in each period all players but one have a singleton action space).) The *future* f^t at date t is the vector of current and future actions: $f^t \equiv (a^t, \ldots, a^T)$. Player i has the von Neumann-Morgenstern payoff function

$$u_i(a^0, a^1, \ldots, a^T) \equiv u_i(h^t, f^t)$$

for all t.

In chapter 3 we defined a (subgame-) perfect equilibrium as a profile of strategies $\sigma_i^* = \{\sigma_i^{t*}(h^t)\}_{t=0,\ldots,T}$ that form a Nash equilibrium for any history: For all t, h^t, i, and σ_i,

$$E_{f^t}(u_i(h^t, f^t)|(\sigma_i^*, \sigma_{-i}^*)) \geq E_{f^t}(u_i(h^t, f^t)|(\sigma_i, \sigma_{-i}^*)).$$

(E_{f^t} indicates the expectation over the futures f^t, where the distribution over the futures is determined by the mixed-strategy profile in the conditional.)

As discussed in the introduction to this chapter, a Markov strategy for player i may be conditioned on less than player i's information. We are thus led to consider *summaries* or *partitions* of the history $\{H^t(h^t)\}_{t=0,\ldots,T}$, which, for each date, are mappings from the set of histories into a set of disjoint and exhaustive subsets of the set of possible histories at that date. Suppose for instance that there are four possible histories, h, h', h'', and h''', at the beginning of date 2. One partition is $H^2(h) = H^2(h') = A$, $H^2(h'') = B$, and $H^2(h''') = C$, in which the first two histories are lumped in the same summary. The partition can also be written $\{(h, h'), (h''), (h''')\}$.

While summarizing the history, a partition must not be too coarse. That is, at each date, the players must be able to recover the strategic elements of the ensuing subgame from the element of the partition to which h^t belongs:

Definition 13.3   A partition $\{H^t(\cdot)\}_{t=0,\dots,T}$ is *sufficient* if, for all t, h^t, and $\tilde{h}^t$ such that $H^t(h^t) = H^t(\tilde{h}^t)$, the subgames starting at date t after histories h^t and $\tilde{h}^t$ are strategically equivalent:
(i) The action spaces (defined conditionally on actions taken from date t on) are identical: For all i, $\tau \geq 0$, and $a^t, \dots, a^{t+\tau-1}$,

$$A_i^{t+\tau}(h^t, a^t, \dots, a^{t+\tau-1}) = A_i^{t+\tau}(\tilde{h}^t, a^t, \dots, a^{t+\tau-1}).$$

(ii) The players' von Neumann-Morgenstern utility functions conditional on h^t and $\tilde{h}^t$ are representations of the same preferences: $\exists \lambda_i(\cdot, \cdot) > 0$ and $\mu_i(\cdot, \cdot, \cdot)$ such that, for all f^t,

$$u_i(h^t, f^t) = \lambda_i(h^t, \tilde{h}^t)u_i(\tilde{h}^t, f^t) + \mu_i(h^t, \tilde{h}^t, f_{-i}^t),$$

where $f_{-i}^t \equiv (a_{-i}^t, \dots, a_{-i}^T)$.

Of course, the entire history $(H^t(h^t) = (h^t))$ is a sufficient partition, but it may be too fine in that it contains information that is not relevant to the subgame.

Definition 13.4   The *payoff-relevant history* is the minimal (i.e., coarsest) sufficient partition.

Note that, by construction, the payoff-relevant history is uniquely defined. In our example, if the subgames starting at date 2 after histories h, h', and h'' (but not h''') are strategically equivalent, the partition $\{(h, h'), (h''), (h''')\}$ is sufficient but not minimal. The coarsest sufficient partition is $\{(h, h', h''), (h''')\}$.

Remark on Infinite-Horizon Games   The above definition of payoff-relevant history is not quite restrictive enough for infinite-horizon games. A Markov strategy of a stationary game should be independent of the calendar time, t. To achieve this, it suffices to include calendar time in the history. Markov strategies are then independent of time if the state, but not time, affects current and future payoffs and action spaces. The analysis then carries through trivially.[11]

11. Similarly, if the game is cyclical, Markov strategies are independent of calendar time except for the location within the cycle.

Definition 13.5 A Markov perfect equilibrium (MPE) is a profile of strategies σ that are a perfect equilibrium and are measurable with respect to the payoff-relevant history ($H^t(h^t) = H^t(\tilde{h}^t) \Rightarrow \forall i, \sigma_i^t(h^t) = \sigma_i^t(\tilde{h}^t)$).[12]

13.2.2 Existence

Theorem 13.2 Suppose either that $T < \infty$, or that $T = \infty$ and the objective functions are continuous at infinity (see chapter 4—for instance, the present discounted value of per-period payoffs is continuous at infinity if the discount factor is less than 1 and if per-period payoffs are uniformly bounded). Then there exists an MPE.

Proof The proof for a finite horizon is trivial: At date T, select a Nash equilibrium that is the same for all histories h^T in the same payoff-relevant history $H^T(h^T)$ (because, for all histories with the same payoff-relevant history, the last-period subgames are strategically equivalent and the sets of Nash equilibria are the same). Folding back, the subgame at $T - 1$ becomes a one-period game, and one can select a Nash equilibrium that depends only on the payoff-relevant history. And so on, by backward induction.

The proof for $T = \infty$ has two steps. The first step was developed in chapter 4: Associate with G^∞ the T-period truncation game G^T in which the players are forced to take a fixed ("null") action after date T. Such games are finite-horizon games and, from the finite-horizon proof admit an MPE $\{\sigma_i^{t,T}\}$. Then take a subsequence converging to some

$$\sigma = \{\sigma_i^t\}_{\substack{i=1,\dots,I \\ t=0,\dots,+\infty}}$$

when T goes to ∞ (see chapter 4), and check that σ is a perfect equilibrium of G^∞. The second step consists of showing that σ is an MPE of G^∞. We will shortly see that limits of MPE are not always MPE. Whether this holds depends on whether the payoff-relevant history becomes coarser or finer in the limit. Here we take the limit of date-t strategies $\sigma_i^{t,T}$ as the horizon T goes to ∞. Intuitively, when T grows to $T' > T$, there is "at least as much relevant history to remember" at date $t \leq T$, because parts of the history might have an influence on the game at dates $T + 1, \dots, T'$ even though they did not have any at dates $t, \dots, T$. That is, the payoff-relevant history at t cannot become coarser in the limit, so the limit of Markov strategies is itself Markov. ∎

12. One can use additional considerations to obtain stronger versions of MPE. For example, the Markov restriction can be combined with the iterated elimination of strictly dominated strategies. The point here is that a past variable that is payoff relevant only if some player plays a strictly dominated strategy in the subgame ought not to be treated as part of the state. The elimination of a (conditionally) strictly dominated strategy results in fewer Markov strategies, which in turn may lead to a new round of deletion of strictly dominated strategies, and so forth.

Existence of a Pure-Strategy MPE in Games of Perfect Information
(technical)

The previous existence result, as in the Nash case, allows mixed strategies. Some researchers have tried to find classes of games in which a pure-strategy MPE exists. One such class is the class of finite games of perfect information (in which all information sets are singletons, which in particular implies that players play sequentially). Let $t = 0, 1, \ldots, T$, where $T < \infty$, and let $i(t)$ denote the player who has the move at date t (knowing the moves $h^t = (a^0, \ldots, a^{t-1})$ up to date t). Proving existence when action spaces are finite is straightforward: At date T, let player $i(T)$ pick one of his optimal actions (the same for each history with the same payoff-relevant history). Folding back, do the same for player $i(T-1)$ and so forth.

Unfortunately, existence of a pure-strategy MPE does not extend to infinite games. Gurvich (1986) exhibits an infinite-horizon game of perfect information without a pure-strategy MPE. Furthermore, even with a finite horizon, infinite action spaces create existence problems. In the case of action spaces that are compact subsets of a Euclidean space, Hellwig and Leininger (1989) identify the following "openness problem": Consider a three-player game; player $i = 1, 2, 3$ chooses at date i an action a_i in the interval $[0, 1]$. Let the preferences of players 2 and 3 be given by, respectively,

$$u_2 = -(a_2 - \tfrac{1}{2})^2 + a_3(a_1 - \tfrac{1}{2})$$

and

$$u_3 = a_3[1 - (a_2 + \tfrac{1}{2})].$$

(Player 1's preferences turn out to be irrelevant.) Note that the payoff-relevant history at dates 2 and 3 is a_1 and a_2, respectively. Player 3's optimal pure-strategy response is

$$s_3^*(a_2) = \begin{cases} 1 & \text{if } a_2 < \tfrac{1}{2} \\ \in [0, 1] & \text{if } a_2 = \tfrac{1}{2} \\ 0 & \text{if } a_2 > \tfrac{1}{2}. \end{cases}$$

Now consider player 2. From the first term in his objective function, he wants to choose a_2 as close as possible to $\tfrac{1}{2}$. But the second term and player 3's reaction imply that whether $a_2 = \tfrac{1}{2}$ or $\tfrac{1}{2} + \varepsilon$ or $\tfrac{1}{2} - \varepsilon$ matters considerably. Figure 13.1 depicts player 2's payoff as a function of a_2 for $a_1 > \tfrac{1}{2}$ and for $a_1 < \tfrac{1}{2}$. Player 2's payoff for $a_2 = \tfrac{1}{2}$ depends on how player 3's indifference is resolved. If $s_3^*(\tfrac{1}{2}) = 0$, then B in figure 13.1 belongs to the bold and broken lines. If $s_3^*(\tfrac{1}{2}) = 1$, then A belongs to the bold line and C to the broken line.

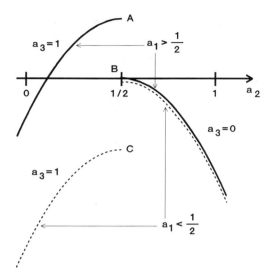

Figure 13.1

Now, for player 2's optimal choice to be well defined (i.e., not face an openness problem) when $a_1 > \frac{1}{2}$, one needs $s_3^*(\frac{1}{2}) = 1$; similarly, for the optimal choice to be well defined when $a_1 < \frac{1}{2}$, one needs $s_3^*(\frac{1}{2}) = 0$. Thus, existence of a pure-strategy equilibrium requires that player 3's response to $a_2 = \frac{1}{2}$ depend on the date-3 payoff-irrelevant variable a_1. We thus conclude that no pure-strategy MPE exists. (The same argument should convince the reader that there is no mixed-strategy MPE either.)

On the other hand, there exists a pure-strategy perfect equilibrium. (As an exercise, find it.) This is actually a more general result: a finite-horizon game of perfect information that satisfies the following conditions for each t has a pure-strategy perfect equilibrium (Goldman 1980; Harris 1985; Hellwig and Leininger 1987):

player $i(t)$'s action space is a compact subset of a Euclidean space,
player $i(t)$'s utility function is continuous in $(a^0, a^1, \ldots, a^T)$, and
player $i(t)$'s action space is a closed-valued and continuous correspondence $A_{i(t)}(a^0, \ldots, a^{t-1})$.

We leave it to the reader to check that, with finite action spaces that approximate the interval $[0, 1]$, there exists a pure-strategy MPE that is close to (converges to when the grid converges to the continuum) the perfect equilibrium mentioned above.

To avoid the nonexistence problem, Hellwig and Leininger (1989) assume that the state variable k^{t+1} at date $t + 1$ depends only on k^t and $a_{i(t)}$, and they make a strong assumption on preferences—namely, that they be (forwardly) recursively separable in state-action pairs,

$$u_{i(t)} = v_t^t(k^t, a_{i(t)}, v_{t+1}^t(k^{t+1}, a_{i(t+1)}, v_{t+2}^t(\cdots))),$$

and that v_τ^t be either independent or strictly increasing in $v_{\tau+1}^t$ ($\tau \geq t$).[13] This class of preferences generalizes the altruistic preferences found in bequest games (Phelps and Pollak 1968) in which the players are generations t ($i(t) = t$) that live for one period, choose the level of capital k^{t+1} (that is, $a^t = k^{t+1}$) to give to their heirs, and have utility

$$u_t = v_t(f(k^t) - k^{t+1}, v_{t+1}(f(k^{t+1}) - k^{t+2}, v_{t+2}(\cdots))).$$

They also make a regularity assumption that rules out another cause of nonexistence associated with the action spaces. To paraphrase them, this regularity assumption guarantees that the effect of a small change in the state variable k^t on the subsequent state variable k^{t+1} can be exactly neutralized by a suitable small change in the choice variable $a_{i(t)}$. This assumption, together with recursive separability of preferences, guarantees existence of a pure-strategy MPE.

13.2.3 Robustness to Payoff Perturbations (technical)

The Markov concept emphasizes the influence of a few key variables to the exclusion of minor ones, and has bite only if a small number of past variables affect the current and future action spaces and objective functions. Perturbing the objective functions is likely to make the entire history payoff relevant and to deprive the Markov concept of its power, as discussed in the introduction to this chapter.

Consider a finite extensive form and identify games with the vector u of payoffs of all players at the terminal nodes of the tree. Let U denote the set of possible games (i.e., payoffs). Define distances between two games u and $\tilde{u}$ as

$$\|\tilde{u} - u\| = \max_{i,s} |\tilde{u}_i(s) - u_i(s)|$$

and distances between two strategies σ and $\tilde{\sigma}$ as

$$\|\tilde{\sigma} - \sigma\| = \max_{i,t,h^t,a_i^t} |\tilde{\sigma}_i^t(a_i^t|h^t) - \sigma_i^t(a_i^t|h^t)|.$$

With a game u is associated a payoff-relevant history H^u. Let $U(H)$ denote the subset of payoffs in U that give rise to payoff-relevant history H.

Consider a game u and an MPE σ of u; and let u^n denote a sequence of small perturbations of game u ($\lim_{n \to +\infty} \|u^n - u\| = 0$). Note that the payoff-relevant history of u^n can be finer than that of u. (Indeed, it is easy to see that "for almost all payoffs v" the payoff-relevant history for game v is the identity, i.e., the entire history itself.) Does there exist a sequence of

---

13. With recursively separable utilities, all interactions between future actions and past ones take place through tomorrow's state variable. Such utility functions have received much attention in (one-player) optimal-growth theory (see, e.g., Beals and Koopmans 1969).

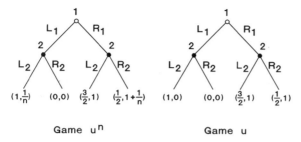

Figure 13.2

MPE σ^n of games u^n such that $\lim_{n \to +\infty} \|\sigma^n - \sigma\| = 0$? Not necessarily, as figure 13.2 demonstrates. In game u^n, the only perfect (and therefore Markov perfect) equilibrium has player 2 playing L_2 if L_1 and playing R_2 if R_1, and player 1 playing L_1. However, in the limit game u, player 1's action a_1 is payoff irrelevant in period 2. Hence, in an MPE of the limit game, player 2's strategy is the same whether player 1 plays L_1 or R_1, and hence player 1 plays R_1.

Game u is, in a sense, exceptional. Not only is a_1 payoff irrelevant; also, for any a_1, player 2 is indifferent between his pure strategies. (Note that the Markov concept has less appeal than usual in such a game. For example, one might argue that player 2, who is indifferent, would seek revenge by playing R_2 if player 1 played L_1 and would reward player 1 by playing L_2 if player 1 played R_1.)

Definition 13.6   A game u is *essential* with respect to MPE if, for any $\varepsilon > 0$, there exists $\zeta > 0$ such that, for all $\tilde{u}$ satisfying $\|\tilde{u} - u\| < \zeta$ and all MPE σ of game u, there exists an MPE $\tilde{\sigma}$ of game $\tilde{u}$ satisfying $\|\tilde{\sigma} - \sigma\| < \varepsilon$.

In other words, a game is essential with respect to MPE if its Markov perfect equilibria are robust to payoff perturbations. We already encountered the notion of essentiality in chapter 12. It was noted there that almost all strategic-form games are essential with respect to Nash equilibrium. We cannot make direct use of this result, however, because Markov perfect equilibrium is a refinement of Nash equilibrium and also because the genericity concepts in strategic-form and extensive-form games do not coincide (see section 12.1). However, it can be shown that the game u in figure 13.2 is indeed exceptional:

Theorem 13.4 (Maskin and Tirole 1989)   Fix a finite, multi-stage tree and consider the set $U(H)$ of games with payoff-relevant history H. Almost all games in $U(H)$ (that is, almost all payoffs consistent with payoff-relevant history H) are essential with respect to the MPE concept and have a finite number of MPE.

Theorem 13.4 shows that, starting from a game in which some past variables have no influence on future payoffs and perturbing the game slightly, one can almost always choose equilibria of the perturbed game in which past variables that (now) have a small influence on future payoffs have only a small influence on equilibrium strategies. (Maskin and Tirole extend the proposition to an infinite horizon and define Markov equilibrium for games of incomplete or imperfect information.)

13.3 Differential Games[†††]

13.3.1 Definition

Differential games are continuous-time stochastic games for which control theorists have analyzed a subclass of solutions in the set of Markov perfect equilibria. Introduced by Isaacs (1954), these games have (unfortunately for social scientists) been developed mainly in the context of zero-sum two-person games. That the sum of the payoffs equals 0 was perceived as a decent approximation for various tactical problems studied in that literature, but it considerably limits the scope of their application to economics. Some progress has been made, however, on the non-zero-sum front. (For more material on differential games, see Basar and Olsder 1982, Blaquiere 1971, and Isaacs 1965. Applications to economics include Levine and Thepot 1982, Reinganum 1982, Simaan and Takayama 1978, and the papers quoted in note 14.) Let us first described the framework.

Time t varies continuously from 0 to $T \le \infty$. Let $k^t \equiv (k^t_1, \ldots, k^t_n)$, a vector in real Euclidean space $\mathbb{R}^n$, denote the position, or state, or payoff-relevant history of the game at date t, and let it obey a system of first-order differential equations,

$$\frac{dk^t_j}{dt} = h^t_j(k^t, a^t), \quad j = 1, \ldots, n, \tag{13.2}$$

where $a^t = (a^t_1, \ldots, a^t_I)$ is the vector of actions chosen by all players at date t. Player i's action a^t_i belongs to some Euclidean space. (One can think of the continuous-time game as a limit of discrete-time games in which the players know the state at the beginning of date t and simultaneously choose actions.) The state at date 0 is given and is equal to $k(0)$.

The payoff functions are

$$u_i = \int_0^T g^t_i(k^t, a^t)\,dt + v^T_i(k^T), \tag{13.3}$$

where the terminal payoff, v^T_i, may depend on the state at the end of the game. A two-player game is zero-sum if $u_1 + u_2 \equiv 0$.

Differential game theorists impose the Markov restriction that strategies depend only on time and the state, and, more strongly, only on the state for stationary games, i.e., games in which h_j^t does not depend on t and the g_i^t take the form $e^{-rt}g_i(\cdot, \cdot)$.

The path of the state is then given by the differential equations

$$\frac{dk_j^t}{dt} = h_j^t(k^t, s^t(k^t)) \equiv \tilde{h}_j^t(k^t), \tag{13.4}$$

with initial condition

$$k^0 = k(0). \tag{13.5}$$

We postpone the discussion of whether these differential equations have a single solution until subsection 13.3.4.

To characterize Nash or perfect equilibria in Markov strategies of a differential game, it is tempting to borrow the techniques of dynamic programming, namely the maximum principle and the Pontryagin conditions (again, we postpone the discussion about whether it is proper to do so). With $s_{-i} = \{s_{-i}^t(k^t)\}$ given as the strategies of player i's rivals, the evolution of the state variables as a function of player i's actions is

$$\frac{dk_j^t}{dt} = h_j^t(k^t, a_i^t, s_{-i}^t(k^t)) \equiv \hat{h}_j^t(k^t, a_i^t). \tag{13.6}$$

If $\hat{g}_i^t(k^t, a_i^t) \equiv g_i^t(k^t, a_i^t, s_{-i}^t(k^t))$, player i's control problem is to maximize

$$\int_0^T \hat{g}_i^t(k^t, a_i^t)\, dt + v_i^T(k^T)$$

subject to equations 13.6 and 13.5 and to $a_i^t \in A_i^t(k^t)$ (player i's date-t action set).

13.3.2 Equilibrium Conditions

As mentioned above, each player's choice of an optimal strategy is a control problem in which the player takes into account the influence of his actions on the state, both directly and indirectly through the influence of the state on the strategies of the player's opponents. Subject to the technical caveats of subsection 13.3.4, it is easy to extend the conditions of Pontryagin et al. (1962) to multi-player situations. Starr and Ho (1969) restrict their attention to equilibria in which the equilibrium payoffs are continuous and almost-everywhere-differentiable functions of the state variables. This restriction obtains naturally for control problems in smooth environments, but it imposes a significant restriction in games: It might be that each player's strategy, and thus each player's payoff, changes discontinuously with the state because of the self-fulfilling expectation that the other players use discontinuous strategies, as we will see in section 13.4. Perhaps the continuity restriction can be justified by the claim that the "endogenous discontinuities" that it prohibits require excessive coordination, or are not robust to the addition of a small amount of noise in the players' observations. We are unaware of formal arguments along these lines.

The technical advantage of restricting attention to smooth equilibria is that necessary conditions can then be derived by means of the variational methods of optimal-control theory. Assume that player i wishes to choose s_i to maximize u_i, subject to the state-evolution equation (13.6) and the initial condition (13.5).

Introducing co-state variables λ_i^t, which are the vectors of $\{\lambda_{ij}^t\}_{j=1,\ldots,n}$, we define $\mathscr{H}_i^t$, the Hamiltonian for player i, as

$$\mathscr{H}_i^t(k^t, a^t, \lambda_i^t) = g_i^t(k^t, a^t) + \sum_j \lambda_{ij}^t h_j^t(k^t, a^t). \tag{13.7}$$

MPE strategies $s_i = \{s_i^t(k^t)\}$ must satisfy the generalized Hamilton-Jacobi equation

$$s_i^t(k^t) \in \arg\max_{a_i} \mathscr{H}_i^t(k^t, a_i, a_{-i}^t, \lambda_i^t) \tag{13.8}$$

as well as, for all j,

$$\frac{d\lambda_{ij}^t}{dt} = -\frac{\partial \mathscr{H}_i^t}{\partial k_j^t} - \sum_{\substack{\ell=1,\ldots,I \\ \ell \neq i}} \left(\frac{\partial \mathscr{H}_i^t}{\partial a_\ell^t}\right)' \frac{\partial s_\ell^t}{\partial k_j^t}, \tag{13.9}$$

where $\partial s_\ell^t/\partial k_j^t$ is the vector of partial derivatives of player ℓ's strategy (which is assumed to be piecewise C^1) with respect to the jth component of the state, with the convention that the derivative of the scalar $\mathscr{H}_i^t$ with respect to the vector a_ℓ^t (i.e., $\partial \mathscr{H}_i^t/\partial a_\ell^t$) is a column vector. They must also satisfy the appropriate transversality condition (e.g., when $T < \infty$, $\lambda_{ij}^T = \partial v_i^T/\partial k_j^T$). Notice that for a one-player game the second term in equation 13.9 vanishes, and the conditions reduce to the familiar ones. In the multi-player case, this second term captures the fact that player i cares about how his opponents will react to changes in the state. Because of the cross-influence term, the evolution of the shadow price of the jth state variable for player i, λ_{ij}^t, is determined by a system of partial differential equations, instead of by ordinary differential equations as in the one-player case. As a result, very few differential games can be solved in closed form. An exception is the linear-quadratic case; see subsection 13.3.3.

Another way of approaching the problem is to work with the value functions $V_i^t(k)$. One has

$$\frac{\partial V_i^t}{\partial k_j^t} = \lambda_{ij}^t$$

and

$$\frac{\partial V_i^t}{\partial t} = \max_{a_i^t} \left\{ \mathcal{H}_i^t \left(k^t, a_i^t, a_{-i}^t, \frac{\partial V_i^t}{\partial k^t} \right) \right\}.$$

A differential game is said to be *normal* (see Starr and Ho 1969) if it is possible to find a unique instantaneous Nash equilibrium $\hat{a}^t$ for the payoffs $\mathcal{H}_i^t$ for all k, λ, and t and if integrating the equations

$$\frac{\partial V_i^t}{\partial t} = \mathcal{H}_i^t \left(k^t, \hat{a}^t, \frac{\partial V_i^t}{\partial k^t} \right)$$

and

$$\frac{dk_j^t}{dt} = h_j^t(k^t, \hat{a}^t)$$

backward from all points on the terminal surface yields feasible trajectories. Starr and Ho prove that linear-quadratic differential games are normal.

13.3.3 Linear-Quadratic Differential Games

Linear-quadratic games are games for which the equations of motion are linear in the state and control variables and the objective functions are quadratic in the state and control variables. They were first studied by Case (1969) and by Starr and Ho (1969). The MPE strategies given by the first-order conditions 13.8 and 13.9 can be obtained numerically. Hoping that a linear-quadratic model is a good Taylor approximation to more general games, many researchers have computed the differential-game equilibria of such models.[14]

For notational simplicity, we assume that the payoffs and the state-evolution equation are independent of time, so that the system is autonomous. The linear-quadratic case is then

$$u_i = \int_0^T \left(\tfrac{1}{2} k' Q_i k + \tfrac{1}{2} \sum_{\ell=1}^I a_\ell' R_{i\ell} a_\ell + \sum_{\ell=1}^I r_{i\ell}' a_\ell + q_i' k + f_i \right) e^{-rt} \, dt$$

$$+ \tfrac{1}{2} k^{T'} S_i k^T, \tag{13.10}$$

and

$$\frac{dk}{dt} = Ak + \sum_{\ell=1}^I B_\ell a_\ell, \tag{13.11}$$

where the dependence of k and $a_i = s_i(k)$ on time is suppressed. Here, a prime denotes a transpose. Q_i, S_i, and A are $n \times n$ matrices, where n is the

---

14. For instance, Pindyck (1977) analyzes a game between the Federal Reserve System and the U.S. Government. Fershtman and Muller (1984), Fershtman and Kamien (1987), and Hanig (1986, chapter 4) apply linear-quadratic games to duopoly markets with slow adjustment of capacity, goodwill, or price variables. Other applications study the arms race. Clemhout and Wan (1979) present further examples.

dimension of the state variable. $R_{i\ell}$ is an $m_\ell \times m_\ell$ matrix where m_ℓ is the dimension of player ℓ's action space. B_ℓ is an $n \times m_\ell$ matrix, $r_{i\ell}$ is an m_ℓ-vector, q_i is an n-vector, and f_i is a real number. r is the instantaneous rate of interest. Matrix R_{ii} is assumed to be negative definite to ensure that player i's optimal control is well defined. Since the quadratic form $x'Cx$ equals $x'[(C + C')/2]x$, we can take the matrices Q_i, $R_{i\ell}$, and S_i to be symmetric without loss of generality.

The "current Hamiltonian" for player i is[15]

$$\mathcal{H}_i = \tfrac{1}{2}k'Q_ik + \tfrac{1}{2}\sum_{\ell=1}^{I} a'_\ell R_{i\ell}a_\ell + \sum_{\ell=1}^{I} r'_{i\ell}a_\ell + q'_ik + f_i$$

$$+ \lambda'_i\left(Ak + \sum_{\ell=1}^{I} B_\ell a_\ell\right).$$

The optimal-control vector a_i for player i maximizes $\mathcal{H}_i$:

$$a_i \equiv -R_{ii}^{-1}(r_{ii} + B'_i\lambda_i). \tag{13.12}$$

The co-state variables vary according to the current version of equation 13.9:

$$\frac{d\lambda_i}{dt} \equiv r\lambda_i - (Q_ik + q_i + A'\lambda_i) - \sum_{\ell\neq i} \frac{\partial s_\ell}{\partial k}(R_{i\ell}a_\ell + r_{i\ell} + B'_\ell\lambda_i), \tag{13.13}$$

where $\partial s_\ell/\partial k$ is the matrix whose jth row is $\partial s_\ell/\partial k_j$.

One then tries to find a solution for which the co-state variables, and thus the strategies from equation 13.12, are affine functions of the state variables. That is, one looks for $n \times n$ matrices Λ_i and n-vectors γ_i such that, for all i,

$$\lambda_i = \Lambda_ik + \gamma_i. \tag{13.14}$$

Equation 13.14 then yields

$$a_i = (-R_{ii}^{-1}B'_i\Lambda_i)k - R_{ii}^{-1}(r_{ii} + B'_i\gamma_i). \tag{13.15}$$

Differentiating equation 13.14, eliminating $\dot{k}$ using equation 13.11, and using equation 13.15 yields an affine function of k in the left-hand side of equation 13.13. Similarly, substituting equations 13.14 and 13.15 into the right-hand side of equation 13.13 yields another affine function of k. Identifying the coefficients of k and the constant coefficients on the two sides of the resulting identity, one finds that, if there exists a solution of the form $\lambda_i = \Lambda_ik + \gamma_i$, then Λ_i and γ_i must satisfy the "Riccati equations":

---

15. The Hamiltonian in equation 13.7 is in terms of present value. Accordingly, we will adjust equation 13.9 to account for the fact that λ_i is a vector of current—not present discounted—shadow prices. Equation 13.13 will thus include a term that represents the interest on the shadow price. Arrow and Kurz (1970) discuss this formulation in a single-player context.

$$\Lambda_i A + A'\Lambda_i + Q_i - r\Lambda_i + \sum_{\ell \neq i} \Lambda'_\ell B_\ell (R_{\ell\ell}^{-1}) R_{i\ell} R_{\ell\ell}^{-1} B'_\ell \Lambda_\ell$$

$$- \sum_{\ell \neq i} \Lambda'_\ell B_\ell (R_{\ell\ell}^{-1})' B'_\ell \Lambda_i - \sum_\ell \Lambda_i B_\ell R_{\ell\ell}^{-1} B'_\ell \Lambda_\ell = 0 \qquad (13.16)$$

and

$$r\gamma_i - A'\gamma_i - q_i + \sum_\ell \Lambda_i B_\ell R_{\ell\ell}^{-1} (r_{\ell\ell} + B'_\ell \gamma_\ell)$$

$$+ \sum_{\ell \neq i} \Lambda'_\ell B_\ell (R_{\ell\ell})^{-1} B'_\ell \gamma_i$$

$$- \sum_{\ell \neq i} \Lambda'_\ell B_\ell (R_{\ell\ell})^{-1} (r_{i\ell} - R_{i\ell}^{-1} R_{\ell\ell}^{-1} (r_{\ell\ell} + B'_\ell \gamma_\ell)) = 0.$$

$$(13.17)$$

These quadratic equations can be solved numerically. In many applications, they are actually much less complex than they appear. For instance, many games have no cross-terms in actions $R_{i\ell}$ ($i \neq \ell$); each player i chooses a one-dimensional action, which affects "his" level of capital k_i, and so forth.

One may wonder whether the linear-quadratic solution derived above indeed forms a perfect equilibrium. In the infinite-horizon case ($T = +\infty$), Papavassilopoulos et al. (1979) have shown that if a stability condition on the matrices is satisfied then the strategies are indeed perfect—for more detail see the original article or chapter 2 of Hanig 1986.[16]

Judd (1985) offers an alternative to the strong functional-form assumptions typically invoked to obtain closed-form solutions to differential games. His method is to analyze the game in the neighborhood of a parameter value that leads to a unique and easily computed equilibrium. In his examples of patent races, he looks at patents with values of almost 0. Obviously, if the value of a patent is exactly 0, in the unique equilibrium the players do no research and development and their values are 0. Judd proceeds to expand the system about this point, neglecting all terms over third order in the value of the patent. Judd's method gives only local results, but it solves an "open set" in the space of games, as opposed to conventional techniques that can be thought of as solving a lower-dimensional subset of games.

13.3.4   Technical Issues

The focus on Markov strategies is guided not only by a subjective notion of what strategies are reasonable, but also by technical considerations associated with continuous time. As Anderson (1985) observes, "general" (that is, functions of the whole history) continuous-time strategies need not lead to a well-defined outcome path for the game, even if the strategies and

---

16. In the finite-horizon case, the above linear-quadratic solution is unique in the space of strategies that are analytic functions of the state variables. See Papavassilopoulos and Cruz 1979.

the outcome path are restricted to be continuous functions of time. Anderson offers the example of a continuous-time game in which two players simultaneously choose actions and there is no state variable. Consider the continuous-time strategy "play at each time t the limit as $\tau \to t$ of what the opponent has played at times τ previous to t." This limit is the natural analogue of the discrete-time strategies "match the opponent's last action." If the players have chosen matching actions at all times before t and the history is continuous, there is no problem in computing what should be played at t. However, there is not a unique way of extending the outcome path beyond time t. Knowing play before t determines the outcome at t but is not sufficient to extend the outcome path to any open interval beyond t. (As a result of this problem, Anderson opts to study the limits of discrete-time equilibria instead of working with continuous time.)

Restricting strategies to be Markov does not avoid the following difficulty: To make things simple, assume that the function g_i^t and h_j^t are defined and continuously differentiable over the whole Euclidean space. Even so, the differential equations 13.4 with the initial condition 13.5 may not have a unique solution. The right-hand side of equation 13.4 may not be Lipschitz-continuous in k^t unless the strategies are continuously differentiable.[17] Alas, the class of continuously differentiable strategies may be too small even in a single-player environment (i.e., in a control problem), as continuously differentiable strategies may be dominated by piecewise continuously differentiable (piecewise-C^1) ones, for instance. Control theorists thus often restrict their attention to piecewise-C^1 strategies; to make sure that the differential equations define a unique path of the state variables, they must verify that the optimal response to a piecewise-C^1 strategy can be chosen piecewise C^1.

In order for the Pontryagin conditions for player i to be applied, the class of allowable strategies must be "sufficiently large" to include all the perturbations required to obtain these conditions. In particular, piecewise-C^1 strategies must be allowed. But, as we already noted, allowing a large class of controls conflicts with the traditional assumption in control theory that the right-hand side of the differential equations governing the evolution of the state variables is C^1. Since at least the class of piecewise-C^1 functions must be allowed for player i's strategy s_i, $\hat{h}_j^t$ may be discontinuous, and applying extensions of traditional control theory to discontinuous evolution equations requires some assumptions about the relationship between the manifolds of discontinuity (which are not always satisfied in examples).

---

17. The function $\tilde{h}_j^i(k^t)$ is Lipschitz-continuous in k^t if, for any k^t and any $\hat{k}^t$ in a neighborhood of k^t,

$$|\tilde{h}_j^i(k^t) - \tilde{h}_j^i(\hat{k}^t)| \le L|k^t - \hat{k}^t| \text{ for some } L > 0.$$

The Lipschitz property plays a crucial role in the existence of a unique solution to differential equations. See Smart 1974.

For sufficient conditions for piecewise-C^1 strategies satisfying the first-order conditions of the players' control problems to form a Nash equilibrium in a zero-sum differential game, see (e.g.) Berkovitz 1971.

Once the class of allowable strategies is fixed, the issue of the existence of an MPE in this class arises. As we mentioned, one may verify that the solution to the first-order conditions satisfies some sufficient conditions. Sufficient conditions to obtain existence without characterizing the equilibrium strategies are known only for special differential games, such as the linear-quadratic games discussed in subsection 13.3.3 and the zero-sum games to which we will turn shortly.

13.3.5 Zero-Sum Differential Games (technical)

Two-player zero-sum games have the convenient property that the sets of perfect-equilibrium and Nash-equilibrium outcomes coincide (see exercise 4.10). The existence of a Markov perfect equilibrium in a class of strategies (e.g., piecewise-C^1 strategies) then results from standard theorems of the existence of Nash equilibria for zero-sum games if such theorems apply. (In particular, a Nash equilibrium exists if the strategy spaces are compact, convex subsets of linear topological spaces and if u_i is upper semi-continuous and concave in s_i and lower semi-continuous and convex in s_j.)

A first approach to proving the existence of a Nash equilibrium relies on the fact that a pure-strategy Nash equilibrium in open-loop strategies (i.e., strategies that depend on time but not on the state) is also a pure-strategy Nash equilibrium in closed-loop strategies (i.e., strategies that are contingent both on time and the state).[18] Sufficient conditions for the existence of a pure-strategy open-loop Nash equilibrium are that the functions h_j^t and g_i^t are linear in the state and actions, that the game has fixed finite duration T, and that the action spaces are compact, convex, and independent of time and state (see, e.g., Fichefet 1970).

This simple method of proving the existence of a Nash equilibrium in closed-loop strategies via the existence of a Nash equilibrium in open-loop strategies breaks down in stochastic differential games, because, with a random evolution of the state variables, a player is not able to perfectly predict the value of the state variables at each date, and thus the best response to even an open-loop strategy of one's opponent is a closed-loop strategy. For instance, researchers have studied differential games with deterministic payoffs as in equation 13.3 but with stochastic differential equations for the evolution of the state variables:

$$dk_j^t = h_j^t(k^t, a^t)\, dt + \gamma^t(k^t)\, dB_j^t,$$

---

18. The point is that player i can perfectly foresee the evolution of the state as a function of s_i since the other player is using a pure strategy. This property also holds for more than two players and for non-zero-sum games.

where B^t is a Brownian motion in $\mathbb{R}^n$. To prove the existence of equilibrium, they use Hamiltonian methods. See, e.g., Elliott 1976, and, for extensions to non-zero-sum games, Uchida 1978 (also with a Brownian perturbation of the differential equations). Wernerfelt (1988) discusses piecewise-constant jump processes.

13.4 Capital-Accumulation Games[†††]

Capital-accumulation games offer a useful illustration of differential-game techniques. Following Spence (1979), we will consider continuous-time, infinite-horizon, stationary, duopoly versions of the capital-accumulation games. Here the control variable a_i is firm i's rate of investment in its own capital k_i. With current capital stocks $k = (k_1, k_2)$ (we continue to suppress the time superscripts), there is an equilibrium in the product market—i.e., outputs $q_i(k)$ and prices $p_i(k)$—with firm i's profit, net of production costs, equal to $R_i(k_1, k_2)$. (We implicitly make a Markov assumption that current pricing and production are not affected by payoff-irrelevant aspects of past play.) The firms also pay a maintenance cost m_i per unit of capital. If we let $C_i(k_i, a_i)$ denote the firm's cost of investing at rate a_i when its capital stock is k_i, the firm's instantaneous net profit is

$$g_i(k, a) = R_i(k) - m_i k_i - C_i(k_i, a_i).$$

We assume that $\partial^2 R_i / \partial k_i^2 < 0$ (concave profit function), $\partial R_i / \partial k_j < 0$ (firms dislike increases in their opponent's capital), and $\partial^2 R_i / \partial k_i \partial k_j < 0$ (the marginal productivity of capital decreases with the rival's capital— that is, the capital levels are strategic substitutes as defined in section 12.3). We also assume that the revenue functions R_i and their derivatives are bounded above and below. For tractability, one of two specifications— irreversible investment and reversible investment—has been adopted for the investment function in the literature.

Irreversible investment The levels of capital never decrease: $dk_i / dt = a_i$. The unit cost of investment is assumed to be 1 up to a maximum investment rate $\bar{a}_i$ and infinite thereafter (so, overall, the investment cost is convex). In other words, the instantaneous action space is $A_i = [0, \bar{a}_i]$ and the cost is $C_i(k_i, a_i) = a_i$. The upper bound on investment prevents firms from investing at an "infinite speed," which allows the game to be truly dynamic.

Reversible investment Firm i's capital level depreciates at rate ρ_i: $dk_i / dt = a_i - \rho_i k_i$. For tractability, the reversible-investment case is studied with a quadratic, rather than a discontinuous, cost of investment: Either $C_i(k_i, a_i) = c_i a_i^2 / 2$ (the cost depends on gross investment) or $C_i(k_i, a_i) = c_i (a_i - \rho_i k_i)^2 / 2$ (the cost depends on net investment).

13.4.1   Open-Loop, Closed-Loop, and Markov Strategies

We begin with the case of irreversible investment. The per-period payoffs are assumed to be $[R_i(k_i, k_j) - m_i k_i - a_i]$, and we impose an upper bound on investment $\bar{a}_i$. Thus, $dk_i/dt = a_i \in [0, \bar{a}_i]$. For concreteness, let both firms enter the market at time $t = 0$ without any capital (but in a perfect equilibrium, equilibrium behavior is defined from any initial levels of capital).

In a first step, we assume that the firms maximize their time-average payoffs, so that only the eventual steady-state capital levels matter. Because the marginal productivity of capital is bounded, and capital requires maintenance, no firm will choose an infinite capital stock. The time-averaging specification has the peculiar feature that the investment path leading to a given steady state does not affect the payoffs, but it allows a simple analysis of strategic investment. (We will discuss the case of a strictly positive rate of interest below.)

We define the "Cournot reaction curves" $r_i(\cdot)$ for this time-averaging specification by $\partial R_i(r_i(k_j), k_j)/\partial k_i = m_i$. Under our assumptions, the reaction curves are downward sloping. We will assume that they have a unique (stable) intersection $C = (k_1^c, k_2^c)$ as in figure 13.3 (which depicts the symmetric case).

Let us first examine the *"precommitment"* or *"open-loop"* equilibria. (See chapter 4 for the notion of open-loop equilibrium.) In a precommitment equilibrium, firms simultaneously commit themselves to entire time paths of investment. Thus, the precommitment equilibria are really static, in that there is only one decision point for each firm. The precommitment equilibria are just like Cournot-Nash equilibria, but with a larger strategy space. In the capital-accumulation game, the precommitment equilibrium is exactly the same as if both firms built their entire capital stocks at the start (because of no discounting). In the resulting "Cournot" equilibrium, each firm invests to the point at which the marginal productivity of capital equals m_i, given the steady-state capital level of its opponent. There are many different paths which lead to this steady state, all of which are precommitment equilibria. For example, each firm's strategy could be to invest as quickly as possible to its Cournot level. We can highlight the similarity of this solution to a Cournot equilibrium by defining the "steady-state reaction curves" that give each firm's desired steady-state capital level as a function of the steady-state capital level of the opposing firm. The precommitment equilibrium is at C in figure 13.3. As we saw in chapter 4, the use of the precommitment concept transforms an apparently dynamic game into a static one. As a modeling strategy, this transformation is ill advised. As Kreps and Spence (1984) note, "one should not allow precommitment to enter by the back door.... If it is possible, it should be explicitly modeled ... as a formal choice in the game."

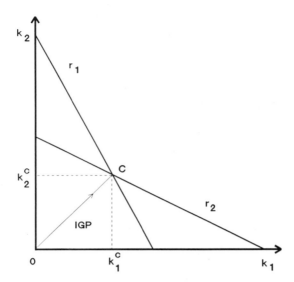

Figure 13.3

Let us next allow firm i's investment at time t to depend on the capital stocks as well as on time. The capital stocks are the state variables. A "closed-loop" equilibrium is a Nash equilibrium in state-dependent ("closed-loop") strategies. The first thing we should point out is that precommitment ("open-loop") equilibria are closed-loop equilibria. If the strategy of a firm's opponent depends only on time (and there are no random disturbances in the system), the firm can without loss restrict itself to a strategy that depends only on time: If the firm's optimal closed-loop strategy is given, the path of the system (the capital stocks at every time) is completely determined. We can then construct an open-loop strategy that calls for the same rate of investment at every point in time as the closed-loop strategy does. In the jargon of optimal control, this is called "synthesizing the feedback control."

Thus simply expanding the strategy space to allow dependence on history does not remove the "static" precommitment equilibria. Moreover, many other implausible outcomes are closed-loop equilibria. For example, firm 1 can threaten to "blow the game up" by building huge amounts of capacity if firm 2 dares to invest beyond some small level. In view of firm 1's threat, firm 2's best response may well be to acquiesce and accept a very small long-run market share. This is the now-familiar "perfection" problem—firm 1 is making a threat it would not choose to carry out were its bluff to be called. Of course, a firm may be willing and able to commit itself to such a threat, using a "doomsday machine" (perhaps a contract with a third player—see Schelling 1960 and Gelman and Salop 1983) preset to inflict some terrible harm on itself if it backs down. The point is the same here as with the open-loop strategies: To the extent that such commitments are possible, they should be included in the formal model. Given such a model, we would expect that each decision a firm makes is part of an optimal plan for the remainder of the game. We will further require that bygones be bygones except to the extent that past choices influence the current and future competitive environment. To impose these requirements we restrict attention to Markov perfect equilibria.

Neither of the two equilibria discussed so far is perfect. In the second equilibrium, if firm 2 were to disregard firm 1's threat and undertake nonnegligible investment, firm 1 would not wish to build the threatened level of capacity. In other words, the given strategies do not form an equilibrium starting from states with nonnegligible k_2. The first (precommitment) equilibrium is imperfect in a less obvious way: If both firms invest as quickly as possible, then generally one of them (say firm 1) will get to its Cournot capital level before the other. The specified strategies then say that firm 1 should stop investing while firm 2 continues on to its Cournot level. But consider what would happen if firm 1 were to deviate by investing past its Cournot level by a small amount before stopping. Firm 2's strategy says that, "no matter what," firm 2 will invest up to its Cournot level. But if firm 1 has already invested past k_1^c, then the best thing for firm 2 to do would be to stop on its reaction curve. The given strategies do not form a Nash equilibrium starting from states in which $k_1 > k_1^c$; thus, they are not perfect.

The above argument suggests that the firm with the greater investment speed (or the firm with a "head start" in investing—the model can be extended to allow unequal entry times) can invest "strategically" (that is, "overinvest" relative to C) in order to reduce the investment of the other firm. Because we have assumed that such "overinvestment" is locked in (there is no depreciation or disinvestment), the best the follower can do when presented with the *fait accompli* of overinvestment is take it as given when making its own decisions. With a large enough discrepancy in speeds, the "leader" can act as a Stackelberg leader and choose its preferred point on the "follower's" reaction curve.

To see this, consider figure 13.4, which depicts a perfect equilibrium. The arrows indicate the direction of motion of the state: horizontal if only firm 1 is investing, vertical if only firm 2 is investing, diagonal if each firm is investing as quickly as possible, and " + " if neither firm invests (because of the linearities, the optimal strategies are "bang-bang"). Note that we have defined choices at every state, and not just those along the equilibrium path—this is necessary in order to test for perfection. Looking at figure 13.4, we see that unless firm 1 has a head start, it cannot enforce its Stackelberg level. But if firm 1 starts with more capital than firm 2, it invests as fast as it can until either it reaches its Stackelberg level of capital or firm 2 reaches its reaction curve (the Stackelberg level is defined as in chapter

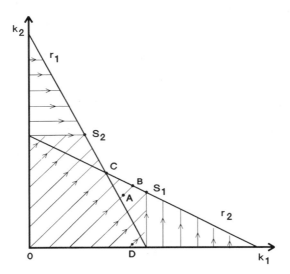

Figure 13.4

3: it maximizes $R_1(k_1, r_2(k_1)) - m_1 k_1$ with respect to k_1. The Stackelberg point is denoted S_1 in figure 13.4). If it reaches the Stackelberg level before firm 2 reaches its reaction curve, firm 2 then continues investing up to r_2. If for some reason firm 1's capital stock already exceeds its Stackelberg level, it stops immediately. The situation is symmetric above the 45° line in figure 13.4, which corresponds to states in which firm 2 has a head start. Thus, this equilibrium demonstrates how an advantage in investment speed or initial conditions can be exploited. The conditions of the growth phase (whose got there first, the costs of adjustment, etc.) have a permanent impact on the structure of the industry. This model also illustrates the importance of using the concept of perfect equilibrium to rule out empty threats.

It turns out that the equilibrium pictured in figure 13.4 is not unique; there are many others. To understand why, consider point A in figure 13.4, which is close to firm 2's reaction curve and past firm 1's reaction curve. The strategies specify that, from A on, both firms invest until r_2 is reached. However, both firms would prefer the *status quo* at A. Firm 1 in particular would not want to invest even if firm 2 stopped investing; it just invests in self-defense to reduce firm 2's eventual capital level. Both firms' stopping at A is an equilibrium in the subgame starting at A, enforced by the credible threat of going to B (or close to B) if anyone continues investing past A. Thus, the Markov restriction does not greatly restrict the set of equilibria in the investment game.

The existence of these early-stopping equilibria naturally depends on each firm's being able to respond quickly to its rival's investment. Sklivas (1986) studies the capital-accumulation game in (infinite-horizon) discrete time.[19] He shows that the early-stopping equilibria do exist as in the continuous-time model, but that the set of such equilibria shrinks and eventually vanishes when the upper bounds on the investment levels $\bar{a}_1$ and $\bar{a}_2$ become large. With the possibility of quick investment, a firm can get very close to its reaction curve within a period, which implies that its rival does not have time to react. The existence of early-stopping equilibria thus relies on the information lag being short relative to the speed of investment.

McLean and Sklivas (1988) consider the finite-horizon, discrete-time game. They show that backward induction has very strong implications: There exists an essentially unique MPE outcome (there may exist two MPE, but they differ only in the last move of one firm). This unique equilibrium converges to the Spence solution (stopping on the upper envelope of the reaction curves) as the horizon goes to ∞ and the discount factor goes to 1. In the last period, a firm that has not reached its reaction curve invests. In the penultimate period, firms invest if they are under their reaction curves as they know that the last period will have investment no matter what, and so forth. Studying the finite-horizon case thus highlights the bootstrap nature of the early-stopping equilibria and illustrates a similarity with the infinite-horizon, possibly cooperative equilibria and the finite-horizon noncooperative equilibria of repeated games (see chapter 5).

The early-stopping equilibria suggest two further refinements which may be in the Markov spirit. First, in the early-stopping equilibria, small causes do not have small effects. A lack of coordination or a small investment mistake drives the equilibrium away from the stopping curve to the upper envelope of the reaction curves. In the spirit of subsection 13.2.3 (which proved that, generically, MPE are robust to small payoff perturbations), one might require that the strategies be fairly insensitive to small variations in the state. However, in general games one cannot find a continuous equilibrium selection. Second, one might require infinite-horizon MPE to be limits of finite-horizon MPE. The idea would be that the modeler may know that the players know that the game ends at some date T, and T is known to the players but not to the modeler. It is not known what conditions imply that all infinite-horizon MPE are limits of finite-horizon MPE.

A word on the case of discounting with irreversible investment: Fudenberg and Tirole (1983) note that when the follower invests as quickly as possible to its reaction curve, it is no longer optimal for the leader to invest as quickly as possible. This is a simple point in optimal control. Suppose, for instance, that in figure 13.4 both firms were engaged in the investment path from D to S_1. Then firm 1 would be better off staying a bit

---

19. In discrete time, with discount factor δ between the periods, the reaction curves are defined by $\partial R_i(r_i(k_j), k_j)/\partial k_i = 1 - \delta + m_i$.

longer near its reaction curve. Its optimal control, in view of firm 2's strategy, is to adopt a "two-switchpoints" or "S-curve" investment strategy. That is, firm 1 may invest as fast as possible, then stop investing, and finally resume investment until the state reaches firm 2's reaction curve. The state then follows an S curve. Naturally, preemption motives may dominate, so that the optimal path may involve zero or one switch points. (For a more thorough analysis, see Nguyen 1986.)

13.4.2 Differential-Game Strategies

Hanig (1986, chapter 3) and Reynolds (1987a,b) have obtained interesting results in the reversible-investment case by applying differential-game techniques. They use the linear-quadratic specification of subsection 13.3.3:

$$R_i(k_i, k_j) = [d - b(k_i + k_j)]k_i,\text{[20]} \tag{13.18}$$

$$\frac{dk_i}{dt} = a_i - \rho k_i, \tag{13.19}$$

and either

$$C_i(k_i, a_i) = \tfrac{1}{2}c(a_i)^2 + \tilde{c}a_i \quad \text{(Reynolds)} \tag{13.20}$$

or

$$C_i(k_i, a_i) = \tfrac{1}{2}c(a_i - \rho k_i)^2 + \tilde{c}a_i \quad \text{(Hanig)} \tag{13.21}$$

(where d, b, c, and $\tilde{c}$ are all strictly positive).

Let us assume that $m_i = 0$ for simplicity (a symmetric maintenance cost can be included in d).

The Cournot-Nash level can be computed in the following heuristic way for the specification 13.21 (we leave it to the reader to check that $k^c = [d - \tilde{c}(r + \rho)]/[3b + c\rho(r + \rho)]$ for specification 13.20): Suppose that the firms are in a steady state at Cournot level (k^c, k^c), and let firm 1, say, increase its investment rate by 1 during a period of time dt, and then revert to its previous investment policy once time dt has elapsed. Because in a steady state $a_i = \rho k_i$, the cost of this investment is $\tilde{c}(dt)$. *If this investment does not affect firm 2's investment policy* (this is the open-loop assumption underlying the steady-state Cournot levels), the extra revenue for firm 1 is

$$\int_0^\infty (d - 3bk^c)(e^{-\rho s}dt)e^{-rs}\,ds = (d - 3bk^c)\frac{dt}{r + \rho},$$

since firm 1's marginal revenue is $d - bk_2 - 2bk_1 = d - 3bk^c$, and the remaining proportion of the extra investment s units of time after it is

---

20. For conditions under which such revenue functions may emerge from Cournot competition, see Fudenberg and Tirole [1983]; on their emerging from price competition under capacity constraints k_i, see chapter 5 of Tirole 1988.

made is $e^{-\rho s}$. At the Cournot level, one must have $\tilde{c} = (d - 3bk^c)/(r + \rho)$ or $k^c = (d - \tilde{c}(r + \rho))/3b$.

Hanig and Reynolds solve equations 13.16 and 13.17 for this game. They find that there exists a unique linear equilibrium, and that the investment strategy $a_i = s_i(k_i, k_j)$ decreases (linearly) with firm j's capital level, k_j. They show that the capital levels (k_1, k_2) converge to a steady state (k_1^*, k_2^*).[21] In the symmetric case, the symmetric steady-state level k^* strictly exceeds the Cournot level derived above. Thus, the two firms engage in a sort of "symmetric Stackelberg behavior," as at the Cournot level each firm has an incentive to increase its capital at least a bit. If its rival didn't react, such a move would affect the firm's profit only to the second order from the previous reasoning; but in the perfect equilibrium, the rival responds to a higher level of capital by reducing its investment. Because of the existence of adjustment costs, the current increase in capital has commitment power, because it would be very costly for the firm to move back quickly to its Cournot level. (Interestingly enough, k^* does not converge to k^c when the adjustment cost c tends to 0. There is thus a "discontinuity" at $c = 0$. In contrast, k^* converges to k^c when the adjustment cost tends to infinity. Reader: Find out why.)

The investment path to the steady state also has interesting features. In particular, a firm may "overshoot" its steady-state capital level (see figure 13.5, which is drawn from Hanig 1986). This result is along the lines of the

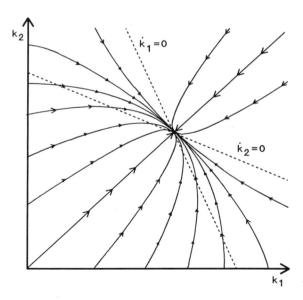

Figure 13.5

---

21. As in the open-loop case (Fershtman and Muller 1984), the steady state is independent of the initial levels of capital.

result found in the irreversible-investment case that a firm may invest beyond its Cournot level to reduce its rival's capital level. Here the firm does not reduce its rival's capital level in the long run, but reduces it in the medium run.

Last, Hanig shows that, in the asymmetric case in which c_1 is very large and c_2 very small, the perfect-equilibrium steady state is close to the Stackelberg level with firm 1 as the leader (as firm 1's capital level has more commitment power).

Exercises

Exercise 13.1** Figure 13.2 shows that, unlike the Nash-equilibrium and perfect-equilibrium correspondences, the MPE correspondence does not have a closed graph in U. Does it have a closed graph in $U(H)$? (Hint: Think of an MPE as a perfect equilibrium of a game with a different information structure.)

Exercise 13.2** Consider the nonmyopic Cournot tâtonnement model. Firms 1 and 2 alternate in choosing quantities q_1 and q_2. Quantities are fixed for two periods, i.e., $q_1^{2n+2} = q_1^{2n+1}$ and $q_2^{2n+1} = q_2^{2n}$ for all n. Payoffs are $\sum_{t=0}^{\infty} \delta^t g_i(q_i^t, q_j^t)$ where $\partial^2 g_i / \partial q_i^2 < 0$, $\partial g_i / \partial q_j < 0$, and $\partial^2 g_i / \partial q_j \, \partial q_i < 0$. Firms adopt Markov strategies: $q_i = r_i(q_j)$.

(a) Check that the "reaction functions" r_i are nonincreasing.

(b) Suppose that $g_i = q_i(1 - q_i - q_j)$. Look for a linear MPE $r_i(q_j) = a - bq_j$. Compare its steady state to the Cournot level ($\frac{1}{3}$). How does it vary with δ? Qualitatively compare the solution against that obtained by Hanig and Reynolds in the capital-accumulation game (see subsection 13.4.2).

(c) Still in the quadratic model of question b, note by backward induction that the reaction functions are linear in the *finite-horizon* (T) model, and that the function that maps the slope and the intercept of r_{t+1}^T and r_{t+2}^T into the slope and intercept of r_{t-1}^T and r_t^T is a contraction mapping in the space of linear functions with slopes in $[-\frac{1}{2}, 0]$ and intercepts in $[0, 1]$. Conclude that the reactions functions r_t^T converge to the infinite-horizon solution derived in part b of this exercise when T tends to $+\infty$. (See Maskin and Tirole 1987 for the answers.)

Exercise 13.3** Two firms play the Cournot game (see chapter 1) repeatedly. Let a_i^t denote firm i's output at date t, and let $a^t \equiv a_1^t + a_2^t$. A fraction $\varepsilon > 0$ of the good sold at date t is recycled (once). The consumers do not receive income when the good they consumed is recycled (a recycling industry purchases the old units at price 0—this ensures that consumers are myopic). The inverse demand curve at date t is $p^t = 1 - a^t - \varepsilon a^{t-1}$. Production by the duopolists is costless. Assume that ε is "small."

(a) What is the payoff-relevant variable in this game?

(b) Write the first-order condition for an MPE. (Use dynamic programming, and introduce valuation functions $V_i(a^{t-1})$.)

(c) Look for a symmetric equilibrium with quadratic valuation function, so $dV/da^{t-1} = -\alpha + \beta a^{t-1}$. Show that $\beta(3 - 2\delta\beta) = \varepsilon^2$ and $\alpha\varepsilon = \beta(1 - \delta\alpha)$ (where δ is the discount factor).

Exercise 13.4** Consider the following three-period game, which is due to Harris (1990): In period 1, two gamblers A and B pick $a \in [0, 1]$ and $b \in [0, 1]$ respectively. In period 2, two greyhounds C and D each receive an injection of size $a + b$, which changes their attitude toward the race. They pick $c \in [0, 1]$ and $d \in [0, 1]$, which are the times in which they complete the course. In period 3 each of two referees E and F must declare a winner, picking $e \in \{C, D\}$ and $f \in \{C, D\}$ respectively. In each period choices are made simultaneously, and players in later periods observe the actions taken by players in earlier periods.

Gambler A obtains a payoff of $1 - a$ if greyhound C is declared the winner by both referees, and $-1 - a$ otherwise. Similarly, gambler B obtains $1 - b$ if both referees declare D to be the winner, and $-1 - b$ otherwise. In other words, A wants C to win, B wants D to win, and both want a result. They would also like to keep their contributions to the injection as small as possible. The payoff to greyhound C is $2c$ if $e = C$ and $1 - (a + b)(1 - c)$ if $e = D$. That is, the form of his payoff depends on whether he or the other greyhound is declared the winner by referee E, but either way he would prefer to run the race as slowly as possible. Also, he would prefer to be first rather than second provided that he does not have to run too fast. The payoff to greyhound D is $2d$ if $e = D$ and $1 - (a + b)(1 - d)$ if $e = C$; like greyhound C, he is interested only in the verdict of referee E. Lastly, referee E gets payoff d if he declares C to be the winner, and c if he declares D to be the winner. Referee F's payoffs are identical. Show that there exists no MPE. (There actually exists no subgame-perfect equilibrium.)

Exercise 13.5** Two firms costlessly produce an infinitely durable good at dates $0, 1, \ldots$. At the beginning of date t, the existing stock of the good is X^t. The firms simultaneously choose outputs a_1^t and a_2^t and the date-t rental price is $1 - X^{t+1}$, where $X^{t+1} \equiv X^t + a_1^t + a_2^t$. The discount factor is δ. Look for a symmetric MPE in which the buyers have rational expectations (date-t price is equal to date-t rental value plus discounted date-$(t + 1)$ price), the strategies are linear in the state variable, and the valuation functions are quadratic. (For more details, see Carlton and Gertner 1989.)

References

Amir, R. 1989. A lattice-theoretic approach to a class of dynamic games. *Computers Math. Applic.* 17: 1345–1349.

Amit, I., and A. Halperin. 1989. Sharing a common product. Mimeo, Hebrew University, Jerusalem.

Anderson, R. 1985. Quick response equilibria. Mimeo, University of California, Berkeley.

Arrow, K., and M. Kurz. 1970. *Public Investment, the Rate of Return, and Optimal Fiscal Policy*. Johns Hopkins University Press.

Basar, T., and G. Olsder. 1982. *Dynamic Noncooperative Game Theory*. Academic Press.

Beals, R., and T. Koopmans. 1969. Maximizing stationary utility in a constant technology. *SIAM Journal of Applied Mathematics* 17: 1001–1015.

Bénabou, R. 1989. Optimal price dynamics and speculation with a storable good. *Econometrica* 57: 41–80.

Berkovitz, L. 1971. Lectures on differential games. In *Differential Games and Related Topics*, ed. H. Kuhn and G. Szegö. North-Holland.

Bernheim, D., and D. Ray. 1989. Markov-perfect equilibria in altruistic growth economies with production uncertainty. *Journal of Economic Theory* 47: 195–202.

Blaquiere, A. 1971. An introduction to differential games. In *Differential Games and Related Topics*, ed. H. Kuhn and G. Szegö. North-Holland.

Carlton, D., and R. Gertner. 1989. Market power and mergers in durable-good industries. *Journal of Law and Economics* 32: S203–S232.

Case, J. H. 1969. Toward a theory of many player differential games. *SIAM Journal of Control* 7: 179–197.

Clemhout, S., and H. Y. Wan, Jr. 1979. Interactive economic dynamics and differential games. *Journal of Optimization Theory and Applications* 27: 7–28.

Cyert, R., and M. DeGroot. 1970. Multiperiod decision models with alternating choice as a solution to the duopoly problem. *Quarterly Journal of Economics* 84: 410–429.

Dana, R. A., and L. Montrucchio. 1987. Dynamic complexity in duopoly games. *Journal of Economic Theory* 40: 40–56.

Duffie, D., J. Geanakoplos, A. Mas-Colell, and A. McLennan. 1988. Stationary Markov equilibria. Mimeo, Stanford University.

Dutta, P., and R. Sundaram. 1988. The tragedy of the commons? A characterization of stationary perfect equilibria in dynamic games. Discussion paper 397, Columbia University.

Dutta, P., and R. Sundaram. 1990a. Stochastic games of resource allocation: Existence theorems for discounted and undiscounted models. Working paper 241, University of Rochester.

Dutta, P., and R. Sundaram. 1990b. How different can strategic models be? Non-existence, chaos and underconsumption in Markov perfect equilibria. Working paper 242, University of Rochester.

Eaton, J., and M. Engers. 1990. Intertemporal price competition. *Econometrica* 58: 637–660.

Elliott, R. 1976. The existence of the value in stochastic differential games. *SIAM Journal of Control and Optimization* 14: 85–94.

Fershtman, C., and M. Kamien. 1987. Dynamic duopolistic competition with sticky prices. *Econometrica* 55: 1151–1164.

Fershtman, C., and E. Muller. 1984. Capital accumulation games of infinite duration. *Journal of Economic Theory* 33: 322–339.

Fichefet, J. 1970. Quelques conditions d'existence de points de selle pour une classe de jeux differentiels de durée fixée. In *Colloque sur la Théorie Mathématique du Contrôle Optimal*, CBRM (Vander, Louvain).

Friedman, J. 1986. *Game Theory with Applications to Economics*. Oxford University Press.

Fudenberg, D., and J. Tirole. 1983. Capital as commitment: Strategic investment to deter mobility. *Journal of Economic Theory* 31: 227–256.

Gelman, J., and S. Salop. 1983. Judo economics: Capacity limitation and coupon competition. *Bell Journal of Economics* 14: 315–325.

Gertner, R. 1986. Dynamic duopoly with price inertia. In Ph.D. thesis, Department of Economics, Massachusetts Institute of Technology.

Goldman, S. 1980. Consistent plans. *Review of Economics Studies* 47: 533–537.

Guesnerie, R., and J.-J. Laffont. 1984. A complete solution to a class of principal-agent problems with an application to the control of a self-managed firm. *Journal of Public Economics* 25: 329–369.

Gurvich, V. 1986. A stochastic game with complete information and without equilibrium situations in pure stationary strategies. *Communications of the Moscow Mathematical Society* 171–172.

Halperin, A. 1990. Price competition and inflation. Mimeo, Department of Economics, Massachusetts Institute of Technology.

Hanig, M. 1986. Differential gaming models of oligopoly. Ph.D. thesis, Massachusetts Institute of Technology.

Harris, C. 1985. Existence and characterization of perfect equilibrium in games of perfect information. *Econometrica* 53: 613–628.

Harris, C. 1988. Dynamic competition for market share: An undiscounted model. Discussion paper 30, Oxford University (Nuffield College).

Harris, C. 1990. The existence of subgame-perfect equilibrium with and without Markov strategies: A case for extensive-form correlation. Mimeo, Oxford University (Nuffield College).

Harris, C., and J. Vickers. 1987. Racing with uncertainty. *Review of Economic Studies* 54: 1–22.

Hellwig, M., and W. Leininger. 1987. On the existence of subgame-perfect equilibrium in infinite-action games of perfect information. *Journal of Economic Theory* 43: 55–75.

Hellwig, M., and W. Leininger. 1989. Markov-perfect equilibrium in games of perfect information. Mimeo, University of Bonn.

Isaacs, R. 1954. Differential games, I, II, III, IV. Reports RM-1391, 1399, 1411, and 1486, Rand Corporation.

Isaacs, R. 1965. *Differential Games*. Wiley.

Judd, K. 1985. Closed-loop equilibrium in a multi-stage innovation race. Mimeo.

Judd, K. 1990. Cournot vs. Bertrand: A dynamic resolution. Mimeo, Hoover Institution.

Kirman, A., and M. Sobel. 1974. Dynamic oligopoly with inventories. *Econometrica* 42: 279–287.

Kohlberg, E. 1976. A model of economic growth with altruism between generations. *Journal of Economic Theory* 13: 1–13.

Kreps, D., and A. M. Spence. 1984. Modelling the role of history in industrial organization and competition. In *Contemporary Issues in Modern Microeconomics*, ed. G. Feiwel. Macmillan.

Lancaster, K. 1973. The dynamic inefficiency of capitalism. *Journal of Political Economy* 81: 1098–1109.

Leininger, W. 1986. The existence of perfect equilibria in a model of growth with altruism between generations. *Review of Economic Studies* 53: 349–368.

Levhari, D., and L. Mirman. 1980. The great fish war. *Bell Journal of Economics*, pp. 322–344.

Levine, J., and J. Thepot. 1982. Open loop and closed loop in a dynamic duopoly. In *Optimal Control and Economic Analysis*, ed. G. Feichtinger. North-Holland.

Loury, G. 1990. Tacit collusion in a dynamic duopoly with indivisible production and cumulative capacity constraints. Mimeo, Kennedy School of Government, Harvard University.

Maskin, E., and J. Tirole. 1987. A theory of dynamic oligopoly. III. Cournot competition. *European Economic Review* 31: 947–968.

Maskin, E., and J. Tirole. 1988a. A theory of dynamic oligopoly. I. Overview and quantity competition with large fixed costs. *Econometrica* 56: 549–570.

Maskin, E., and J. Tirole. 1988b. A theory of dynamic oligopoly. II. Price competition. *Econometrica* 56: 571–600.

Maskin, E., and J. Tirole. 1989. Markov equilibrium. Mimeo, Harvard University.

McLean, R., and S. Sklivas. 1988. Capital accumulation in an intertemporal duopoly. Discussion paper 145, Columbia University.

Mertens, J.-F., and T. Parthasarathy. 1987. Equilibria for discounted stochastic games. CORE research paper 8750, Université Catholique de Louvain.

Nguyen, D. 1986. Capital investment in a duopoly as a differential game. Mimeo, City University of New York.

Papavassilopoulos, G., and J. Cruz. 1979. On the uniqueness of Nash strategies for a class of analytic differential games. *Journal of Optimization Theory and Applications* 27: 309–314.

Papavassilopoulos, G. P., et al. 1979. On the existence of Nash strategies and solutions to coupled Riccati equations in linear-quadratic games. *Journal of Optimization Theory and Applications* 28: 49–76.

Parthasarathy, T. 1982. Existence of equilibrium stationary strategies in discounted stochastic games. *Sankya* 44: 114–127.

Phelps, E., and R. Pollak. 1968. On second-best national savings and game equilibrium growth. *Review of Economic Studies* 35: 185–199.

Pindyck, R. 1977. Optimal economic stabilization policies under decentralized control and conflicting objectives. *IEEE Transactions on Automatic Control* 22: 517–530.

Pontryagin, L. S., V. G. Boltyanskii, R. V. Gamkrelidze, and E. F. Mischenko. 1962. *The Mathematical Theory of Optimal Processes.* Tr. K. N. Trirogoff. Wiley.

Reinganum, J. 1982. A dynamic game of R&D: Patent protection and competitive behavior. *Econometrica* 50: 671–688.

Reynolds, S. 1987a. Capacity investment, preemption, and commitment in an infinite horizon model. *International Economic Review* 28.

Reynolds, S. 1987b. Capital accumulation and adjustment costs: A dynamic game approach. Mimeo, University of Arizona.

Rieder, U. 1979. Equilibrium plans for nonzero sum Markov games. In *Game Theory and Related Topics*, ed. O. Moeschlin and D. Pallasche. North-Holland.

Schelling, T. 1960. *The Strategy of Conflict.* Harvard University Press.

Shapley, L. 1953. Stochastic games. *Proceedings of the National Academy of Sciences* 39: 1095–1100.

Sobel, M. 1971. Noncooperative stochastic games. *Annals of Mathematical Statistics* 42: 1930–1935.

Simaan, M., and T. Takayama. 1978. Game theory applied to dynamic duopoly problems with production constraints. *Automatica* 14: 161–166.

Sklivas, S. 1986. Capital as a commitment in discrete time. Mimeo, Columbia University.

Smart, D. R. 1974. *Fixed Point Theorems.* Cambridge University Press.

Spence, A. M. 1979. Investment strategy and growth in a new market. *Bell Journal of Economics* 10: 1–19.

Starr, A., and Y. C. Ho. 1969. Nonzero-Sum differential games, *Journal of Optimization Theory and Applications* 3: 183–206.

Sundaram, R. 1989. Nash equilibrium in a class of symmetric dynamic games: An existence theorem. *Journal of Economic Theory* 47: 153–177.

Tirole, J. 1988. *The Theory of Industrial Organization* MIT Press.

Uchida, K. 1978. On existence of *n*-person nonzero sum stochastic differential games. *SIAM Journal of Control and Optimization* 16: 142–149.

Villas-Boas, M. 1990. Dynamic duopolies with non-convex adjustment costs. Mimeo, Massachusetts Institute of Technology.

Wernerfelt, B. 1988. On existence of a Nash equilibrium point in *N*-person non-zero sum stochastic jump differential games. *Optimal Control Applications and Methods* 9: 449–456.

Whitt, W. 1980. Representation and approximation of noncooperative sequential games. *SIAM Journal of Control and Optimization* 1: 35–48.

14 Common Knowledge and Games

14.1 Introduction[††]

The idea of common knowledge—that players know that their opponents know that they know...—is a useful tool for understanding how the equilibria of a game depend on its information structure. This chapter gives a formal definition of common knowledge of an event, and illustrates its implications in a few examples of games.

Section 14.2 gives two equivalent definitions of common knowledge. The first is the recursive one just mentioned: An event is common knowledge if players know this event, know that the other players know this event, and so on *ad infinitum*. The second is perhaps less natural but is simpler to apply: To be common knowledge, an event must be implied by (that is, be a superset of) the element in the finest common coarsening of the players' partitions that contains the state of nature. We apply the definitions to the celebrated "dirty face" example.

Section 14.3 illustrates the difference between knowledge (that everyone knows) and common knowledge (that everyone knows that everyone knows that...) of the payoff functions in games. The point of that section is to emphasize some strong implications of common knowledge. We first add strategies and payoffs for the players to the "dirty face" information structure introduced in section 14.2, and show how the set of equilibria is changed by the introduction of an extra player who publicly announces something that everyone knew was true but that was not common knowledge. This example illustrates the strikingly different implications of knowledge and common knowledge for games. The second illustration is the archetypal result in the theory of asset pricing that trade among asymmetrically informed players should not occur if the allocation determined before they receive their private information is Pareto optimal. This result is shown to be closely related to the fact that, starting from a common prior on states of nature, players cannot disagree on the posterior probability of an event if the posterior probabilities of the event are common knowledge.

Section 14.4 asks whether a Nash equilibrium of a given game is close to one of a perturbed game with an information structure very "close" to that of the original game. Because this is a lower-hemicontinuity property, it can hold at most for generic games. (See sections 1.3, 12.1, and 13.2.) But even for generic games, one must use an appropriate definition of what it means for two information structures to be close. The "electronic mail" example developed in section 14.4 shows that such a definition must be stringent. However, it turns out that a simple generalization of common knowledge, "almost common knowledge," ensures generic lower hemicontinuity. Take a given game with payoffs u and a Nash equilibrium σ, and consider a perturbed game in which it is very likely that payoffs are u but in which payoffs may differ from u with small probability. Roughly, game u is almost common knowledge if each player puts high probability on payoffs' being u, puts high probability on other players' putting high probability on payoffs' being u, and so on *ad infinitum*. For generic games u, the perturbed game has an equilibrium near σ if payoffs u are almost common knowledge.[1]

Our focus throughout is on the players' knowledge about payoff functions and other exogenous data of the game. There is also an extensive literature characterizing various equilibrium concepts in terms of the players' beliefs about the others' *strategies*, about their opponents' beliefs, and so forth (see, e.g., Aumann 1987, Brandenburger and Dekel 1987, and Tan and Werlang 1988; Brandenburger 1990 and Brandenburger and Dekel 1990 are recent surveys).

14.2 Knowledge and Common Knowledge[2] ††

Before defining common knowledge we must give a definition of knowledge. That is, when will we say that an agent "knows" something? As throughout the book, we represent agent i's beliefs by an information partition H_i. Formally, we suppose that the exogenous uncertainty in the model is represented by a finite set Ω of nature's moves and a common prior distribution p, and that all of player i's information about ω is represented by the element (or event) $h_i(\omega)$ of H_i that contains ω. The interpretation is that player i knows that the true state is some $\omega' \in h_i(\omega)$, but he does not know which one it is. In particular, player i's information partition represents all the information he has about the information of other players, about their information about his information, and so on. (See the discussion of the notion of type in chapter 6.) We suppose that all states in Ω have positive prior probability; states with probability 0 are dropped from the description of the state space.

Player i's posterior beliefs about the state when knowing that $h_i(\omega) = h_i$ are given by

$$p(\omega \mid h_i) = p(\omega) \Big/ \sum_{\omega' \in h_i} p(\omega') = p(\omega)/p(h_i).$$

We then say that player i knows event E at ω if he knows that the true state lies in E—that is, if $h_i(\omega) \subseteq E$. The event "player i knows E," denoted $K_i(E)$,

---

1. These perturbations are closely related to those in the Fudenberg-Kreps-Levine paper discussed in section 11.4, except that here we do not restrict our attention to "personal types." Fudenberg, Kreps, and Levine show that with the more general elaborations discussed in this chapter, any Nash equilibrium is near-strict. We should also point out that in a sequence of elaborations $\tilde{E}$ that approach E in the sense of section 11.4, the payoff functions of E become almost common knowledge.
2. Our presentation of this material draws heavily on the excellent survey by Binmore and Brandenburger (1989).

is then $\{\omega | h_i(\omega) \subseteq E\}$. Since information partitions must satisfy $\omega \in h_i(\omega)$, if player i knows E, then E is true.[3,4] With this formulation of knowledge, more precise information corresponds to knowing a *smaller* set: Knowledge here is the ability to rule out some of the states that were possible *ex ante*. In particular, if a player knows that the true state is in E, he knows that the true state is in any superset of E.

The event "everyone knows E," denoted $K_{\mathscr{I}}(E)$, is then the set

$$\left\{\omega \left| \bigcup_{i \in \mathscr{I}} h_i(\omega) \subseteq E \right. \right\}.$$

Then, because all players know the information partitions, player i knows that everyone knows E if $h_i(\omega) \subseteq K_{\mathscr{I}}(E)$, and the event that everyone knows that everyone knows E is

$$K_{\mathscr{I}}^2(E) = \left\{\omega \left| \bigcup_{i \in \mathscr{I}} h_i(\omega) \subseteq K_{\mathscr{I}}(E) \right. \right\}.$$

Then event $K_{\mathscr{I}}^{\infty}(E)$ is the intersection of all sets of the form $K_{\mathscr{I}}^n(E)$, which is a decreasing sequence of events in the sense that $K_{\mathscr{I}}^{n+1}(E) \subseteq K_{\mathscr{I}}^n(E)$ for all n.

Definition 14.1 Event E is common knowledge at ω if $\omega \in K_{\mathscr{I}}^{\infty}(E)$.

If E is common knowledge, then any statement of the form "player i knows that players j and k know that player i knows that m knows...E" is true. The term "common knowledge" was first used to describe the infinite regress of "I know that you know" by Lewis (1969), who attributed the basic idea to Schelling (1960). Aumann (1976) proposed the notion independently, and gave a characterization of common knowledge in terms of the *meet* of the individual agents' partitions; we discuss this below. It is interesting to note that Littlewood (1953) developed some examples of common-knowledge-type reasoning without developing a formal definition of the concept.

The definition of common knowledge takes for granted a state space and information partitions that incorporate *all* of the agents' initial uncertainty about the structure of the game. This framework makes the information

---

3. This is called the "axiom of knowledge." Other axioms implied by the partition formulation are $K_i(E) = K_i K_i(E)$ (player i knows E if and only if he knows that he knows it) and $\sim K_i(\sim K_i(E)) \subseteq K_i(E)$ (if player i does not know that he does not know E, then he knows E). Although the partition model of knowledge is standard for decision theory, with other interpretations of knowledge this model may be too strong. Bacharach (1985), Brown and Geanakoplos (1988), Geanakoplos (1989), Rubinstein and Wolinsky (1989), Samet (1987), and Shin (1987) discuss common knowledge when knowledge is modeled by more general "knowledge operators" K_i that need not be derived from partitions. This work is discussed in the survey by Binmore and Brandenburger (1989).

4. If some states had probability 0, then it might be that $h_i(\omega)$ is not contained in E, yet player i assigns E posterior probability 1. See Brandenburger and Dekel 1987a for a discussion of how to extend the definitions of knowledge and common knowledge to models in which some states have prior probability 0.

partitions common knowledge in an informal sense. Otherwise, if (say) player 2 did not know whether player 1's information when ω occurred was h_1' or h_1'', we would need to add additional states of the world to Ω to model the different beliefs player 2 might have; since the players' beliefs derive from a common prior, there would have to be positive probability that each of player 2's beliefs was in fact correct.

Returning to the formal definition of common knowledge, we can easily check that if everyone knows E, then E must be true, that is, $K_{\mathcal{I}}(E) \subseteq E$. Thus, as we iterate the (everyone knows) operator, the set of states included cannot grow, and if E is common knowledge and the state space is finite, there must be a finite n such that $K_{\mathcal{I}}^n(E) = K_{\mathcal{I}}^\infty(E)$.

As an illustration of this definition, consider the following examples, which are variants of an example originally developed by Littlewood (1953).

Example 14.1: Dirty Face without a Sage

Suppose there are three players and eight states, written in binary notation as 000, 001, 010, 011, etc., and that all states have equal prior probability. Player 1 knows the second and third components of the state but not the first, player 2 knows the first and third but not the second, and player 3 knows the first and second but not the third. In terms of the information partitions, this means that H_1 has the four elements $\{000, 100\}$, $\{001, 101\}$, $\{010, 110\}$, and $\{011, 111\}$. In a famous story we develop below, the ith component of the state is 0 if player i's face is clean and 1 if player i's face is dirty; each player observes the others' faces but not his own.

With this information structure, everyone knows the event $E^* =$ "at least one player's face is dirty"—that is, "not 000"—if there are at least two dirty faces; if there is only one dirty face, then the player whose face is dirty doesn't know if there are zero or one dirty faces. So

$$K_{\mathcal{I}}(E^*) = \{111, 110, 101, 011\} \equiv E^{**}.$$

Then $K_{\mathcal{I}}^2(E^*) = K_{\mathcal{I}}(E^{**}) = 111$: At 101, for example, player 1 cannot rule out the state 001, which is not in E^{**}. Finally, $K_{\mathcal{I}}^3(E^*) = K_{\mathcal{I}}(111) = \varnothing$, since no player can distinguish 111 from the state in which his face is clean and the others are dirty. Thus, there is no ω at which E^* is common knowledge. In fact, theorem 14.1 below makes it easy to check that the only event that is common knowledge is the whole state space Ω.

Example 14.2: Dirty Face with a Sage

Next, we modify the information partitions in example 14.1 so that if all three faces are clean, all players are informed of this in public by a sage. Then player 1's partition H_1 has the five elements $\{000\}$, $\{100\}$, $\{001, 101\}$, $\{010, 110\}$, and $\{011, 111\}$. Now the event 000 is common knowledge when it occurs, as is the complementary event $K_{\mathcal{I}}(E^*) = E^*$, so $K_{\mathcal{I}}^2(E^*) = E^*$, and so on.

In examples 14.1 and 14.2 it was fairly easy to determine when a state was common knowledge by applying definition 14.1 directly. This is not always the case. Aumann (1976) gives an equivalent definition of common knowledge that provides a simple algorithm for determining the commonly known information without explicitly iterating the (everyone knows) operator. To present this definition, we first recall that the *meet* $\mathcal{M}$ of a collection of partitions H_i is the finest common coarsening of the partitions. We let $M(\omega)$ be the element of $\mathcal{M}$ containing ω. The requirement that $\mathcal{M}$ be a common coarsening means that it is not more informative than any of the H_i; that is, for all players i and all ω,

$$h_i(\omega) \subseteq M(\omega).$$

$\mathcal{M}$ is the finest common coarsening if there does not exist another common coarsening $\mathcal{M}'$ with $M'(\omega) \subseteq M(\omega)$ for all ω and strict inclusion— $M'(\hat{\omega}) \subset M(\hat{\omega})$—for at least some $\hat{\omega}$.[5]

The idea of "reachability" provides a simple algorithm for computing the meet, and some intuition for understanding theorem 14.1 below. It is easy to see that $\omega' \in M(\omega)$ if there exists a chain $\omega_0 \equiv \omega, \omega_1, \omega_2, \ldots, \omega_m \equiv \omega'$ such that for all $k \in \{0, \ldots, m-1\}$ there exists a player $i(k)$ such that $h_{i(k)}(\omega_k) = h_{i(k)}(\omega_{k+1})$. In words, there exists a chain of states from ω to ω' such that two consecutive states are in the same information set of some player. One can also check that ω' belongs to $M(\omega)$ only if ω' is *reachable* from ω in the above sense.

Theorem 14.1 (Aumann 1976)   Let $\mathcal{M}$ be the meet of the individual players' partitions. Event E is common knowledge at ω if and only if $M(\omega) \subseteq E$.

The intuition for theorem 14.1 is as follows: If there exists ω' reachable from ω through $\omega_1, \ldots, \omega_{m-1}$, then player $i(0)$ cannot exclude the possibility that ω_2 is consistent with the information of player $i(1)$, who in turn cannot exclude the possibility that ω_3 is consistent with the information of player $i(2)$, who.... Hence, someone believes that someone believes that...that someone believes that ω' is possible, and thus no event E that excludes ω' can be common knowledge. In contrast, if any chain of "i knows that j knows that..." is "trapped" in E (that is, if $M(\omega) \subseteq E$), everyone knows that everyone knows that...the state is in E.

Figure 14.1 gives an example, where $\Omega = (1, 2, 3, 4)$, $H_1 = \{(1, 2), (3, 4)\}$, $H_2 = \{(1), (2, 3), (4)\}$, and $\omega = (2)$. Then player 2 cannot rule out state 3, and at state 3 player 1 cannot rule out state 4. Since player 1 cannot rule out state 1,

$$M(2) = \Omega = (1, 2, 3, 4),$$

---

5. Actually, if this inclusion holds for some $\hat{\omega}$ it must also hold for some $\tilde{\omega} \in \Omega \backslash M'(\hat{\omega})$. Proving this is exercise 14.4.

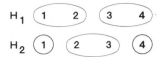

Figure 14.1

and so all that is commonly known at state 2 is the original set of possible states.

Proof of Theorem 14.1   First we claim that $M(\omega)$ is common knowledge at every $\omega' \in M(\omega)$:

$$K_{\mathscr{I}}(M(\omega)) = \left\{\omega \,\middle|\, \bigcup_{i \in \mathscr{I}} h_i(\omega) \subseteq M(\omega)\right\} = M(\omega),$$

since $\mathscr{M}$ is a coarsening of each H_i, and so $K_{\mathscr{I}}^n(M(\omega)) = M(\omega)$ for all n, and $K_{\mathscr{I}}^{\infty}(M(\omega)) = M(\omega)$. Next if E contains $M(\omega)$, then, since $M(\omega)$ is common knowledge, so is E.

Conversely, if E is common knowledge at ω,

$$M(\omega) \subseteq E.$$

To see this, suppose that there exists $\omega' \in M(\omega)$ such that $\omega' \notin E$. Because $\omega' \in M(\omega)$, and because of the reachability criterion, there exists a sequence $k = 0, \ldots, m$ with associated states of nature $\omega_0, \ldots, \omega_m$ and information sets $h_{i(k)}(\omega_k)$ such that $\omega_0 = \omega$, $\omega_m = \omega'$, and $\omega_k \in h_{i(k)}(\omega_{k+1})$. But at information set $h_{i(m)}(\omega_{m-1})$, player $i(m)$ does not know event E; working backward on k, we see that event E cannot be common knowledge.   ∎

If $E = K_{\mathscr{I}}(E)$, we say that event E is a *common truism* (Binmore and Brandenburger 1989), a *public event* (Milgrom 1981), or *self-evident* (Samet 1987). Clearly a common truism is common knowledge whenever it occurs.[6] Moreover, the proof above shows that the common truisms are precisely the elements of $\mathscr{M}$ and unions of elements of $\mathscr{M}$, and so any commonly known event must be the consequence of a common truism. Note how easy this makes it to determine which events can be common knowledge in example 14.1: Since no player knows the state of his own face, the only common truism is the whole state space.

14.3   Common Knowledge and Equilibrium[††]

We now analyze two games in which common knowledge has strong implications. Throughout our discussion, we assume that the structure of the game is common knowledge in an informal sense. Applying formal

---

6. Iterating the (everybody knows) operator: $K_{\mathscr{I}}(E) = K_{\mathscr{I}}^2(E) = \cdots$ and therefore $E = K_{\mathscr{I}}^{\infty}(E)$.

definitions of common knowledge to the structure of the game leads to technical and philosophical problems that we prefer not to address.

14.3.1   The Dirty Faces and the Sage

To illustrate how the equilibrium strategies in a state of nature (more precisely, the projection of the set of equilibrium profiles onto the state) can vary with what is common knowledge, we return to the examples 14.1 and 14.2, which gave the story of the clean and dirty faces. To make this into a game, suppose that in each of $T + 1$ periods ($t = 0, 1, \ldots, T$, where $T \geq 2$) the three players simultaneously decide whether or not to blush, and that their actions are revealed at the end of each period (so this is a multi-stage game with observed actions). Each player can blush in at most one period. Each player receives payoff δ^t if he blushes in period t when his face is dirty, 1 if he does not blush when his face is clean, -100 if he blushes when his face is clean, and -1 if he does not blush when his face is dirty. Thus, no player will blush unless he is quite certain that his face is dirty. We assume that the discount factor δ is smaller than 1 (so that a player blushes immediately when he knows that his face is dirty), that each player's face is equally likely to be clean or dirty, and that the states of the three faces are independently distributed.

We begin with the information structure of example 14.1, in which each player's information is the state of the other two players' faces. We claim that the unique Nash equilibrium is for no player ever to blush, even if all faces are actually dirty. To see that this is an equilibrium, note that players will learn nothing from their opponents' play and that, as long as no player deviates, each player's posterior beliefs about his own face equals the prior, which is that clean and dirty are equally likely. To see that not blushing is the only Nash equilibrium of the game, let t_0 denote the first date at which at least one player (player i, say) blushes with positive probability. At date t_0 player i has not learned anything, because no one blushes before t_0. Therefore, his posterior belief that his face is dirty is still 0.5, and he gets a very negative expected payoff by blushing at t_0.

Now consider the information structure of example 14.2. This can be explained by saying that there is a sage who will announce at the beginning of the first period that at least one face is dirty if and only if this is in fact the case. With this information structure there is no longer a Nash equilibrium where players never blush when all their faces are dirty.

To see this, we proceed by induction on the number of dirty faces. When exactly one face is dirty, the sage announces that there is at least one dirty face; the player with a dirty face sees two clean faces and thus will blush in the first period (because of discounting). Since all players know that their opponents know the game structure, all players know that a player with a dirty face would blush if there were exactly one dirty face. Thus, if no one blushes in the first period, everyone knows that there are at least two dirty faces; this fact is common knowledge in the formal sense, because the structure of the game is common knowledge in the informal sense. Continuing with the induction: If there are exactly two dirty faces, two players each see one clean face, and these two players will blush in the second period. Thus, if no one blushes in the second period, all players know that all three faces are dirty, and all players blush in the third period. More generally, it is easy to see that, with I players, if all faces are dirty, all blush at date $I - 1$.

This example is sometimes analyzed by considering only the change the announcement makes in the state where all faces are dirty. Since even without the sage's announcement all the players know that there is at least one dirty face when all faces are dirty, the only change the announcement makes *in this state* is to make common knowledge a fact that was previously known to all the players. An alternative interpretation is that, in any state where only one face is dirty, the sage's announcement gives the player with the dirty face payoff-relevant information he did not have before. This observation can be generalized: With a fixed state space Ω and a prior p, if E is not common knowledge at ω with partitions $\{H_i\}$ but E is common knowledge at ω with partitions $\{\hat{H}_i\}$, there must be a player j and a state $\hat{\omega}$ such that player j's *knowledge* at $\hat{\omega}$ is different under H_j and $\hat{H}_j$. That is, there must be an event $\tilde{E}$ such that $h_j(\hat{\omega}) \nsubseteq \tilde{E}$ but $\hat{h}_j(\hat{\omega}) \subseteq \tilde{E}$. (Proving this is exercise 14.5.)

14.3.2 Agreeing to Disagree[†††]

The first and best-known result obtained with the formal definition of common knowledge is Aumann's proof that rational players cannot "agree to disagree" about the probability of a given event. The intuition for this is that if a player knows that his opponents' beliefs are different from his own, he should revise his beliefs to take the opponents' information into account. Of course, this intuition doesn't make sense if the player thinks his opponents are simply crazy; it requires that he believe that his opponents process information correctly and that the difference in the beliefs reflects some objective information. More formally, Aumann's result requires that the players' beliefs be derived by Bayesian updating from a common prior distribution.

Theorem 14.2 (Aumann 1976)   Suppose that it is common knowledge at ω that player i's posterior probability of event E is q_i and that player j's posterior probability of E is q_j. Then $q_i = q_j$.

Proof   Let $\mathcal{M}$ be the meet of all the players' partitions, and let $M(\omega)$ be the element of $\mathcal{M}$ that contains ω. Write $M(\omega) = \bigcup_k h_i^k$, where each h_i^k is an element of player i's partition H_i. Since player i's posterior probability of event E is common knowledge, it is constant on $M(\omega)$, and hence

$$q_i = p(E \cap h_i^k)/p(h_i^k) \text{ for all } k.$$

Hence,

$$p(E \cap h_i^k) = q_i p(h_i^k),$$

and summing over k yields

$$p(E \cap M(\omega)) = q_i p(M(\omega)).$$

Applying the same reasoning to player j shows that

$$p(E \cap M(\omega)) = q_j p(M(\omega)),$$

so $q_i = q_j$.   ∎

The proof of the theorem makes explicit use of the assumption of a common prior over Ω. It should be clear that this assumption is necessary: When the priors differ, each player, rather than ascribe any difference in an opponent's beliefs to "real" information, is free to ascribe such a difference to the opponent's having used the "wrong" prior.[7]

As Aumann observes, the theorem also requires the assumption that the players' partitions are common knowledge in the informal sense that the model fully describes the players' information. Intuitively, if player i did not know how player j arrived at his posterior beliefs, player i would not know how to evaluate the fact that player j's beliefs differed from his own. Somewhat more formally, if the partitions were not common knowledge (say, because player i did not know player j's partition), we would need to enlarge the state space to assign positive probability to each H_j that player i believes has positive probability. On this new expanded state space, the theorem would hold as before, again with the supposition that the players have a common prior.

Aumann also gives an example to show that the result fails if the players merely know each other's posteriors, as opposed to the posteriors' being common knowledge. In his example, Ω has four equally likely elements, ω_1, ω_2, ω_3, and ω_4, players 1's partition is $H_1 = \{(\omega_1, \omega_2), (\omega_3, \omega_4)\}$, and player 2's partition is $H_2 = \{(\omega_1, \omega_2, \omega_3), (\omega_4)\}$. Let E be the event (ω_1, ω_4). Then at ω_1, player 1's posterior probability of E is

$$q_1(E) = p[(\omega_1, \omega_4) | (\omega_1, \omega_2)] = \tfrac{1}{2},$$

and player 2's posterior probability of E is

$$q_2(E) = p[(\omega_1, \omega_4) | (\omega_1, \omega_2, \omega_3)] = \tfrac{1}{3}.$$

Moreover, player 1 knows that player 2's information is the set $(\omega_1, \omega_2, \omega_3)$, so player 1 knows $q_2(E)$. Player 2 knows that player 1's information is either (ω_1, ω_2) or (ω_3, ω_4), and either way player 1's posterior probability of E is

---

7. Here and in his 1987 paper, Aumann appeals to the "Harsanyi doctrine" to support the assumption of a common prior.

$\frac{1}{2}$, so player 2 knows $q_1(E)$. Thus, each player knows the other player's posterior, yet the two players' posteriors differ. The explanation is that the posteriors are not common knowledge. In particular, player 2 does not know what player 1 thinks $q_2(E)$ is, as $\omega = \omega_3$ is consistent with player 2's information, and in this case player 1 believes there is probability $\frac{1}{2}$ that $q_2(E) = \frac{1}{3}$ (if $\omega = \omega_3$) and probability $\frac{1}{2}$ that $q_2(E) = 1$ (if $\omega = \omega_4$).

Geanakoplos and Polemarchakis (1982) observe that Aumann's result does not address the issue of how and when the players' posterior beliefs might come to be common knowledge in the first place. They assume that the players communicate only by announcing their posterior beliefs. (In particular, players are not allowed to communicate information sets.) Geanakoplos and Polemarchakis analyze the process in which two agents take turns announcing their posterior distributions cooperatively (i.e., truthfully) to each other; if the state space is finite, this process converges in finite time. Or, as in the title of their paper, "we can't disagree forever."

Geanakoplos and Polemarchakis note that the fact that agents' posteriors about an event E eventually converge (and thus become common knowledge) does not imply that the players know as much about the event as they would know if they pooled their information. That is, even though the agents' posteriors are equal, they need not have the same information. A counterexample has four equally likely states, $\omega = 00, 10, 01,$ and 11; player 1 knows the first component of ω, and player 2 knows the second component. Consider the posterior probabilities for the event $E = \{00, 11\}$ when the state is 00. On the basis of only his own information, each player's posterior probability for E is always $\frac{1}{2}$. When player 1 announces that his posterior for E is $\frac{1}{2}$, this does not give any information to player 2—for instance, when the second component is 0, player 2 knows that ω is either 00 or 10; if it is 00, player 1 knows that ω is either $00 \in E$ or $01 \notin E$, and so assigns E probability $\frac{1}{2}$; if it is 10, player 1 knows that ω is either $10 \notin E$ or $11 \in E$, and again player 1's posterior for E is $\frac{1}{2}$. Thus, player 2's revised posterior for E is still $\frac{1}{2}$, and his announcing this gives no information to player 1.

14.3.3   No-Speculation Theorems[†††]

Results on agreeing to disagree are closely related to the results about the impossibility of risk-averse agents' taking opposing sides of the same purely speculative bet.[8] Intuitively, if player 1 is risk averse and accepts an even-odds bet that a coin will come up heads, he must assign heads a probability greater than $\frac{1}{2}$, whereas if player 2 is risk averse and accepts the other side of this bet—that the coin will come up tails—then player 2 thinks the probability of heads is less than $\frac{1}{2}$, so the players would be "agreeing to disagree."

8. Rubinstein and Wolinsky (1989) present a theorem that encompasses both types of results.

There are two sorts of no-speculation theorems in the literature: "equilibrium" theorems, which assert that speculation cannot occur in equilibria of various games, and "common-knowledge" theorems, which assert that it cannot be common knowledge that all players expect to gain from speculation. Although the first no-speculation theorems were equilibrium theorems, we will present a common-knowledge theorem first, as it is more closely related to the theme of this chapter.

To distinguish speculative trading from trading for other purposes, we decompose the state of nature ω into two parts: $\omega = (x, z)$, where the players' *ex post* utilities and initial endowments depend only on x and where z is a vector of signals $z = (z_1, \ldots, z_I)$, possibly correlated with the payoff-relevant uncertainty x. Thus, player i's information is

$$h_i(\omega) = h_i(x, z) = \{(x', z') \text{ such that } z'_i = z_i\}.$$

A *net trade* is a map y from the set of states Ω to consumption bundles in a set B, where $y_i(\omega)$ is player i's net trade at ω; y is *feasible* if it lies in the (exogenous) set Y. Player i's endowment is $e_i(x)$, and his utility of $y(\cdot)$ at ω is

$$\tilde{u}_i(y_i(\omega) + e_i(x), x) \equiv u_i(y_i(\omega), x).$$

Suppose as usual that the players have a common prior distribution $p(\cdot)$ on Ω.

Theorem 14.3 (Milgrom and Stokey 1982) Suppose that traders are weakly risk averse (i.e., u_i is concave in y_i) and that $\hat{y} \equiv 0$ is Pareto optimal in the set of all feasible net trades. If y is feasible, and if it is common knowledge at ω and that each player weakly prefers y to $\hat{y}$, then every player is indifferent between y and $\hat{y}$; if all players are strictly risk averse (that is, u_i is strictly concave), then $y = \hat{y}$.

Proof If it is common knowledge at ω' that all players at least weakly prefer y to $\hat{y}$, then for all ω'' in $M(\omega')$

$$E[u_i(y_i(\omega), x) | h_i(\omega'')] \geq E[u_i(\hat{y}_i(\omega), x) | h_i(\omega'')]. \tag{14.1}$$

We claim that equation 14.1 must hold with exact equality for all players i. To see this, define $y^*(\omega)$ by

$$y^*(\omega) = y(\omega) \text{ if } \omega \in M(\omega')$$
$$y^*(\omega) = \hat{y}(\omega) \text{ if } \omega \in M^c(\omega'), \tag{14.2}$$

where $M^c(\omega') = \Omega \backslash M(\omega')$ is the complement of $M(\omega')$. Now, since $h_i(\omega'') \subseteq M(\omega')$ for every $\omega'' \in M(\omega')$, each player can deduce whether $y^*(\omega) = y(\omega)$ or $y^*(\omega) = \hat{y}(\omega)$. Player i's *ex ante* expected utility of y^* is

$$E[u_i(y^*(\omega), x)]$$

$$= \sum_{h_i \in M(\omega')} p(h_i) E[u_i(y_i(\omega), x) | h_i]$$

$$+ \sum_{h_i \in M^c(\omega')} p(h_i) E[u_i(\hat{y}_i(\omega), x) | h_i]$$

$$\geq E[u_i(y(\omega), x)], \tag{14.3}$$

where the last inequality comes from substituting equation 14.1 for player i's utility conditional on $h_i(\omega'') \in M(\omega')$. Moreover, if equation 14.1 holds strictly for some player j, so does equation 14.3, which contradicts the assumption that $\hat{y}$ was Pareto optimal *ex ante*. If traders are strictly risk averse and the endowments are Pareto optimal, then no other allocation is Pareto indifferent for all players.   ∎

To see the role of common knowledge in the theorem, consider an economy in which there are two equally likely states, ω_1 and ω_2, and two players, 1 and 2, with $H_1 = \{(\omega_1), (\omega_2)\}$ and $H_2 = \{(\omega_1, \omega_2)\}$. There is a single consumption good, and endowments and utility functions are independent of the state (so that, in our previous notation, $\omega \equiv z$). Players are risk neutral, so $Eu_i(y_i(\omega), x) = Ey_i(\omega)$, and $\hat{y}(\omega) \equiv 0$ for all ω is Pareto optimal. Consider the feasible net trade given by

$$y_1(\omega_1) = 1, \; y_2(\omega_1) = -1, \; y_1(\omega_2) = -2, \; y_2(\omega_2) = 2.$$

At state ω_1 both players strictly prefer y to $\hat{y}$ on the basis of their information $(E[y_1(\omega) | h_1(\omega_1)] = 1$, and $E[y_2(\omega) | h_2(\omega_1)] = \frac{1}{2}(-1) + \frac{1}{2}(2) = \frac{1}{2})$, and so the conclusion of the theorem does not hold. Of course, it is not common knowledge at ω_1 that both players prefer y to $\hat{y}$, as $M(\omega_1) = (\omega_1, \omega_2)$ contains $h_1(\omega_2)$ and, conditional on $h_1(\omega_2)$, player 1 prefers $\hat{y}$ to y. Thus, inequality 14.1 does not apply, which is why the proof fails.

Intuitively, a situation like this could not arise in equilibrium, as it involves player 1's "fooling" player 2. For example, if we suppose that the players vote on moving from $\hat{y}$ to y after receiving their information, and that either player can veto a move, then it is not an equilibrium for player 1 to vote for y in state ω_1 and for $\hat{y}$ in state ω_2, and for player 2 to vote for y: The outcome of this profile would be y in state ω_1 and $\hat{y}$ in state ω_2, and player 2 would do better to vote for $\hat{y}$.

The first formal result along these lines was reported by Kreps (1977), who gave credit to Stiglitz (1971). Kreps shows that in a rational-expectations equilibrium of an asset market with asymmetric information but common priors on the distribution of uncertainty, and with risk-neutral traders, traders cannot be better off than if they held onto their initial shares and refused to participate in the market. In other words, a trader with superior information cannot benefit from his information, because other traders anticipate that he will buy when the asset is likely to do well and sell when prospects are less favorable. That is, with common priors, the trading game is a zero-sum game. This result can be called the "no-speculation" result. The slightly stronger "no-trade" result does not quite hold here; risk-neutral traders can trade at commonly agreed fair prices, and thus we can assert only that traders can do as well by not trading, not that they will not trade.

One way to obtain "equilibrium" no-speculation results is to suppose that the equilibrium strategies are common knowledge in an informal sense, as are the structure (information and strategy spaces) of the game and the fact that each player's equilibrium strategy maximizes his expected payoff. Consider some arbitrary "trading game" of announcements, bids, etc., supposing only that there is a fixed "no-trade" outcome, independent of the sequence of play, and that each player i can, for each $h_i \in H_i$, play a strategy that ensures the no-trade outcome (for himself) when his information is h_i. Let s^* be an equilibrium profile in the (unmodeled) trading game, and let $y = y(s^*(\cdot))$ be the corresponding net trade. Then, since s_i^* is a best response to each s_{-i}^*, inequality 14.1 is satisfied at all $\omega \in \Omega$, so that equations 14.2 and 14.3 hold where $M(\omega')$ is replaced by Ω; hence the conclusion that all players must be indifferent between y and the null trade $\hat{y}$.

Although the absence of speculation in equilibrium can thus be inferred from no-common-knowledge-of-speculation by assuming that the strategies are themselves common knowledge, we should point out that the equilibrium version of the result can also be obtained directly without formal use of the concept of common knowledge: In any Nash equilibrium, equation 14.1 holds for all $\omega \in \Omega$.

Tirole (1982) extends the Kreps-Stiglitz result in another direction than Milgrom and Stokey. Studying an intertemporal (finite or infinite horizon) asset market where traders are risk neutral and their time-varying private information stems from a filtration, he shows that with a finite number of traders the price of the asset must be at each instant equal to the expected present discounted value of dividends for any "interior trader," that is, for any trader who is not constrained by short-sale constraints or by the impossibility of buying more than 100 percent of the shares. Thus, there cannot exist any "bubble" (difference between market price and market fundamental) for any interior trader. The link with the no-speculation result is that it can be shown that if a bubble existed in a given period, at least one (interior) trader would have an intertemporal trading strategy that would make him strictly better off than if no trade occurred. And, again, the asset market is a zero-sum game and each trader can guarantee himself his no-trade payoff.[9]

9. The no-bubble result fails if there are an infinite number of overlapping finite-lived generations.

14.3.4 Interim Efficiency and Incomplete Contracts[†††]

The version of the Milgrom-Stokey result we gave allowed the set of feasible contracts Y to be arbitrary; in particular, the feasible set need not include all complete contingent contracts. In this case, though, it is not always natural to suppose that, as theorem 14.3 requires, the initial allocation is *ex ante* Pareto optimal, as the players may not be able to meet and contract before receiving any private information.

It is interesting, then, to note that the conclusion of theorem 14.1 holds on the weaker assumption that the initial allocation $\hat{y}$ is *interim efficient* in the sense of Holmström and Myerson (1983). If v is the vector of *ex post* verifiable variables (which might be a subvector of x, for instance), let $e(z, v)$ denote the contingent endowment if no further trading occurs[10] and let $y(z, v)$ denote a v-contingent trade when players receive private information z_i. The no-trade allocation is interim efficient if there exists no $y(\cdot, \cdot)$ such that for all i, z, and z_i'

$$E(u_i(y(z, v), x) \mid z_i) \geq E(u_i(0, x) \mid z_i) \tag{14.4}$$

and

$$E(u_i(y(z, v), x) \mid z_i) \geq E(u_i(y((z_i', z_{-i}), v), x) \mid z_i). \tag{14.5}$$

Inequalities 14.4 and 14.5 express individual rationality and incentive compatibility. In particular, inequality 14.5 reflects the fact that information z_i is private and must be elicited. It is clear that, for any trading or bargaining process, the no-trade allocation will not be altered by further contracting as long as each trader can guarantee himself the no-trade allocation. In other words, interim efficient allocations are "strongly renegotiation-proof."[11]

14.4 Common Knowledge, Almost Common Knowledge, and the Sensitivity of Equilibria to the Information Structure[†††]

This section discusses how the Nash equilibria of a game can vary with "small" changes in its information structure. Changes in the information

---

10. $e(z, v)$ may stem from a previous contract that, for instance, sets up a revelation mechanism for eliciting the private signals z (see chapter 7). But note the following subtlety: We implicitly assume that $e(z, v)$ is not affected by the process of bargaining for further trade. For instance, the revelation mechanism that elicits z may have several equilibria; although this is not an issue when only one player has private information, unique implementation with several informed players requires some care (see the references cited in section 7.2). If the revelation game has multiple equilibria, then even if one equilibrium is interim efficient, it might be the case that trade occurs, i.e., that the initial contract is renegotiated. Renegotiation may be enforced by the threat that a "bad equilibrium" prevails if the initial contract is not replaced by a new one. Also, if the revelation game is not in dominant strategies, the change in beliefs during the bargaining phase may destroy incentive compatibility of the revelation game.
11. They are strongly renegotiation-proof (at least if traders are strictly risk averse) in two ways. First, for a given bargaining process for renegotiation, they are not renegotiated in *any*

structure can change what each player knows, and thus change what is common knowledge, so the notion of exact common knowledge will not be very useful here. As we will see, though, the closely related notion of "almost common knowledge" is very useful indeed.

We start with two examples to illustrate the possibility that apparently small perturbations in the information structures starting from common knowledge of the payoffs may change the equilibrium set considerably. That is, some equilibria of the game in which payoffs are common knowledge are not near any equilibrium of the perturbed game, even if with high probability all players know that the payoffs are as in the original game. This lack of lower hemi-continuity should not be surprising: We saw that small changes in the payoffs of players may eliminate equilibria; a small probability that the payoffs might be different can have the same effect. The simplest version of this point is that, if a player is slightly uncertain about his opponent's payoffs, he may be unwilling to play a strategy that is weakly dominated. As a slightly more complicated example, the game illustrated in figure 14.2a has a component of equilibria where player 2 plays A with probability 1 and player 1 plays A with probability at least $\frac{5}{6}$, and also a pure-strategy equilibrium where both players play B; the equilibrium where both play A with probability 1 has the highest payoffs. However, playing A is weakly dominated for player 1. Now consider a perturbed version of the game in which with probability $1 - \varepsilon$ the state is ω_1 and the payoffs are as in figure 14.2a, and with probability ε the state is ω_2 and the payoffs are as in figure 14.2b. Moreover, player 1 is uncertain which state prevails—he has the trivial partition $H_1 = \{\omega_1, \omega_2\}$—but player 2 knows the state. Then, in state ω_2, player 2 plays B, as B strictly dominates A, and so player 1, not knowing the state, will not play the weakly dominated strategy A. Hence, although (A, A) is an equilibrium if it is common knowledge the payoffs are as in figure 14.2a, (A, A) is not an equilibrium of the perturbed game.

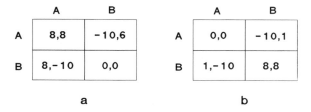

a   b

Figure 14.2

Nash equilibrium of the renegotiation game. Second, this holds for any renegotiation process. Interim efficiency is necessary and sufficient for strong renegotiation-proofness. (However, it is not in general necessary for the allocation to be "weakly renegotiation-proof" (in that there exists *some* equilibrium of the renegotiation game for which it is not renegotiated). One can define a weaker notion of efficiency—"weak interim efficiency"—to characterize weakly renegotiation-proof allocations. See Maskin and Tirole 1989.)

Subsection 14.4.1 develops two subtler examples of this lack of lower hemi-continuity. Example 14.3 considers a situation where, as above, the payoffs are given by either figure 14.2a or figure 14.2b, but the information structure considered is more complex. In particular, there is a state with high prior probability in which *both* players know that the payoffs are as in figure 14.2a. Nevertheless, even in this state, both players must play B in equilibrium: Although both players know that the payoffs are as in figure 14.2a, this is not common knowledge.

The payoffs in figure 14.2a are not generic, and the argument concerning this figure focused on eliminating equilibria where the weakly dominated strategy A is played. This raises the question of whether a similar lack of lower hemi-continuity can arise when each payoff matrix being considered is generic in the space of strategic forms. In example 14.4, the payoff matrices considered are generic and the perturbation in the information structure eliminates a strict equilibrium. Whether this is viewed as a failure of lower hemi-continuity for generic payoffs depends on whether the perturbation in the information structure is indeed small. This turns out to be a subtle question (as the example has an infinite state space), and there are several seemingly reasonable ways to define what it means for two information structures on an infinite state space to be close. Subsection 14.4.2 develops one such notion, based on the idea of almost common knowledge, and shows that it yields a result with generic lower hemi-continuity.

14.4.1 The Lack of Lower Hemi-Continuity

Example 14.3
Suppose that players 1 and 2 play the game depicted in figure 14.2a. As remarked above, this game has two components of Nash equilibria: the pure-strategy equilibrium (B, B) and any profile where player 2 plays A with probability 1 and player 1 plays A with probability at least $\frac{5}{6}$. Now suppose we want to model a situation in which players 1 and 2 both know that the payoffs are as in figure 14.2a but player 1 assigns positive probability to player 2's believing that the payoffs may actually be as in figure 14.2b. Then (since players 1 and 2 have a common prior over nature's moves) nature must assign positive probability to player 1's not being fully informed of nature's move. One such game is depicted in figure 14.3.

In this game, nature has four possible moves: ω_1, ω_2, ω_3, and ω_4. In ω_1 and ω_2 the payoffs are as in figure 14.2a; in ω_3 and ω_4 the payoffs are as in figure 14.2b. The players' information partitions are $H_1 = \{(\omega_1, \omega_2), (\omega_3, \omega_4)\}$ and $H_2 = \{\omega_1, (\omega_2, \omega_3), \omega_4\}$. Player 1 always knows the payoffs, and player 2 may or may not know them. Furthermore, player 1 does not know whether player 2 knows. In state ω_1 the players know that the payoffs are as in figure 14.2a, and player 2 can infer that player 1 knows the payoffs, but player 1 does not know whether player 2 knows the payoffs. In ω_2 the

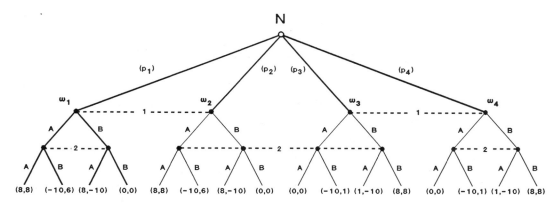

Figure 14.3

payoffs are as in figure 14.2a, and player 1 is informed of this but player 2 is not; again player 1 does not know whether player 2 is informed. In state ω_3 the payoffs are as in figure 14.2b, player 1 is informed, and player 2 is not; in ω_4 the payoffs are as in figure 14.2b and both players know this; in ω_3 and ω_4 player 1 does not know whether player 2 is informed. Thus, if all states have positive prior probability, the only common-knowledge event is the whole state space.

Note well that there are more states in Ω than there are possible payoff matrices: The state must describe not only the payoffs, and each player's information about the payoffs, but also each player's information about his opponent's information, and so on. Note as well how we changed the model to incorporate the uncertainty of player 1: We added additional states until the players' beliefs could again be described by a common prior over a common state space. To precisely describe a real-world game in this way could easily require a large (or even infinite) state space. The hope is that small-state-space models can be a good approximation to the game of interest.

Returning to the example, let p_i denote the prior probability of ω_i. Then, if $p_2 < p_3$, the only Nash equilibrium of the game is for both players to play B in every state. The proof of this proceeds state by state. First, in state ω_4, both players know the payoffs, and for each of them it is a dominant strategy to play B, as it is for player 1 in state ω_3. Let $q = p_2/(p_2 + p_3)$ be player 2's posterior probability of ω_2 when his information is (ω_2, ω_3). Given that player 1 plays B in state ω_3, when player 2 is told (ω_2, ω_3) he receives at most $q8 + (1 - q)(-10)$ from playing A and at least $q(-10) + (1 - q)8$ from playing B. Since $p_2 < p_3$, $q < \frac{1}{2}$, and player 2 must play B. Next, when player 1 is told (ω_1, ω_2) he knows that the payoffs are as in figure 14.2a, so that (A, A) is Pareto optimal, yet he also knows that there is positive probability that the state is ω_2 and so player

2 will play B, and hence player 1 plays B. Finally, given that player 1 plays B in ω_1 and ω_2, player 2 plays B when told ω_1.

The foregoing is true regardless of the absolute probabilities of ω_2 and ω_3, and regardless of the relative probability of ω_1 and ω_2. In particular, the conclusions hold at each p^n in the sequence $p_1^n = 1 - 4/n$, $p_2^n = 1/n$, $p_3^n = 2/n$, and $p_4^n = 1/n$, which converges to the limit $p_1 = 1$ where (A, A) is a Nash equilibrium. Moreover, when n is large, in state ω_1 both players know that the payoffs are as in figure 14.2a, player 2 knows that player 1 knows this, and player 1 is "almost certain" that player 2 knows this, yet the set of equilibria differs from what it would be if state ω_1 had probability 1 and hence were common knowledge. Put differently: The set of equilibria with "almost common knowledge" of the payoffs (defined formally in subsection 14.4.2) differs from that where the payoffs are known with certainty. Though this may be troubling, it should not be a surprise: We saw in chapter 1 that the Nash correspondence need not be lower hemi-continuous in the payoffs, and this example simply illustrates a lack of lower hemi-continuity in the prior distribution. Note also that the payoffs in figure 14.2a are not generic, which is why a very small chance of ω_2 could force player 1 to play B. From the result on *generic* lower hemi-continuity of the Nash correspondence in finite games (see section 12.1), we would expect that if the players were uncertain which of a finite number of payoff matrices prevailed, and if the underlying state space Ω were finite, the Nash correspondence would be lower hemi-continuous in the neighborhood of common knowledge of a generic game, so that for generic payoffs "almost common knowledge" and common knowledge should have the same implications. Subsection 14.4.2 presents a version of this result. First, though, we give an example to show that the preceding intuition relies on the restriction to finite games.

Example 14.4: Electronic-Mail Game

Consider Rubinstein's (1989) updated version of Gray's (1978) "coordinated attack problem." Here the payoff matrices are as in figure 14.4. Note that in figure 14.4a the Pareto-optimal equilibrium (A, A) is strict. The information structure (represented in figure 14.5) is as follows: In state 0 the payoffs are as in figure 14.4b. In states $1, 2, \ldots$, the payoffs are as in

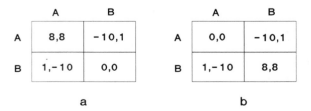

a   b

Figure 14.4

figure 14.4a. Player 1's partition is the sets

$$(0), (1, 2), (3, 4), \ldots, (2n - 1, 2n), \ldots.$$

Player 2's partition is

$$(0, 1), (2, 3), \ldots, (2n, 2n + 1), \ldots.$$

The prior probability of state 0 is $\frac{2}{3}$, and the probability of state $n \geq 1$ is $\varepsilon(1 - \varepsilon)^{n-1}/3$. The interpretation of this information structure is that if player 1 learns that the payoffs are as in figure 14.4a he sends a message to player 2, who does not know *a priori* which matrix is relevant. The message has probability ε of not being received; if player 2 does receive the message, he sends a response, which in turn has probability ε of not being received; if player 1 receives player 2's response, he sends a second response, acknowledging his receipt of player 2's response, and so on. In the original version of this game the messages were carried by horsemen through a valley occupied by enemy forces who might intercept them[12]; the updated version has the messages sent by an electronic-mail system that sometimes fails. It is important that sending the messages is not a strategic decision on the part of the players, but rather an exogenous process that determines their initial information.

Figure 14.5

12. Halpern (1986) gives the following account of the coordinated-attack problem:

Two divisions of an army are camped on two hilltops overlooking a common valley. In the valley awaits the enemy. It is clear that if both divisions attack the enemy simultaneously they will win the battle, whereas if only one division attacks it will be defeated. The divisions do not initially have plans for launching an attack on the enemy, and the commanding general of the first division wishes to coordinate a simultaneous attack (at some time the next day). Neither general will decide to attack unless he is sure that the other will attack with him. The generals can only communicate by means of a messenger. Normally, it takes the messenger one hour to get from one encampment to the other. However, it is possible that he will get lost in the dark or, worse yet, be captured by the enemy. Fortunately, on this particular night, everything goes smoothly. How long will it take them to coordinate an attack?

Suppose the messenger sent by general A makes it to general B with a message saying "Let's attack at dawn." Will general B attack? Of course not, since general A does not know he got the message, and thus may not attack. So general B sends the messenger back with an acknowledgement. Suppose the messenger makes it. Will general A attack? No, because now general B does not know he got the message, so he thinks general A may think that he (B) didn't get the original message, and thus not attack. So A sends the messenger back with an acknowledgement. But of course, this is not enough either. I will leave it to the reader to convince himself that no amount of acknowledgements sent back and forth will ever guarantee agreement. Note that this is true even if the messenger succeeds in delivering the message every time. All that is required in this reasoning is the *possibility* that each messenger doesn't succeed.

If the state is $n > 0$, then n messages have been sent and $n - 1$ of them received. For instance, if $n = 2k$, player 2 knows that the payoffs are as in figure 14.4a and that player 1 knows the payoffs, player 1 knows that player 2 knows this, player 2 knows that player 1 knows, and so on, for any string of length less than n. That is, $n \in K_{\mathcal{I}}^{n-1}$ $(n > 0)$. Nevertheless, there is no finite n for which "$n > 0$" is common knowledge, as figure 14.5 and the reachability criterion make clear.

One might think that for small ε it would be possible for the players to coordinate on (A, A) when the payoff matrix is that of figure 14.4a (that is, when $n > 0$); however, this is not the case. The argument is an extension of the argument in the last example. Player 1 plays B in state 0, as this is a dominant strategy. Since player 1 plays B in state 0, and the probability of 0 given $(0, 1)$ is greater than $\frac{1}{2}$, playing B in states $(0, 1)$ gives player 2 a payoff of at least 4, whereas playing A gives at most -1, so player 2 plays B. Since the probability of state 1 given $(1, 2)$ is

$$q = \frac{\varepsilon}{\varepsilon + \varepsilon(1 - \varepsilon)} > \tfrac{1}{2},$$

player 1, who in states $(1, 2)$ knows that player 2 plays B with probability greater than $\frac{1}{2}$, plays B in states $(1, 2)$, as he obtains at most $(1 - q)8 + q(-10) < 0$ from playing A and at least 0 from playing B. Given that player 1 plays B in states $(1, 2)$, in states $(2, 3)$ player 2 plays B, as the probability of 2 given $(2, 3)$ exceeds $\frac{1}{2}$, so player 1 plays B in states $(3, 4)$, and the proof continues by induction to show that the two players play B in any state.

If ε is very small, then conditional on $n \neq 0$ a large number of messages are likely to be sent and received. However, for any $\varepsilon > 0$, there is no equilibrium where players play (A, A) whenever the payoffs are as in figure 14.4a, even through (A, A) is a strict equilibrium when payoffs are certain to be as in figure 14.4a.

Whether example 14.4 represents a failure of lower hemi-continuity depends on whether we view the ε error probability as a "small" change in the information structure, and also on whether we test for convergence of the strategies as maps from the state space to actions or instead test for convergence of the probability distribution over the payoff-relevant outcomes. We begin with the second point. Eddie Dekel-Tabak has suggested a way of defining convergence of strategies under which the equilibrium strategies in the perturbed games are in fact the *same* as the strategies when $\varepsilon = 0$. First, compactify the state space by adding the point ∞, corresponding to an infinite number of messages, so that the state space is $\bar{\Omega} = \Omega \cup \{\infty\}$. When an infinite number of messages are sent and received (which has probability 0 under the prior distribution), it is common knowledge that the payoffs are as in figure 14.4a. Thus, on this expanded state space, one equilibrium is for both players to play **B** when ω is finite and **A** when $\omega = \infty$. On this view, since the set of equilibrium strategies in the example does not vary with the parameter ε, there is no failure of lower hemi-continuity regardless of whether or not the information structures converge as $\varepsilon \to 0$. (But recall that in example 14.3 the set of equilibrium *strategies* in state ω_1 did change.)

Although this viewpoint may help illuminate the structure of the problem, it does not address the fact that equilibrium *payoffs* when $\varepsilon > 0$ are strictly lower than the payoffs when $\varepsilon = 0$. This suggests looking at convergence in the space of probability distributions over outcomes. Since the equilibrium probability distributions are changed by introducing the error probability, the question of lower hemi-continuity then depends on whether the error probability represents a small perturbation.

The next subsection introduces the concept of "almost common knowledge," and shows that the set of equilibria is lower hemi-continuous when the perturbations are deemed to be small only if the unperturbed payoffs remain almost common knowledge. With that approach, example 14.4 does not display lack of lower hemi-continuity, because the perturbation is not judged to be small.

But there is an equally intuitive way of defining a small perturbation so that example 14.4 does qualify. This definition is based on extending the topology of weak convergence to the probability distributions in the compactified state space $\bar{\Omega}$. (To extend this topology, note that $\bar{\Omega}$ is isomorphic to the set $\{2, 1, \frac{1}{2}, \frac{1}{3}, \ldots, 1/n, \ldots, 0\}$ by the transformation $x(n) = 1/n$ for $n > 0$, $x(\infty) = 0$, and $x(0) = 2$. Probability distributions over $\bar{\Omega}$ are thus a subset of the set $\mathscr{P}$ of probability distributions over the interval $[0, 2]$, which we endow with the weak topology on distributions over $[0, 2]$.[13]) Many sequences of distributions on $\bar{\Omega}$ converge in the weak topology to the distribution $\{p(0) = \frac{2}{3}, p(\infty) = \frac{1}{3}\}$, which corresponds to common knowledge. One sequence is that of the example. Another is defined by

$$p^\varepsilon(0) = \tfrac{2}{3},$$

$$p^\varepsilon(2k + 1) = \varepsilon(1 - \varepsilon)^k(1 - 2\varepsilon)^k/3$$

and

$$p^\varepsilon(2k + 2) = 2\varepsilon(1 - \varepsilon)^{k+1}(1 - 2\varepsilon)^k/3$$

---

13. p^n converges to p in the weak topology if, for each continuous function f on $[0, 2]$,

$$\int_0^2 f(x) \, dp^n(x)$$

converges to

$$\int_0^2 f(x) \, dp(x).$$

for $k \geq 0$. The interpretation of this information structure is that player 2's messages are twice as likely to be intercepted as player 1's.

With this notion of closeness, the reason the example displays a lack of lower hemi-continuity is that the set of equilibria of the "limit" game where payoffs are common knowledge contains both the limits of equilibria of perturbed games with equal error probabilities and the limits of equilibria of games where player 2's error probability is twice as large as player 1's. And with this latter information structure, it is a Nash equilibrium for player 1 to play A in all states $n > 0$, and for player 2 to play A in all states $n > 1$: Now, when player 1 has partition $(1, 2)$, his posterior probability of state 2 is

$$\frac{2\varepsilon(1 - \varepsilon)}{\varepsilon + 2\varepsilon(1 - \varepsilon)} \cong \frac{2}{3}$$

for small ε, and player 1 is willing to play A if he expects player 2 to play A in state 2, even though he expects player 2 to play B in state 1. Thus, player 2 is willing to play A when his partition is $(2, 3)$, and so on.

14.4.2 Lower Hemi-Continuity and Almost Common Knowledge (technical)

Subsection 14.4.1 presented examples of ways in which the equilibrium correspondence can fail to be lower hemi-continuous. One response is to parallel chapter 1 and ask for lower hemi-continuity of the ε-equilibrium correspondence; another is to identify conditions for the Nash correspondence itself to be lower hemi-continuous. We will consider the two responses in turn, beginning with the idea of using ε equilibrium. As we will see, there are two distinct versions of ε equilibrium: *"ex ante"* and *"interim."* Stronger conditions are required for lower hemi-continuity of the interim version.

The ε equilibrium we discussed in chapter 1 corresponds to *ex ante* ε equilibrium in the present setting. Recall from chapter 1 that, for a family of (finite) strategic-form games with the same strategy space S and with payoff functions $u_i(\cdot, \lambda)$ that vary continuously in λ, a Nash equilibrium for payoffs λ is an ε Nash equilibrium for payoffs λ^n, where $\varepsilon \to 0$ as $\lambda^n \to \lambda$. (Recall that an ε equilibrium is a profile where no player can increase his payoff by more than ε by deviating.) That is, lower hemi-continuity is restored by looking at ε equilibria. Since changing the prior distribution on a fixed state space changes only the payoffs to each strategy and not the strategy space itself, this result extends immediately to changes in the prior distribution. Formally, fix a finite state space Ω, partitions H_i, H_i-measurable strategies $\sigma_i \in \mathscr{S}_i$ from Ω to the space of (probability mixtures) on S_i, and payoff functions u_i on $S \times \Omega$, and let $G(p)$ be the strategic-form game corresponding to prior distribution p over Ω. If profile σ is a Nash equilibrium of $G(p)$, then σ is an ε Nash equilibrium for $G(p^n)$, with $\varepsilon \to 0$ as

$p^n \to p$. In particular, if under prior p there is probability 1 that $u_i(s, \omega)$ equals some fixed $\bar{u}_i(s)$ for all s and all i, so that the payoff functions are common knowledge under p, and under p^n each player assigns high probability to the payoffs being given by $\bar{u}$, then any Nash equilibrium for $G(p)$ is an ε Nash equilibrium of $G(p^n)$. The reason is that if the probability that u differs from $\bar{u}$ is very small, then each player loses very little in terms of expected payoff (in the usual game) by playing a response that is suboptimal when u differs from $\bar{u}$. Thus, in an ε equilibrium of $G(p)$, players can make "big" mistakes at unlikely information sets. To emphasize this point, ε equilibria of the strategic-form game are called *ex ante* ε equilibria.

The electronic-mail game of example 14.4 has an infinite number of states, so the observation in the preceding paragraph does not apply. However, the following profile is an *ex ante* ε' equilibrium of that game, where ε' is of the same order as the probability ε that a message is lost: "Player 1 plays B when $h_1 = 0$ and plays A in all other states. Player 2 plays B when $h_2 = (0, 1)$ and plays A in all other states." Given player 1's strategy, player 2's strategy is exactly optimal: When player 2's information is $(0, 1)$, player 1 is likely to play B; when player 2's information is anything else, player 1 is certain to play A. Player 1's strategy is exactly optimal when his information is state 0 or when player 2 knows that the state is greater than 2. Player 1's choice of A given information $h_1 = (1, 2)$ is not optimal; however, since this event has probability $[\varepsilon + \varepsilon(1 - \varepsilon)]/3$, player 1's strategy is almost optimal when ε is small.

More generally, if s is a Nash-equilibrium profile when payoffs are known to be given by $\bar{u}(\cdot)$, then s is an *ex ante* ε Nash equilibrium, with ε small, if there is probability near 1 that each player believes payoffs are very likely to be given by $\bar{u}(\cdot)$. This is true whether the state space is finite or infinite (Monderer and Samet 1988; Stinchcombe 1988).

Though this result provides one resolution of the lack of lower hemicontinuity, it is not completely satisfying, because the notion of *ex ante* ε equilibrium is so weak. Instead, one might wish to use the concept of *interim* ε equilibrium: Profile σ is an *interim* ε equilibrium[14] if, for all players i and all states ω, strategy $\sigma_i(\omega)$ comes within ε of maximizing player i's expected payoff conditional on $\omega' \in h_i(\omega)$. If we let E denote the expectation operator, the formal condition is

$$E[u_i(\sigma_i(\omega'), \sigma_{-i}(\omega'))|h_i(\omega)] \geq E[u_i(s_i, \sigma_{-i}(\omega'))|h_i(\omega)] - \varepsilon$$

for all ω and i, and for all $s_i \in S_i$. Clearly, every interim ε equilibrium is an *ex ante* ε equilibrium; the converse is true for $\varepsilon = 0$ (so long as all states have positive probability) but not for positive ε. As we noted above, a player might make a big mistake in an unlikely state in an *ex ante* ε equilibrium.

---

14. Monderer and Samet call it an *ex post* ε equilibrium. We prefer "interim" because in information economics *"ex post"* refers to the situation in which the state of the world is revealed to all.

Monderer and Samet (1989) provide conditions for the lower hemi-continuity of the interim ε equilibrium in terms of "almost common knowledge" of the payoffs, which requires that all players be "pretty sure" that their opponents are "pretty sure" about the payoffs, and so on..., as opposed to knowing that their opponents know them.

Monderer and Samet say that player i "r-believes E" at ω if his posterior probability $p(E|h_i(\omega))$ is greater than or equal to r. The event "player i r-believes E" is denoted $B_i^r(E)$; this is the set $\{\omega|p(E|h_i(\omega)) \geq r\}$. When all states have strictly positive prior probability, 1-belief is equivalent to knowledge.[15] Event E is "common r-belief" if everyone believes E has probability at least r, everyone believes there is probability at least r that everyone believes E has probability at least r, and so on.[16] Common 1-belief is the same as common knowledge; common r-belief for r large corresponds to almost common knowledge. Stinchcombe (1988) independently proposed a closely related notion of almost common knowledge that is slightly weaker. His approach can be interpreted as defining common (r, n)-belief in E to require that statements of the form "everyone believes there is probability at least r that there is probability at least r...that E is true" hold so long as they involve n or fewer instances of "everyone believes." For instance, if all players know E but no player knows that his opponents know it, then E is common $(1, 1)$-belief; common r-belief in the Monderer-Samet sense is common (r, ∞)-belief in Stinchcombe's sense. Stinchcombe then says that payoffs are almost common knowledge if they are common (r, n)-belief for r near 1 and n near infinity.

The two definitions of almost common knowledge are equivalent with a finite state space (because the r-knowledge operator stops after a finite number of steps), but they can differ in games with a countable state space, as is shown by example 14.4. Here, payoffs become almost common knowledge in Stinchcombe's sense as the number of messages sent increases: If $n + 1$ messages have been sent, it is common $(1, n)$-belief that the state is not 0. However, no matter how many messages have been sent, the payoffs

---

15. With a continuum of states and a smooth prior, no individual state has positive probability, and one may wish to distinguish between knowledge and 1-belief. For instance, if a number is picked at random from the interval $[0, 1]$, players 1-believe that the number is irrational, but they do not "know" it in the sense we have defined.

16. Starting from $^1B_i^r(E) \equiv B_i^r(E)$ and $^1B_{\mathscr{I}}^r(E) \equiv \bigcap_{i \in \mathscr{I}} {}^1B_i^r(E)$, let

$$^nB_i^r(E) \equiv \{\omega|p(^{n-1}B_{\mathscr{I}}^r(E)|h_i(\omega)) \geq r\}$$

and

$$^nB_{\mathscr{I}}^r(E) \equiv \bigcap_{i \in \mathscr{I}} {}^nB_i^r(E).$$

Then E is common r-belief at ω if $\omega \in {}^{\infty}B_{\mathscr{I}}^r(E)$.

As with common knowledge, there is an equivalent, "Aumann-style" definition of common r-belief. An event E is a common r-truism if $E \subseteq B_{\mathscr{I}}^r(E)$. That is, when E occurs, every player assigns a probability of at least r to its occurrence. An event E' is a common r-belief at ω if there exists a common r-truism E such that $\omega \in E$ and $E \subseteq B_{\mathscr{I}}^r(E')$.

never become almost common knowledge in the Monderer-Samet sense, as for no $r > (1 - \varepsilon)/(2 - \varepsilon)$ are the payoffs common r-belief.[17]

In contrast, the payoffs in example 14.3 become almost common knowledge as the probability of ω_1 goes to 1: Let $p_1^n \to 1$ be the prior probability of ω_1. At state ω_1, the event $E = \{\omega_1\}$ is common p_1^n-belief (because $E \subseteq B_{\mathscr{I}}^{p_1^n}(E)$), so the game is almost common knowledge for n large. (Theorem 14.5 extends this observation by showing that as the prior probability of an event E goes to 1, the probability that the event is almost common knowledge goes to 1 as well.)

Monderer and Samet generalize theorem 14.2 by showing that if the posterior probabilities of event E are common r-belief, then any two posteriors can differ by at most $2(1 - r)$.[18] As we said above, they also use their notion of almost common knowledge to extend the result about the lower hemi-continuity of *ex ante* ε equilibria to the interim version.

Monderer and Samet consider games G with a finite number of payoff functions $u^\ell(\cdot)$, where $\ell = 1, \ldots, L$. The payoffs in state ω are $u(\cdot, \omega) = u^{\lambda(\omega)}(\cdot)$, where Ω is either finite or countably infinite. Let $G^\ell = \{\omega | \lambda(\omega) = \ell\}$ be the set of all states ω at which the payoffs are given by u^ℓ. Payoffs u^ℓ are common r-belief at ω if the event G^ℓ is common r-belief at ω. For each ℓ, let σ^ℓ be a Nash equilibrium for common-knowledge payoffs u^ℓ, and define $\sigma^*: \Omega \to \Sigma$ by $\sigma^*(\omega) = \sigma^{\lambda(\omega)}$. This function assigns each ω a Nash equilibrium for the payoffs $\lambda(\omega)$. If the payoffs are common knowledge at each ω, then σ^* is a Nash equilibrium of the overall game G.[19]

Monderer and Samet show that for each Nash profile σ^ℓ of a common-knowledge game u^ℓ there exists an interim ε equilibrium of any game in which it is almost common knowledge that payoffs are u^ℓ, such that the players play strategy profile σ^ℓ with probability close to 1.

---

17. $(1 - \varepsilon)/(2 - \varepsilon)$ is the conditional probability that a player attaches to his not receiving a message that was sent by the other player (as opposed to the other player's not receiving his previous message), given that he did not receive a new message. One way of showing that no $r > (1 - \varepsilon)/(2 - \varepsilon)$ makes the payoffs common r-belief is to use the iterative definition of common r-belief. Another is to note that for E to be a common r-truism (that is, $E \subseteq B_{\mathscr{I}}^r(E)$), r must be less than $(1 - \varepsilon)/(2 - \varepsilon)$: Let n_0 denote the lowest element in E; if $n_0 > 1$, one of the players at state n_0 puts probability $1/(2 - \varepsilon)$ on E being false; and at $n_0 = 1$, player 2 puts probability $2/(2 + \varepsilon)$ on E being false.

18. The precise meaning of the posterior probabilities that an event E is common r-belief is the following: Fix I posterior beliefs, $q_1, \ldots, q_I$, for event E. Let

$$E' = \{\omega | p(E | h_i(\omega)) = q_i \text{ for all } i \in \mathscr{I}\}.$$

Posterior beliefs $(q_1, \ldots, q_I)$ are common r-beliefs at ω if E' is common r-belief at ω. The result is then

$$\max_{i,j} |q_i - q_j| \leq 2(1 - r).$$

The method of proof is to show that if E'' is a common r-truism contained in E' (whose existence is ascertained in note 16), each q_i cannot differ from the probability of E conditional on E'' by more than $1 - r$.

19. Every Nash equilibrium of G is not necessarily such an σ^*, unless there is a single ω in each G^ℓ; otherwise, correlation over the equilibria can be introduced through the public signal ω.

Theorem 14.4 (Monderer and Samet 1989)   Fix an $r \in (0.5, 1]$, and set $q = p[\omega|$ for some ℓ, payoffs u^ℓ are common r-belief at $\omega]$. Then for any selection $\jmath^*$ from the set of common-knowledge-of-payoffs Nash equilibria, there is a profile $\jmath$ of G such that

$$p[\omega | \jmath(\omega) = \jmath^*(\omega)] \geq q$$

and such that $\jmath$ is an interim ε equilibrium for all

$$\varepsilon > 4(1 - r) \max_{i, \ell, \ell', \sigma, \sigma'} |u_i^\ell(\sigma) - u_i^{\ell'}(\sigma')|.$$

In particular, for any $\varepsilon > 0$, there are $\bar{r} < 1$ and $\bar{q} < 1$ such that for all $r \geq \bar{r}$ and $q \geq \bar{q}$ there exists an interim ε equilibrium $\jmath$ such that $p[\omega | \jmath(\omega) = \jmath^*(\omega)] > 1 - \varepsilon$.

Proof [20]   Let $E^\ell = \{\omega | G^\ell$ is common r-belief at $\omega\}$. If $r > \frac{1}{2}$, there can be at most one E^ℓ that player i r-believes has occurred. Set $\Omega_i = \bigcup_\ell B_i^r(E^\ell)$, and let Ω_i^c be the complement of Ω_i. For $\omega \in \Omega_i$, specify that player i plays the strategy σ_i^ℓ corresponding to the payoffs he r-believes to be true. Let

$$K = \max_{i, \ell, \ell', \sigma, \sigma'} |u_i^\ell(\sigma) - u_i^{\ell'}(\sigma')|.$$

We claim first that, at any $\omega \in B_i^r(E^\ell)$, σ_i^ℓ is a $4K(1 - r)$-optimal interim response to any $\jmath_{-i}$ with $\jmath_j(\omega) = \sigma_j^\ell$ for all $\omega \in B_j^r(E^\ell)$, all ℓ, and all j. To see this, note that at $\omega \in B_i^r(E^\ell)$ player i assigns probability at least r to the payoffs' being given by u^ℓ (he himself r-believes G^ℓ), and he also assigns probability at least r to $\omega \in B_j^r(E^\ell)$ for all $j \neq i$ (for $\omega \in E^\ell$, player j r-believes E^ℓ), so he assigns probability at least r to $\jmath_{-i}(\omega) = \sigma_{-i}^\ell$. Since σ_i^ℓ is a best response to σ_{-i}^ℓ, we have

$$E(u_i(\sigma_i^\ell, \jmath_{-i}(\omega)) | h_i(\omega))$$

$$\geq u_i^\ell(\sigma_i^\ell, \sigma_{-i}^\ell) - 2K(1 - r) \geq u_i^\ell(\sigma_i', \sigma_{-i}^\ell) - 2K(1 - r)$$

(since σ_i^ℓ is a best response to σ_{-i}^ℓ in game G^ℓ)

$$\geq E(u_i(\sigma_i', \jmath_{-i}(\omega)) | h_i(\omega)) - 4K(1 - r).$$

Note also that $p[\omega | \jmath(\omega) = \jmath^*(\omega)] \geq q$, as required.

It remains only to define $\jmath_i(\omega)$ for $\omega \notin \Omega_i$. For this, it suffices to look for a Bayesian equilibrium in the game where players are constrained to follow the strategy $\jmath_i$ defined above in Ω_i. (Such an equilibrium exists from Glicksberg's existence theorem in chapter 1.)   ∎

Corollary   If for each ℓ σ^ℓ is a *strict* equilibrium, then for any $\varepsilon > 0$ there are $\bar{r}$ and $\bar{q}$ such that, for all $r > \bar{r}$ and $q > \bar{q}$, if the payoffs are com-

---

20. Monderer and Samet use a slightly different proof.

mon r-belief with probability q, there is an *exact* equilibrium σ with $p[\omega \mid \sigma(\omega) = \sigma^*(\omega)] > 1 - \varepsilon$.

As we remarked above, the almost-common-knowledge condition is not satisfied in the electronic-mail example, so the theorem and its corollary do not apply. However, in the truncated version of the game where player 2 does not respond after receiving n messages, the state $2n$ is common $(1 - \varepsilon)$-belief where it occurs: Player 2 knows that the state is $2n$, and player 1 assigns it probability $1 - \varepsilon$. Hence, from the corollary, there is an exact equilibrium where both players play A when the state is $2n$.

This truncated example brings us to our final point: On a finite state space, if event C has probability close to 1 there is high probability that it is a common r-belief for r large.

Theorem 14.5 Consider an event C on a finite state space Ω and a sequence of prior distributions p^n on Ω such that $p^n(C) \to 1$. Then there are sequences $q^n \to 1$ and $r^n \to 1$ such that under p^n there is probability q^n that C is a common r^n-belief.

Remark To apply this theorem to the existence of interim ε Nash equilibria, fix an $\hat{r}$ close to 1 and let the event C be "all players $\hat{r}$-believe they know the true payoffs."

The proof of the theorem uses the following lemma:

Lemma 14.1 If $p(C) \geq q$, then

$$p(B_i^r(C)) \geq \frac{q - r}{1 - r}.$$

Thus, if event C is likely *ex ante*, there is high probability that player i will believe that it is likely conditional on his information.

Proof of Lemma Let $\mu_i(\omega) = p(C \mid h_i(\omega))$. Then

$$p(C) = \mathrm{E}\mu_i(\omega)$$

$$= p[\mu_i(\omega) \geq r]\mathrm{E}[\mu_i \mid \mu_i \geq r] + (1 - p[\mu_i(\omega) \geq r])\mathrm{E}[\mu_i \mid \mu_i < r].$$

Then

$$p[\mu_i(\omega) \geq r] = \frac{p(C) - \mathrm{E}[\mu_i \mid \mu_i < r]}{\mathrm{E}[\mu_i \mid \mu_i \geq r] - \mathrm{E}[\mu_i \mid \mu_i < r]}$$

$$\geq \frac{q - r}{1 - r}. \qquad \blacksquare$$

Proof of Theorem 14.5 For $q \in (0, 1)$, define the function $\imath(q) = q - \sqrt{1 - q}$. For fixed n, set $D^0 = C^0 = C$ and $q^0 = p^n(C)$. Recursively define

$$C^m = \bigcap_{i=1}^{I} B_i^{\imath(q^{m-1})}(D^{m-1}),$$

$$D^m = C^m \cap D^{m-1},$$

$$q^m = q^{m-1} - \frac{I\sqrt{1 - q^{m-1}}}{1 + \sqrt{1 - q^{m-1}}}.$$

(Note that q^m is a decreasing sequence.) We claim that $p^n(D^m) \geq q^m$; that $D^m \subseteq D^{m-1}$; that for some M, $D^{M+1} = D^M$; and that with probability q^M the event D^M is common $\imath(q^M)$-belief.

The first claim is true for D^0. For $m > 0$, use lemma 14.1 to conclude that if $p^n(D^{m-1}) \geq q^{m-1}$ then, for each player i,

$$p^n(B_i^{\imath(q^{m-1})}(D^{m-1})) \geq \frac{q^{m-1} - \imath(q^{m-1})}{1 - \imath(q^{m-1})} = \frac{1}{1 + \sqrt{1 - q^{m-1}}},$$

and so the probability of $[\bigcap_{i=1}^{I} B_i^{\imath(q^{m-1})}(D^{m-1})] \cap D^{m-1}$ is at least

$$[1 - I(1 - p^n(B_i^{\imath(q^{m-1})}(D^{m-1})))] + q^{m-1} - 1 = q^{m-1} - \frac{I\sqrt{1 - q^{m-1}}}{1 + \sqrt{1 - q^{m-1}}}.$$

(Recall that for any sets A and B, $p(A \cap B) = p(A) + p(B) - p(A \cup B) \geq p(A) + p(B) - 1$.) That $D^m \subseteq D^{m-1}$ results from the definition of D^m. Since the D^m are nested, and Ω is finite, there exists an M such that $D^{M+1} = D^M$. Let $r = \imath(q^M)$. By definition of D^{M+1},

$$D^{M+1} = D^M = \bigcap_{i=1}^{I} B_i^r(D^M),$$

so that D^M is evident r-belief. Because $D^M \subseteq C$ and because D^M has probability at least q^M, C is r-common belief with probability q^M. Finally, note that with a finite state space, M is bounded above by the number of states $\overline{M}$, and that from the difference equation giving q^m, $q^{\overline{M}}$ ($\leq q^M$) converges to 1 when $p^n(C)$ goes to 1; similarly, $r = q^M - \sqrt{1 - q^M}$ converges to 1. ∎

Because many games do not have strict equilibria, we now turn to the question of lower hemi-continuity (with exact equilibria) for general games. We will consider a finite state space Ω, and a finite pure-strategy set S_i for each player. As in chapter 12, we can define a distance between two strategy profiles σ and $\tilde{\sigma}$:

$$\|\tilde{\sigma} - \sigma\| = \max_{\substack{i \in \mathscr{I} \\ s_i \in S_i}} |\sigma_i(s_i) - \tilde{\sigma}_i(s_i)|.$$

We consider a sequence of priors $p^n(\cdot)$ on Ω, and a finite collection of games $\{G^\ell\}_{\ell=1}^{L}$ with payoffs $\{u^\ell\}_{\ell=1}^{L}$, and assume that, as $n \to \infty$, it becomes very likely that one of these games is almost common knowledge.

Theorem 14.6 Consider a game with a finite state space Ω, partitions H_i, and L possible payoff functions u^ℓ. Consider a sequence of priors p^n such that, for some sequence $r^n \to 1$,

$$p^n(\omega \,|\, \exists G^\ell \text{ with } G^\ell \text{ common } r^n\text{-belief at } \omega) \to 1.$$

Suppose further that, for each player i and each $\ell \in \{1, \ldots, L\}$, $B_i^{r^n}(G^\ell)$ is a single information set h_i^ℓ. Then, for a generic choice of $\{G^\ell\}_{\ell=1}^L$, for any $\varepsilon > 0$ and any selection of equilibria $\sigma^{\ell *}$ of (common-knowledge) games G^ℓ, there exists N such that, for $n > N$, there exists a Bayesian equilibrium of the game with prior p^n such that

$$\omega \in \bigcap_{i \in \mathscr{I}} B_i^{r^n}(G^\ell) \Rightarrow \|\sigma(\cdot \,|\, \omega) - \sigma^{\ell *}\| < \varepsilon.$$

Remark We have no reason to believe that the restriction to a single information set is necessary; we use it to simplify the proof.

Proof Assume that $B_i^{r^n}(G^\ell)$ is a single information set for all i, ℓ. Let σ_i^ℓ denote the strategy of player i when $\omega \in B_i^{r^n}(G^\ell)$, let $\sigma^\ell = (\sigma_i^\ell)_{i \in \mathscr{I}}$, let $\sigma_i^c(\omega)$ denote player i's strategy for $\omega \notin \bigcup_\ell B_i^{r^n}(G^\ell)$, let $\sigma^c = (\sigma_i^c)_{i \in \mathscr{I}}$, and let $\sigma^{-\ell} = (\sigma^1, \ldots, \sigma^{\ell-1}, \sigma^{\ell+1}, \ldots, \sigma^L, \sigma^c)$. Fixing $\sigma^{-\ell}$, we define a I-player common-knowledge game $G(\sigma^{-\ell})$ among the types who are almost sure that the game is G^ℓ. Because at information set $B_i^{r^n}(G^\ell)$ player i is almost sure that the payoffs are u^ℓ and is almost sure that $\omega \in \bigcap_{j \neq i} B_j^{r^n}(G^\ell)$, player i's payoff in $G(\sigma^{-\ell})$, as a function of σ^ℓ, is very close to $u_i^\ell(\sigma^\ell)$ for n large. From Wu and Jiang's theorem (section 12.1), for a generic choice of u^ℓ, for any equilibrium $\sigma^{\ell *}$ of u^ℓ, and for n sufficiently large, there exists an equilibrium $\sigma^\ell = \hat{\sigma}^\ell(\sigma^{-\ell})$ of the common-knowledge game $G(\sigma^{-\ell})$ such that $\|\sigma^\ell - \sigma^{\ell *}\| < \varepsilon$. Furthermore, σ^ℓ is continuous in $\sigma^{-\ell}$. We now extend $\hat{\sigma}^\ell$ to a function on the space of strategy profiles Σ by defining $\hat{\sigma}^\ell(\sigma) = \hat{\sigma}^\ell(\sigma^{-\ell})$.

After constructing $\hat{\sigma}^\ell(\cdot)$ for each ℓ, we define $\hat{\sigma}_i$ on $C_i \equiv \Omega \backslash \bigcup_\ell B_i^{r^n}(G^\ell)$ (sets which have vanishingly small probability). There we simply require player i to optimize against $(\sigma^1, \ldots, \sigma^L, \sigma_{-i}^c)$. Let $\sigma_i^c = \hat{\sigma}_i^c(\sigma)$ denote player i's optimal response in those states; it is nonempty, it is compact-and-convex-valued, and it has a closed graph.

Let $\Sigma_\varepsilon^\ell = \{\sigma^\ell \,|\, \|\sigma^\ell - \sigma^{\ell *}\| < \varepsilon\}$ and $\Sigma^c = \{\sigma^c\}$. The correspondence $\hat{\sigma}$: $(\sigma^1, \ldots, \sigma^L, \sigma^c) \to \hat{\sigma}(\sigma)$ maps $\Sigma_\varepsilon^1 \times \cdots \times \Sigma_\varepsilon^L \times \Sigma^c$ into itself, and has a fixed point from Kakutani's theorem. By construction, this fixed point is a Bayesian equilibrium and lies within ε of $\sigma^{\ell *}$ when $\omega \in \bigcap_{i \in \mathscr{I}} B_i^{r^n}(G^\ell)$. ∎

Remark The hypotheses of the theorem are satisfied if each G^ℓ is common r^n-belief at exactly one ω^ℓ, as the total probability of $\{\omega^\ell\}_{\ell \in L}$ converges to 1. Thus, the theorem applies to the information structure of the truncated (finite-message) version of the electronic-mail game, and so the equilibrium correspondence there is lower hemi-continuous at the common-knowledge limit even if the payoff matrices given in figure 14.4 (which have strict equilibria) are replaced by matrices with essential equilibria.

The nice mathematical properties of games with a finite state space do not mean that infinite-state-space models are irrelevant—indeed, the standard requirement that the state space represent all the players' uncertainty about one another's information may naturally lead to state spaces that are even uncountably infinite. Mertens and Zamir (1985) and Brandenberger and Dekel (1987b) explicitly construct the "universal type space" required to capture general uncertainty of this kind, and find that it is quite large indeed. Mertens and Zamir observe that the universal type space can be approximated by a finite one when closeness is measured by the weak topology. However, as our discussion of example 14.4 shows, the set of equilibria is not continuous in that topology, so that the finite-state-space "approximation" can have a very different set of equilibria. In practical applications one works with finite type spaces for reasons of tractability. The sensitivity of even the Nash-equilibrium set to low-probability infinite-state perturbations is another reason to think seriously about the robustness of one's conclusions to the information structure of the game.

Exercises

Exercise 14.1** Consider the following version of the undiscounted, three-times-repeated prisoner's dilemma. There are three equally likely states of the world, ω_1, ω_2, and ω_3, with $H_1 = \{(\omega_1, \omega_2), (\omega_3)\}$ and $H_2 = \{(\omega_1), (\omega_2, \omega_3)\}$. In state ω_1 both players' payoffs are the sum of the stage-game payoffs shown in figure 14.6. In states ω_2 and ω_3, player 1's payoff is still the sum of the stage-game payoffs above, but player 2 must play the strategy "tit for tat" (which is to play C in the first period, and in periods 2 and 3 to play the same way player 1 did the period before). Show that the profile "both players play D every period," which is the unique equilibrium when it is common knowledge that the state is ω_1, is *not* a Nash-equilibrium outcome of the overall game—i.e., that there is no Nash equilibrium when in state ω_1 both players play D every period. Show that there is an equilibrium where both players play C in the first period in *every* state.

Exercise 14.2* Let Ω be the set of integers from 1 to 11. (a) Let $H_1 = \{(0, 3, 6, 9), (1, 4, 7, 10), (2, 5, 8, 11)\}$ and $H_2 = \{(0, 4, 8), (1, 5, 9), (2, 6, 10), (3, 7, 11)\}$. What is the meet of these partitions?
(b) What if $h_1(\omega)$ is the set of all ω' such that $[\omega'/2] = [\omega/2]$, where $[x]$ denotes the greatest integer less than or equal to x, and $h_2(\omega)$ is the set of all ω' such that $[\omega'/4] = [\omega/4]$?

	C	D
C	2,2	−1,3
D	3,−1	0,0

Figure 14.6

Exercise 14.3* Consider a game of incomplete information (see chapter 6) with prior $p(\cdot)$ over types θ. Suppose that p has full support ($p(\theta) > 0$ for all θ). Describe the information structure (Ω, H), and determine the meet.

Exercise 14.4** Prove the claim made in footnote 5.

Exercise 14.5** Prove that if E is common knowledge at ω under partitions $\{\hat{H}_i\}$ but E is not common knowledge at ω under partitions $\{H_i\}$, then there is a player j and a state $\hat{\omega}$ such that player j's knowledge is different at $\hat{\omega}$. (See the last paragraph of subsection 14.3.1.)

References

Aumann, R. 1976. Agreeing to disagree. *Annals of Statistics* 4: 1236–1239.

Aumann, R. 1987. Correlated equilibrium as an expression of Bayesian rationality. *Econometrica* 55: 1–18.

Bacharach, M. 1985. Some extensions to a claim of Aumann in an axiomatic model of knowledge. *Journal of Economic Theory* 37: 167–190.

Binmore, K., and A. Brandenburger. 1989. Common Knowledge and Game Theory. Mimeo.

Brandenburger, A. 1990. Knowledge and equilibrium in games. *Journal of Economic Perspectives*, forthcoming.

Brandenburger, A., and E. Dekel. 1987a. Common knowledge with probability 1. *Journal of Mathematical Economics* 16: 237–245.

Brandenburger, A., and E. Dekel. 1987b. Hierarchies of beliefs and common knowledge. Mimeo.

Brandenburger, A., and E. Dekel. 1987c. Rationalizability and correlated equilibria. *Econometrica* 55: 1391–1402.

Brandenburger, A., and E. Dekel. 1990. The role of common knowledge assumptions in game theory. In *The Economics of Information, Games, and Missing Markets*, ed. F. Hahn. Oxford University Press.

Brown, D., and J. Geanakoplos. 1988. Common knowledge without partitions. Mimeo, Yale University.

Geanakoplos, J. 1988. Common knowledge, Bayesian learning and market speculation with bounded rationality. Unpublished manuscript , Yale University.

Geanakoplos, J. 1989. Game theory without partitions, and applications to speculation and consensus. Mimeo, Yale University.

Geanakoplos, J., and H. Polemarchakis. 1982. We can't disagree forever. *Journal of Economic Theory* 28: 192–200.

Gray, J. 1978. Notes on data base operating systems. Research report RJ 2188, IBM.

Halpern, J. 1986. Reasoning over knowledge: An overview. In *Theoretical Aspects of Reasoning about Knowledge*, ed. J. Halpern. Morgan Kaufmann.

Holmström, B., and R. Myerson. 1983. Efficient and durable decision rules with incomplete information. *Econometrica* 51: 1799–1820.

Kreps, D. 1977. A note on fulfilled expectations' equilibria. *Journal of Economic Theory* 14: 32–43.

Lewis, D. 1969. *Conventions: A Philosophical Study*. Harvard University Press.

Littlewood, J. E. 1953. *Mathematical Miscellany*, ed. B. Bollobas.

Maskin, E., and J. Tirole. 1989. The principal-agent relationship with an informed principal: The case of common values. Mimeo, Harvard University and Massachusetts Institute of Technology; forthcoming in *Econometrica*.

Mertens, J.-F., and S. Zamir. 1985. Formulation of Bayesian analysis for games with incomplete information. *International Journal of Game Theory* 10: 619–632.

Milgrom, P. 1981. An axiomatic characterization of common knowledge. *Econometrica* 49: 219–222.

Milgrom, P., and N. Stokey. 1982. Information, trade and common knowledge. *Journal of Economic Theory* 26: 177–227.

Monderer, D., and D. Samet. 1989. Approximating common knowledge with common beliefs. *Games and Economic Behavior* 1: 170–190.

Rubinstein, A. 1989. The electronic mail game: Strategic behavior under 'almost common knowledge.' *American Economic Review* 79: 385–391.

Rubinstein, A., and A. Wolinsky. 1989. Remarks on the logic of "agreeing to disagree"-type results. London School of Economics ICERD paper TE/89/188.

Samet, D. 1987. Ignoring ignorance and agreeing to disagree. Discussion paper no. 749, KGSM, Northwestern University.

Schelling, T. 1960. *The Strategy of Conflict*. Harvard University Press.

Shin, H. 1987. Logical structure of common knowledge, I and II. Mimeo, Nuffield College, Oxford University.

Stiglitz, J. 1971. Information and capital markets. Mimeo.

Stinchcombe, M. 1988. Approximate common knowledge. Mimeo, University of California, San Diego.

Tan, T., and S. Werlang. 1988. The Bayesian foundations of solution concepts of games. *Journal of Economic Theory* 45: 370–391.

Tirole, J. 1982. On the possibility of speculation under rational expectations. *Econometrica* 50: 1163–1182.

Index